Python Essentials for Biomedical Data Analysis:
An Introductory Textbook

Julhash U. Kazi

# Python Essentials for Biomedical Data Analysis: An Introductory Textbook

Julhash U. Kazi
Department of Laboratory Medicine
Lund University
Lund, Sweden

ISBN 978-3-031-85599-3 ISBN 978-3-031-85600-6 (eBook)
https://doi.org/10.1007/978-3-031-85600-6



This Springer imprint is published by the registered company Springer Nature Switzerland AG
The registered company address is: Gewerbestrasse 11, 6330 Cham, Switzerland

# Preface

Welcome to *Python Essentials for Biomedical Data Analysis: An Introductory Textbook*. This book serves as an introduction for graduate students and early-career researchers in the biomedical field who want to use Python for data analysis. In today's world, the ability to analyze and interpret complex datasets is an important skill, particularly in biomedicine, where working with data can drive new progress in health and medicine.

This book begins by introducing the basics of Python, a programming language known for its simplicity and flexibility. We recognize that many readers may be new to coding, so the early chapters focus on fundamental concepts and syntax, offering a gentle introduction. Our primary aim is not to turn you into expert programmers but to equip you with a basic understanding of Python. This knowledge will enable you to run pre-written programs effectively and create simple pipelines for executing sequences of applications.

After covering the basics of Python, we move on to practical applications for managing, visualizing, and analyzing biomedical data. Each chapter builds on the previous one, ensuring your skills and confidence grow steadily. Throughout the book, you will find examples and exercises related to biomedical scenarios. These examples may help to develop understanding and skills that can be directly applied to tasks and problems in biomedical research.

Key topics include statistical analysis, machine learning, image processing, genomics data analysis, pharmacokinetics, pharmacodynamics, and an introduction to natural language processing. The book covers a range of areas reflecting the broad scope of biomedical research, while demonstrating how Python can be used to meet various needs.

Additionally, each chapter begins with clearly stated learning goals; offers example code, available on GitHub as Jupyter Notebooks with accompanying data, where necessary, at ► https://github.com/sn-code-inside/BioPy; and concludes with relevant exercises and questions to assess learning outcomes. Whether you are taking your first steps into data analysis or looking to expand your current skills, our goal is for this book to serve as a useful and encouraging guide.

**Julhash U. Kazi**
Lund, Sweden

**Competing Interests** The author has no competing interests to declare that are relevant to the content of this manuscript.

**Disclaimer** The language of the human-generated text was corrected with the assistance of artificial intelligence (AI) tools [GPT-3.5 and GTP-4 from OpenAI], using the following prompts: "Could you check the English in the following text?" or "Correct English language in the following text:". GitHub Co-Pilot was used to check the correctness of the codes with the following prompt: "Check the code and suggest any improvement if required".

Every section was carefully reviewed, revised, and approved by the author to ensure that the text is appropriate concerning content and scientific accuracy, and the integrity of the original content is maintained. The author takes full responsibility for the content of the final publication.

Although every code block in this book has been carefully reviewed, occasional formatting errors may still occur. We recommend that readers refer to the accompanying Jupyter Notebooks on GitHub for the original and most up-to-date code.

# About this Book

This book introduces graduate students and early-stage biomedical researchers to Python as a practical tool for data analysis, emphasizing both basic programming skills and applications in biomedicine. It begins with the basics of Python syntax and core concepts, aiming to give readers enough understanding to run pre-written programs and develop basic data-processing pipelines. Each chapter starts with learning goals, includes code examples available on GitHub (at ▶ https://github.com/sn-code-inside/BioPy), and ends with exercises. Building upon each new concept in a structured progression, the book discusses data management, visualization, and analysis scenarios, ranging from statistical analysis and machine learning to genomics and introductory natural language processing. By highlighting the versatility of Python across diverse topics such as image processing, pharmacokinetics, and pharmacodynamics, this adaptable resource aims to stay current with the rapid evolution of data science, offering updatable content to remain relevant over time.

# Contents

## Supplementary Information

# Introduction to Python

Contents




J. U. Kazi, *Python Essentials for Biomedical Data Analysis: An Introductory Textbook*,
https://doi.org/10.1007/978-3-031-85600-6_1

The ability to process, analyze, and visualize large datasets is a key skill in biomedicine. Python has become a widely used programming language in this field due to its simplicity, flexibility, and extensive library support. This chapter introduces Python and its possible application in biomedical data analysis. Originally developed by Guido van Rossum in the late 1980s, Python has evolved into one of the most popular programming languages globally. Its interpreted nature and emphasis on readability make it especially accessible to beginners. The chapter provides a brief overview of Python's development and use across biomedical applications. It also describes the procedure for setting up the Python environment, from installation and choosing integrated development environments (IDEs) to creating and managing virtual environments. Practical steps for installing libraries, verifying the setup, and troubleshooting common issues are also covered.

**Learning Goals**

This chapter introduces the basics of Python programming and its importance in biomedical data analysis. You will learn the history and development of Python, how to set up the Python environment, and how to use integrated development environments (IDEs) and notebooks. These tools will facilitate hands-on practice and make the learning process more accessible and practical.

## 1.1 Overview of Python Programming

Python is a high-level,[1] interpreted programming language, widely recognized for its readability and ease of use. Its design philosophy prioritizes code clarity, benefiting both novice and experienced programmers. One of Python's key advantages is its portability, allowing programs to run on various platforms without modification. As an interpreted language, Python processes code line-by-line, simplifying debugging and testing without the need for prior compilation. Python is also dynamically typed, meaning variable types are determined during runtime, eliminating the need for explicit declarations.

1 High-level programming languages are designed to simplify the process of programming by using syntax that is closer to human language, which makes them easier to read, write, and understand. These languages abstract away many of the complex details of the computer's hardware, allowing you to focus more on solving problems relevant to your field, without needing deep knowledge of the underlying computer system. This level of abstraction also makes programs written in high-level languages portable, meaning they can often run on different types of computers without modification.

Python supports various programming approaches, including object-oriented,[2] procedural,[3] and functional[4] programming [1]. The language comes with a standard library that offers modules and functions for tasks like file operations, system management, and Internet protocols. In addition, Python's ecosystem which is enriched by a large and active community, provides access to numerous third-party libraries and frameworks. Its straightforward syntax and wide capabilities make Python particularly well suited for biomedical data analysis, facilitating efficient data processing, analysis, and visualization.

### 1.1.1 Python Is an Interpreted Programming Language

As an interpreted programming language, Python executes programs directly from the source code without first compiling the entire program into machine code (◘ Fig. 1.1). When a Python program is run, the interpreter reads the source code, translates it into bytecode, a lower-level, machine-readable format, and then executes the bytecode almost immediately using the Python virtual machine (PVM). This process facilitates rapid testing and debugging, as errors are reported in real time, allowing modifications to be integrated into the workflow. This immediate execution style is particularly beneficial in research settings, where scripts are often adjusted on the fly to analyze complex datasets. Additionally, Python's bytecode handling provides an efficient means to rerun scripts.

The bytecode generated by Python is stored in the `__pycache__` directory within the script's directory as `.pyc` files. Each bytecode file is stamped with a timestamp to ensure that only the most up-to-date version of the script is executed. When the original Python scripts are modified, new bytecode files are generated, replacing the older versions. This process simplifies subsequent executions by avoiding the need to recompile unchanged parts of the code. However, if the system's permissions restrict the interpreter from writing bytecode files, Python compensates by holding and translating the code directly in memory each time the program runs. Despite this fallback, the efficiency of Python's execution remains intact. Finally, the PVM interprets bytecode instructions for the CPU at runtime. [2]. While this can be slower than

2 Object-oriented programming (OOP) is a paradigm where programs are structured around objects, which represent real-world entities or concepts. Each object contains data (attributes) and methods (functions) that operate on the data. OOP encourages the organization of code into reusable, modular components, making it easier to maintain and scale. In Python, OOP is implemented through classes and objects.

3 Procedural programming is a programming paradigm based on the concept of procedure calls, where a program is structured as a sequence of instructions or functions. In this approach, the code is written as a series of steps that manipulate data, with the main focus on how tasks are performed. Python supports this paradigm through the use of functions to break down complex tasks into smaller, manageable parts.

4 Functional programming is a paradigm that views computation as the evaluation of mathematical functions and emphasizes immutability, avoiding changes to state and mutable data. This approach emphasizes the use of pure functions, which consistently yield the same output for identical inputs and operate without causing any side effects.

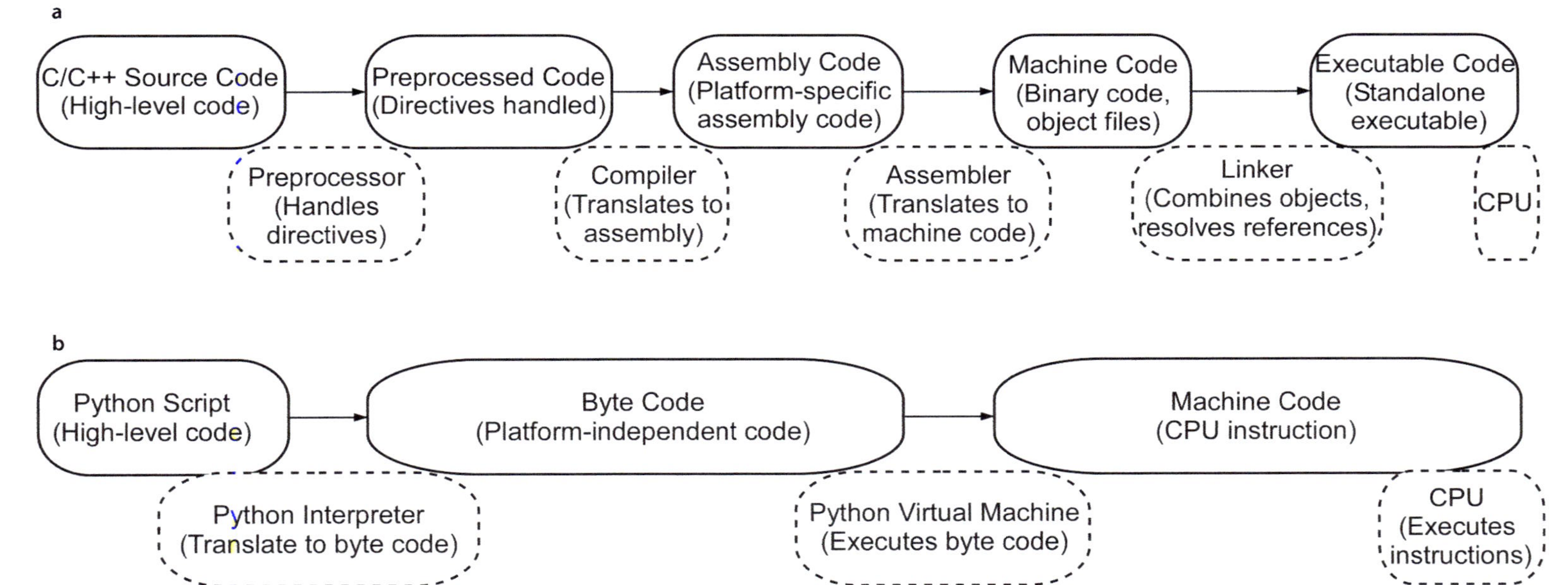


**Fig. 1.1** The workflows of code compilation and execution in C/C++ and Python. (**a**) In C/C++, the process begins with the source code, which is preprocessed to handle directives. The preprocessed code is then compiled into assembly code, which is subsequently translated into machine code by an assembler. Finally, a linker combines all object files into a single executable that the CPU can run. (**b**) In Python, the process starts with a script that is interpreted into bytecode, a lower-level, platform-independent code. This bytecode is then executed by the PVM, which dynamically converts it into machine-specific instructions for the CPU

compiled languages, it provides portability and flexibility, supporting the iterative and exploratory style of programming common in scientific research.

### 1.1.2 The Evolution of Python Programming

Guido van Rossum, a Dutch programmer, created Python as a hobby project during the Christmas holidays of 1989 while working at Centrum Wiskunde & Informatica (CWI) in the Netherlands. He named the language after the British comedy series "Monty Python's Flying Circus," one of his favorites. Python was designed to combine the readability and ease of writing of its predecessors, such as ABC[5] and Modula-3,[6] with enhanced power and functionality. Its design philosophy, captured in "The Zen of Python" [3], emphasizes code readability and simplicity through a collection of aphorisms that reflect the language's aesthetic and philosophical approach to programming. Python's syntax enables programmers to convey complex concepts concisely, often requiring fewer lines of code than languages like C++ or Java, while supporting various programming approaches.

The first beta version of Python was released in early 1991, followed by the first major version, Python 1.0, in January 1994 (◘ Fig. 1.2). This version established the foundation for future development, introducing key features such as exception handling, functions, and core data types. Python 2.0, released in October 2000, brought significant enhancements to the language and its standard library. Notably, it introduced list comprehensions—a feature borrowed from Haskell[7]—and improved Unicode support, making Python more accessible for non-English language environments. The release of Python 3.0 in December 2008 marked a major turning point; it was intentionally not backward compatible with Python 2, allowing the language to break free from legacy design issues. Changes included expanded Unicode support (with Unicode strings as the default) and transforming the print statement into a built-in function, improving consistency and readability.

In recent years, Python has widely been used in fields such as machine learning and data science. Several tools were developed to address these areas, including

---

5 ABC is a high-level, general-purpose programming language and environment developed at CW in the early 1980s. It was designed primarily as a teaching language, with a focus on simplicity and ease of use for nonexpert programmers. ABC's syntax was structured to be clear and concise, using indentation rather than braces or keywords to define code blocks, which later influenced the design of Python.

6 Modula-3 is a systems programming language that evolved from the Modula-2 language, designed by Niklaus Wirth in the late 1980s. It extended Modula-2 s approach of rigorous type safety and modularity while adding support for object-oriented programming, including features like classes, objects, and inheritance. Modula-3 was significant for its design goal of creating robust programs with safety and reliability, incorporating exception handling and garbage collection to manage memory and errors automatically.

7 Haskell is a purely functional programming language with static typing, named in honor of the logician Haskell Curry. It is known for its strong emphasis on immutability, higher-order functions, and type inference. Haskell promotes writing code that avoids side effects, which means functions consistently produce the same outputs given the same inputs without altering the program's state. Due to these properties, Haskell is often used in academic research, teaching, and certain areas of industry that require rigorous mathematical precision.

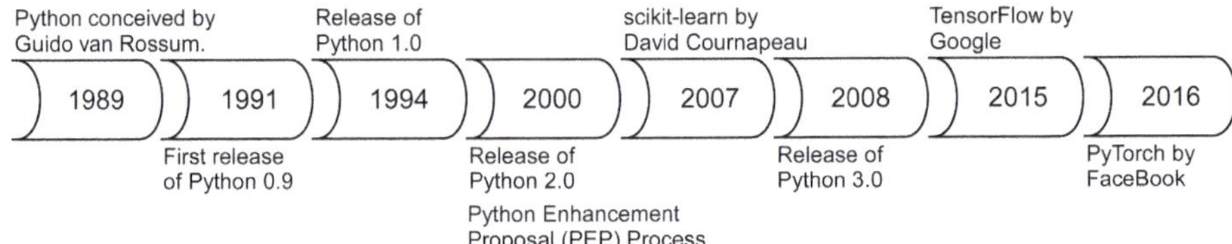


**Fig. 1.2** Key milestones in Python language development. A timeline of key milestones in the history of Python programming, illustrating its evolution from inception to becoming a major tool in data science and machine learning. The timeline begins in 1989 when Guido van Rossum first conceived Python, followed by the release of Python 0.9 in February 1991. Major subsequent releases include Python 1.0 in January 1994, Python 2.0 in October 2000, and Python 3.0 in December 2008. Additionally, key developments within the Python community include the establishment of the Python Enhancement Proposal (PEP) process in 2000, and the introduction of influential libraries like scikit-learn by David Cournapeau in 2007, TensorFlow by Google in 2015, and PyTorch by Facebook in 2016

scikit-learn in 2007 by David Cournapeau, TensorFlow in 2015 by Google, and PyTorch in 2016 by Facebook. According to TIOBE, an organization that tracks software quality, Python became the most popular programming language in 2024–2025, surpassing both C and Java [4].

## 1.2 Importance of Python in Biomedical Data Analysis

Biomedical data analysis plays a major role in research and discovery [5]. The ability to analyze and interpret large datasets helps us gain a better understanding of complex biological systems, study disease mechanisms, and develop new treatments [6]. Python, with its ecosystem of libraries and tools, has become a widely used resource for biomedical researchers, allowing them to process, visualize, and examine data with enhanced efficiency and precision.

### 1.2.1 Python in Genomic Data Analysis

Python has been widely used in genomics, particularly in the analysis of vast amounts of DNA sequence data generated by high-throughput technologies such as next-generation sequencing (NGS). The sheer volume and complexity of NGS datasets require robust computational tools for analysis [7]. Python libraries such as Biopython [8] and scikit-allel [9] are widely used for genomic data processing tasks, including sequence manipulation, annotation, and variant detection. These tools facilitate the discovery and characterization of genetic variants associated with human diseases, advancing the field of precision medicine. Precision medicine relies on the ability to tailor therapeutic interventions to the genetic profiles of individual patients, thereby enhancing treatment efficacy and patient outcomes.

### 1.2.2 Python in Drug Discovery and Development

Python tools are used in drug discovery and development to navigate large volumes of chemical and biological data. Specialized libraries, such as RDKit for cheminformatics and PyMOL for molecular visualization, enable researchers to analyze compound structures, predict biological activity, and optimize drug properties for efficacy and safety [10, 11]. Additionally, Python's adaptability and integration with machine learning algorithms enhance the modeling of complex biological interactions and improve prediction of therapeutic potential.

### 1.2.3 Python in Clinical Data Analysis

In clinical research, Python is used to analyze patient data, including electronic health records (EHRs). With Python libraries such as pandas and NumPy, researchers can efficiently process large-scale clinical datasets, while machine learning frameworks like TensorFlow and scikit-learn allow for the development of predictive models [12, 13]. These models can assist in predicting patient outcomes, improving diagnoses, and recommending treatment plans. Furthermore, Python's natural language processing (NLP) capabilities allow the extraction of useful information from unstructured clinical notes, that can be used for analysis of patient health trends.

### 1.2.4 Python in Image Analysis

Python is widely used for imaging data analysis, particularly in medical fields like oncology. Libraries such as OpenCV and scikit-image [14, 15] are used to automate the analysis of imaging data, from standard X-rays to advanced MRI scans and single-cell imaging techniques. These tools provide precise and reproducible processing of images, improving diagnostic accuracy and providing updates on disease progression. Python-based imaging analysis is instrumental in detecting and quantifying tumor regions, tracking disease progression, and evaluating the effectiveness of treatments, thereby supporting improved therapeutic decision-making.

## 1.3 The Python Environment

A Python environment is a configurable virtual space where Python applications are developed and executed. It includes the Python interpreter, a defined set of installed libraries, and the necessary configuration settings tailored to meet the requirements of a specific project or application. This setup allows projects to run independently from one another, avoiding conflicts between dependencies and ensuring that different projects can maintain their own isolated environments. Managing Python environments in this way facilitates more effective handling of project dependencies and enhances reproducibility across diverse applications.

### 1.3.1 Key Characteristics of a Python Environment

The key characteristics of a Python environment include isolation, reproducibility, and manageability:

- *Isolation:* Each Python environment operates independently, with its own set of dependencies and libraries. This isolation ensures that changes made in one environment do not affect others, helping to reduce conflicts between different project requirements.
- *Reproducibility:* By locking the versions of the libraries used in a project, Python environments ensure that the project can be replicated and executed with the exact same configuration elsewhere, whether on a different machine or by another developer. This is important for maintaining consistency across development and production stages.
- *Manageability:* Python environments simplify the management of project-specific dependencies, allowing developers to update or modify libraries without affecting other projects or the global Python installation. This flexibility makes handling dependencies more efficient and organized.

### 1.3.2 Python Installation

We outline here steps to install and set up Python on your computer. Installing Python is an initial task for anyone starting their journey in biomedical data analysis. The following guide ensures that you install Python correctly, equipping you with the necessary tools to begin your work efficiently.

**Preinstallation check:** Before proceeding with the installation, it is important to verify whether Python is already installed on your system. This helps avoid redundancy and ensures a clean setup.

Linux and macOS:

1. Open the "Terminal" application.
2. Type `python --version` or `python3 --version` and press Enter.

Windows:

1. Open the "Command Prompt" (search for "cmd" in the Start menu).
2. Type `python --version` and press Enter.

If Python is not installed or has not been added to the system's "PATH," the command will return an error. If it is installed, the command will display the currently installed Python version.

**Choosing the right version:** For most applications, Python 3 is recommended over Python 2, as Python 3 includes the latest features and improvements. Follow these steps to download and install the correct version of Python:

1. Visit the official Python website at ► www.python.org.
2. Navigate to the "Downloads" section, where the latest Python release is available for various operating systems.

3. Select and download the version compatible with your operating system. Ensure you download Python 3.xx, as it is the most current and widely supported version.

**Installation via Anaconda Navigator:** For users focused on data science, machine learning, or scientific computing, Anaconda Navigator provides a user-friendly alternative to manual installation. Anaconda Navigator is a graphical interface included in the Anaconda distribution, designed to simplify the management of Python environments and packages. It includes the latest version of Python, numerous popular libraries, and Jupyter Notebook—a tool for interactive data exploration and visualization. Anaconda also simplifies the complexities of package management and environment configuration.

Installation Steps:

1. Visit ▶ www.anaconda.com and navigate to the "Download" section.
2. Select the appropriate version for your operating system, and download the Anaconda installer.
3. Run the installer and follow the on-screen instructions to install Anaconda, which will also install Python and the necessary tools.

### 1.3.3 Integrated Development Environments (IDEs) for Python

Integrated development environments (IDEs) are tools that may enhance the productivity and efficiency of Python developers. IDEs combine various components of software development, such as code editing, testing, debugging, and version control, into a single interface. This integration simplifies the process of writing, testing, and debugging Python code, thereby streamlining development workflows. ◘ Table 1.1 provides a comparative overview of some of the most popular IDEs used for Python programming, including PyCharm, Visual Studio Code (VS Code), Spyder, and Thonny, as well as the more simplified, interactive environment provided by Jupyter Notebook.

### 1.3.4 Virtual Environments for Python

Several tools have been developed to facilitate the creation and management of isolated Python environments, each offering unique features and capabilities. The most commonly used tools include `venv`, `virtualenv`, and `conda`:

**venv:** Introduced in Python 3.3 and integrated into the Python Standard Library, `venv` is designed to create lightweight virtual environments. It offers a simple, direct approach to environment management, making it a standard choice for most Python projects. Each environment created with `venv` contains its own Python binary and a separate directory for installing packages, ensuring isolation from the global Python environment. This separation allows different projects to use different versions of the same packages without interference. `venv` is ideal for projects where external system

**Table 1.1** Popular IDEs for Python

| IDE | Overview | Key features | Suitability |
|---|---|---|---|
| PyCharm | Developed by JetBrains, PyCharm is one of the most comprehensive IDEs for Python. It offers a rich set of features including code analysis, a graphical debugger, an integrated unit tester, and support for web development with Django | PyCharm provides intelligent code assistance, smart code navigation, and fast and safe refactoring. It integrates with major version control systems and supports scientific tools like Jupyter notebook, Anaconda as well as numerous scientific packages | Ideal for professional developers working on large projects or anyone looking for an all-encompassing development tool |
| VS Code | VS Code, developed by Microsoft, is a streamlined yet robust source code editor that operates on your desktop. It features built-in support for JavaScript, TypeScript, and Node.js, and offers a vast array of extensions for Python through its extensive ecosystem | VS Code features include debugging support, intelligent code completion (IntelliSense), snippets, and code refactoring. It is highly customizable, allowing users to adjust themes, keyboard shortcuts, and preferences to suit their needs | A fast and highly customizable editor that can be expanded with extensions to suit a wide range of Python development needs |
| Spyder | Spyder is an open-source IDE specifically designed for scientists, engineers, and data analysts working with the Python language. It features integration with many of the major scientific packages | Includes an advanced editing, interactive testing, debugging, and introspection features. It also has a variable explorer that allows you to interact with and modify your data directly from the GUI | Best suited for those involved in scientific and technical computing who require an IDE that integrates deeply with Python's scientific stack |
| Thonny | Thonny is a simple Python IDE useful for beginners. It provides a clean and straightforward interface that focuses on teaching beginners the basics of programming in Python | Offers easy access to Python expressions, built-in debugger, and simple interface for code errors and debugging. It is light on resources and does not overwhelm the user with features | An excellent starting point for programming novices or educational environments where simplicity and ease of use are paramount |
| Jupyter Notebook | Jupyter Notebook is an open-source web application that enables the creation and sharing of documents featuring live code, equations, visualizations, and narrative text. It is especially useful for projects involving data visualization and machine learning | Supports over 40 programming languages including Python. It allows for the integration of rich media, visualizations, widgets, and narratives that combine source code and its output | Excellent choice for data scientists and researchers focusing on data analysis, statistical modeling, and machine learning |

packages are not required, and dependency management is confined to the Python ecosystem.

**virtualenv:** The `virtualenv` is a third-party tool that predates `venv` and is particularly useful for Python versions earlier than 3.3, which do not include `venv`. It offers more flexibility and additional features compared to `venv`. Like `venv`, it creates isolated environments with independent Python binaries and libraries but provides isolation and compatibility with multiple Python versions. `virtualenv` is useful in environments that require managing different Python versions simultaneously or where greater isolation from the system Python is necessary.

**Conda:** Conda is an open-source package and environment management system developed by Anaconda, Inc. Unlike `venv` and `virtualenv`, which are limited to Python, Conda supports multiple programming languages, making it ideal for projects that involve mixed-language dependencies. Conda simplifies package management and deployment in scientific computing environments by handling both the environment setup and package installation. Its cross-language capability and platform support (Windows, macOS, and Linux) make it particularly suited for data science and complex scientific projects requiring a wide array of dependencies.

### 1.3.5 Creating a Virtual Environment Using venv

Setting up a virtual environment using Python's built-in `venv` module is straightforward, and the process is largely similar across Linux, macOS, and Windows platforms.

- **Creating a virtual environment**

1. Create a folder where the virtual environment will be established.
2. Open your terminal (Linux/macOS) or command prompt (Windows).
3. Navigate to the directory where you want to create your virtual environment using the `cd` command.
4. To create a virtual environment named `env` (you can choose any name), run the command:

```
python -m venv env
or
python3 -m venv env
```

This command creates a directory named "env" (or your chosen name) in the current directory, containing the Python interpreter, standard library, and various supporting files.

- **Activating the virtual environment**
  1. Use the following command to activate the virtual environment:

  On Linux and macOS:

```
source env/bin/activate
```

On Windows:

```
env\Scripts\activate
```

  2. Once activated, your command prompt will change to indicate that you are in the virtual environment, displaying the name of the environment (e.g., (env)).

- **Working within the virtual environment**
  1. With the virtual environment activated, you can now install packages using `pip`, which will be isolated to this environment.
  2. Any Python commands you run will use the interpreter and packages installed in the virtual environment.

- **Deactivating the virtual environment**

To deactivate a virtual environment after you have finished working in it, simply run the following command:

```
deactivate
```

Each virtual environment is independent, requiring the activation of the correct one to access its packages and interpreter. Using `venv` is a simple and effective way to manage project-specific dependencies without affecting the global Python setup.

### 1.3.6 Creating a Virtual Environment Using Anaconda

Anaconda is a popular distribution of the Python and R programming languages, primarily aimed at simplifying package management and deployment in data science, machine learning, and scientific computing. There are several reasons to consider Anaconda as a tool for creating virtual environments and managing packages:

- Comprehensive package management: Anaconda simplifies the management of packages and dependencies. It uses Conda, an open-source package and environ-

ment management system, which makes it easy to install, run, and update complex scientific and analytical software packages.
- A wide array of pre-built packages: Anaconda provides a vast repository of precompiled packages designed for data science, machine learning, and scientific computing. This ready-to-use suite minimizes the challenges associated with software installation and dependency management, allowing users to commence their projects more efficiently and focus on development rather than setup.
- Virtual environment management: Anaconda allows the creation of isolated environments for different projects. These environments contain unique sets of dependencies and libraries, ensuring that different projects run smoothly without interference from one another. This isolation maintains consistency and stability across diverse project requirements.
- Cross-platform compatibility: Anaconda's ability to operate across Windows, macOS, and Linux platforms makes it versatile. This cross-platform compatibility is useful for teams working in heterogeneous computing environments or individuals who need to ensure their projects function on multiple operating systems.
- Ease of use: Anaconda Navigator, a graphical user interface (GUI) included in Anaconda, provides a user-friendly way to manage packages, environments, and updates without needing to use command-line commands. This GUI is designed to make navigation and administration straightforward, especially for users who prefer graphical interfaces to command-line operations.
- Community and support: Anaconda is supported by a large community of developers and users who continuously enhance its ecosystem with new tools, features, and documentation. This community support helps resolving technical issues, exchanging knowledge, and discovering best practices in data science and analytics.
- Integration with cloud services: Anaconda can be integrated with various cloud providers, which facilitates the deployment of data analysis solutions at scale. Such type of integration allows for the effective handling of large datasets and the deployment of complex models, facilitating a smoother transition from development to production environments.
- Educational resources: Anaconda offers a wealth of educational materials, including tutorials, webinars, and detailed documentation. These resources are designed to help both novice and advanced users sharpen their skills in data science, machine learning, and related fields, promoting continual learning and professional development.

Given these features, Anaconda is a robust choice for researchers, data scientists, and developers who require a reliable and flexible environment for their scientific computing needs. The ability to manage multiple environments and packages makes it particularly useful in academic and research settings where multiple projects may be ongoing concurrently. In this book, we will predominantly use Anaconda to generate examples.

**Setting up Anaconda:** Download and install Anaconda from ▶ www.anaconda.com, selecting the version that is compatible with your operating system (described previously in ▶ Sect. 1.3.2 as well). After completing the installation, launch the Anaconda Navigator to access a variety of IDEs and tools through a user-friendly interface. This centralized access point facilitates the exploration and use of numerous resources that Anaconda offers.

**Creating a virtual environment using Anaconda:** Utilize the environment manager within Anaconda Navigator to create a new virtual environment. This feature allows you to select the specific Python version and the packages required for your project. Anaconda handles the installation and configuration of these selected packages, ensuring that your environment is set up with all the necessary tools.

**Activating and using a virtual environment:** Activate your virtual environment through Anaconda Navigator or use the command line by typing `conda activate environment_name`. Once activated, all operations such as running Python scripts or installing additional packages are confined to this environment. This isolation protects your main Python installation from any changes or conflicts that might arise from project-specific dependencies.

**Managing environments:** Anaconda Navigator provides a graphical interface for management of your virtual environments. This interface simplifies the process of switching between different environments, making it easy to manage multiple projects simultaneously. By leveraging Anaconda and its robust environment management capabilities, you enhance the organization of your workflow and maintain the integrity of your Python projects. This method not only aids in keeping projects well organized but also simplifies collaboration and sharing with other developers.

### 1.3.7 Installing Python Libraries

One of Python's standout features is its vast collection of libraries, which is particularly beneficial for data analysis and scientific computing. Python libraries such as NumPy, Pandas, Matplotlib, and SciPy are indispensable for tasks including data manipulation, statistical analysis, and data visualization.

*In a standard Python environment:* Most Python packages can be easily installed using the `pip` package manager with a simple command:

```
pip install package-name
```

This command fetches the latest version of the package from Python Package Index (PyPI) and installs it, along with any required dependencies, into your Python environment.

*In an Anaconda environment (Conda environment):* Anaconda environments come preinstalled with many packages optimized for data science and scientific computing. To manage additional libraries, users can use the `conda` package manager:

```
conda install package-name
```

This command ensures compatibility with other installed packages by checking the best available version from the Anaconda repository, reducing the risk of dependency conflicts.

However, if a specific package is not available in Conda's repositories, you can still use pip within a Conda environment:

```
pip install package-name
```

While `conda` and `pip` can coexist, it is generally recommended to prefer `conda` for package installation when possible, as it better handles dependency resolution in complex environments. If you use `pip`, it is advisable to run `conda update --all` afterward to ensure compatibility between the installed packages.

### 1.3.8 Verifying the Setup

After completing the setup of your Python environment, it is important to verify that everything is functioning as expected. This step ensures you are ready to proceed with your Python projects without any unforeseen issues. Here is how to verify your Python setup:

- **Check Python installation**
  1. Open your command line or terminal.
  2. Type `python -version` or `python3 -version` and press Enter.
  3. You should see the Python version number that you installed. If not, review the installation steps to ensure Python is correctly installed and added to your system's PATH.
- **Test virtual environment**
  1. Activate your virtual environment as previously described.
  2. While the environment is active, install a simple package like `requests` using `pip install requests`.
  3. After installation, start a Python interpreter by typing `python` in the command line or terminal.
  4. Import the installed package by typing `import requests` in the Python interpreter. If no error messages appear, the environment and package installation are functioning correctly.
- **Validate Anaconda installation (if applicable)**
  1. Open Anaconda Navigator from your applications or start menu.
  2. Ensure you can see the list of applications like Jupyter Notebook, Spyder, etc.
  3. Optionally, launch Jupyter Notebook or another IDE from the Navigator and execute a simple Python command, such as `print("This is Anaconda!")`.

- **Test IDE setup (if applicable)**
  1. Open your chosen IDE or notebook (e.g., PyCharm, Spyder, Visual Studio Code, or Jupyter Notebook).
  2. Create a new Python file or notebook.
  3. Write a simple Python script, such as `print("Testing Notebook!")`, and run it. Successful execution without errors confirms that your IDE is correctly set up.

- **Check package management**
  1. In your active environment, try installing a different package using either `pip` or `conda`, depending on your setup.
  2. After installation, repeat the import test in the Python interpreter.

Completing these steps assures that your Python setup, including the base installation, virtual environments, Anaconda, and your IDE, is correctly configured and ready for use.

### 1.3.9 Troubleshooting Common Setup Issues

Setting up Python and its environments can sometimes lead to common issues. This section provides solutions to typical problems, ensuring a smoother setup experience.

- **Python is not recognized in command line or terminal**
  - *Issue:* After installing Python, the `python` command is not recognized in your command line or terminal.
  - *Solution:* Ensure that Python is correctly installed and that its path is added to the system's environment variables. On Windows, you have the option to add Python to PATH during the installation process.

- **Issues with virtual environment activation**
  - *Issue:* The virtual environment does not activate, or an error occurs during activation.
  - *Solution:* Verify that you are in the directory where the virtual environment is located. Use the correct activation command (`source env/bin/activate` on Linux/macOS, `env\Scripts\activate` on Windows for the environment created as `env`).

- **Problems installing packages with pip**
  - *Issue:* Errors occur when installing packages using `pip`.
  - *Solution:* Confirm that you are using the latest version of `pip` by upgrading it with `pip install --upgrade pip`. Ensure you have an active internet connection and that the package name is spelled correctly.

- **Conda command is not found**
  - *Issue:* The `conda` command is not recognized after installing Anaconda.

   - *Solution:* Ensure Anaconda is properly installed. It should be added to your system's PATH, or you can use Anaconda Prompt or Navigator to initiate `conda` commands.

- **Library compatibility and dependency errors**
   - *Issue:* Installing a new package disrupts an existing environment due to compatibility issues.
   - *Solution:* Use virtual environments to isolate dependencies. Pay attention to the versions of packages you install and consult the package documentation for version compatibility.

- **Slow performance or timeout errors**
   - *Issue:* Slow package installation or timeouts occur.
   - *Solution:* This could be due to a slow Internet connection or issues with the PyPI servers. Consider using a different PyPI mirror or trying the installation at another time.

- **Seeking help**
   - Learn to effectively seek assistance from online resources such as Stack Overflow, Python's documentation, and community forums.

## 1.4 Exercises and Questions

We encourage you to complete the following exercises and attempt the questions to reinforce your understanding of the material covered in this chapter.

### Exercises

Practice the following exercises to become comfortable with key functionalities:

1. Install Python and Set Up a Virtual Environment:
   - Download and install the latest version of Python from the official website.
   - Create a virtual environment using the `venv` module and activate it.
   - Document any issues you encounter during the installation or setup process, and describe how you resolved them.
2. Exploring IDEs:
   - Install two different IDEs mentioned in ► Sect. 1.3.3 (e.g., PyCharm and VS Code).
   - Compare their basic features.
3. Library installation and testing:
   - Using your active virtual environment, install the *Pandas* and *Matplotlib* libraries using both `pip` and `conda`.
   - Write a simple script in a Jupyter Notebook that imports these libraries.

### Answer the Following Questions

1. Describe the evolution of Python as a programming language. Include key milestones that have contributed to its popularity today.
2. What are the key characteristics of a Python environment that make it suitable for data analysis in biomedicine? Discuss at least two characteristics.
3. Compare and contrast the use of the `venv` module and Anaconda for creating virtual environments in Python. What are the basic advantages of each method?
4. Why is it important to verify your Python setup? Outline the steps you would take to ensure your Python environment is properly configured.

**Acknowledgement** The language of the human-generated text was corrected with the assistance of artificial intelligence (AI) tools [GPT-3.5 and GTP-4 from OpenAI]. GitHub Co-Pilot was used to check the correctness of the codes. The text underwent subsequent human revision to ensure its accuracy.

## References

1. Severance C. Guido van Rossum: The early years of Python. Computer. 2015;48(2):7–9. https://doi.org/10.1109/MC.2015.45.
2. Jiang C, Hua B, Ouyang W, Fan Q, Pan Z. PyGuard: finding and understanding vulnerabilities in Python virtual machines. In: 2021 IEEE 32nd international symposium on software reliability engineering (ISSRE), 25–28 October 2021; 2021. https://doi.org/10.1109/ISSRE52982.2021.00055.
3. Peters T. The Zen of Python: github.com/python/peps/blob/main/peps/pep-0020.rst. 2004. Available from: https://github.com/python/peps/blob/main/peps/pep-0020.rst.
4. TIOBE. TIOBE Index for May 2025. 2025. Available from: https://www.tiobe.com/tiobe-index/.
5. Luo J, Wu M, Gopukumar D, Zhao Y. Big data application in biomedical research and health care: a literature review. Biomed Inform Insights. 2016;8:1–10. https://doi.org/10.4137/BII.S31559.
6. Li R, Romano JD, Chen Y, Moore JH. Centralized and federated models for the analysis of clinical data. Annu Rev Biomed Data Sci. 2024;7(1):179–99. https://doi.org/10.1146/annurev-biodatasci-122220-115746.
7. Consortium I. Deciphering the impact of genomic variation on function. Nature. 2024;633(8028):47–57. https://doi.org/10.1038/s41586-024-07510-0.
8. Cock PJ, Antao T, Chang JT, Chapman BA, Cox CJ, Dalke A, et al. Biopython: freely available Python tools for computational molecular biology and bioinformatics. Bioinformatics. 2009;25(11):1422–3. https://doi.org/10.1093/bioinformatics/btp163.
9. Miles A, pyup.io bot, Rodrigues MF, Ralph P, Kelleher J, Schelker M, et al. Cggh/scikit-allel: V1.3.13. Zenodo, 17 September 2024. https://doi.org/10.5281/zenodo.13772087
10. RDKit. Open-source cheminformatics software. RDKit; 2024. rdkit.org.
11. Schrödinger L, DeLano W. The PyMOL molecular graphics system. Schrödinger, LLC; 2010. pymol.org.
12. Abadi M, Agarwal A, Barham P, Brevdo E, Chen Z, Citro C, et al. TensorFlow: large-scale machine learning on heterogeneous systems. 2015. tensorflow.org.
13. Pedregosa F, Varoquaux G, Gramfort A, Michel V, Thirion B, Grisel O, et al. Scikit-learn: machine learning in python. J Mach Learn Res. 2011;12:2825–30.
14. Bradski G. The OpenCV library. Dr Dobb's J Softw Tools. 2000;4:2236121.
15. van der Walt S, Schonberger JL, Nunez-Iglesias J, Boulogne F, Warner JD, Yager N, et al. scikit-image: image processing in Python. PeerJ. 2014;2:e453.

# Python Basics

## Contents




J. U. Kazi, *Python Essentials for Biomedical Data Analysis: An Introductory Textbook*,
https://doi.org/10.1007/978-3-031-85600-6_2

This chapter introduces the main aspects of Python programming relevant to biomedical data analysis. It begins with the basic syntax and operations that form the core of Python programming, providing a foundation for writing code. Key topics include variables, data types, and data structures, which are used to organize, store, manipulate, and manage information in complex biomedical datasets. The chapter also covers control structures, including loops and conditionals, which enable programs to make decisions and automate repetitive tasks. Together, these concepts offer a starting point for further exploration of advanced Python programming in biomedical applications.

**Learning Goals**

Learners will develop an understanding of Python's syntax and operations, including printing outputs, using comments, and performing arithmetic operations. They will gain familiarity with variables and data types, enabling them to define and manipulate variables across various data types such as strings, Booleans, lists, tuples, sets, and dictionaries. This chapter will introduce learners to control structures, including conditional statements, loops, and comprehensions, for managing program flow. Moreover, they will learn to implement error handling within control structures, equipping them to manage unexpected outcomes effectively.

## 2.1 Basic Syntax and Operations

The syntax of Python is renowned for its readability and simplicity—characteristics that have contributed to its widespread adoption [1]. Unlike many programming languages that require complex structures and repetitive code, the syntax of Python is straightforward and almost English-like. This section introduces basic Python syntax and operations, including printing output, performing arithmetic operations, and working with variables. These elements work together and can be used for various programming tasks.

### 2.1.1 Printing Output

**Notebook: Section 2.1.1. Printing Output**

This notebook provides code examples demonstrating how to print output in Python.

Link to the GitHub repository:

▸ https://github.com/sn-code-inside/BioPy

Go to: Chap. 2—▸ Sect. 2.1.1

One of the most basic functions in Python, as in many programming languages, is displaying output. In Python, this is achieved using the `print()` function. The purpose of `print()` function is to send data to the standard output device, typically the screen. It is capable of outputting strings, numbers, and even the results of complex expressions and function calls. For example: `print("Print this text!")`, when executed, Python evaluates the contents within the parentheses of the `print()` function and displays it. Here, the message `Print this text!` is displayed.

**Printing strings:** A string[1] can be directly passed to the `print()` function. Strings must be enclosed in quotes (either single `' '` or double `" "` quotes). For example: `print("Learning Python is fun!")` will result in an output `Learning Python is fun!`

**Printing numbers and expressions:** The `print()` function can also display numbers and arithmetic expressions. The command `print(5 + 10)` will return `15`.

**Printing multiple items:** We can print multiple items, for example, a string and a number, separated by commas within the `print()` function. Python will output these items separated by spaces by default.

For example: `print("The answer is:", 42)` returns `The answer is: 42`.

**Note:** In this example, we used the number 42 directly, but you can print any value or variable. The point is to show how you can print multiple items at once, not that you must use the exact numbers or words from the example.

**String concatenation:** We can concatenate[2] strings and variables within the `print()` function using the + operator. For example, the following code will print a name after `"I am"`. String concatenation is useful for creating new strings (discussed below), but it is not always the convenient way to combine values for printing.

```
name = "Jerry"
print("I am " + name)
```

Output:

```
I am Jerry
```

---

1 A string is a sequence of characters used to store and manipulate text-based information. Strings are enclosed within single quotes (' '), double quotes (" "), or triple quotes (""" """ or ''' ''') which allow for the inclusion of both single and double quotes within the text. They are immutable, meaning once a string is created, the characters within it cannot be changed. Python provides a rich set of methods to perform operations on strings, such as searching, slicing, replacing, and formatting, making strings one of the most utilized data types for handling text in programming tasks.

2 Concatenation refers to the process of joining two or more strings together into one continuous string. This is typically achieved using the "+" operator, which when applied between strings, appends the second string directly to the end of the first. For example, if we have two strings, string1 = "Hello," and string2 = "world!", concatenating these using string1 + string2 would result in the single string "Hello, world!". This operation is not limited to just two strings; multiple strings can be concatenated in the same way, creating longer sequences of text.

**Why print() Matters?**

In programming, especially when learning, it is important to have a way to see the output of your code. The `print()` function is a simple tool for debugging and understanding how data is being manipulated within your program. It is often used to display the state of variables, track the flow of execution, and as a tool to help the programmer understand how their code is working. The `print()` function in Python is a primary way of sending messages and data from a program to the user, making it an indispensable part of the Python toolkit.

### 2.1.2 Comments

Comments are used to explain the code, making it easier to understand the purpose and functionality of various sections or lines. In Python, comments are created using the # symbol. Comments are used to describe what the code is doing, the logic behind certain sections, or to specify any assumptions or important information. Such annotations are useful not just for others who may read the code later, but also for our future selves, particularly when revisiting complex sections. Furthermore, temporarily commenting out certain lines of code is a common practice for debugging. It allows us to isolate sections of code to identify issues without deleting them.

To insert a comment in Python, simply prefix the line with the # symbol. Everything following this symbol on the same line is treated as a comment and is ignored by the Python interpreter. For instance, writing `# This is a comment` will ensure that this piece of text has no impact on the code's execution.

There are several types of comments:

*Single-line comments:* As the name suggests, these extend only over a single line, e.g.,

```
# This is a single-line comment.
```

They are commonly used for brief annotations.

*Multi-line comments:* Unlike some other programming languages, Python does not have a specific syntax for multi-line comments. Instead, we can use the # symbol at the start of each line to create a block of comments:

```
# This is a multi line comment
# It spans multiple lines
# Each line starts with a #
```

*Docstrings:* Although not traditional comments, docstrings (enclosed in triple quotes, `""" """`) describe the purpose of a function or class. They resemble comments in their use for code documentation and are interpreted by certain tools to generate documentation automatically:

```
"""
Write our comments here.
"""
```

As a best practice, comments should be clear, concise, and relevant. Avoid redundant comments that do not add value or clarity. Moreover, it is important to update comments alongside the code they describe to maintain consistency. Effective commenting makes the code more accessible and maintainable, aiding both future readers and the original developers.

### 2.1.3 Arithmetic Operations

**Notebook: Section 2.1.3. Arithmetic Operations**
This notebook provides code examples demonstrating simple arithmetic operations.
Link to the GitHub repository:
► https://github.com/sn-code-inside/BioPy
Go to: Chap. 2—► Sect. 2.1.3

Arithmetic operations are widely used in data analysis and scientific computing [2]. They provide the means to perform calculations, analyze data, and solve numerical problems. These operations are necessary for manipulating datasets, creating algorithms, and developing models.

Several basic operations include:

- Addition (+): Used to add two numbers.

```
print(5 + 3)
```

Output:

```
8
```

- Subtraction (−): Used to subtract one number from another.

```
print(10 - 2)
```

Output:

```
8
```

- Multiplication (*): Used to multiply two numbers.

```
print(4 * 2)
```

Output:

```
8
```

- Division (/): Used to divide one number by another.

```
print(16 / 2)
```

Output:

```
8.0
```

Note: avoid division by zero. When we divide two numbers in Python, the result is automatically formatted as a floating-point number.[3] For instance, dividing 6 and 5 by 2, as in `6 / 2` and `5 / 2`, yields `3.0` and `2.5`. However, if we require a full number as a result, discarding any fractional part, we can use the integer division operator (//). For example, `10 // 3` will return `3`, efficiently truncating the decimal portion.

- Exponentiation (**): Used to raise a number to the power of another number:

```
print(2 ** 3)
```

Output:

```
8
```

3 A floating-point number is a numerical data type used in programming to represent real numbers that can contain fractional parts. These numbers are called "floating-point" because the decimal point can move (or "float"), allowing the representation of a wide range of values with varying degrees of precision. In most programming languages, floating-point numbers are represented using types like float or double, which differ in the precision and range they cover. This system allows computers to approximate real numbers efficiently, though it can sometimes introduce small rounding errors due to the way numbers are stored, which is an aspect to consider in high-precision calculations.

### 2.1.4 Assignment Operators



**Notebook: Section 2.1.4. Assignment Operators**
This notebook provides code examples demonstrating various assignment operators.
Link to the GitHub repository:
▸ https://github.com/sn-code-inside/BioPy
Go to: Chap. 2—▸ Sect. 2.1.4

Assignment operators in Python are used for assigning values to variables that enables the storage and manipulation of data [3]. These operators facilitate the organization and management of data, improving the ability to perform complex analyses and operations on datasets.

The basic assignment operator is denoted by =, where the variable name is placed on the left side and the value to be assigned on the right side:

```
count = 100
```

will assign an integer[4] value to "count"

```
name = "Aspirin"
```

will assign a string value to "name"

```
concentration = 1.5
```

will assign a floating-point number to "concentration"

Compound assignment operators combine arithmetic operations with assignment, streamlining code and reducing redundancy:

- Addition assignment += operator adds the right operand to the left operand and assigns the result to the left operand. For example, `x += 2` is equivalent to `x = x + 2`

4 An integer is a data type used to represent whole numbers without any fractional or decimal component. Unlike floating-point numbers, which can contain decimal points and are used for more precise calculations, integers are typically used when dealing with counts, indexes, and other scenarios where only whole numbers are required.

```
total_dose = 0        # Initialize total_dose to 0
total_dose += 50      # Increment total_dose by 50
print(total_dose)
```

Output:

```
50
```

- Subtraction assignment -= operator subtracts the right operand from the left operand and assigns the result to the left operand. For example, `x -= 2` is equivalent to `x = x - 2`

```
total_dose = 50        # Initialize total_dose to 50
total_dose -= 10       # Decrement total_dose by 10
print(total_dose)
```

Output:

```
40
```

- Multiplication assignment *= operator multiplies the left operand with the right operand and assigns the result to the left operand. For example, `x *= 2` is equivalent to `x = x * 2`:

```
total_dose = 3        # Initialize total_dose to 3
total_dose *= 2       # Multiplying total_dose by 2
print(total_dose)
```

Output:

```
6
```

- Division assignment /= operator divides the left operand by the right operand and assigns the result to the left operand. For example, `x /= 2` is equivalent to `x = x / 2`.

```
total_dose = 3        # Initialize total_dose to 3
total_dose /= 2       # Dividing total_dose by 2
print(total_dose)
```

Output:

```
1.5
```



- Exponentiation assignment `**=` operator raises the left operand to the power of the right operand and assigns the result to the left operand. For example, `x **= 2` is equivalent to `x = x ** 2`.

```
total_dose = 3          # Initialize total_dose to 3
total_dose **= 2        # Square total_dose
print(total_dose)
```

Output:

```
9
```

Assignment operators are often utilized in biomedical data analysis for tasks such as initializing variables to store data, iteratively updating values during data processing, and managing the state of variables in simulation models. For instance, in pharmacokinetics, assignment operators might be used to update the concentration of a drug in the bloodstream over time. The use of assignment operators facilitates advanced data manipulation and analysis.

### 2.1.5 Comparison Operators

**Notebook: Section 2.1.5. Comparison Operators**

This notebook provides code examples demonstrating various comparison operators.

Link to the GitHub repository:

► https://github.com/sn-code-inside/BioPy

Go to: Chap. 2—► Sect. 2.1.5

Comparison operators evaluate measurements against reference values or thresholds [4]. These operators can be used to assess a patient's response to treatment and, thereby, compare whether outcomes are above or below expected levels. Moreover, they help in categorizing data based on specific criteria, such as age groups or risk factors, thereby aiding in the programming logic and decision-making processes that are useful in clinical and research environments.

Here is an overview of the primary comparison operators in Python:

- **Equal to** == operator verifies if the values of two operands are equal. If so, the condition becomes true.

```
print(2 == 3)
```

Output:

```
False
```

- **Not equal to** != operator verifies if the values of two operands are not equal. If the values are not equal, the condition becomes true.

```
print(7 != 2)
```

Output:

```
True
```

- **Greater than** > operator verifies if the value of the left operand is greater than the value of the right operand. If so, the condition becomes true.

```
print(5 > 3)
```

Output:

```
True
```

- **Less than** < operator verifies if the value of the left operand is less than the value of the right operand. If so, the condition becomes true.

```
print(5 < 3)
```

Output:

```
False
```

- **Greater than or equal to** >= operator verifies if the left operand is greater than or equal to the right operand.

```
print(8 >= 8)
```

Output:

```
True
```

- **Less than or equal to** <= operator verifies if the left operand is less than or equal to the right operand.

```
print(8 <= 8)
```

Output:

```
True
```

When comparing values to `None`, it is advisable to avoid using == and !=. Instead, use `is` or `is not`. These operators offer a more precise and intended meaning for comparisons with singletons[5] like `None`. This practice is important for ensuring accurate logic flows in programs, particularly when handling variables that may be set to `None` under certain conditions.

## 2.2 Variables and Data Types

Variables and data types enable efficient data organization and manipulation. Variables serve as containers for storing data, while data types define the kind of data being stored, such as integers, strings, or Booleans. In this section, we will explore how to declare and work with variables, the common data types in Python, and how these elements interact to manage and process data in various applications.

5 A singleton refers to an object of which only one instance exists within a given scope, typically within an application or system. This concept is commonly implemented in software design to ensure that a class has only one instance while providing a global point of access to that instance. Singletons are used in various scenarios where exactly one object is needed to coordinate actions across the system, such as in managing configurations, handling database connections, or implementing logging functionalities. The singleton pattern ensures controlled access and consistent behavior from different parts of an application.

### 2.2.1 Introduction to Variables

**Notebook: Section 2.2.1. Introduction to Variables**
This notebook provides code examples demonstrating various variables.
Link to the GitHub repository:
▶ https://github.com/sn-code-inside/BioPy
Go to: Chap. 2—▶ Sect. 2.2.1

Variables in Python are used to store data [5]. They can be seen as containers or storage locations in the memory of computer, where data can be held and manipulated during program execution. The primary purpose of variables is to label and keep track of data values with descriptive names, making the code more readable and manageable. They enable the dynamic handling of data, facilitate calculations, and help control the flow of a program.

Python has specific rules and conventions for naming variables:

- Names must start with a letter or an underscore. For example, `_name` and `name` are valid, but `1name` is not valid.
- Names can only contain letters, numbers, and underscores; spaces or special characters are not allowed.
- Variable names are case-sensitive. This means `gender`, `Gender`, and `GENDER` are three different variables.
- Python's reserved words like `if`, `else`, `for`, `while`, and others that have special meanings in Python should not be used as variable names.
- Choose names that are descriptive of the purpose of the variable (e.g., `patient_age` instead of just `pa`).

Assigning a value to a variable in Python is straightforward. We simply use the equals = sign to assign a value to a variable. For example, `height = 170` assigns the value `170` to the variable `height`. Python also allows for variable reassignment, where the value of a variable can be updated or changed to a different value or type. If we then execute `height = 180`, the value of `height` is now `180`.

Variables are extensively used in calculations. They can represent numbers, strings, or more complex data structures and can be manipulated using various operators and functions. For example:

- Arithmetic calculations:

```
price_per_item = 10
quantity = 100
total_cost = price_per_item * quantity
print(total_cost)
```

Output:

```
1000
```



In this example, we use variables to store values and perform an arithmetic operation. The variable `price_per_item` is assigned the value 10, representing the cost of one item, while `quantity` is assigned the value 100, representing the number of items being purchased. These two variables are then multiplied together in the line `total_cost = price_per_item * quantity` to calculate the total cost. The result (10 * 100 = 1000) is stored in the variable `total_cost`, and the `print(total_cost)` statement outputs this value to the screen.

- String concatenation:

```
first_name = 'Tom'
last_name = 'Hanks'
full_name = first_name + " " + last_name
print(full_name)
```

Output:

```
Tom Hanks
```

Here we used string concatenation to combine two separate strings. The variable `first_name` is assigned the value `'Tom'`, and the variable `last_name` is assigned the value `'Hanks'`. To create the full name, the `full_name` variable is assigned the result of concatenating `first_name` and `last_name` using the + operator, with a space (`" "`) added in between to separate the two names. The `print(full_name)` statement then outputs the combined result, which is `'Tom Hanks'`, to the screen. This demonstrates how strings can be joined together in Python to form a complete message or name using variables.

- Storing and manipulating data:

```
heart_rates = [72, 78, 85, 95, 76, 88, 82, 80]
average_heart_rate = sum(heart_rates) / len(heart_rates)
print(average_heart_rate)
```

Output:

```
82.0
```

In this example, we store and manipulate data using a list of heart rates. The variable `heart_rates` is assigned a list of values representing individual heart rate measurements. To calculate the average heart rate, we use the `sum()` function to get the total of all the heart rates and divide it by the number of entries in the list, which is determined using the `len()` function. The result is then stored in the variable `average_heart_rate`. Finally, the `print(average_heart_rate)` statement outputs the calculated average, which is 82.0, demonstrating how Python can be used to easily store, manipulate, and compute values from a dataset.

In biomedical data analysis, variables might represent data points like blood pressure readings, medication dosages, patient identifiers, and much more. Proper use of variables allows for efficient and accurate calculations and manipulations necessary in data analysis.

### 2.2.2 Basic Data Types

Basic data types are the building blocks used to construct and manipulate data [6]. These include integers (`int`) for whole numbers, floating-point numbers (`float`) for decimals, strings (`str`) for text, and Booleans (`bool`) for true/false values. Each type serves a specific purpose, allowing programmers to accurately represent and work with different kinds of data. The "type()" function can be used to determine the data type in a variable. For example, if "var_x = 15", then "type(var_x)" will return "int", suggesting "var_x" is an integer.

**Integers (`int`):** Used for whole numbers without a fractional part, integers are ideal for counts, indexes, and scenarios where decimal precision is not needed.

**Floating-point numbers (`float`):** These represent real numbers with a decimal point, suitable for measurements, probabilities, and calculations requiring fractional values.

**Strings (`str`):** Sequences of characters used to store and manipulate text, enclosed in single quotes (' '), double quotes (" "), or triple quotes (''' ''' or """ """). Single and double quotes are commonly used for simple text assignments, while triple quotes are beneficial for strings containing both single and double quotes.

**Booleans (`bool`):** Represent logical values, either `True` or `False`. These are often the result of conditions or comparisons. Numerically, `True` is equivalent to `1`, and `False` is equivalent to `0`. Generally, non-zero and non-null values are interpreted as `True`, while `zero`, `None`, and empty collections (such as an empty list `[ ]`) are considered `False`.

### 2.2.3 Basics of String Manipulations

> **Notebook: Section 2.2.3. Basics of String Manipulations**
> This notebook provides code examples demonstrating various basics of string manipulations.
> Link to the GitHub repository:
> ▶ https://github.com/sn-code-inside/BioPy
> Go to: Chap. 2—▶ Sect. 2.2.3

Strings can be manipulated using built-in methods to perform operations like concatenation (joining strings together), repetition (repeating the same text), indexing (marking each character), slicing (extracting portions of strings), and replacement (changing specific parts of a string). Strings also support methods for case conversion, trimming whitespace, and finding substrings or characters. These capabilities make string manipulation a useful skill in tasks ranging from simple data cleaning to complex text analysis. These methods automate the manipulation of textual data, handling diverse text-processing tasks effectively.

Concatenation (also discussed sections in print and variables) is used to join multiple strings:

```
string1 = "String"              # Define the first string
string2 = "concatenation!"      # Define the second string

# Concatenating strings by adding a space between them
concatenated_string = string1 + " " + string2

# Print the concatenated string
print(concatenated_string)
```

Output:

```
String concatenation!
```

Repetition method for creating repeated text:

```
string = "Repeat"           # Define the string to be repeated

# Repeating the string 3 times
repeated_string = string * 3

# Print the repeated string
print(repeated_string)
```

Output:

```
RepeatRepeatRepeat
```

Each character in a Python string has an index, starting from 0 for the first character. We can access individual characters using these indices:

```
string = "Python"                          # Define the string

# Accessing characters by index
first_character = string[0]          # Access the first character 'P'
third_character = string[2]          # Access the third character 't'

# Print the accessed characters with descriptive messages
print("First character:", first_character)
print("Third character:", third_character)
```

Output:

```
First character: P
Third character: t
```

Python also supports negative indexing, where −1 refers to the last character, −2 to the second last, and so on:

```
string = "Python"                      # Define the string
# Accessing characters using negative indices
last_character = string[-1]        # Access the last character 'n'
second_last_character = string[-2]
# Access the second-to-last character 'o'

# Print the accessed characters with descriptive messages
print("Last character:", last_character)
print("Second last character:", second_last_character)
```

Output:

```
Last character: n
Second last character: o
```

Slicing is used to obtain a substring from the string. The syntax is `string[start:end]`, where `start` is inclusive and `end` is exclusive:

```
string = "Python"                  # Define the string

# Slicing the string to get a substring
substring = string[1:4]        # Extracts "yth" from "Python"

# Print the sliced substring with a descriptive message
print("Substring:", substring)
```

Output:

```
Substring: yth
```



We can also use an optional **step** parameter in the slicing syntax (`string[start:end:step]`) and omit `start` and/or `end` (`string[:end:step]` or `string[::step]`):

```
string = "Python"                                   # Define the string

# Slicing the string to get every second character
every_second_character = string[::2]  # "Pto" (every second character)

# Reversing the string using slicing
reverse_string = string[::-1]  # "nohtyP" (reverses the string)

# Print the results of the slicing operations with descriptive messages
print("Every second character:", every_second_character)
print("Reversed string:", reverse_string)
```

Output:

```
Every second character: Pto
Reversed string: nohtyP
```

Python provides numerous methods for string manipulation, such as `upper()` for converting to uppercase, `lower()` for lowercase, `strip()` for removing whitespace, `replace()` for replacing substrings, and `split()` for dividing a string into a list of substrings.

The `upper()` method converts all characters in a string to uppercase:

```
string = "Learning Python is fun"          # Define the initial string
# Using the .upper() method to convert the entire string to uppercase
uppercase_string = string.upper()
# Print the converted uppercase string
print(uppercase_string)
```

Output:

```
LEARNING PYTHON IS FUN
```

Conversely, the `lower()` method converts all characters in a string to lowercase:

```
string = "Learning Python is fun"      # Define the initial string
# Using the .lower() method to convert the entire string to lowercase
lowercase_string = string.lower()
# Print the converted lowercase string
print(lowercase_string)
```

# Output:

```
learning python is fun
```

The `strip()` method removes any leading and trailing whitespaces (including tabs, newlines) from the string:

```
string = "Learning Python is fun" # Define the initial string with extra spaces
# Using the .strip() method to remove all leading and trailing spaces
stripped_string = string.strip()
# Print the string after stripping whitespace
print(stripped_string)
```

Output:

```
Learning Python is fun
```

The `replace()` method replaces a specified substring with another substring:

```
string = "Learning Python is fun"
# Replace the substring 'Python' with 'Programming'
replaced_string = string.replace("Python", "Programming")
# Print the string after replacement
print(replaced_string)
```

Output:

```
Learning Programming is fun
```

The `split()` method divides a string into a list of substrings based on a specified delimiter (default is whitespace):

2

```
string = "Learning Python is fun"  # Define the initial string
# Using the .split() method without specifying a delimiter to
# split the string by whitespace
split_list = string.split()
# Print the list of words
print(split_list)
```

Output:

```
['Learning', 'Python', 'is', 'fun']
```

Splitting by a specific delimiter:

```
# Define the initial string containing comma-separated items
string = "apple,banana,cherry"

# Using the .split() method with ',' as the delimiter
split_list = string.split(',')
# Print the list of separated words
print(split_list)
```

Output:

```
['apple', 'banana', 'cherry']
```

String formatting methods, like the `format()` function or f-strings (e.g., `f"Hello, {name}!"`), are used for creating strings with dynamic content:

Using the `format ()` Method

```
# Basic usage of the format() method
name = " Jerry "
# Substitute {} with the value of 'name'
greeting = "Hello, {}!".format(name)
print(greeting)
```

Output:

```
Hello, Jerry!
```

```
# More complex example with multiple placeholders
name = " Jerry "
age = 30
# Substitute the first {} with 'name' and the second with 'age'
intro = "My name is {} and I am {} years old.".format(name, age)
print(intro)
```

Output:

```
My name is Jerry and I am 30 years old.
```

Note: In this case, the method automatically matches each argument to the corresponding placeholder in the order they are given. However, you can also specify the order manually by using numbered placeholders like {0} and {1}, which refer to the position of each argument in the tuple passed to the method.

```
# Formatting with positional arguments
product = "book"
price = 19.991
# Using positional arguments to format strings.
#{0} is reused, {1:.2f} formats the price to 2 decimal places.
order = "I bought a {0} for ${1:.2f}. It's a great {0}!".
format(product, price)
print(order)
```

Output:

```
I bought a book for $19.99. It's a great book!
```

Using f-strings (formatted string literals): An f-string is a way to create formatted strings in Python by placing an f before the quotation marks. You can include variables or expressions inside curly braces {} within the string, and Python will automatically replace them with their values.

```
# Basic example with f-strings
name = "Jerry"
# Directly insert the value of 'name' into the string
greeting_f = f"Hello, {name}!"
print(greeting_f)
```

Output:



```
Hello, Jerry!
```

```
# Complex example with multiple variables
name = "Jerry"
age = 30
intro_f = f"My name is {name} and I am {age} years old."
# Insert 'name' and 'age' into the string
print(intro_f)
```

Output:

```
My name is Jerry and I am 30 years old.
```

```
# Inline expressions within f-strings
hours = 7
minutes = 45
time_message = f"The current time is {hours}:{minutes:02d} AM."
# Format 'minutes' to always show two digits
print(time_message)
```

Output:

```
The current time is 7:45 AM.
```

Note: Inside the f-string, {hours} is replaced with the value of hours, and {minutes:02d} is replaced with the value of minutes, formatted to always show two digits (for example, "05" instead of "5").

```
# Using f-strings with calculations
total_cost = 150.75
discount = 20.25
# Perform the calculation inside the f-string and format the
# result to two decimal places
final_price = f"After a discount of ${discount},"\
f"the total cost is ${total_cost - discount:.2f}."
print(final_price)
```

Output:

```
After a discount of $20.25, the total cost is $130.50.
```

### 2.2.4 Boolean Expressions and Logical Operators

**Notebook: Section 2.2.4. Boolean Expressions and Logical Operators**
This notebook provides code examples demonstrating Boolean expressions and logical operators.

Link to the GitHub repository:
► https://github.com/sn-code-inside/BioPy
Go to: Chap. 2—► Sect. 2.2.4

Boolean expressions are operations that yield a `True` or `False` outcome. They are often used in control flow statements, such as `if` conditions. Python uses logical operators like `and`, `or`, `not` to combine or invert Boolean expressions. For instance, `True` and `False` evaluates to `False`, `True` or `False` evaluates to `True`, and `not True` evaluates to `False`.

Here is an example code demonstrating these concepts:

Boolean values from comparisons:

```
a = 5
b = 3
print(a > b)  # True, since 5 is greater than 3
```

Output:

```
True
```

Numerical treatment of True and False:

```
print(True + True)  # True is treated as 1
```

Output:

```
2
```

```
print(True * 10)    # True is treated as 1
```

Output:

```
10
```

```
print(False * 10)   # False is treated as 0
```



Output:

```
0
```

Non-zero/non-null values as True:

```
print(bool(10))  # True, because a non-zero number is considered True
```

Output:

```
True
```

```
print(bool("Python"))  # True, because a non-empty string is considered True
```

Output:

```
True
```

Zero, none, and empty collections as False:

```
print(bool(0))   # False, because zero is treated as False
```

Output:

```
False
```

```
print(bool(None))  # False, because None is a null value, thus False
```

Output:

```
False
```

```
print(bool([]))  # False, because an empty list is considered False
```

Output:

```
False
```

Boolean expressions with logical operators:

```
print(True and False)  # False, because the and operator returns
# True only if both operands are True
```

Output:

```
False
```

```
print(True or False)  # True, because the or operator returns
# True if at least one operand is True
```

Output:

```
True
```

```
print(not True)  # False, because not negates True to False
```

Output:

```
False
```

### 2.2.5 Type Conversion

**Notebook: Section 2.2.5. Type Conversion**
This notebook provides code examples demonstrating type conversion in Python.
Link to the GitHub repository:
► https://github.com/sn-code-inside/BioPy
Go to: Chap. 2—► Sect. 2.2.5

Type conversion is often needed for data preprocessing, such as converting strings read from a file into numerical data for calculation, or changing numerical IDs into strings for output formatting. Proper use of type conversion helps ensure that data is handled accurately and without errors.

**Implicit type conversion**, also known as automatic type conversion, is performed by Python automatically when it is necessary. This typically happens in operations involving multiple data types where Python converts one type to another to avoid errors.

- *Adding an integer to a floating-point number:*
- `5` (int) + `3.2` (float) automatically converts the integer to a float, resulting in `8.2` (float)

- *Combining a string with a number using print:*
- `print("The value is " + 42)` will cause an error, but `print("The value is", 42)` works because print implicitly converts the number into a string.

**Explicit type conversion**, or type casting, involves converting the data type of an object to another data type explicitly by the programmer. This is done using functions like `int()`, `float()`, `str()`, etc.

- *Converting a float to an integer:*
- `int(5.7)` will result in `5`.

- *Converting an integer to a string:*
- `str(100)` will result in `'100'`.

Before performing type conversion, you need to understand the data types you are working with and the expected outcome. Particularly when casting user inputs or data from uncertain sources, validate the data to ensure it can be safely and meaningfully converted. Leverage Python's built-in functions for conversion and avoid creating unnecessary custom conversion logic. Exercise caution when converting from a more informative type, such as a `float`, to a less informative type like an `int`, as this can lead to a loss of precision. Attempting to convert a value to an incompatible type, such as converting the string `abc` to an integer, results in a `ValueError`. Furthermore, misunderstandings about implicit type conversion can lead to bugs, especially in complex expressions.

### 2.2.6 Complex Data Types

Complex data types commonly used in biomedical data analysis include lists, tuples, sets, and dictionaries. Lists and tuples both represent ordered sequences of data, such as time series measurements from a patient. However, tuples are immutable, which helps maintain fixed data integrity. Sets are ideal for handling unique elements, eliminating duplicates, which is useful for managing distinct sets of genes or symptoms. Dictionaries are useful for mapping relationships, such as linking patient IDs to their medical records, due to their key-value pairing which allows for fast retrieval and linking of related information.

### 2.2.7 Lists

**Notebook: Section 2.2.7. Lists**

This notebook provides code examples demonstrating complex data types—list.

Link to the GitHub repository:

► https://github.com/sn-code-inside/BioPy

Go to: Chap. 2—► Sect. 2.2.7

Lists are created by enclosing elements within square brackets [ ], with the elements separated by commas. These elements can be of various types, including numbers, strings, or even other lists. To access elements in a list, Python uses their index, which starts at 0 for the first element. In addition, Python supports negative indexing, allowing access to elements from the end of the list, starting with −1 for the last element. This indexing system provides a flexible way to reference list contents from both the beginning and the end.

Creating a list with different types of elements:

```
# A list containing an integer, a string, a float, and another list
my_list = [1, "Python", 3.14, [2, 4, 6]]
print(my_list)
```

Output:

```
[1, 'Python', 3.14, [2, 4, 6]]
```

Accessing elements using positive indexing:

```
first_element = my_list[0]  # Access the first element (1)
second_element = my_list[1]  # Access the second element ("Python")
print("First element:", first_element)
print("Second element:", second_element)
```

Output:

```
First element: 1
Second element: Python
```

Accessing elements using negative indexing:

```
last_element = my_list[-1]  # Access the last element ([2, 4, 6])
# Access the second to last element (3.14)
second_last_element = my_list[-2]
print("Last element:", last_element)
print("Second to last element:", second_last_element)
```

Output:

```
Last element: [2, 4, 6]
Second to last element: 3.14
```

### 2.2.8 List Operations

**Notebook: Section 2.2.8. Lists Operations**
This notebook provides code examples demonstrating complex data types—list operations.
Link to the GitHub repository:
▶ https://github.com/sn-code-inside/BioPy
Go to: Chap. 2—▶ Sect. 2.2.8

Lists are dynamic and, therefore, allow various modifications through different operations and methods. Some common operations include concatenation, which combines multiple lists, and replication, which repeats a list a specified number of times. There are also several useful methods for manipulating lists including `append()` to add an element to the end of a list, `remove()` to delete a specified element, `sort()` to arrange the elements in ascending or custom order, and `reverse()` to flip the order of the elements in the list. Furthermore, slicing is a feature that enables access to a specific range of elements within the list, allowing for efficient extraction and manipulation of list data.

List concatenation:

```
list1 = [1, 2, 3]              # Define the first list
list2 = [4, 5, 6]              # Define the second list
combined_list = list1 + list2  # Concatenate list1 and list2
print("Concatenated List:", combined_list)
```

Output:

```
Concatenated List: [1, 2, 3, 4, 5, 6]
```

The + operator is used to combine two lists into one extended list. This does not modify the original lists but instead creates a new list that includes elements from both original lists in the order they appear.

List replication:

```
list1 = [1, 2, 3]                 # Define the first list
replicated_list = list1 * 2       # Replicate list1 two times
print("Replicated List:", replicated_list)
```

# Output:

```
Replicated List: [1, 2, 3, 1, 2, 3]
```

The * operator is used to repeat a list a specified number of times. This is useful when we need multiple copies of the list elements in a single list.

To add an element:

```
list1 = [1, 2, 3]          # Define the initial list
list1.append(7)            # Append the number 7 to the end of list1
print("After append:", list1)
```

Output:

```
After append: [1, 2, 3, 7]
```

The method `.append(item)` directly modifies the original list by adding `item` to its end.

To remove an element:

```
list1 = [1, 2, 7, 3, 7]  # Define the initial list
list1.remove(7)  # Remove the first occurrence of the number 7 from list1
print("After remove:", list1)
```

Output:

```
After remove: [1, 2, 3, 7]
```

The method `.remove(value)` targets the first instance of value in the list and removes it. If the value is not found, Python raises a `ValueError`.

To sort the list:

```
unsorted_list = [3, 1, 4, 2]   # Define the initial unsorted list
# Sorts the list in place in ascending order
unsorted_list.sort()
# Print the Sorted List
print("Sorted List:", unsorted_list)
```

Output:

```
Sorted List: [1, 2, 3, 4]
```

The method `.sort()` sorts the list items in ascending order. This method changes the order of the list items permanently. Sorting order can be customize by passing arguments to the `sort()` method. For example, the argument `reverse = True` will sort the list in descending order, and the argument `key = function` allows specifying a function of one argument that is used to extract a comparison key from each list element (e.g., `key = str.lower`, `key = len`).

Sorting in descending order:

```
unsorted_list.sort(reverse=True)
print("Descending Sorted List:", unsorted_list)
```

# Output:

```
Descending Sorted List: [4, 3, 2, 1]
```

To reverse the order of list elements:

```
unsorted_list = [3, 1, 4, 2]  # Define an unsorted list
# Reverses the order of elements in the list
unsorted_list.reverse()
print("Reversed List:", unsorted_list)
```

# Output:

```
Reversed List: [2, 4, 1, 3]
```

The method `.reverse()` directly modifies the list by inverting the order of all elements, so that the first element becomes the last and the last becomes the first.

### 2.2.9 Slicing a List

**Notebook: Section 2.2.9. Slicing a List**
This notebook provides code examples demonstrating complex data types—slicing a list.

Link to the GitHub repository:
▶ https://github.com/sn-code-inside/BioPy
Go to: Chap. 2—▶ Sect. 2.2.9

Basic slicing syntax is `list[start:end]` extracts elements from `start` to `end - 1`. If `start` is omitted, it defaults to the beginning of the list; if `end` is omitted, it goes to the end:

```
# Create a list for slicing
slice_list = [1, 10, 20, 30, 40, 50, 60, 70, 80, 90]
```

```
# Slicing - getting elements from index 2 to 5
sub_list = slice_list[2:6]
print("Sliced List (2 to 5):", sub_list)
```

Output:

```
Sliced List (2 to 5): [20, 30, 40, 50]
```

```
# Slicing - getting elements from the beginning to index 3
start_slice = slice_list[:4]
print("Sliced from Start to 3:", start_slice)
```

Output:

```
Sliced from Start to 3: [1, 10, 20, 30]
```

```
# Slicing - getting elements from index 5 to the End
end_slice = slice_list[5:]
print("Sliced from 5 to End:", end_slice)
```

Output:

```
Sliced from 5 to End: [50, 60, 70, 80, 90]
```



### 2.2.10 Iterating Over Lists

**Notebook: Section 2.2.10. Iterating Over Lists**
This notebook provides code examples demonstrating complex data types—iterating over lists.
Link to the GitHub repository:
▶ https://github.com/sn-code-inside/BioPy
Go to: Chap. 2—▶ Sect. 2.2.10

Lists can be iterated over using a `for` loop. This is a common way to access and perform operations on each element within a list. During each iteration of the loop, the variable defined in the `for` statement takes on the value of the next element in the list. This feature is particularly useful for tasks like processing data, performing calculations on list elements, or even creating new lists based on the existing ones.

Basic iteration over a list:

```
numbers = [10, 12, 23] # Define a list of numbers
# Iterate over the list using a for loop
for num in numbers:
    print(num)
```

Output:

```
10
12
23
```

The `for` loop goes through each element in the list named `numbers`. The variable `num` takes on the value of each element sequentially from the first to the last.

Performing operations on each element:

```
numbers = [2, 4, 5] # Define a list of numbers
# Iterate over the list and perform an operation (e.g., squaring
# each number)
for num in numbers:
    squared = num ** 2  # Square the current number
    print(f"Square of {num} is {squared}")
```

Output:

```
Square of 2 is 4
Square of 4 is 16
Square of 5 is 25
```

As the `for` loop progresses through each item in the list, it calculates the square of the current number `num` by raising it to the power of two (`num ** 2`).

Creating a new list from an existing list:

```
# Define the original list with integers
original_list = [11, 2, 10, 4, 5]
# Initialize an empty list to hold modified values
modified_list = []
# Iterate over each number in the original list
for num in original_list:
    # Double the current number and append it to the new list
    modified_list.append(num * 2)
# Print the new list that contains doubled values
print("Doubled List:", modified_list)
```

Output:

```
Doubled List: [22, 4, 20, 8, 10]
```

Each element in the `original_list` is accessed sequentially through the loop, where it is doubled (`num * 2`). The result is then added to the `modified_list` using the `.append()` method.

Iterating with index and element:

```
items = ['apple', 'banana', 'cherry']  # Define a list of fruit items
# Use enumerate() to iterate over the list, to obtain both the
# index and the element
for index, item in enumerate(items):
    # Print the index and the corresponding item from the list
    print(f"Item {index}: {item}")
```

Output:

```
Item 0: apple
Item 1: banana
Item 2: cherry
```

The `enumerate()` function allows us to loop over an iterable (like lists, tuples, strings, etc.) and has an automatic counter. It adds an index to each item of the iterable and returns it as an enumerate object. The enumerate object is an iterator that produces pairs containing a count (from start, which defaults to 0) and the corresponding value from the iterable. For example, `for index, item in enumerate(items)` allows us to access the index (0, 1, 2) and the value (item) from `items` during each iteration.

Nested loops for multidimensional lists:

```
# Define a 2D list (list of lists) representing a matrix
matrix = [[1, 2, 3], [4, 5, 6], [7, 8, 9]]

# Iterating over each sublist in the matrix
for sublist in matrix:
    # Iterating over each element in the sublist
    for element in sublist:
        # Print each element followed by a space,
        #without moving to a new line immediately
        print(element, end=' ')
    # Print a newline after each sublist to separate rows of the matrix
    print()
# This results in each row of the matrix being printed on a new line
```

Output:

```
1 2 3
4 5 6
7 8 9
```

Nested loops are loops within loops (discussed further in ▶ Sect. 2.3.1). They are used to iterate over the elements of multidimensional lists (or other complex data structures). In the context of a 2D list, for example, the outer loop iterates over each sublist, while the inner loop iterates over each element within the current sublist. This allows us to access and manipulate each element within the multidimensional array.

### 2.2.11 Tuples

**Notebook: Section 2.2.11. Tuples**

This notebook provides code examples demonstrating complex data types—tuples.

Link to the GitHub repository:

▶ https://github.com/sn-code-inside/BioPy

Go to: Chap. 2—▶ Sect. 2.2.11

Tuples are immutable collections in Python. Therefore, once tuples are created, their elements cannot be changed, added to, or removed. This immutability contrasts sharply with lists, which are mutable and allow for modification after creation. Tuples are defined by enclosing their elements within parentheses `( )`. For example:

```
my_tuple = (1, 'a', 3.14)
```

Once a tuple is created, any attempt to modify its contents, such as changing an element or appending new elements, will result in a `TypeError`. Tuples are useful when a collection of items should remain unchanged throughout the program, ensuring that the data is write-protected. They are ideal for storing data that requires protection from modifications, thus maintaining data integrity. Because tuples are immutable, they are generally faster to access than lists, making them suitable for optimizing performance in situations where the data is static and only read operations are required.

Creating a tuple:

```
my_tuple = (1, "Python", 3.14)
print("This is a tuple:", my_tuple)
```

Output:

```
(1, 'Python', 3.14)
```

Attempting to change a tuple's element (will raise an error):

```
my_tuple = (1, "Python", 3.14)
try:
    my_tuple[1] = "world"
except TypeError as e:
    print(e)
```

Output:

```
'tuple' object does not support item assignment
```

Using tuples over lists:



```
# Using a tuple for constant values
days_of_week = ('Monday', 'Tuesday', 'Wednesday', 'Thursday',
'Friday', 'Saturday', 'Sunday')
# Iterating over a tuple (read-only operation)
for day in days_of_week:
    print(day)
```

Output:

```
Monday
Tuesday
Wednesday
Thursday
Friday
Saturday
Sunday
```

### 2.2.12 Sets

**Notebook: Section 2.2.12. Sets**

This notebook provides code examples demonstrating complex data types—sets.

Link to the GitHub repository:

▶ https://github.com/sn-code-inside/BioPy

Go to: Chap. 2—▶ Sect. 2.2.12

Sets are mutable collections of unordered, unique elements. They can be created by placing items inside curly braces `{ }` or by using the `set()` function on an iterable. Since sets are unordered, their elements do not have a defined order and cannot be accessed by index. Further, sets automatically enforce uniqueness by removing duplicate entries.

Define a set of fruits using curly braces:

```
# Creating a set with curly braces
fruits = {'apple', 'banana', 'cherry'}
print("Fruits Set:", fruits)
```

Output:

```
Fruits Set: {'apple', 'banana', 'cherry'}
```

Creating a set from a list to remove duplicates:

```
# Define a list with duplicate numbers
numbers_list = [5, 3, 2, 3, 4, 2, 4, 5]
# Convert the list to a set, automatically removes duplicates
numbers_set = set(numbers_list)
print("Numbers Set:", numbers_set)
```

Output:

```
Numbers Set: {2, 3, 4, 5}
```

### 2.2.13 Basic Set Operations

**Notebook: Section 2.2.13. Basic Set Operations**
This notebook provides code examples demonstrating complex data types—basic set operations.

Link to the GitHub repository:
▶ https://github.com/sn-code-inside/BioPy
Go to: Chap. 2—▶ Sect. 2.2.13

Sets support several basic operations that enable efficient handling and manipulation of grouped data. These operations are used to perform mathematical set operations programmatically.

Union (|): The union operation combines all elements from two sets, omitting duplicates. It results in a set that contains every element from both sets, with each element appearing only once.

Union of two sets:

```
set_a = {1, 2, 3}
set_b = {3, 4, 5}
set_union = set_a | set_b
print("Union:", set_union)
```

Output:

```
Union: {1, 2, 3, 4, 5}
```

Intersection (&): Intersection retrieves only the elements common to both sets. This operation is useful for finding shared data between datasets.

Intersection of two sets:

```
set_a = {1, 2, 3}
set_b = {3, 4, 5}
set_intersection = set_a & set_b
print("Intersection:", set_intersection)
```

Output:

```
Intersection: {3}
```

Difference (−): The difference method returns the elements that are present in the first set but not in the second. This operation is useful for identifying items that are unique to one set.

Difference of two sets:

```
set_a = {1, 2, 3}
set_b = {3, 4, 5}
set_difference = set_a - set_b
print("Difference:", set_difference)
```

Output:

```
Difference: {1, 2}
```

Symmetric difference (^): The symmetric difference operation returns the elements that are in either of the sets but not in both. In other words, it produces all items from both sets, excluding those that are common to both.

Symmetric difference of two sets:

```
set_a = {1, 2, 3}
set_b = {3, 4, 5}
set_sym_diff = set_a ^ set_b
print("Symmetric difference:", set_sym_diff)
```

Output:

```
Symmetric Difference: {1, 2, 4, 5}
```

### 2.2.14 Methods to Modify Sets

**Notebook: Section 2.2.14. Methods to Modify Sets**

This notebook provides code examples demonstrating complex data types—methods to modify sets.

Link to the GitHub repository:

▸ https://github.com/sn-code-inside/BioPy

Go to: Chap. 2—▸ Sect. 2.2.14

As sets are mutable data structures, elements can be added or removed after creating a set. Python provides several methods to modify sets.

`add()`: The `add()` method inserts a single element into the set. If the element is already in the set, the set does not change.

Adding an element to the set:

```
fruits = {'apple', 'banana', 'cherry'}
fruits.add('mango')  # Adds 'mango' to the set
print("After add:", fruits)
```

Output:

```
After add: {'apple', 'banana', 'mango', 'cherry'}
```

`remove()`: The `remove()` method removes a specific element from the set. The method raises a `KeyError` if the element is not preset.

Removing an element from the set:

```
fruits = {'apple', 'banana', 'cherry'}
fruits.remove('banana')  # Removes 'banana' from the set
print("After remove:", fruits)
```

Output:

```
After remove: {'apple', 'cherry'}
```

Attempt to remove an item which is not in the set:

```
fruits = {'apple', 'banana', 'cherry'}
try:
    fruits.remove('mango')
# Attempt to remove 'mango', which is not in the set
except KeyError as e:
    print(f"Error: {e}")
```

Output:

```
Error: 'mango'
```

`discard()`: The `discard()` method serves a purpose similar to `remove()`. However, unlike `remove()`, if the specified element does not exist in the set, `discard()` does nothing and does not raise any error. Use `discard()` when the presence of the element in the set is not important, thus eliminating the need for additional error handling.

Removing an element from the set using discard():

```
fruits = {'apple', 'banana', 'cherry'}
fruits.discard('banana')  # Successfully removes 'banana'
print("After discard:", fruits)
```

Output:

```
After discard: {'apple', 'cherry'}
```

```
fruits = {'apple', 'banana', 'cherry'}
fruits.discard('mango')
# Does nothing and no error is raised, even though 'mango' is not
# in the set
print("After discard:", fruits)
```

Output:

```
After discard: {'apple', 'cherry', 'banana'}
```

`pop()`: The `pop()` method removes and returns an arbitrary element from a set. Since sets are unordered, there is no specific first or last element to remove; `pop()` simply removes one of the elements, but the exact element cannot be predicted unless the order of insertion has been tracked. This method is often used in algorithms where elements need to be processed one by one until the set is empty, and the order of processing is irrelevant. If `pop()` is called on an empty set, a `KeyError` is raised, as there are no elements to remove.

Removing an element from the set using pop():

```
fruits = {'apple', 'banana', 'cherry'}
# Removes and returns a random element from the set
element = fruits.pop()
print('Element:', element)
print('Set:', fruits)
```

Outputs:

```
Element: apple
Set: {'cherry', 'banana'}
```

`clear()`: The `clear()` method removes all elements from the set, resulting in an empty set.

Clearing the set:

```
fruits = {'apple', 'banana', 'cherry'}
fruits.clear()  # Removes all elements, resulting in an empty set
print("After clear:", fruits)
```

Output:

```
After clear: set()
```

`update()`: The method `update()` updates the set by adding elements from another set or any iterable (such as a list, tuple, etc.). Using `update()` with another set is functionally identical to using the `|=` operator. Both methods add all elements from the specified set to the original set, ignoring duplicates. However, the `update()` method can also take any iterable as an argument, not just sets, effectively expanding its utility. This flexibility allows `update()` to incorporate elements from lists, tuples, or even strings into the existing set.

Update a set using another set:

```
fruits = {'apple', 'banana'}
fruits.update({'banana', 'orange', 'grape'})
# Duplicate 'banana' is ignored
print(fruits)
```

Output:

```
{'apple', 'grape', 'orange', 'banana'}
```

Equivalent to:

```
fruits = {'apple', 'banana'}
fruits |= {'banana', 'orange', 'grape'}
print(fruits)
```

Output:

```
{'apple', 'grape', 'orange', 'banana'}
```

Update a set using a list:

```
fruits = {'apple', 'banana'}
fruits.update(['banana', 'orange', 'grape'])
# Duplicate 'banana' is ignored
print(fruits)
```

Output:

```
{'apple', 'grape', 'orange', 'banana'}
```

`intersection_update()`: This method updates the set, keeping only elements that are found in both the original set and a specified iterable or another set. The `&=` operator performs the same operation but may not work with all iterables.

intersection_update a set using another set:

```
fruits = {'apple', 'banana'}
fruits.intersection_update({'banana', 'orange', 'grape'})
print(fruits)
```

Output:

```
{'banana'}
```

Equivalent to &=:

```
fruits = {'apple', 'banana'}
fruits &= {'banana', 'orange', 'grape'}
print(fruits)
```

Output:

```
{'banana'}
```

Update a set using a list:

```
fruits = {'apple', 'banana'}
fruits.intersection_update(['banana', 'orange', 'grape'])
print(fruits)
```

Output:

```
{'banana'}
```

Attempting to use &= with a list:

```
fruits = {'apple', 'banana'}
try:
    fruits &= ['banana', 'orange', 'grape']
except TypeError as e:
    print("Error:", e)
```

Outputs:

```
Error: unsupported operand type(s) for &=: 'set' and 'list'
```



`difference_update()`: The `difference_update()` method modifies the set by removing all elements found in another set or any iterable (like lists or tuples). The -= operator is used to perform a similar operation as `difference_update()`, but it strictly requires the operand to be a set. It cannot handle other iterables like lists directly.

Using difference_update with another set:

```
fruits = {'apple', 'banana'}
fruits.difference_update({'banana'})
print(fruits)
```

Output:

```
{'apple'}
```

Using difference_update with a list:

```
fruits = {'apple', 'banana'}
fruits.difference_update(['banana'])
print(fruits)
```

Output:

```
{'apple'}
```

Correct use of -= with a set:

```
fruits = {'apple', 'banana'}
fruits -= {'banana'}
print(fruits)
```

Output:

```
{'apple'}
```

`symmetric_difference_update()`: The `symmetric_difference_update()` method updates the set by keeping only the elements that are found in either the set itself or a specified iterable (another set, list, etc.), but not in both. This method is an in-place operation, which means it modifies the set directly and does not return a new set. The `^=` operator performs the same function as `symmetric_difference_update()` when used between two sets. It updates the left-hand set, keeping only the elements that are unique to each set (i.e., those not shared between the two).

Using symmetric_difference_update() with a set:

```
fruits = {'apple', 'banana', 'cherry'}  # Define the original set
# Update the set using symmetric_difference_update with another set
fruits.symmetric_difference_update({'banana', 'orange', 'apple'})
print(fruits)
```

Output:

```
{'cherry', 'orange'}
```

Equivalent operation using ^= operator:

```
fruits = {'apple', 'banana', 'cherry'} # Reset fruits set
# Using ^= operator with another set
fruits ^= {'banana', 'orange', 'apple'}
print(fruits)
```

Output:

```
{'cherry', 'orange'}
```

Using symmetric_difference_update() with a list:

```
fruits = {'apple', 'banana', 'cherry'} # Reset fruits set
# Update using a list
fruits.symmetric_difference_update(['banana', 'orange'])
print(fruits)
```

Output:



```
{'cherry', 'apple', 'orange'}
```

### 2.2.15 Dictionaries

**Notebook: Section 2.2.15. Dictionaries**
This notebook provides code examples demonstrating complex data types—dictionaries.
Link to the GitHub repository:
▶ https://github.com/sn-code-inside/BioPy
Go to: Chap. 2—▶ Sect. 2.2.15

Dictionaries are vital data structures in Python that store collections of key-value pairs. Each key-value pair maps a key to its corresponding value, making dictionaries ideal for fast data retrieval and manipulation. Keys in dictionaries are unique, ensuring that each key maps distinctly to a specific value. Keys must be immutable, which means they can be types such as strings, numbers, or tuples—provided the tuples contain only immutable elements.

Dictionaries are denoted by curly braces `{ }` with key-value pairs separated by colons (`:`):

```
biodata = {'name': 'John Doe', 'age': 30, 'occupation': 'Biologist'}
```

To access values from a dictionary, square brackets `[ ]` are used with the key to retrieve the corresponding value:

```
print("Name:", biodata['name'])
```

Output:

```
Name: John Doe
```

To add or modify values in a dictionary, assign a value directly to a key. If the key does not exist, it will be added to the dictionary:

```
biodata = {'name': 'John Doe', 'age': 30, 'occupation': 'Biologist'}
biodata['city'] = 'New York'  # Adds a new key-value pair
biodata['age'] = 31  # Updates the existing key 'age'
print("Biodata:", biodata)
print("Age:", biodata['age'])
```

Output:

```
Biodata: {'name': 'John Doe', 'age': 31, 'occupation': 'Biologist',
'city': 'New York'}
Age: 31
```

### 2.2.16 Dictionary Methods

**Notebook: Section 2.2.16. Dictionary Methods**
This notebook provides code examples demonstrating complex data types—dictionary methods.
Link to the GitHub repository:
▶ https://github.com/sn-code-inside/BioPy
Go to: Chap. 2—▶ Sect. 2.2.16

Like other data types, Python provides various built-in methods for manipulating dictionaries. Common methods include `clear()`, `copy()`, `keys()`, `fromkeys()`, `get()`, `items()`, `pop()`, `popitem()`, `setdefault()`, and `update()`. Below are examples demonstrating the practical usage of these methods:

The clear() method removes all items in a dictionary:

```
biodata = {'name': 'John Doe', 'age': 30, 'occupation': 'Biologist'}
biodata.clear()
print(biodata)
```

Output:

```
{}
```

The copy() method copies all items to a new dictionary:

```
biodata = {'name': 'John Doe', 'age': 30, 'occupation': 'Biologist'}
copy_biodata = biodata.copy()
print("New dictionary:", copy_biodata)
```

Output:

```
New dictionary: {'name': 'John Doe', 'age': 30, 'occupation': 'Biologist'}
```

The method fromkeys() is used to create a new dictionary with the same initial value for all keys:

```
keys = ['name', 'age', 'occupation']
new_dict = dict.fromkeys(keys, "None")
print("New dictionary:", new_dict)
```

Output:

```
New dictionary: {'name': 'None', 'age': 'None', 'occupation': 'None'}
```

The get() method retrieves a specific item using a key:

```
biodata = {'name': 'John Doe', 'age': 30, 'occupation': 'Biologist'}
print("Age:", biodata.get('age'))
```

Output:

```
Age: 30
```

View all items in a dictionary using items() methods. The list() function is used to convert the dictionary object to a list:

```
biodata = {'name': 'John Doe', 'age': 30, 'occupation': 'Biologist'}
print("All items:", list(biodata.items()))
```

Output:

```
All items: [('name', 'John Doe'), ('age', 30), ('occupation', 'Biologist')]
```

Retrieve all keys in a dictionary using keys() method. Similarly the list() function is used to convert the dictionary object to a list:

```
biodata = {'name': 'John Doe', 'age': 30, 'occupation': 'Biologist'}
print("All keys:", list(biodata.keys()))
```

Output:

```
All keys: ['name', 'age', 'occupation']
```

Getting all values from a dictionary by using value() method:

```
biodata = {'name': 'John Doe', 'age': 30, 'occupation': 'Biologist'}
print("All values:", list(biodata.values()))
```

Output:

```
All values: ['John Doe', 30, 'Biologist']
```

Removing and returning a value using a key:

```
biodata = {'name': 'John Doe', 'age': 30, 'occupation': 'Biologist'}
print("Value removed:", biodata.pop('occupation'))
print("Current dictionary:", biodata)
```

Output:

```
Value removed: Biologist
Current dictionary: {'name': 'John Doe', 'age': 30}
```

Inserting a default value if key not present:

```
biodata = {'name': 'John Doe', 'age': 30}
print("New data:", biodata.setdefault('occupation', 'Chemist'))
print("Updated dictionary:", biodata)
```

Output:

```
New data: Chemist
Updated dictionary: {'name': 'John Doe', 'age': 30, 'occupation': 'Chemist'}
```

Removing and returning the last item:

```
biodata = {'name': 'John Doe', 'age': 30, 'occupation': ' Chemist'}
print("Removed item:", biodata.popitem())
print("Current dictionary:", biodata)
```

Output:

```
Removed item: ('occupation', 'Chemist')
Current dictionary: {'name': 'John Doe', 'age': 30}
```

Updating a dictionary using another dictionary:

```
biodata = {'name': 'John Doe', 'age': 30, 'occupation': 'Biologist'}
biodata.update({'age': 40, 'education': 'MS'})
print("Updated dictionary:", biodata)
```

Output:

```
Updated dictionary: {'name': 'John Doe', 'age': 40, 'occupation':
'Biologist', 'education': 'MS'}
```

### 2.2.17 Iterating over Dictionaries

**Notebook: Section 2.2.17. Iterating Over Dictionaries**
This notebook provides code examples demonstrating complex data types—iterating over dictionaries.

Link to the GitHub repository:
► https://github.com/sn-code-inside/BioPy
Go to: Chap. 2—► Sect. 2.2.17

In Python, dictionaries are flexible data structures that allow for various methods of iteration. It is possible to iterate over keys, values, or both keys and values together.

*Iterating over keys:* By default, when we iterate over a dictionary, we iterate over its keys. This is the most direct method to access each key in the dictionary:

```
biodata = {'name': 'John Doe', 'age': 30, 'occupation': 'Biologist'}
for key in biodata:
    print(key)
```

Outputs:

```
name
age
occupation
```

*Iterating over values:* If we need to access the values directly, we can iterate over the values of a dictionary using the `values()` method:

```
biodata = {'name': 'John Doe', 'age': 30, 'occupation': 'Biologist'}
for value in biodata.values():
    print(value)
```

Outputs:

```
John Doe
30
Biologist
```

Iterating over key-value pairs: To iterate over both keys and values at the same time, we use the `items()` method. This method returns each item in the dictionary as a tuple containing the key and its corresponding value, providing a convenient way to access both during iteration:

```
biodata = {'name': 'John Doe', 'age': 30, 'occupation': 'Biologist'}
for key, value in biodata.items():
    print(f"{key}: {value}")
```

Outputs:

```
name: John Doe
age: 30
occupation: Biologist
```

## 2.3 Control Structures: Loops and Conditionals

Control structures like loops and conditionals are required for directing the flow of a program. Loops allow repetitive execution of a block of code until a specific condition is met, enabling efficient handling of repetitive tasks. The most common loop in Python is the “for” loop, which we have used on several occasions in previous sections. In this section, we will elaborate on the concept of control structures.

### 2.3.1 Introduction to Control Structures



Control structures guide the flow of execution within a program. They are used in data analysis for tasks such as filtering data, analyzing patterns, and automating repetitive processes. In such a way, they enable programs to respond dynamically under varying conditions and repeat code blocks as needed.

Key functions of control structures include decision-making, repetitive execution, program flexibility, and complexity management. Control structures allow a program to select from different execution paths based on varying conditions, which is required for decision-making. They also automate repetitive operations, reducing the amount of code needed. Furthermore, control structures provide a framework that dictates how and when specific code blocks are executed, adding flexibility to the program's functionality. Finally, they simplify complexity by organizing code in a structured way, making it easier to manage and maintain.

Conditional statements within the control structure are the backbone of decision-making in Python programming and include `if`, `elif`, and `else` statements, which execute different code blocks based on specific conditions. The `if` statement is the fundamental structure for condition-based execution, allowing a code block to run only if a specified condition is met. The `elif` and `else` statements work in conjunction with `if` to handle multiple conditions, providing alternative code blocks when the initial `if` condition is not met, providing alternative path so that the program can respond to various scenarios.

Loops are commonly implemented using `for` and `while` constructs. These can also be nested to handle more complex scenarios. The `for` loop iterates over a sequence, such as a list, tuple, or string, executing a code block for each item. This is especially useful for processing items in a data structure and performing operations like data transformation or aggregation systematically. The `while` loop executes a code block as long as a specified condition remains true, useful in cases where the number of iterations is unknown, such as waiting for an event or processing data until a certain state is reached. Nested loops involve placing one loop inside another, require for handling complex data structures, like multidimensional arrays. They allow multilayered data processing tasks, such as comparing elements across datasets or processing grid-based data.

Control flow tools such as `break`, `continue,` and `pass` in loops provide additional control over loop execution. Regardless of the iteration condition `break` exits the loop immediately. If the loop condition holds, `continue` skips the remainder of the current loop iteration and moves to the next. Within a loop or conditional statement `pass` acts as a placeholder allowing for syntactical presence without affecting execution flow.

### 2.3.2 Usage of Conditional Statements

**Notebook: Section 2.3.2. Usage of Conditional Statements**
This notebook provides code examples demonstrating the usage of conditional statements.

Link to the GitHub repository:
▶ https://github.com/sn-code-inside/BioPy
Go to: Chap. 2—▶ Sect. 2.3.2

Conditional statements enable scripts to make complex decisions, which is useful for handling the diverse scenarios encountered in biomedical data analysis and healthcare applications. Let's break it down using some simple examples.

`if` *statement:* The `if` statement is used to execute a block of code only if a specified condition is true. The syntax of `if` condition is followed by an indented block of code:

```
age = 21
if age > 18:
    print("Adult")
```

Output:

```
Adult
```

In this code block, the `if` statement checks if the `age` variable is greater than 18. If the condition is met, it prints "Adult." Otherwise, if `age` is less than or equal to 18, it does nothing.

`elif` *statement:* The `elif` statement is short for "else if." It allows for multiple conditions to be checked, one after another, following an initial `if` or another `elif` statement.

For example, in the previous code block, we only checked if the `age` variable was more than 18 and printed "Adult." However, the code does nothing if the `age` is exactly 18. We can use an `elif` condition to check if `age` is 18 and print a different statement:

```
age = 18
if age > 18:
     print("Adult")
elif age == 18:
     print("Just turned adult")
```

Output:

```
Just turned adult
```



Since we updated the age variable to 18, the `if` condition (`age > 18`) does not meet, so the code proceeds to check the `elif` condition. The `elif` condition (`age == 18`) is met, so it prints "`Just turned adult`." If neither the `if` nor `elif` conditions are met, the code will return nothing.

`else` *statement:* The `else` statement executes a block of code if none of the preceding conditions are true. The syntax of else follows an `if` or `elif` statement.

In our previous examples, the code does not provide an output if the `age` is below 18. To handle this case, we can use an `else` statement that executes when the `age` is less than 18 and prints "Minor."

```
age = 16
if age > 18:
     print("Adult")
elif age == 18:
     print("Just turned adult")
else:
     print("Minor")
```

Output:

```
Minor
```

Since the provided `age` is 16, both the `if` and `elif` conditions fail, so the `else` statement executes and prints "`Minor`."

Conditional statements are widely used in Python to build functions. Functions are blocks of code that can be executed separately to perform a specific task. The following example demonstrates the use of `if`, `elif`, and `else` statements within a function:

```
def check_age(age):
    if age > 18:
       print("Adult")
    elif age == 18:
       print("Just turned adult")
    else:
       print("Minor")
# Example usage of the function
check_age(20)  # This should print "Adult"
check_age(18)  # This should print "Just turned adult"
check_age(16)  # This should print "Minor"
```

In this example, the `check_age` function uses conditional statements to check the age and print a corresponding message.

*Nested conditional statements:* Nested conditional statements involve placing an `if`, `elif`, or `else` statement within another `if` or `elif` block. This approach is particularly useful for evaluating a series of conditions where one is a subset or a specific case of another. For example, in a medical context, we might first want to check if a patient is an adult and then, within that adult category, check for a specific medical condition.

The syntax for nested conditionals adheres to the standard `if-elif-else` structure but with additional `if-elif-else` blocks inserted within the primary ones. It is important to indent each level correctly to ensure the code is readable and the logic flow is clear.

Consider a scenario where we want to check a patient's age category and then, for adults, further check if they have a specific medical condition. We can create a function for this specific task:

```
def patient_check(age, medical_condition):
    if age >= 18:
        # This is the first level of nesting
        if medical_condition == "Hypertension":
            print("Adult patient with Hypertension")
        elif medical_condition == "Diabetes":
            print("Adult patient with Diabetes")
        else:
            print("Adult patient with no specified condition")
    else:
        print("Not an adult, no further checks required")


# Example usage of the function
patient_check(25, "Hypertension")  # Adult patient with Hypertension
patient_check(20, "Diabetes")      # Adult patient with Diabetes
patient_check(25, "Other") # Adult patient with no specified condition
patient_check(16, "None")
# Not an adult, no further checks required
```

In this example, the `patient_check` function takes two arguments: `age` and `medical_condition`. It outputs a response based on the age and specified medical condition, using nested conditional statements to organize the checks logically.

### 2.3.3 Looping Structures

**Notebook: Section 2.3.3. Looping Structures**

This notebook provides code examples demonstrating the looping structures.

Link to the GitHub repository:

▶ https://github.com/sn-code-inside/BioPy

Go to: Chap. 2—▶ Sect. 2.3.3

Looping structures automate repetitive tasks, particularly in the context of data-intensive fields. They provide efficient processing and analysis of large datasets, allowing for the extraction of infromation and the automation of routine data handling tasks. Here, we will briefly discuss `for` and `while` loops with examples.

**For loops:** A "`for`" loop iterates over a sequence such as a list, tuple, or string, or any iterable object, executing a block of code for each item. For example, `for i in range` (5): `print(i)` prints numbers 0–4.

Iterating over a list using a `for` loop is straightforward. For instance:

```
my_list = [1, 2, 3, 4, 5]
for item in my_list:
    print(item)
```

Output:

```
1
2
3
4
5
```

In this example, the `item` variable in the loop statement reads each element from `my_list` and prints them one by one in each iteration.

Similar to lists, tuples can be iterated item by item:

```
my_tuple = ('a', 'b', 'c')
for item in my_tuple:
    print(item)
```

Output:

```
a
b
c
```

A for loop can be used to iterate character by character of a string:

```
my_string = "Python"
for char in my_string:
    print(char)
```

Output:

```
P
y
t
h
o
n
```

For loops can incorporate conditional logic to perform actions like filtering, such as extracting even numbers from a list, or managing multidimensional data structures with nested for loops, such as iterating through a matrix.

We can combine for loops with conditional logic if-elif-else structures to perform conditional operations on each element in a sequence:

```
# Filtering values from a list.
numbers = [1, 2, 3, 4, 5, 6]
even_numbers = []
for number in numbers:
    if number % 2 == 0:
        even_numbers.append(number)
print(even_numbers)
```

Output:

```
[2, 4, 6]
```

In this example, we start by creating a list containing numbers and aim to separate the even numbers using a for loop and conditional statements. We create an empty list called even_numbers to store the results. Next, we iterate over the numbers list using a variable, number, which represents each item in the list as it is processed. The conditional statement within the loop checks if the number is even, using the modulo operator %, which returns the remainder of the division. If the remainder is zero, the number is even. We then use the append() method to add the even number to the even_numbers list. Only numbers that meet the condition (i.e., are even) are appended to the list.

Furthermore, a for loop can be placed within another for loop, creating what is known as a nested for loop. Nested loops are particularly useful when working with multidimensional data structures:

```
# Iterating through a matrix.
matrix = [[1, 2, 3], [4, 5, 6], [7, 8, 9]]
for row in matrix:
    for element in row:
        print(element, end=' ')
    print()  # For a new line after each row
```

Output:

```
1 2 3
4 5 6
7 8 9
```

In this example, the outer `for` loop iterates through each `row` in the matrix (a list of lists). The inner `for` loop then iterates through each `element` within that row, printing each element on the same line. After completing each row, the `print()` function is called to move to a new line. This will print each element of the matrix row by row.

**While loops:** The while loop is yet another type of loops that repeats a block of code as long as a condition remains true. For example, `count = 0` followed by `while count <5: print(count); count += 1` prints numbers 0–4. Care must be taken to avoid infinite loops by updating loop conditions or using the `break` statements to exit based on specific conditions. An optional `else` block can follow a `while` loop to execute additional code when the loop condition becomes false, except if the loop is exited with a `break`, which bypasses the `else` block.

An infinite loop occurs when the loop's condition remains true indefinitely without modification. To prevent this, ensure the loop condition is updated within the loop. The `break` statement can also be used to exit the loop when a specific condition is met:

```
# Example of a break condition:
count = 0
while True:  # Potentially infinite loop
    if count >= 5:
        break  # Break out of the loop when count reaches 5
    print(count)
    count += 1
```

Output:

```
0
1
2
3
4
```

An `else` block can follow a `while` loop and will execute when the loop's condition becomes false naturally (i.e., without a `break`). If the loop is exited with a `break` statement, the `else` block is skipped.

Example with else:

```
count = 0
max_count = 5
while count < max_count:
    print("Count is:", count)
    count += 1
else:
    print("Loop finished. Count reached", max_count)
```

Output:

```
Count is: 0
Count is: 1
Count is: 2
Count is: 3
Count is: 4
Loop finished. Count reached 5
```

In this example, we initialize `count` to 0 and set `max_count` to 5. The `while` loop runs as long as the count is less than `max_count`. Within each iteration, it prints the current value of `count` and then increments the `count` by 1. Once the `count` reaches 5, the condition `count < max_count` becomes false, causing the loop to end naturally. At this point, the else block executes, printing "Loop finished. Count reached 5." The else block is executed only because the loop finished without being interrupted by a break statement.

Example with break and else:

```
count = 0
max_count = 5
while count < max_count:
    print("Count is:", count)
    count += 1
    if count == 3:
        print("Breaking out of the loop")
        break
else:
    # This block will not be executed as the loop is broken
    print("Loop finished. Count reached", max_count)
```

Output:

```
Count is: 0
Count is: 1
Count is: 2
Breaking out of the loop
```

In this code, we begin by setting count to 0 and max_count to 5. The while loop runs as long as the count is less than max_count. Each time the loop iterates, it prints "Count is:" followed by the current value of count. After printing, the count is incremented by 1. Inside the loop, there is an if statement that checks if count equals 3. When the count reaches 3, the message "Breaking out of the loop" is printed, and the break statement immediately exits the loop. Because the break statement stops the loop before it completes all iterations, the else block does not execute. The else block only runs if the loop completes naturally, without being interrupted by the break. Therefore, the final output after the break statement is simply "Breaking out of the loop," and the else message "Loop finished. Count reached 5" is skipped.

### 2.3.4 Use of Control Flow Tools

**Notebook: Section 2.3.4. Use of Control Flow Tools**

This notebook provides code examples demonstrating the use of control flow tools.

Link to the GitHub repository:

▶ https://github.com/sn-code-inside/BioPy

Go to: Chap. 2—▶ Sect. 2.3.4

The proper use of control flow tools like break, continue, and pass can enhance the functionality and efficiency of loops and conditional statements in Python, particularly in complex tasks. These tools provide greater flexibility and control, allowing users to precisely dictate how and when specific blocks of code are executed.

**The** break **statement** is used to exit a loop prematurely. When executed, it immediately terminates the loop it is in, regardless of the loop's original stopping condition. This is particularly useful when a specific condition, different from the loop's main condition, needs to be met to stop the iteration.

Let's consider we are iterating over patient data and need to immediately exit the loop if a critical condition is met, such as encountering an alarmingly high blood pressure reading.

Example code:

```
patient_blood_pressure_readings = [120, 125, 130, 180, 110, 115]
for reading in patient_blood_pressure_readings:
    if reading > 170:
        print(f"Critical reading encountered: {reading}")
        break
    print(f"Reading {reading} is normal")
```

Output:

```
Reading 120 is normal
Reading 125 is normal
Reading 130 is normal
Critical reading encountered: 180
```

In this example, the "for" loop iterates over a list of blood pressure readings. The break statement is used inside an if condition to check for a critical reading (above 170). When a critical reading is encountered, the break statement is executed, which immediately stops the loop and exits it. This demonstrates how the break statement can be used to respond to specific conditions within a loop, providing a tool for controlling the flow of our program.

**The** `continue` **statement** is used within loops to skip the remaining portion of the loop's code block for the current iteration. After a `continue` statement is encountered, the loop does not terminate; instead, it immediately proceeds to the next iteration. This feature is particularly useful for bypassing specific values or conditions within a loop.

Consider an example where we are iterating through a list of test results. Using "continue," we can skip any null or invalid entries and proceed directly to the next item in the list for further processing.

Example code:

```
test_results = [85, 90, None, 95, 'invalid', 100]
for result in test_results:
    if result is None or not isinstance(result, int):
        # Skip null or non-integer results
        continue
    print(f"Processing result: {result}")
```

Output:

```
Processing result: 85
Processing result: 90
Processing result: 95
Processing result: 100
```

In this example, the loop iterates over each item in `test_results`. When it encounters a `None` value or a non-integer value (like the string "`invalid`"), the `continue` statement is executed. This causes the loop to immediately skip the remaining part of its body for that iteration and proceed to the next item in the list. Only valid integer results are processed and printed.

**The "**`pass`**" statement** is basically a placeholder; it represents a null operation. When executed, it results in no action. Its primary use is to maintain the integrity of the code structure in scenarios where a statement is syntactically necessary, but the developer does not wish to execute any code at that moment. This is particularly useful during the development phase, where we might want to outline the structure of our code but have not implemented the logic yet.

Example use cases:

*As a placeholder in a loop:* If we have a loop structure but have not yet implemented the logic inside it, we can use "`pass`" to avoid syntax errors and maintain the loop structure:

```
for i in range(5):
    # Future logic goes here
    pass
# This loop does nothing for now but can be filled in later
```

*In conditional statements:* Similarly, in conditional statements like "`if`," "`elif`," or "`else`," where we might not be ready to write the full logic, "pass" can be used.

The following code will not do anything:

```
x = 10
if x > 5:
    pass  # Placeholder for future logic
else:
    print("x is 5 or less")
# The if block does nothing for now, but can be implemented later
```

But assigning `x = 3` will print the message `x is 5 or less.`

*In function definitions:* When defining functions that we intend to implement later, pass can be used to create a function skeleton:

```
def future_function():
    pass  # Future implementation goes here
# This function currently does nothing.
```

**Combining with conditional logic:** Control flow tools can be combined with loops and conditional statements to create complex logic in a program. Often, loops are used with conditional statements, and `break` and `continue` are employed to control the flow based on certain conditions.

Using `break` in a `while` loop: In this example, a `while` loop is combined with an `if` condition. The `break` statement is used to exit the loop once a specific criterion is met:

```
count = 0
max_value = 10
while count < max_value:
    count += 1
    if count == 5:
        print("Breaking out of loop at count =", count)
        break
    print("Current count:", count)
```

Output:

```
Current count: 1
Current count: 2
Current count: 3
Current count: 4
Breaking out of loop at count = 5
```

In this script, the `while` loop iterates until count reaches the value of 10. However, the `if` statement checks if `count` is equal to 5. Once this condition is met, the `break` statement is executed, causing the loop to terminate early. This example demonstrates how we can use conditional logic within a loop to control when the loop should exit.

Here is another example where a `for` loop is used along with conditional logic to filter out specific items from a list:

```
numbers = [1, 2, 3, 4, 5, 6, 7, 8, 9, 10]
for number in numbers:
    if number % 2 == 0:
        continue # Skip the even numbers
    print("Odd number:", number)
```

Output:

```
Odd number: 1
Odd number: 3
Odd number: 5
Odd number: 7
Odd number: 9
```

In this example, the `for` loop iterates over a list of numbers. The `if` statement checks if the number is even. This is done using the modulus operator `%`, which calculates the remainder of the `number` divided by 2. If the number is found to be even (i.e., `number % 2 == 0` is `True`), the `continue` statement is executed. The `continue` statement immediately stops the current iteration and moves the control back to the beginning of the loop for the next iteration. Therefore, the `print` statement following this condition is skipped for even numbers.

### 2.3.5 Comprehensions

**Notebook: Section 2.3.5. Comprehensions**

This notebook provides code examples demonstrating comprehensions.

Link to the GitHub repository:

► https://github.com/sn-code-inside/BioPy

Go to: Chap. 2—► Sect. 2.3.5

Comprehensions provide an elegant and efficient way to create collections such as lists, sets, and dictionaries. They simplifies the process of generating and manipulating these collections, making it beneficial in data analysis and data processing. Comprehensions enable the transformation, filtering, and generation of new collections from existing datasets using a straightforward and intuitive syntax. This not only enhances the readability of the code but also improves efficiency, facilitating various data processing operations.

**List comprehensions** provide a concise way to create lists. They are commonly used to generate new lists by applying operations to each element of another sequence or iterable, or to form a subset of elements that meet a specific condition.

The basic syntax of a list comprehension is:

```
[expression for item in iterable]
```

Here, "`expression`" is the current item in the iteration, but it is also the outcome, which we can manipulate before it ends up like an item in the new list. The "`iterable`" is a collection of objects—for example, a list or tuple.

Creating a list of squares for numbers from 0 to 9 will be as follows:

```
squares = [x**2 for x in range(10)]
print(squares)
```

Output:

```
[0, 1, 4, 9, 16, 25, 36, 49, 64, 81]
```

Using a conditional statement to create a list of even numbers from 0 to 9:

```
evens = [x for x in range(10) if x % 2 == 0]
print(evens)
```

Output:

```
[0, 2, 4, 6, 8]
```

Applying a function—suppose we have a function that converts Celsius to Fahrenheit. We can use a list comprehension to apply this function to a list:

```
def celsius_to_fahrenheit(c):
    return (9/5) * c + 32


temperatures_c = [0, 10, 20, 30]
temperatures_f = [celsius_to_fahrenheit(c) for c in
temperatures_c]
print(temperatures_f)
```

Output:

```
[32.0, 50.0, 68.0, 86.0]
```

Using a nested list comprehension, we can easily access and manipulate a matrix represented as a list of lists. For example, if we have a matrix as defined below, we can flatten this matrix into a single list where each element is sequentially accessed from the nested lists. This is done using a nested list comprehension, which iterates through each row and then each number within the row:

```
matrix = [[1, 2, 3], [4, 5, 6], [7, 8, 9]]
flattened = [num for row in matrix for num in row]
print(flattened)
```

Output:

```
[1, 2, 3, 4, 5, 6, 7, 8, 9]
```

List comprehensions offer several advantages such as conciseness, which reduces the amount of code needed to create a new list; readability, allowing the intent of the code to be quickly understood in simple cases; and better performance, generally being faster than equivalent for-loops. However, they also have disadvantages, including readability issues in complex cases where they can be hard to read if overused or if the logic is too complex, and they can be more challenging to debug compared to traditional loops. List comprehensions are widely used due to their syntactical simplicity and efficient processing, they should be used carefully to ensure that code remains readable and maintainable.

**Dictionary comprehensions** are a concise way to create dictionaries. They follow a similar principle to list comprehensions, providing a shorter syntax when we want to create a new dictionary based on the values of an existing iterable. Dictionary comprehensions consist of an expression pair `(key: value)` followed by a `for` statement inside curly braces `{ }`.

The basic syntax of dictionary comprehension is expressed as `{key: value for item in iterable}`, which allows the creation of dictionaries directly from iterables, which facilitates the direct creation of dictionaries from iterables. Dictionary comprehensions can also incorporate conditional statements, acting as filters to refine the elements that are included in the final dictionary. This capability allows for selective inclusion based on specific criteria, increasing the flexibility and utility of dictionary comprehensions. Moreover, dictionary comprehensions are not limited to a single iterable; they can iterate over multiple iterables simultaneously. This versatility enables the combination of different data sources into a single dictionary, which can be particularly useful in scenarios where key-value pairs are derived from distinct datasets.

Basic usage:

```
squares = {x: x*x for x in range(6)}
print(squares)
```

Output:

```
{0: 0, 1: 1, 2: 4, 3: 9, 4: 16, 5: 25}
```

Conditional logic:

```
even_squares = {x: x*x for x in range(10) if x % 2 == 0}
print(even_squares)
```

Output:

```
{0: 0, 2: 4, 4: 16, 6: 36, 8: 64}
```

Using two iterables:

```
keys = ['a', 'b', 'c']
values = [1, 2, 3]
dictionary = {k: v for k, v in zip(keys, values)}
print(dictionary)
```

Output:

```
{'a': 1, 'b': 2, 'c': 3}
```

Creating a dictionary from a string:

```
word = "dictionary"
letter_count = {letter: word.count(letter) for letter in word}
print(letter_count)
```

Output:

```
{'d': 1, 'i': 2, 'c': 1, 't': 1, 'o': 1, 'n': 1, 'a': 1, 'r': 1, 'y': 1}
```

Dictionary comprehensions can make code more efficient and readable, particularly when dealing with complex data structures or transformations. However, it is important to balance their use with code readability, especially when working in collaborative environments.

**Set comprehensions** are a concise way to create sets using a single line of code, similar to the list comprehensions but for sets. Set comprehensions allow the construction of sets in a way that is both syntactically clearer and often more efficient than using traditional `for` loops.

The basic syntax of a set comprehension is:

```
{expression for item in iterable if condition}
```

where `expression` is the current item in the iteration, but it also could be any other valid Python expression. The `item` represents each item in the `iterable`. The `iterable` can be a sequence, collection, or object that can be iterated over. The `condition`, which is optional, is like a filter that only accepts the items that are `True`.

The basic syntax of a set comprehension is:

```
# Create a set of squares for numbers from 0 to 9
squares = {x**2 for x in range(10)}
print(squares)
```

Output:

```
{0, 1, 64, 4, 36, 9, 16, 49, 81, 25}
```

Set comprehension with condition:

```
# Create a set of even numbers between 0 and 9
even_numbers = {x for x in range(10) if x % 2 == 0}
print(even_numbers)
```

Output:

```
{0, 2, 4, 6, 8}
```



Set comprehension with complex expression:

```
# Create a set of tuples (number, square of number)
number_square = {(x, x**2) for x in range(4)}
print(number_square)
```

Output:

```
{(1, 1), (3, 9), (2, 4), (0, 0)}
```

Using set comprehension to remove duplicates:

```
# Remove duplicates from a list
list_with_duplicates = [1, 2, 2, 3, 4, 4, 4, 5]
unique_elements = {x for x in list_with_duplicates}
print(unique_elements)
```

Output:

```
{1, 2, 3, 4, 5}
```

Set comprehensions are beneficial for several reasons. They are more concise and readable than traditional loops, often perform faster for similar tasks, and support a functional programming style, which makes code cleaner. Sets are unordered and automatically remove duplicates, which is ideal for ensuring element uniqueness without extra coding. They are particularly useful for data transformation and filtering, simplifying operations where the uniqueness of elements is important.

### 2.3.6 Error Handling in Control Structures

**Notebook: Section 2.3.6. Error Handling in Control Structures**

This notebook provides code examples demonstrating error handling in control structures.

Link to the GitHub repository:

▶ https://github.com/sn-code-inside/BioPy

Go to: Chap. 2—▶ Sect. 2.3.6

Error handling in control structures is required to ensure the robustness and reliability of code. Control structures, such as loops and conditionals, direct the flow of execution within a program. By incorporating error handling techniques, such as try-except blocks, users can anticipate and manage errors or exceptional conditions that might disrupt this flow. This proactive approach not only prevents crashes but also provides opportunities to gracefully handle unexpected situations, ensuring the program operates smoothly under various conditions.

**Exception handling with `try`, `except`, and `finally`:** Exception handling is a mechanism that allows users to respond appropriately to errors that may occur during the execution of a program. The primary constructs used for exception handling are `try`, `except`, and `finally`. We will discuss each of these constructs and provide example codes to illustrate their usage.

`try` block: The `try` block lets us test a block of code for errors. The code inside the try block is executed, and if any error occurs, it is caught, and the control is passed to the except block.

`except` block: The `except` block enables to handle the error. It allows one to specify different types of exceptions to catch only specific errors or use a general `except` block to catch any error. If the error matches the exception specified in the `except` block, the code in this block is executed.

`finally` block: The `finally` block, if present, will be executed regardless of whether an exception is raised or not. It is typically used to perform clean-up actions, like closing files or releasing resources, which should be done whether the operation was successful or not.

Using `try` and `except` without `finally`: This is common for handling exceptions without needing any cleanup or final actions regardless of success or failure:

```
try:
    print("Trying to convert input to an integer.")
    number = int(input("Enter a number: "))
except ValueError:
    print("That's not a number!")
```

Output after entering x:

```
Trying to convert input to an integer.
Enter a number: x
That's not a number!
```

In this example, the `try` block attempts to convert user input to an integer. If the conversion fails (e.g., if the user enters a non-numeric string, for example, “x”), the `ValueError` is caught by the `except` block.

Using `try` and `finally` without `except`: This is less common but can be used to ensure that certain cleanup code is always executed, but user does not want to handle the exception in the current function (letting it propagate up the call stack):

```
try:
    file = open("example_data/data.txt")
    data = file.read()
finally:
    file.close()
    print("File closed.")
```

Output:

```
File closed
```

Here, the file is always closed, whether an exception occurred or not. However, if an exception does occur (e.g., if the file does not exist), it is not caught here and will propagate upward.

Using `try`, `except`, `else`, and `finally`: The `else` block is an additional block that can be used with these constructs. It is executed if the try block does not raise an exception:

```
try:
    print("Trying to divide numbers.")
    result = 10 / 2
except ZeroDivisionError:
    print("Divided by zero.")
else:
    print("Division successful:", result)
finally:
    print("Execution completed.")
```

Output:

```
Trying to divide numbers.
Division successful: 5.0
Execution completed.
```

In this case, the `else` block prints a success message if no exception is raised, while the `finally` block executes regardless of whether an exception occurred.

Another example with a function:

```
def divide(x, y):
    try:
        result = x / y
    except ZeroDivisionError:
        print("Sorry, division by zero is not allowed.")
    else:
        print("The result is", result)
    finally:
        print("Executing finally clause.")
```

```
# Test the function
divide(2, 1)   # Normal division
divide(2, 0)   # Division by zero
```

Output:

```
The result is 2.0
Executing finally clause.
Sorry, division by zero is not allowed.
Executing finally clause.
```

In this example, the `try` block contains the code that might raise an exception, in this case, division of two numbers. The `except ZeroDivisionError` block catches the `ZeroDivisionError` that occurs if the denominator (`y`) is zero. The `else` block (which is optional) runs if no exceptions were raised in the `try` block. The `finally` block runs regardless of whether an exception was raised or not, making it suitable for clean-up activities. Note that in the second call to `divide(2, 0)`, the `ZeroDivisionError` is caught, and the appropriate message is printed. In both calls, the `finally` clause is executed, demonstrating that it runs irrespective of the occurrence of an exception.

**Handling specific exceptions:** Catching specific exceptions is a recommended practice for error handling. This approach involves specifying the exact type of exception that the code is expected to handle. By doing this, we ensure that our exception handling is targeted and precise, avoiding the pitfalls of a general except block which can inadvertently catch and obscure unexpected exceptions. Here is an example to illustrate this concept:

```
def divide_numbers(num1, num2):
    try:
        result = num1 / num2
    except ZeroDivisionError:
        print("Error: Division by zero is not allowed.")
    except TypeError:
        print("Error: Non-numeric types cannot be divided.")
    else:
        print("The result is:", result)
# Example usage
divide_numbers(10, 2)  # This will execute without any errors
divide_numbers(10, 0)  # This will trigger the ZeroDivisionError
divide_numbers("10", 2)  # This will trigger the TypeError
```

Output:

```
The result is: 5.0
Error: Division by zero is not allowed.
Error: Non-numeric types cannot be divided.
```

In the divide_numbers function, two specific types of exceptions are caught: ZeroDivisionError, which is raised when a division or modulo operation is attempted with zero as the divisor, and TypeError, which occurs when an operation or function is applied to an object of an inappropriate type. The former is used to handle cases where num2 is zero, while the latter deals with situations where either num1 or num2 is not a numeric type. Moreover, there is an else block that follows the except blocks, which executes only if no exceptions are raised in the try block. This block is optional but useful for executing code that should run only if the try block is successful. By specifying these exceptions, the divide_numbers function can provide more meaningful error messages and handle each error case appropriately, thus improving the code's reliability and maintainability.

**Multiple exception blocks:** Having multiple except blocks within a try-except structure allows for handling different types of exceptions separately. This practice is particularly beneficial when different exceptions require distinct handling logic. By specifying multiple except blocks, each tailored to a specific exception type, we can manage errors more effectively and provide clearer, more specific responses to different error conditions. Here is an example to illustrate the use of multiple exception blocks:

```
def calculate_operation(a, b, operation):
    try:
        if operation == "add":
            return a + b
        elif operation == "subtract":
            return a - b
        elif operation == "multiply":
            return a * b
        elif operation == "divide":
            return a / b
        else:
            raise ValueError("Unsupported operation")
    except ZeroDivisionError:
        return "Cannot divide by zero"
    except TypeError:
        return "Both operands must be numbers"
    except ValueError as ve:
        return str(ve)
# Example usage
print(calculate_operation(10, 5, "add"))       # Performs addition
print(calculate_operation(10, 0, "divide"))   # Handles ZeroDivisionError
print(calculate_operation(10, "five", "add")) # Handles TypeError
print(calculate_operation(10, 5, "mod"))      # Handles ValueError
```

Output:

```
15
Cannot divide by zero
Both operands must be numbers
Unsupported operation
```

In the `calculate_operation` function, multiple except blocks are employed to handle different exceptions: the first block catches `ZeroDivisionError`, occurring when an attempt is made to divide by zero, as in a / b where b is zero; the second block addresses `TypeError`, raised when either operand a or b is not a number, thus ensuring both operands are numeric; and the third block manages `ValueError`, invoked for an unsupported operation, using an exception instance (`ve`) to capture and relay the exception's message. This method of using multiple except blocks for handling various exceptions allows for more detailed and tailored error handling in Python, providing specific responses for different error types and improving the robustness and clarity of the code.

## 2.4 Exercises and Questions

Here we include exercises and questions to complement the material covered in the chapter.

### Exercises

The following exercises will enhance our understanding of the materials we covered in this chapter.

1. Write a Python script that prints "Learning Python is fun!" and includes a comment saying, "This is my first Python script!"
2. Perform the following operations and print the results:
   (a) Add 15 and 7.
   (b) Multiply 5 by 3.
   (c) Raise 4 to the power of 2.
3. Print the result of whether 10 is less than 20.
4. Use comparison operators (==, !=, <, >, <=, >=) to compare the following pairs of numbers and print the result of each comparison:
   (a) 5 and 3.
   (b) 10 and 10.
   (c) 7 and 8.
5. Determine if given clinical measurements fall within a normal range. Given the normal heart rate range for adults is between 60 and 100 beats per minute (bpm), write a script to check if a patient's heart rate of 75 bpm is normal. Use <= and >= operators for the comparison. Print whether the heart rate is "Normal" or "Abnormal."
6. Use multiple comparison operators in a single statement. Assign values to age = 30 and heart_rate = 85. Determine if the age is greater than 18 and heart_rate is less than 100 simultaneously. Print "Checkup Required" if both conditions are true, else print "Checkup Not Required."
7. Apply comparison operators to a real-world dataset. Imagine a dataset patient_bmi = [22, 27, 31, 29, 24, 19] representing body mass index (BMI) of several patients. Write a script to count how many patients are overweight. Consider BMI over 25 as overweight. Print the count of overweight patients.

8. To apply conditional statements in a practical health-related scenario, write a script that uses if-elif-else structures to categorize patients based on age and BMI into different health risk categories.
9. To practice using for loops in data filtering, create a program that iterates over a list of patient records, extracting and printing out only those records where certain conditions, like a specific ailment or age range, are met.
10. To combine loops and conditional statements for data cleaning and analysis, develop a script to calculate the average heart rate from a list of heart rate measurements, excluding any outliers or erroneous data points.
11. To practice variable assignment and basic arithmetic operations, create variables to store patient age, name, and heart rate. Perform basic arithmetic operations using these variables.
12. To enhance skills in string manipulation and data extraction, let's consider a string with patient information; extract and format specific details like patient name and diagnosis.
13. To understand list operations and their application in handling sequential data, create a list of lab test results, and perform operations like adding new results, sorting, and slicing the list.
14. To demonstrate the use of tuples for immutable data, use tuples to store fixed information about a medication (e.g., name, dosage, frequency), and display it.
15. To learn dictionary operations and their use in organizing complex data like patient records, create a dictionary to store patient information and update, retrieve, and manipulate data within it.

## Questions to Be Answered

1. What is the purpose of using comments in a Python script, and how do they affect the execution of the code?
2. Explain the difference between the "/" and "//" operators in Python. Provide an example where each might be used.
3. How does Python determine the data type of a variable, and what happens when we use the "`type()`" function on a variable?
4. Provide an example of how we would concatenate two strings in Python and discuss a situation where string immutability might impact this operation.
5. Describe a real-world application where Boolean expressions and logical operators ("`and`", "`or`", "`not`") could be utilized effectively.
6. What are the pitfalls of implicit type conversion in Python, and how can explicit type conversion be enforced?
7. Compare and contrast lists and tuples in Python. When would we choose to use a tuple over a list?
8. How would we extract every third element from a list starting from the second element? Provide the Python code to accomplish this.
9. Write a Python snippet that iterates over a dictionary and prints each key-value pair in the format "`Key: [key] has value: [value]`."
10. Discuss how error handling can be integrated within a loop structure to continue processing data even if an error occurs during one iteration. Provide a Python code example illustrating this scenario.

**Acknowledgement** The language of the human-generated text was corrected with the assistance of artificial intelligence (AI) tools [GPT-3.5 and GTP-4 from OpenAI]. GitHub Co-Pilot was used to check the correctness of the codes. The text underwent subsequent human revision to ensure its accuracy.

## References

1. Jones P. Python: the fundamentals of Python programming. CreateSpace Independent Publishing Platform; 2016.
2. Python. Arithmetic operators. 2024. https://python-reference.readthedocs.io/en/latest/docs/operators/#arithmetic-operators.
3. Python. Assignment operators. 2024. https://python-reference.readthedocs.io/en/latest/docs/operators/#assignment-operators.
4. Python. Relational operators. 2024. https://python-reference.readthedocs.io/en/latest/docs/operators/#relational-operators.
5. Liang Y. Introduction to Python programming and data structures. Pearson Education Limited; 2022.
6. McKinney W. Python for data analysis: data wrangling with Pandas, NumPy, and Jupyter. O'Reilly Media; 2022.

# Working with Biomedical Data: Basic Data Handling

## Contents




J. U. Kazi, *Python Essentials for Biomedical Data Analysis: An Introductory Textbook*,
https://doi.org/10.1007/978-3-031-85600-6_3

This chapter provides a guide to key techniques for managing biomedical data. It begins with an overview of various biomedical data types, suggesting the unique attributes of clinical, imaging, genomic, and transcriptomic data, and explores the challenges inherent in handling these complex datasets. The chapter emphasizes the importance of understanding biomedical data formats, focusing on both common and proprietary types, as well as the need for format compatibility and accessibility. It covers data import techniques, including methods for reading flat files, navigating hierarchical data, and managing large datasets, with specialized methods for handling medical imaging, genomic sequences, and transcriptomic data. This chapter also details effective export methods, including strategies for handling large-scale and hierarchical datasets, and methods to export processed imaging and genomic data.



**Learning Goals**

The goal of this chapter is to provide students with a basic understanding of the different types and formats of biomedical data and the unique challenges they present in data handling. Students will develop practical skills in importing and exporting biomedical data across various formats, including flat files, hierarchical data, medical imaging, and genomic data. They will become familiar with handling large-scale biomedical datasets. Moreover, students will learn to apply basic data handling techniques in biomedical contexts, preparing them for more advanced data analysis tasks.

## 3.1 Overview of Biomedical Data Types

Biomedical data can come in many forms, including but not limited to genomic data, transcriptomic data, proteomic data, metabolomic data, epigenetic data, microbiome data, clinical trial data, electronic health records, imaging data, biobank data, pharmacogenomic data, and digital health data. We briefly provide a simplified description of each data type.

*Genomic data:* This includes data related to genes and DNA sequencing. Genomic datasets can be massive, often involving whole-genome sequencing, which generates large amounts of data. This data helps us understand genetic variations, gene-disease associations, and evolutionary processes [1].

*Transcriptomic data:* This type involves studying the complete set of RNA transcripts that are produced by the genome under specific circumstances or in specific cells, providing information regarding gene function and regulation [2].

*Proteomic data:* Proteomics involves the study of the proteome, the entire set of proteins produced by an organism. Proteomic data includes information about protein expression, structure, functions, and interactions [3]. Proteomic data including site specific phosphorylation data provides better understanding of diseases at the molecular level and helps in drug development.

*Metabolomic data:* This focuses on metabolites, the small molecule substrates, intermediates, and products of metabolism [4]. This data provides information about the metabolic changes associated with diseases.

*Epigenetic data:* This type of data deals with heritable changes in gene expression that do not involve alterations to the DNA sequence itself, such as DNA methylation and histone modification. These modifications play a role in gene regulation and are integral to understanding developmental biology and disease progression [5].

*Microbiome data:* This type of data involves the analysis of the microbiomes, including bacteria, viruses, fungi, and other microorganisms that inhabit various parts of the human body [6]. This data is key to studies on immunity, gastrointestinal diseases, and even mental health.

*Clinical trial data:* These datasets consist of information gathered during clinical trials. They include patient demographics, treatment regimens, clinical outcomes, and side effects [7]. Clinical trial data serves an important role for evaluating the efficacy and safety of new treatments and drugs.

*Electronic health records (EHRs):* EHRs contain patient health information collected over time, including medical history, diagnoses, medications, treatment plans, immunization dates, allergies, radiology images, and laboratory test results [8]. They hold useful information for personalized medicine and epidemiological studies.

*Imaging data:* Medical imaging data, such as MRI, CT scans, X-rays, and ultrasound images, are often used for diagnostics and treatment planning. This data type is usually large and requires specialized software for analysis.

*Biobank data:* Biobanks are repositories that store biological samples, such as blood, tissue, and DNA, for use in research. Data from biobanks includes sample information, donor information, and associated clinical or phenotype data.

*Pharmacogenomic data:* This data type combines pharmacology and genomics to understand how genetic makeup affects an individual's response to drugs. Such type of data are widely used in developing personalized medicine strategies [9].

*Digital health data:* Emerging from wearable technology and other Internet of Things (IoT) devices, digital health data includes information on physical activity, sleep patterns, heart rate variability, and more, facilitating real-time health monitoring and chronic disease management.

## 3.2 Data Handling Challenges

Handling biomedical datasets presents various challenges that may impact research, clinical decisions, and drug development. Accurate and efficient data management may directly support better research outcomes [10]. Effective data handling ensures that data is accessible and usable for a range of research purposes. Careful management of data also influences the validity of research findings, emphasizing the importance of data handling in propelling scientific progress [11].

Data generated from biomedical research can support the development of more personalized and effective approaches to patient care. For example, data on patient outcomes and treatment efficacy may guide clinicians in choosing the most appropriate treatment plans for individual patients [12]. In drug development, the analysis of biological data helps identify drug targets, and supports preclinical studies assessing drug efficacy and safety. Clinical trial data lead to evaluating the performance of drugs, and therefore developing robust data handling pipelines for such type of data may enhance the drug development process [13].

However, several challenges arise in managing biomedical datasets. Biomedical datasets are often large and complex, encompassing various data types, from genetic sequences to clinical trial information. Managing such data demands advanced computational tools and algorithms. The volume and complexity of these datasets make processing and analysis challenging, requiring specialized bioinformatics and data science expertise. Ensuring data quality and integrity is also important. Accurate, consistent, and complete data is required for reliable research findings, particularly because of its implications for patient care. Rigorous data validation and cleaning are necessary to avoid inaccuracies or incomplete information that could lead to erroneous conclusions, compromising patient outcomes and scientific credibility.

Interoperability is another challenge, as integrating and harmonizing data from diverse sources and formats are often required for analysis. Achieving interoperability involves standardizing data formats and creating protocols for data exchange, enabling collaborative research and facilitating data sharing across studies [14]. Privacy and security concerns are critical, as biomedical data often includes sensitive personal information. Strict adherence to privacy laws and ethical standards is required, along with robust security measures to prevent unauthorized access and breaches. Ethical considerations are also paramount in research involving human subjects.

Lastly, handling biomedical data is resource-intensive. The processing, analysis, and storage of large datasets require substantial computational resources and expertise, including the hardware and software infrastructure necessary for data management. The demand for skilled personnel who can effectively analyze and interpret biomedical data adds to the resource burden. This intensity can be a barrier, particularly for smaller institutions or research groups that may lack the necessary infrastructure and personnel to handle extensive datasets.

## 3.3 Understanding Data Formats

Different data formats can affect how data is accessed, shared, and analyzed. Before exploring methods of handling them, it is important to understand some of the most commonly used formats in this domain.

### 3.3.1 Common Biomedical Data Formats

Data formats in the biomedical domain are largely purpose-dependent, and while we discuss several common types below, numerous other specialized formats exist to meet specific requirements.

*Basic flat files:* Comma-separated values (CSV) and tab-separated values (TSV) are simple file formats used to store tabular data, such as spreadsheets or databases. Each line in the file is a data record, with CSV using commas and TSV using tabs as delimiters. These formats are widely used due to their simplicity and readability, making them ideal for exporting and importing small to medium-sized datasets.

*Hierarchical data structures:* JavaScript object notation (JSON) and eXtensible markup language (XML) are formats for storing and transporting data. JSON is a text format that uses key-value pairs and array data types, making it lightweight and easy for humans to read and write. XML is a markup language that defines a set of rules for encoding documents in a format that is both human-readable and machine-readable. Both are used in biomedical research for data that require hierarchical structuring, like nested data from complex experiments or studies.

*Large-scale scientific data:* Hierarchical data format version 5 (HDF5) is designed to store and organize large amounts of heterogeneous data. HDF5 supports complex data types and is ideal for handling vast datasets that require efficient storage and quick access. It is extensively used in biomedical research for storing everything from large-scale imaging data to complex genomic datasets. An alternative to HDF5, Zarr is a format that provides an implementation of chunked, compressed, N-dimensional arrays. This format is particularly well suited for use with cloud storage and parallel computing environments, making it increasingly popular in bioinformatics projects that require scalability and efficient data access across distributed systems.

*Medical imaging data:* Digital imaging and communications in medicine (DICOM) is the standard for the communication and management of medical imaging information and related data. It is used for storing, exchanging, and transmitting medical images, enabling the integration of medical imaging devices such as scanners, servers, workstations, and network hardware.

*Genomic data formats:* FASTA[1] is a text-based format for representing nucleotide sequences or peptide sequences, in which nucleotides or amino acids are represented using single-letter codes. FASTQ[2] is similar to FASTA but also includes quality scores for each nucleotide, making it useful for next-generation sequencing (NGS) data.

### 3.3.2 Proprietary Formats

In biomedical research, several proprietary data formats are often encountered, particularly in relation to specific types of equipment or software. These formats are generally developed and controlled by companies for their own products and may require specialized software to access or convert.

*Instrument-specific formats:* Many laboratory instruments produce data in proprietary formats unique to the manufacturer. For example, different brands of mass spectrometers or microarray scanners use distinct file formats that are not interchangeable. For instance, Thermo Fisher's mass spectrometers produce files in the ".raw" format, which are specific to Thermo's software and instruments.

1 FASTA is a plain text file format used to represent nucleotide sequences (DNA or RNA) or protein sequences. It is widely utilized in bioinformatics for sequence alignment, storage, and analysis. A FASTA file typically contains one or more sequences, each beginning with a description line, also known as a header line, which starts with the ">" symbol. The header is followed by one or more lines representing the sequence itself.

2 FASTQ is a widely used text-based file format for storing both nucleotide sequences (such as DNA or RNA) and their corresponding quality scores. It is primarily used in high-throughput sequencing technologies, such as next-generation sequencing (NGS), where both the sequence data and the associated quality information are required for downstream analysis.

*Software-specific formats:* Bioinformatics and analysis software often use proprietary formats for data input and output. This is particularly common in fields such as molecular modeling and computational chemistry, where software like Schrödinger and ChemDraw utilize their own formats like ".mae" or ".cdx" for storing molecular structures and reactions.

*Conversion and compatibility issues:* Working with proprietary formats can cause various problems related to data conversion and compatibility. Researchers might need to rely on specific software tools or conversion services to make the data accessible in a usable form. This often involves additional costs or workflow complexities, as proprietary formats are not designed for easy interoperability with other systems.

*Licensing and access restrictions:* The use of proprietary formats can also impose licensing and access restrictions, limiting the sharing and collaborative potential of research data. This can hinder scientific progress, especially in fields where data sharing is needed for advancements.

## 3.4 Data Import Techniques

We provide here an overview of techniques for importing various types of biomedical data into Python. We will cover major data types, including CSV, TSV, JSON, XML, HDF5, DICOM, and FASTA, among others, each commonly encountered in biomedical research. In addition, we will discuss efficient methods for importing very large files, addressing memory management and performance optimization techniques.

### 3.4.1 Reading Flat Files

**Notebook: Section 3.4.1. Reading Flat Files**

This notebook provides code examples demonstrating how to read flat files.

Link to the GitHub repository:

► https://github.com/sn-code-inside/BioPy

Go to: Chap. 3—example data—► Sect. 3.4.1

Python provides built-in methods for reading text, CSV, and other file types. However, for more advanced file handling and data manipulation, specialized libraries like Pandas offer convenient methods [15]. While Python's built-in functions such as `open()` and the `csv` module can read and write basic flat files (e.g., `.txt`, `.csv`, `.json`), these methods are limited in comparison to external libraries like Pandas. For easy handling of CSV or TSV files or when extensive data manipulation is needed, Pandas is typically used.

Example codes using Python's built-in functions:

```
# Reading a text file line by line
with open('example_data/data.txt', 'r') as file:
    for line in file:
        print(line.strip())
# strip() removes any leading/trailing whitespace


# Reading a CSV file using the csv module3
import csv
with open('example_data/data.csv', 'r') as file:
    reader = csv.reader(file, delimiter=',')
# Use semicolon, if it is used as the delimiter)
    for row in reader:
        print(row) # Each row is read as a list
```

Pandas simplifies reading flat files with functions like `read_csv()` and `read_table()`, which offer a wide range of parameters for handling custom delimiters, headers, date formats, and more. It also provides robust tools for managing missing data, allowing users to fill, forward-fill, or drop missing values. Furthermore, Pandas supports automatic or explicit data type inference, ensuring data integrity. To address formatting issues such as inconsistent date formats, mixed data types, or special characters, Pandas offers functionality for cleaning and standardizing data, making it ready for analysis.

Using Pandas modules

```
# Install pandas4
%pip install pandas


# import pandas as pd alias5
import pandas as pd
```

---

3 The statement import csv is necessary because it brings the csv module into your Python script, allowing you to use its functions and tools to work with CSV files. The csv module is not automatically available in the global scope of your script, so you must explicitly import it before accessing any of its functionalities. Without importing, the program will not recognize commands like csv.reader() or csv.writer(), which are needed for reading from and writing to CSV files. Importing modules in Python is a way to reuse existing code and libraries efficiently.

4 The command %pip to install Pandas is used in Jupyter Notebooks or IPython environments to install the Pandas library. Since Pandas is not included in the standard Python library, we need to install it first before using its functionalities. The %pip magic command ensures that the installation happens in the correct environment linked to the notebook's kernel, which is especially important when working in environments like Conda or virtual environments.

5 The statement import pandas as pd is used to import the Pandas library and assign it the alias pd. This alias is a common convention in the Python community to make the code more concise and readable. Instead of typing out the full name pandas every time we call a function from the library, we can simply use the shorter pd.

```
# Reading a txt or tsv file
df = pd.read_csv('example_data/data.txt', sep = '\t')
print(df.head())


# Reading a CSV file
df = pd.read_csv('example_data/data.csv', sep = ',')
print(df.head())


# Reading a txt or tsv file using read_table
df = pd.read_table('example_data/data.txt')
print(df.head())
```

Pandas `read_csv` and `read_table` are two functions for handling flat files, among many other input functions. These functions take numerous arguments for proper data handling, which can be found in the Pandas documentation [16, 17].

### 3.4.2 Working with Hierarchical Data

> **Notebook: Section 3.4.2. Working with Hierarchical Data**
> This notebook provides code examples demonstrating how to work with hierarchical data.
> Link to the GitHub repository:
> ▶ https://github.com/sn-code-inside/BioPy
> Go to: Chap. 3—example data—▶ Sect. 3.4.2

Python provides libraries like `json` and `xml.etree.ElementTree` to parse JSON and XML files, respectively. These libraries enable the extraction of nested data structures, attribute values, and text elements from complex hierarchical data.

Handling JSON file:

```
# import Python default library json
import json
# Reading a JSON file
with open('example_data/data.json', 'r') as file:
    data = json.load(file)
# Printing the data
print(data)
```

Handling XML file:

```
# Import Python's built-in library for XML parsing
import xml.etree.ElementTree as ET


# Reading an XML file (Ensure the file is in XML format, not CSV)
tree = ET.parse('example_data/data.xml')
root = tree.getroot()


# Iterating through elements
for child in root:
    print(child.tag, child.attrib)
    for subchild in child:
        print(subchild.tag, subchild.text)
```

Converting hierarchical structures (like those in JSON/XML) to a tabular format (like `DataFrame`[6] in pandas) is often necessary for analysis. Techniques include flattening nested structures, handling repeated elements, and mapping hierarchical elements to table columns.

Converting JSON to Pandas DataFrame:

```
# import pandas and json
import pandas as pd
import json
# Load JSON data
with open('example_data/data.json', 'r') as file:
    data = json.load(file)
# Convert the JSON data into a DataFrame
df = pd.DataFrame(data)
# Print the resulting DataFrame
print(df)
```

6 A DataFrame is one of the often used data structures in the Pandas library, designed specifically for data manipulation and analysis in Python. DataFrames can be portrayed as tables, similar to an Excel spreadsheet or a SQL table, where data is organized in rows and columns. Each column in a DataFrame can store different types of data, such as integers, strings, floats, or even date-time values, making it ideal for managing mixed datasets, like those often encountered in biomedical research.

Converting XML to Pandas DataFrame:

```
# Import necessary libraries
import xml.etree.ElementTree as ET
import pandas as pd
# Parse the XML file
tree = ET.parse('example_data/data.xml')
root = tree.getroot()
# Create an empty list to store the extracted data
data = []
# Iterate through the XML structure to extract data
# Here each child of the root considered to be a separate record
for child in root:
    record = {}
    # Extract attributes of the child element
    for subchild in child:
        record[subchild.tag] = subchild.text
        # Store the tag as column and text as value
    data.append(record)
# Convert the list of dictionaries to a DataFrame
df = pd.DataFrame(data)
# Print the resulting DataFrame
print(df)
```



### 3.4.3 Importing Large Datasets

**Notebook: Section 3.4.3. Importing Large Datasets**

This notebook provides code examples demonstrating how to import large datasets.

Link to the GitHub repository:

▶ https://github.com/sn-code-inside/BioPy

Go to: Chap. 3—▶ Sect. 3.4.3

For datasets too large to fit in memory, techniques like chunking and lazy loading can be instrumental. These methods enable efficient data processing without exhausting system resources.

***Chunking:*** Chunking involves dividing a large dataset into smaller, more manageable parts (or chunks) and processing each chunk sequentially. This approach is especially useful when working with large files that cannot be loaded into memory at once.

Pandas provides a convenient way to read large CSV files in chunks using the `chunksize` parameter:

```
import pandas as pd
# Define the size of each chunk
chunk_size = 10000
# Define your data path
file_path = 'path/to/your/large_file.csv'
# Using an iterator to load chunks of the file
chunk_iter = pd.read_csv(file_path, chunksize=chunk_size)
# Print the first few rows of each chunk
for chunk in chunk_iter:
    print(chunk.head())
# Example: Process each chunk and combine them into a final DataFrame
processed_chunks = []
for chunk in pd.read_csv(file_path, chunksize=chunk_size):
    # Perform data processing (e.g., filtering, aggregation, etc.)
    processed_chunk = process_chunk(chunk)
    # Here "process_chunk" is a defined function
    processed_chunks.append(processed_chunk)
# Concatenate all processed chunks into a single DataFrame
final_df = pd.concat(processed_chunks)
```

This example demonstrates how to process large datasets efficiently by using chunking. Here, the data is divided into 1000 small pieces (chunks) and processed sequentially before being combined. This approach allows for handling large files without loading the entire dataset into memory at once, reducing memory usage and improving performance. Each chunk is processed individually, allowing for specific transformations, filtering, or aggregation on a manageable subset of the data. The `process_chunk` function should be defined according to your data processing requirements, or it can be replaced by any other function suited to your analysis goals.

***Lazy loading:*** Lazy loading refers to the technique of loading data on-demand—only when it is needed—instead of loading the entire dataset into memory at once. This method helps minimize memory usage, making it suitable for processing large files efficiently.

Python's generator functions can be used to implement lazy loading. Generators yield items one at a time and only when required, thereby saving memory. We use a generator function in this example. We will discuss more about functions in the following chapters.

```
# Define a generator function to read a large file line-by-line
def read_large_file(file_obj):
    """A generator function to read a large file lazily."""
    while True:
        data = file_obj.readline()
        if not data:
            break
        yield data
# Use the generator to read and process the file
with open('path/to/your/large_file.txt', 'r') as file:
    for line in read_large_file(file):
        # Process each line (e.g., print or analyze)
        print(line.strip())  # Replace with actual data processing
```

This code demonstrates how to read a large file in Python line by line using a generator function, which helps to manage memory efficiently by loading only one line at a time. First, the `read_large_file` function is defined as a generator that reads each line from a given file object using the `readline()` method. It continues reading lines until the end of the file, yielding each line back to the caller without storing the entire file in memory. Next, the code opens the file in read mode using `with open(...) as file`: to ensure the file is closed automatically after processing. Within this context, the code loops through each line generated by `read_large_file(file)`, processing one line at a time (in this example, stripping whitespace and printing it). This line-by-line processing approach, often referred to as "lazy loading," is particularly useful for handling large files that may not fit entirely in memory.

**Working with HDF5 using h5py:** HDF5 is a file format optimized for storing and managing large quantities of data. The `h5py` library provides a simple and efficient Python interface to read and write HDF5 files, enabling handling of large datasets.

First, we need to install the h5py package assuming that it has not been installed before:

```
%pip install h5py
```

Reading HDF5 files with h5py

```
import h5py

# Reading data from an HDF5 file
with h5py.File('data.h5', 'r') as files:
    data = files['my_dataset'][...]
    print(data)
```

This code demonstrates how to read a dataset from an HDF5 file using the `h5py` library. First, the `h5py` library is imported to provide functionality for working with HDF5 files. Then, the file `'data.h5'` is opened in read mode (`'r'`) using `h5py.File`. This line creates a file object `files` that represents the HDF5 file, allowing access to its contents without loading the entire file into memory. Within this context, `data = files['my_dataset'][...]` accesses the dataset named `'my_dataset'` inside `data.h5`. The `[...]` notation loads the entire dataset into memory as a NumPy array. Finally, `print(data)` outputs the loaded data to the console, displaying its contents. This code assumes that `data.h5` exists in the specified location and that `my_dataset` is a valid dataset within that file.

### 3.4.4 Importing Medical Imaging Data

> **Notebook: Section 3.4.4. Importing Medical Imaging Data**
> This notebook provides code examples demonstrating how to import medical imaging data.
> Link to the GitHub repository:
> ► https://github.com/sn-code-inside/BioPy
> Go to: Chap. 3—example data—► Sect. 3.4.4

Medical imaging data often requires specialized tools for processing and analysis. Libraries such as pydicom [18], SimpleITK [19], and OpenCV [20], offer robust capabilities for reading, processing, and visualizing medical images. The choice of library depends on the type of image, the necessary operations, and the overall workflow requirements. We will cover image processing in detail in ► Chap. 9, but here, we will briefly discuss how to open or import imaging data into Python.

DICOM files are the standard format for storing and transmitting medical imaging data. These files contain both image data and associated metadata, such as patient information, study details, and imaging parameters. Python libraries like pydicom make it possible to read DICOM files, extract image data, and access metadata for further analysis and visualization.

Before working with DICOM files in Python, you may need to install the pydicom library:

```
# Install pydicom using pip
%pip install pydicom
```

Working with DICOM files:

```
# Import the necessary library for reading DICOM files
import pydicom

# Example DICOM image was downloaded from the Cancer Imaging
# Archive:
# https://www.cancerimagingarchive.net/collection/cmb-aml/

dicom_file_path = 'example_data/example_image.dcm'

# Read the DICOM file into a pydicom dataset object
ds = pydicom.dcmread(dicom_file_path)

# Check if the dataset contains pixel data (to avoid errors)
if hasattr(ds, 'pixel_array'):
    # Access the pixel data from the DICOM file and store it in a
    # NumPy array
    image_data = ds.pixel_array
    print("Image shape:", image_data.shape)
    # Display the shape of the image data
else:
    raise ValueError("The DICOM file does not contain pixel data.")

# Access basic metadata (examples)
# Use .get() to avoid errors if the attribute is missing
patient_name = ds.get('PatientName', 'N/A')
patient_id = ds.get('PatientID', 'N/A')
study_date = ds.get('StudyDate', 'N/A')
modality = ds.get('Modality', 'N/A')

# Print the extracted metadata
print(f"Patient Name: {patient_name}")
print(f"Patient ID: {patient_id}")
print(f"Study Date: {study_date}")
print(f"Modality: {modality}")

# Additional metadata can be accessed similarly using the appropriate
# DICOM tag names
```

Output:

```
Image shape: (1024, 1024)
Patient Name: MSB-01723
Patient ID: MSB-01723
Study Date: 19600115
Modality: XA
```

To visualize the image data, you can use libraries like matplotlib or PIL (Python imaging library). Below is an example of how to display the first slice of a DICOM image using matplotlib:

```
# If you need to install matplotlib, use "%pip install matplotlib"
import matplotlib.pyplot as plt

plt.imshow(image_data, cmap='gray')
plt.title("DICOM Image")
plt.axis('off')  # Turn off axis labels
plt.show()
```

SimpleITK is a toolkit for medical image analysis [19]. The ReadImage function in SimpleITK can import medical image files and store them in a SimpleITK-specific image format. Although SimpleITK has its built-in image viewer (sitk.Show), viewing is dependent on external applications. To visualize these images using matplotlib, we need to first convert the SimpleITK image to a NumPy array using the GetArrayViewFromImage or GetArrayFromImage function. This conversion ensures compatibility with matplotlib, allowing for visualization and further processing if needed:

```
# Installation
%pip install SimpleITK

import SimpleITK as sitk
import matplotlib.pyplot as plt

# Example image path
image_path = 'example_data/example_image.nii'
image = sitk.ReadImage(image_path)

# Convert the image to a numpy array for visualization with
# matplotlib
image_array = sitk.GetArrayViewFromImage(image)
plt.imshow(image_array, cmap='gray')

# Add a title and display the image
plt.title("Medical Image")
plt.axis('off')  # Turn off axis labels
plt.show()
```

OpenCV (Open-source computer vision library) is widely used for computer vision and general image processing tasks. Although it is not specifically designed for medical imaging, its robust functionality can be applied to tasks such as image filtering, segmentation, and enhancement [20]. The built-in image viewer (imshow) does not display images inline, so in the following example, we use matplotlib to visualize the image:

```
# Installing OpenCV
%pip install opencv-python
import cv2
import matplotlib.pyplot as plt

# Read an image in grayscale mode
image_path = 'example_data/example_image.jpg'
image = cv2.imread(image_path, cv2.IMREAD_GRAYSCALE)

# Display the image inline using matplotlib
plt.imshow(image, cmap='gray')
plt.title("Image")
plt.axis('off')  # Turn off axis labels
plt.show()
```

Choosing the right library depends on the task: pydicom is best for reading and extracting metadata and pixel data from DICOM files; SimpleITK is suitable for complex medical image processing tasks, such as segmentation and registration; and OpenCV is useful for general image processing tasks like noise reduction, feature detection, and edge detection, though it is not specific to medical imaging.

### 3.4.5 Reading Genomic Data

**Notebook: Section 3.4.5. Reading Genomic Data**

This notebook provides code examples demonstrating how to read genomic data.

Link to the GitHub repository:

► https://github.com/sn-code-inside/BioPy

Go to: Chap. 3—example data—► Sect. 3.4.5

Biopython is a widely used tool for biological computation in Python, capable of handling genomic data formats like FASTA and FASTQ [21]. It allows for parsing of these files, extraction of sequence data, and handling quality scores associated with sequencing data:

```
# Install Biopython using pip
%pip install biopython
```

```
# Reading Fasta Files
# Reading Fasta Files
from Bio import SeqIO
```

```
# Example data was taken from
# https://www.bioinformatics.nl/tools/crab_fasta.html
fasta_file = 'example_data/example.fasta'
# Reading sequences from a FASTA file
for seq_record in SeqIO.parse(fasta_file, 'fasta'):
    print(f"ID: {seq_record.id}")
    print(f"Sequence: {seq_record.seq}\n")

# Reading FASTQ Files
from Bio import SeqIO
# Example FASTQ file was taken from
# https://zenodo.org/records/3736457
fastq_file = 'example_data/example.fastq'
# Reading sequences and quality scores from a Fastq file
for seq_record in SeqIO.parse(fastq_file, 'fastq'):
    print(f"ID: {seq_record.id}")
    print(f"Sequence: {seq_record.seq}")
    print(
    f"Quality: {seq_record.letter_annotations['phred_quality']}\n"
    )
```

Other functionalities include sequence alignment, manipulation, and analysis, making it a suitable tool for genomic data analysis.

### 3.4.6 Reading Transcriptomic Data

**Notebook: Section 3.4.6. Reading Transcriptomic Data**
This notebook provides code examples demonstrating how to read transcriptomic data.
Link to the GitHub repository:
▶ https://github.com/sn-code-inside/BioPy
Go to: Chap. 3—example data—▶ Sect. 3.4.6

Transcriptomic data, such as RNA-sequencing (RNA-seq) and single-cell RNA-sequencing (scRNA-seq), captures gene expression levels across different conditions, cell types, or time points. A popular Python tool for handling this kind of data is AnnData, which stands for "Annotated Data." AnnData is a Python library specifically designed to manage large annotated datasets commonly used in genomics and transcriptomics [22]. It offers a scalable way to keep track of both data and metadata, making it especially useful for single-cell analyses.

Transcriptomic data typically consists of raw sequence reads, count matrices, and metadata about samples or cells. The raw data is often preprocessed to generate a count matrix, which forms the basis for most downstream analyses.

The main data structure in AnnData is the AnnData object, which stores expression data in a structured format. This object contains the following:

- .X: the data matrix, typically gene expression counts
- .obs: metadata about observations, such as cell or sample annotations
- .var: information about variables, such as gene names



The library also supports additional data layers, like normalized counts or log-transformed values, which can be stored alongside the primary data matrix.

For handling large datasets, such as those common in single-cell studies, AnnData provides features for on-disk storage and chunked processing. This functionality allows for efficient management of data that exceeds memory capacity.

Example code for using AnnData:

```
# Install AnnData using pip
%pip install anndata

import anndata as ad
import numpy as np

# Example: Creating an AnnData object with random data
data_matrix = np.random.rand(100, 10)  # 100 cells, 10 genes
obs_metadata = {'cell_type': ['type1']*50 + ['type2']*50}
var_genes = {'gene_name': ['gene'+str(i) for i in range(10)]}

# Creating the AnnData object
adata = ad.AnnData(X=data_matrix, obs=obs_metadata, var=var_genes)

# Accessing data
print("Data matrix shape:", adata.X.shape)
print("Cell types:", adata.obs['cell_type'].unique())
print("Gene names:", adata.var['gene_name'].tolist())

# Saving the AnnData object to disk (HDF5 format)
adata.write_h5ad('my_dataset.h5ad')

# Loading AnnData object from disk
adata_loaded = ad.read_h5ad('my_dataset.h5ad')
```

## 3.5 Data Export Techniques

Data export techniques are important for managing the diverse data types and formats encountered in biomedical research. These techniques offer efficient and flexible methods for data sharing, publication, and further analysis, ensuring interoperability across various software and platforms.

### 3.5.1 Exporting to Flat Files

**Notebook: Section 3.5.1. Exporting to Flat Files**
This notebook provides code examples demonstrating how to export flat files.
Link to the GitHub repository:
► https://github.com/sn-code-inside/BioPy
Go to: Chap. 3—► Sect. 3.5.1

Python's built-in `open` function, which is commonly used to open flat files, also provides options to export or save data to these files. For instance, changing the argument in the open function from "`r`" (read) to "`w`" (write) enables you to write data to a file using the `write()` method.

Writing to text files:

```
with open("output.txt", "w") as file:
    file.write("Created by Python")
```

This approach is effective for saving plain text or simple data.

Writing lists or lines: The writelines() can be used to write multiple lines (from lists or strings) directly to a file:

```
lines = ["First line\n", "Second line\n"]
with open("output2.txt", "w") as file:
    file.writelines(lines)
```

The `csv` module in Python's standard library allows you to export data to CSV files:

```
import csv

data = [["Name", "Age"], ["Jerry", 30], ["Tom", 25]]
with open("output.csv", "w", newline="") as file:
    writer = csv.writer(file)
    writer.writerows(data)
```

The Pandas library in Python is instrumental for exporting "DataFrame" objects to CSV or TSV files using "to_csv()" method. This functionality is useful for sharing data with collaborators who may not use Python, as CSV and TSV files are universally accessible across various applications, including spreadsheet software. Pandas provides options to customize the output file, such as selecting which columns to export, whether to include the index, and handling missing values. Formatting options like setting custom headers, choosing delimiters, and specifying encoding types are also available, allowing for tailored output to meet specific needs.

### 3.5.2 Exporting Hierarchical Data



**Notebook: Section 3.5.2. Exporting Hierarchical Data**
This notebook provides code examples demonstrating how to export hierarchical data.
Link to the GitHub repository:
▶ https://github.com/sn-code-inside/BioPy
Go to: Chap. 3—▶ Sect. 3.5.2

Python provide several built-in libraries for exporting, hierarchical data.

JSON format with the `json` library: The `json` module is ideal for exporting structured data, such as dictionaries or lists into JSON format. JSON format is particularly useful for web applications and for transferring data between different systems due to its lightweight and readable structure:

```
import json

data = {"name": "Jerry", "age": 30}
with open("output.json", "w") as file:
    json.dump(data, file)
```

XML format with the `xml.etree.ElementTree` library: For XML data, the `xml.etree.ElementTree` library provides tools for building an XML structure from Python objects.

```
import xml.etree.ElementTree as ET

# Create the root element
root = ET.Element("Users")

# Add a child element (user) with sub-elements
user = ET.SubElement(root, "User")
name = ET.SubElement(user, "Name")
name.text = "Jerry"
age = ET.SubElement(user, "Age")
age.text = "30"

# Convert the XML structure to an ElementTree object
tree = ET.ElementTree(root)

# Save the XML to a file
with open("output.xml", "wb") as file:
    tree.write(file, encoding="utf-8", xml_declaration=True)
```

Pickle format with the `pickle` library: The `pickle` module allows for serialization and deserialization of Python objects, preserving complex data structures and object states. This can be especially useful for storing Python-specific data that might be needed later in the exact format, though it is less interoperable with non-Python systems:

```
import pickle

data = {"name": "Jerry", "age": 30}
with open("output.pkl", "wb") as file:
    pickle.dump(data, file)
```

### 3.5.3 Handling Large-Scale Data Export

**Notebook: Section 3.5.3. Handling Large-Scale Data Export**
This notebook provides code examples demonstrating how to handle large-scale data export.

Link to the GitHub repository:
▶ https://github.com/sn-code-inside/BioPy
Go to: Chap. 3—▶ Sect. 3.5.3

Exporting large-scale datasets requires efficient data writing techniques to optimize performance, conserve memory, and maintain system stability. Strategies for exporting large datasets include writing data in chunks, using buffer memory effectively, and parallelizing writing operations (if supported by the storage system). These techniques help prevent memory overflow and can substantially reduce the time required for data export.

Writing data in chunks: For large datasets, writing data in manageable chunks can prevent memory overload. By processing data in smaller portions, memory is only occupied by the chunk being processed at any time, which is particularly beneficial when dealing with datasets too large to fit into memory:

```
import pandas as pd

# Define a sample large dataset
data = pd.DataFrame({"ID": range(1, 100000001), "Value":
range(1, 100000001)})

# Export the dataset in chunks to a CSV file
chunk_size = 100000  # Number of rows per chunk
for i in range(0, len(data), chunk_size):
    data_chunk = data.iloc[i:i+chunk_size]
    data_chunk.to_csv("large_data.csv", mode="a", index=False,
    header=(i == 0))
```

The mode="a" parameter appends data to the file, ensuring that each chunk is added to the same file.

Using HDF5 format with the h5py library: For very large datasets, the HDF5 file format provides efficient methods to write, store, and access extensive data volumes. Python's h5py library allows the exporting of large datasets to HDF5 format, which supports compression and chunking. This conserves disk space and enhances data access speed, making it an ideal format for high-performance applications:

```
import h5py
import numpy as np

# Create a large dataset in memory
data = np.random.random((10000, 10000))

# Export the dataset to HDF5 format
with h5py.File("large_data.h5", "w") as h5file:
    h5file.create_dataset("dataset", data=data, compression="gzip",
    chunks=(1000, 1000))
```

In this code compression="gzip" enables data compression, which reduces file size.

### 3.5.4 Exporting Processed Imaging Data

**Notebook: Section 3.5.4. Exporting Processed Imaging Data**

This notebook provides code examples demonstrating how to export processed imaging data.

Link to the GitHub repository:

▸ https://github.com/sn-code-inside/BioPy

Go to: Chap. 3—▸ Sect. 3.5.4

Processed medical imaging data often needs to be converted into widely accessible formats, such as JPEG, PNG, or TIFF, to facilitate easier sharing, viewing, and integration into reports, presentations, or publications. OpenCV and PIL offer extensive functionalities for handling, converting, and exporting imaging data to these formats.

Using the PIL library for exporting imaging data: The PIL library provides straightforward methods for converting and saving medical images. Below is an example of how to use PIL to convert imaging data to JPEG format:

```
from PIL import Image
import numpy as np

# Example: Simulated grayscale imaging data as a NumPy array
image_data = np.random.rand(256, 256) * 255  # 256x256 grayscale image
# Convert to a PIL Image object
image = Image.fromarray(image_data.astype('uint8'))

# Save as a JPEG file
image.save("processed_image.jpg", format="JPEG")
```

Using OpenCV for exporting imaging data: OpenCV supports multiple image file formats and offers more advanced image manipulation functions. Below is an example of how to use OpenCV to export imaging data to PNG format:

```
import cv2
import numpy as np

# Example: Simulated color imaging data as a NumPy array
image_data = np.random.rand(256, 256, 3) * 255  # 256x256 RGB image
# Convert to uint8 type required by OpenCV
image_data = image_data.astype('uint8')

# Save as a PNG file
cv2.imwrite("processed_image.png", image_data)
```

Using `matplotlib` for saving image data: Matplotlib is especially useful when you want to add annotations, overlays, or specific colormaps to the image before exporting it:

```
import matplotlib.pyplot as plt
import numpy as np

# Example: Simulated grayscale imaging data as a NumPy array
# 256x256 grayscale image
image_data = np.random.rand(256, 256) * 255
image_data = image_data.astype('uint8')  # Convert to uint8 type

# Display and save the image using matplotlib
plt.imshow(image_data, cmap="gray")  # Display in grayscale
plt.axis("off")  # Turn off axis for a cleaner output

# Save as a PNG file with matplotlib
plt.savefig("processed_image_matplotlib.png", format="png",
bbox_inches="tight", pad_inches=0)
plt.close()
```

### 3.5.5 Exporting Genomic Data

**Notebook: Section 3.5.5. Exporting Genomic Data**

This notebook provides code examples demonstrating how to export genomic data.

Link to the GitHub repository:

▶ https://github.com/sn-code-inside/BioPy

Go to: Chap. 3—▶ Sect. 3.5.5



Biopython is widely used library for exporting genomic data to formats like FASTA and FASTQ. It provides tools for formatting and writing sequence data, ensuring that embedded information, such as sequence identifiers, descriptions, and quality scores (for FASTQ), are preserved. This functionality is vital for sharing sequence data with other researchers or for submitting to genomic data repositories.

Exporting sequences to FASTA format

```
from Bio import SeqIO
from Bio.Seq import Seq
from Bio.SeqRecord import SeqRecord

# Define a sample sequence record
sequence_data = SeqRecord(
    Seq("ATGCGTACGTAGCTAGCTAGCTGACT"),
    id="Sequence1",
    description="Example DNA sequence"
)

# Export the sequence to a FASTA file
with open("output.fasta", "w") as file:
    SeqIO.write(sequence_data, file, "fasta")
```

Exporting sequences to FASTQ format

```
from Bio import SeqIO
from Bio.Seq import Seq
from Bio.SeqRecord import SeqRecord

# Define a sample sequence record with quality scores
sequence_data = SeqRecord(
    Seq("ATGCGTACGTAGCTAGCTAGCTGACT"),
    id="Sequence1",
    description="Example DNA sequence",
    letter_annotations={"phred_quality": [40, 30, 35, 38, 37, 40,
    30, 32, 30, 35, 38, 37, 40, 30, 32, 30, 35, 38, 37, 40, 30,
    32, 30, 35, 38, 37]}
)
```

```
# Export the sequence to a FASTQ file
with open("output.fastq", "w") as file:
    SeqIO.write(sequence_data, file, "fastq")
```

## 3.6 Exercises and Questions

We included exercises and questions to complement the material covered in this chapter.

### Exercises

The following exercises will enhance your understanding of the materials we covered in this chapter.

1. Read a CSV file: Import a CSV file containing patient demographic information using Python and Pandas.
2. Work with JSON data: Load a JSON file that includes patient treatment records and extract specific fields.
3. Import a large dataset: Use efficient techniques to load a large dataset of blood test results without overwhelming system memory.
4. Import medical imaging data: Use a Python library, such as PyDICOM, to read DICOM files for radiological images.
5. Read genomic data: Import a FASTQ file containing genomic sequencing data and display information about the sequences.
6. Import transcriptomic data: Load and preprocess RNA-seq data from a file to check gene expression levels.
7. Export data to a flat file: Write processed clinical trial data back to a CSV file with appropriate headers and formatting.
8. Export hierarchical data: Save complex patient data, including nested records of treatments and outcomes, to a JSON file.
9. Handle large-scale data export: Demonstrate techniques to efficiently export a large dataset.
10. Export processed imaging data: Save enhanced medical images to a TIFF file after applying image processing algorithms.

### Questions to Be Answered

1. What are the primary types of biomedical data, and how do they differ?
2. What are the common challenges associated with handling biomedical data?
3. Describe some common and proprietary formats used in biomedical data.
4. How would you import a large biomedical dataset without compromising system performance?
5. What considerations should be made when importing medical imaging data?
6. What are the steps involved in reading genomic data from a FASTQ file?
7. What techniques can be used to ensure the efficiency and scalability of data import and export processes?
8. Explain how hierarchical data can be exported, and why it might be necessary to use formats like JSON or XML for such tasks.

**Acknowledgement** The language of the human-generated text was corrected with the assistance of artificial intelligence (AI) tools [GPT-3.5 and GTP-4 from OpenAI]. GitHub Co-Pilot was used to check the correctness of the codes. The text underwent subsequent human revision to ensure its accuracy.



## References

1. All of Us Research Program Genomics I. Genomic data in the all of us research program. Nature. 2024;627(8003):340–6. https://doi.org/10.1038/s41586-023-06957-x.
2. Glaves PD, Tugwood JD. Generation and analysis of transcriptomics data. Methods Mol Biol. 2011;691:167–85. https://doi.org/10.1007/978-1-60761-849-2_10.
3. Zhu H, Bilgin M, Snyder M. Proteomics. Annu Rev Biochem. 2003;72:783–812. https://doi.org/10.1146/annurev.biochem.72.121801.161511.
4. Idle JR, Gonzalez FJ. Metabolomics. Cell Metab. 2007;6(5):348–51. https://doi.org/10.1016/j.cmet.2007.10.005.
5. Zhang L, Lu Q, Chang C. Epigenetics in health and disease. Adv Exp Med Biol. 2020;1253:3–55. https://doi.org/10.1007/978-981-15-3449-2_1.
6. Shetty SA, Lahti L. Microbiome data science. J Biosci. 2019;44(5):1–6. https://doi.org/10.1007/s12038-019-9930-2.
7. Kandi V, Vadakedath S. Clinical trials and clinical research: a comprehensive review. Cureus. 2023;15(2):e35077. https://doi.org/10.7759/cureus.35077.
8. Sauer CM, Chen LC, Hyland SL, Girbes A, Elbers P, Celi LA. Leveraging electronic health records for data science: common pitfalls and how to avoid them. Lancet Digit Health. 2022;4(12):e893–e8. https://doi.org/10.1016/S2589-7500(22)00154-6.
9. Hockings JK, Pasternak AL, Erwin AL, Mason NT, Eng C, Hicks JK. Pharmacogenomics: an evolving clinical tool for precision medicine. Cleve Clin J Med. 2020;87(2):91–9. https://doi.org/10.3949/ccjm.87a.19073.
10. Anderson NR, Lee ES, Brockenbrough JS, Minie ME, Fuller S, Brinkley J, et al. Issues in biomedical research data management and analysis: needs and barriers. J Am Med Inform Assoc. 2007;14(4):478–88. https://doi.org/10.1197/jamia.M2114.
11. Luo J, Wu M, Gopukumar D, Zhao Y. Big data application in biomedical research and health care: a literature review. Biomed Inform Insights. 2016;8:1–10. https://doi.org/10.4137/BII.S31559.
12. Cascini F, Santaroni F, Lanzetti R, Failla G, Gentili A, Ricciardi W. Developing a data-driven approach in order to improve the safety and quality of patient care. Front Public Health. 2021;9:667819. https://doi.org/10.3389/fpubh.2021.667819.
13. Oronsky B, Burbano E, Stirn M, Brechlin J, Abrouk N, Caroen S, et al. Data Management 101 for drug developers: a peek behind the curtain. Clin Transl Sci. 2023;16(9):1497–509. https://doi.org/10.1111/cts.13582.
14. Teodoro D, Choquet R, Schober D, Mels G, Pasche E, Ruch P, et al. Interoperability driven integration of biomedical data sources. Stud Health Technol Inform. 2011;169:185–9. https://doi.org/10.3233/978-1-60750-806-9-185.
15. McKinney W. Python for data analysis: data wrangling with Pandas, NumPy, and Jupyter. O'Reilly Media; 2022.
16. pandas.read_table [Internet]. 2024. Available from: https://pandas.pydata.org/docs/reference/api/pandas.read_table.html.
17. pandas.read_csv [Internet]. 2024. Available from: https://pandas.pydata.org/docs/reference/api/pandas.read_csv.html.
18. Mason D. et al. pydicom/pydicom: Pydicom v2.4.0, Zenodo, 2023. https://doi.org/10.5281/zenodo.8034250.
19. Lowekamp BC, DChen DT, Ibáñez L, Blezek D. The Design of SimpleITK, Front Neuroinform. 2013;7:45. https://doi.org/10.3389/fninf.2013.00045.
20. Bradski G. The OpenCV library. Dr Dobb's J Softw Tools. 2000;4:2236121.
21. Cock PJ, Antao T, Chang JT, Chapman BA, Cox CJ, Dalke A, et al. Biopython: freely available Python tools for computational molecular biology and bioinformatics. Bioinformatics. 2009;25(11):1422–3.
22. Virshup I, Rybakov S, Theis FJ, Angerer P, Wolf FA. anndata: Annotated data. bioRxiv. 2021;2021.12.16.473007.

# Biomedical Data Preprocessing

## Contents




J. U. Kazi, *Python Essentials for Biomedical Data Analysis: An Introductory Textbook*,
https://doi.org/10.1007/978-3-031-85600-6_4

This chapter describes several techniques and considerations in biomedical data preprocessing to ensure data quality, integrity, and suitability for analysis. It discusses common challenges in biomedical datasets, including complexity, heterogeneity, and the prevalence of missing or inconsistent data. Methods for assessing and improving data quality are discussed, along with strategies for handling missing data to minimize its impact on analyses. The chapter also covers data transformation techniques such as normalization, encoding, and dimensionality reduction, as well as approaches to manage outliers effectively. Data integration methods are examined to address challenges with duplicate data and metadata management. Finally, privacy and security practices are discussed.

**Learning Goals**

The learning goals for this chapter include building an understanding of the challenges found in biomedical datasets, such as their complexity, heterogeneity, and privacy concerns, and gaining the skills necessary to address them. Students are expected to learn how to assess and improve data quality, identify and solve common data problems, and apply methods such as normalization, standardization, and missing data handling. This chapter also aims to prepare learners to protect sensitive information by using appropriate privacy and security techniques.

## 4.1 Overview of Common Issues with Biomedical Datasets

Biomedical datasets often come with a unique set of challenges that can impact the quality and reliability of the data. Recognizing these issues helps to draw reliable data-driven conclusions.

### 4.1.1 Complexity and Heterogeneity

Biomedical datasets are complex and heterogeneous, comprising diverse data types that are important for research and personalized medicine, but they also introduce analytical and operational challenges [1]. Clinical data, for example, can include patient records, treatment outcomes, diagnostic tests, and demographic information, all of which are highly variable and reflect individual health trajectories [2]. These records often contain longitudinal information, disease progression or treatment responses over time, and can vary widely in how data are collected, healthcare is delivered, and in the level of detail available. Such variability necessitates sophisticated normalization and preprocessing for meaningful comparative and predictive analyses. Genomic data, including DNA sequencing, gene expression profiles, single nucleotide polymorphisms (SNPs), and epigenetic modifications like methylation, add further complexity. This complexity arises from both the volume of data per individual and the complex genetic interactions involved require substantial computational resources for storage, processing, and analysis, alongside advanced statistical and machine learning tools for biological and clinical interpretation [3]. Medical imaging data from MRI, CT scans, X-rays, and ultrasounds introduce additional challenges, with each modality providing unique information in resolution, contrast, and underlying physical principles [4]. Bringing together clinical, genomic, and imaging data to form a complete view of the patient requires advanced computational methods, including machine learning and multimodal integration.



### 4.1.2 Missing and Incomplete Data

Missing or incomplete data is a common challenge in biomedical research and can affect the validity and reliability of studies. These issues arise from various sources, each introducing specific obstacles to data analysis and interpretation [5]. Some of these sources include:

- *Patient drop-out in longitudinal studies*: In studies that follow patients over time, drop-out is a common problem [6]. Participants may discontinue their involvement due to various reasons, such as improved health, side effects, dissatisfaction with the treatment, or even loss of interest. If drop-outs are related to specific patient characteristics or outcomes, this can introduce bias. Handling this issue often requires relevant statistical techniques such as multiple imputation, or sensitivity analyses to assess how the missing data might influence conclusions of the study [7, 8].
- *Equipment failure*: Technical failures in medical or data-recording equipment can lead to significant gaps in data collection. For instance, a malfunction in imaging equipment may result in non-usable images or failure in genomic sequencing tools could lead to incomplete genetic data. These issues are not merely operational but can also affect the longitudinal integrity of the data, where consistency across time points is relevant [9]. Rigorous quality control measures and, if needed, repeated experiments or imaging sessions are important, which can be costly and time-consuming.
- Manual entry of clinical, laboratory, or health-related information is still common and prone to mistakes such as typos, omissions, and misclassifications. Such inaccuracies can lead to incorrect conclusions or misdiagnoses if used for clinical decision-making. Preventative strategies include the use of double-data entry systems, where two independent data entries are compared for discrepancies, and the implementation of software that automatically checks for data anomalies or outliers [10].

The presence of missing or incomplete data can complicate statistical analyses and lead to biases in research findings. Standard statistical methods often prove inadequate as they typically presume the availability of complete datasets [11]. The management of missing data requires careful planning and proper methodologies from the design phase of a study [12]. Managing missing data involves identifying the sources of incomplete data and implementing strategies to mitigate their impact. Such strategies could include designing engaging study protocols to minimize participant dropout, investing in reliable equipment, and ensuring thorough training of personnel in accurate data collection and handling practices [13]. Moreover, the challenge of handling missing and incomplete data necessitates continuous advancements in statistical methods and study designs.

### 4.1.3 Outliers and Noise

Outliers and noise are common issues in biomedical datasets that can affect the analysis and interpretation of data, sometimes leading to misleading conclusions [14]. These phenomena can originate from a variety of sources, each requiring specific strategies for identification and mitigation. Here, we discuss some sources of outliers and noise:

- *Experimental error*: Variations in experimental conditions are a primary source of outliers including but not limited to fluctuations in temperature, inconsistencies in sample handling, or variations in the calibration of instruments. Even minor deviations in experimental protocol can introduce variability in the data, which manifests as outliers. These outliers can skew statistical analyses, leading to erroneous conclusions if not properly managed [15]. Using appropriate experimental design and strictly following standardized protocols, applying statistical methods like robust regression or outlier detection algorithms can help minimize and address such errors.
- *Biological variability*: Biological variability refers to the natural differences in biological responses among subjects in a study. This type of variability is inherent and expected, especially in studies involving human subjects, due to genetic, environmental, and lifestyle differences [16]. However, in certain analyses, extreme cases of biological variability can appear as outliers. Such type of outliers are not errors per se but rather real data points that represent extreme responses. Identifying whether such outliers should be retained or treated depends on the research question and the statistical methods employed. In many cases, these data points can provide information relevant to possible responses and help in understanding mechanisms underlying extreme cases [17].
- Data entry mistakes: As discussed above as well manual data entry can introduce noise through typos, incorrect numbers, misclassified terms, or coding errors in scripts. This artificial noise can mask real patterns in the data. Techniques such as automated error detection algorithms, periodic audits of data integrity, and cross-validation with external data sources can help identify and correct these issues.

Managing outliers and noise is important for producing reliable research. This involves both preventative measures, such as improving the accuracy of data collection and experimental design and corrective measures, like employing statistical techniques tailored for robust outlier handling. Statistical tools such as box plots, scatter plots, and analytical techniques like principal component analysis (PCA) or clustering can help to identify outliers. Once identified, decisions on how to handle these data points, whether to exclude them, adjust them, or otherwise account for them in the analysis, should be made based on the context of the study and the nature of the data.

### 4.1.4 Inconsistencies and Errors

When data are collected from multiple sources, they often contain inconsistencies and errors that can hamper their utility for research and clinical applications. These issues arise from a variety of factors, each contributing to the overall complexity of data management and requiring attention during data collection, integration, and analysis. Below we discuss three common sources of these issues:



*Naming conventions*: Lack of standardized naming conventions across different researchers, institutions, or databases can cause confusion when combining datasets. For example, one dataset may refer to a biomarker with a specific name, while another uses a different abbreviation or alternative nomenclature for the same biomarker. Such discrepancies can lead to challenges in data merging and can even result in erroneous data interpretation if not properly addressed. Adopting common data elements (CDEs) [18] and standards, such as those developed by the Clinical Data Interchange Standards Consortium (CDISC), can help ensure data from different sources are compatible and comparable [19].

*Measurement units*: Inconsistencies in measurement units, such as differences between the metric and imperial systems, or varying scales for laboratory values (e.g., mg/dL vs. mmol/L for glucose levels), can lead to errors in data interpretation. This problem is especially prevalent in global studies where data collection involves sources from countries with different measurement practices. Such inconsistencies can lead to errors in dosing calculations, misinterpretation of diagnostic criteria, and incorrect statistical analyses, resulting in harmful clinical outcomes. The use of automated conversion tools or integrated data management systems that standardize units at the point of data entry can help prevent these errors. Furthermore, protocols should be established to check and convert data into a unified system before analysis.

*Data formats*: Variability in data formats across different sources also presents substantial challenges in data integration and analysis. For example, some datasets may store information in structured databases, while others use unstructured formats, such as text files or images. The lack of uniformity can complicate data processing and integration, requiring extensive preprocessing and transformation efforts. Furthermore, different software tools and platforms might only support certain data formats, necessitating additional conversion steps that can introduce errors or data loss. Implementing interoperability standards, such as Health Level Seven International (HL7) or Fast Healthcare Interoperability Resources (FHIR), can aid in addressing these challenges by providing guidelines for data formatting and exchange [20, 21].

Effectively dealing with inconsistencies and errors in data is required for maintaining the integrity and reliability. This involves strategies encompassing both technological solutions and organizational practices. Technologically, employing advanced data integration tools equipped with capabilities for semantic reconciliation (aligning data meaning across different systems) and format normalization can be useful. Organizationally, fostering collaboration across institutions to agree on common standards and practices can greatly reduce the risk of inconsistencies.

## 4.2 Understanding Data Quality

Data quality evaluates the suitability of a dataset by assessing various dimensions that determine whether the data are reliable and can be effectively used for their intended purposes [22]. High-quality data directly impacts planning and decision-making across various fields.

### 4.2.1 Defining Data Quality in a Biomedical Context

Data quality is not just a beneficial attribute but a requisite for ensuring the reliability, validity, and applicability of research results [23, 24]. It is a multidimensional concept contributing to the overall integrity and utility of data. Dimensions offer a structured framework for evaluating and managing data quality, enabling us to identify and address weaknesses systematically [25]. To provide clarity and focus, data quality dimensions can be broadly grouped into three categories: intrinsic data quality, contextual data quality, and resource-related data quality, facilitating an approach to managing and maintaining high-quality data across various research contexts (Fig. 4.1).

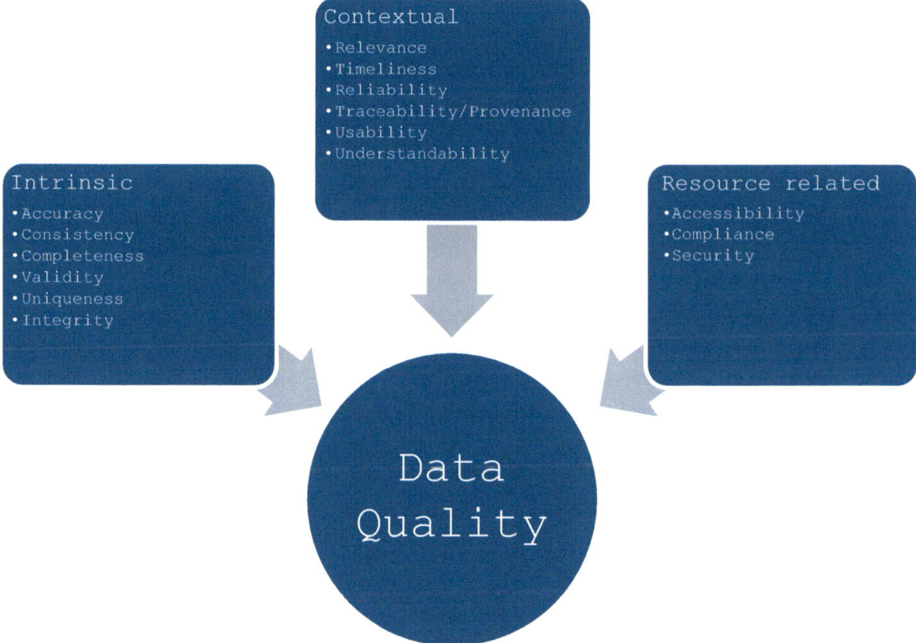


**Fig. 4.1** Dimensions of data quality. Dimensions of data quality can be grouped into three categories including intrinsic, contextual, and resource-related

A. **Intrinsic data quality**: Intrinsic data quality pertains to the inherent properties of the data itself, independent of external factors. It focuses on ensuring that the data values are accurate, complete, and logically sound. The following dimensions fall under this category:
   - **Accuracy** evaluates how closely data values correspond to the true or real-world values they are supposed to represent. For instance, numerical values that correctly reflect physical measurements or exact spelling of names improve data accuracy.
   - **Consistency** ensures that data does not contain contradictions or discrepancies, both within a single dataset and across multiple datasets or systems. A consistent dataset maintains coherence in data points, such as a patient's birth date being the same in multiple records.
   - **Completeness** checks whether all necessary data points and attributes are present. Missing values or fields, such as an address without a zip code or a medical record without a diagnosis, indicate incomplete data, reducing its usefulness.
   - **Validity** assesses whether data values conform to defined formats, rules, or constraints. For example, a field designated for dates should only contain valid date entries, and numerical values should fall within an expected range.
   - **Uniqueness** ensures that records or data points are not duplicated within a dataset, confirming that each entity is represented only once. Duplicate entries, such as having multiple records for the same customer, can lead to confusion and misinterpretation.
   - **Integrity** refers to the logical coherence and correctness of relationships within the data. It checks for appropriate linkages and dependencies, such as referential integrity in a relational database where foreign keys accurately point to corresponding primary keys.

B. **Contextual data quality**: Contextual data quality is concerned with the suitability and applicability of data for a specific context, task, or scenario. It considers whether the data is relevant and useful for the intended use, ensuring it meets the requirements of its application. The following dimensions fall under this category:
   - **Relevance** measures how applicable and useful the data is for its intended purpose. Data that is relevant for one analysis may not be suitable for another. For example, daily patient admission data is relevant for hospital operational decision-making but might be less useful for long-term epidemiological trend analysis.
   - **Timeliness** refers to how current and available the data is relative to its intended use. Time-sensitive applications, such as patient monitoring or real-time drug administration, require up-to-date data, as delayed information can lead to incorrect decisions.
   - **Reliability** evaluates whether data can be trusted, based on its source, collection methods, and history of usage. Reliable data comes from credible sources and stable collection methods, without unexplained fluctuations.

- **Traceability/Provenance** assesses the ability to trace data back to its origins, transformations, and lineage. It ensures that data lineage is documented, making it possible to verify its history and the processes applied to it.
- **Usability** measures how easy it is for users to interpret and utilize the data for analysis, decision-making, or other tasks. Well-structured data with clear naming conventions, proper formatting, and supporting documentation is considered more usable.
- **Understandability** relates to how clearly the data is defined, labeled, and described, including the availability of metadata. Easily understandable data reduces ambiguity and improves communication among stakeholders.

C. **Resource-related data quality**: Resource-related data quality dimensions focus on how data is accessed, stored, and managed. These dimensions address the technical and logistical facets of data management, ensuring that data is readily available and compliant with standards. The key dimensions in this category include:
   - **Accessibility** evaluates how easily authorized users can access and retrieve data. This involves considering permissions, system availability, and how straightforward it is to extract or use the data for analysis. If data is stored in overly complex or restricted systems, its accessibility may be limited.
   - **Compliance** examines whether data handling meets legal, regulatory, or organizational requirements. This includes adherence to data protection laws such as the General Data Protection Regulation (GDPR) and standards like the Health Insurance Portability and Accountability Act (HIPAA). Compliance ensures that data is managed ethically and legally.
   - **Security** ensures that data is protected from unauthorized access, breaches, and tampering. This includes implementing access controls, encryption, and other security measures to safeguard sensitive information such as financial records or personal health data.

### 4.2.2 Common Data Quality Issues

Data quality issues are pervasive in biomedical research and can compromise the validity and reliability of a study [26, 27]. Common challenges include missing values, outliers, inconsistencies, data entry errors, and duplication of records (■ Fig. 4.2) which are also partly discussed earlier.

**Missing data**, often encountered in clinical trials or patient records, can reduce statistical power, distort estimates, and introduce biases, leading to an incomplete understanding of disease patterns or treatment effects. The classification of missing data influences the choice of handling methods, ranging from basic imputation techniques to advanced statistical models.

**Outliers** may arise due to experimental errors, unique patient characteristics, or variations in measurement techniques and can substantially skew results. They can lead to misleading interpretations, such as overestimating the efficacy of a drug or misidentifying risk factors for a disease.

**Fig. 4.2** Data quality issues. Typical challenges encompass missing values, outliers, inconsistencies, data entry mistakes, and record duplication

**Inconsistencies** emerge from data entry errors, variations in data collection methods, or discrepancies across different sources and often require harmonization efforts. Addressing these inconsistencies involves accounting for differences in methodologies and applying efficient data integration techniques.

**Data entry errors and duplication of records** frequently arise from manual data entry, merging multiple sources, or technical errors in data collection and storage systems. For example, duplicated patient records can overestimate the prevalence of a condition or skew population-based statistics. Mitigating these issues necessitates thorough data cleaning and validation processes, often requiring cross-referencing with multiple sources and employing automated error detection algorithms to ensure accuracy.

Addressing these common data quality issues enhances the reliability of biomedical research and ultimately for generating accurate data and meaningful conclusions.

### 4.2.3 Addressing Data Quality Issues

Addressing data quality issues ensures the integrity and applicability of research findings in clinical and public health contexts. This process involves the implementation of robust data management practices, advanced analytical techniques, and encouraging interdisciplinary collaboration [28].

Key strategies include establishing standardized data entry protocols, conducting regular data quality audits, and utilizing appropriate data management systems to maintain high data standards and reduce errors at the source. In parallel, advanced statistical methods and machine learning algorithms can be employed to handle missing data, detect outliers, and correct inconsistencies. These techniques improve data quality by filling gaps, identifying anomalies, and reconciling discrepancies, which in turn improves the reliability of research outcomes.

Interdisciplinary collaboration, such as engaging data scientists, statisticians, and domain experts, facilitates the development and application of tailored methodologies for data cleaning and analysis. This collaborative approach ensures that data quality solutions are both contextually relevant and technically sound.

Together, these strategies form a framework for recognizing and addressing data quality challenges, ultimately supporting the generation of accurate, meaningful, and reproducible research results in the biomedical field.

## 4.3 Handling Missing Data

Here, we discuss methods for effectively handling missing data in biomedical datasets, which are required for ensuring the accuracy and reliability of their analyses.

### 4.3.1 Type of Missing Data

In statistical analysis and data science, missing data is typically categorized into three main types: Missing Completely at Random (MCAR), Missing at Random (MAR), and Missing Not at Random (MNAR) [5]. These categories help researchers understand the nature of missingness in their data and choose appropriate methods for handling it.

MCAR: The probability that data are missing is the same for all observations. In this case, the missing data are a random subset of the data, and are not related to any variables in the dataset, whether observed or unobserved.

MAR: The probability of missingness may differ across groups but is related only to observed data, not to the missing values themselves. Here, the missingness can be systematically related to other observed variables but remains unrelated to the specific values that are missing.

MNAR: The probability of missingness is related to the unobserved value itself or to other unmeasured variables. This is the most challenging scenario because the missingness depends on information that is not available in the dataset. A brief comparison within MCAR, MAR and MNAR is presented in the table below (◘ Table 4.1).

**Table 4.1** Comparison of missing data mechanisms

| Characteristic | MCAR | MAR | MNAR |
|---|---|---|---|
| Probability | The probability of missingness is the same for all observations | The probability of missingness is the same within groups defined by observed data | The missingness is related to the value of the missing data itself |
| Dependency | Missing data does not depend on any values (observed or unobserved) | Missing data depends on observed data but not on the missing data | Missing data depends on unobserved data or the missing data itself |
| Impact on data analysis | Generally, does not bias the results if data are ignored | Can bias results if not properly handled | Often biases results if ignored or improperly handled |
| Handling complexity | Simplest to handle as it requires fewer assumptions | Requires modeling the mechanism of missingness based on observed data | Most complex to handle; often requires modeling the missing data mechanism explicitly |
| Examples | Randomly losing some data entries due to technical issues | Income data missing more frequently for people in certain age groups | People with high debt not reporting their income level |
| Imputation feasibility | Easier to impute since the missingness is unrelated to any data | Feasible with methods that model the relationship with observed variables | Challenging to impute accurately without insights into the reasons for missingness |
| Statistical methods | Simple random sampling methods can be effective | More sophisticated methods like multiple imputation needed | May require specialized statistical models like selection models or pattern-mixture models |

### 4.3.2 Identifying Missing Data

**Notebook: Section 4.3.2. Identifying Missing Data**

This notebook provides code examples demonstrating how to identify missing data.

Link to the GitHub repository:

► https://github.com/sn-code-inside/BioPy

Go to: Chap. 4—example data—► Sect. 4.3.2

The first step involves detecting missing data using visual methods like heatmaps, which can show the presence and distribution of missing values in a dataset. Moreover, programmatic checks in Python, utilizing libraries like Pandas, can automate this detection process. It is important to analyze whether the missingness in the data is random or systematic. Therefore, determining the type of missing data is an important step. This understanding can affect the approach to handling the missing data and the interpretation of the results. Furthermore, evaluating the proportion and distribution of missing data is useful to understand how it might impact the analysis. This involves examining the extent and nature of the missing data across variables.

Determining the type of missing data, whether it is MCAR, MAR, or MNAR, is needed for selecting appropriate methods for handling missing values in statistical analyses. There are several statistical tests and techniques used to infer the likely mechanism of missingness:

**Little's MCAR test**: This is a formal statistical test used to determine if the missing data are MCAR [29]. The test compares the observed patterns of missingness against the patterns that would be expected if the data were MCAR. A significant test result indicates that the data are not MCAR, suggesting a systematic pattern in the missingness [30]. It is important to note that this test does not differentiate between MAR and MNAR.

The MCAR test can be performed using tools like `XeroGraph`. An example code snippet for using `XeroGraph` to perform the MCAR test is shown below:

```
# Install XeroGraph using PIP
%pip install XeroGraph

# Import necessary libraries
import pandas as pd
from XeroGraph import XeroAnalyzer
# Load the dataset
data = pd.read_csv('example_data/data.csv', index_col=0)
print(data.head(5))
# Initialize XeroAnalyzer
xg_test = XeroAnalyzer(data)
# Perform MCAR test
mcar_result = xg_test.mcar()
# Print the MCAR test result
print(f"MCAR Test Result: {mcar_result}")
```

**Tests and strategies for MAR and MNAR**: For MAR and MNAR, there are no definitive tests equivalent to Little's MCAR test. This is primarily because these mechanisms involve conditions where the probability of missingness is related to the observed data alone (MAR) or both the observed and unobserved data (MNAR), making direct testing challenging.

### 4.3.3 Techniques for Handling Missing Data

Selecting the appropriate method for handling missing data is highly context-dependent. The choice should consider the pattern of missingness (MCAR, MAR, or MNAR), the type of data (categorical, continuous, or ordinal), and the specific objectives of the study [31]. Below are several techniques commonly employed to address missing data, each with its advantages and limitations.



**Notebook: Section 4.3.3. Techniques for Handling Missing Data**
This notebook provides code examples demonstrating techniques for handling missing data.
Link to the GitHub repository:
▶ https://github.com/sn-code-inside/BioPy
Go to: Chap. 4—example data—▶ Sect. 4.3.3

A. *Removal of Incomplete Records*

The removal of incomplete records is a method used in handling missing data in datasets. This method involves identifying and eliminating records (rows of data) that contain missing values in one or more fields. There are several approaches to implement this:

**Listwise deletion** is also known as complete case analysis, this method involves removing an entire record if any single value is missing [32, 33]. It is a straightforward approach and easy to implement. However, it can lead to a reduction in the dataset size, especially if missing values are widespread. This reduction can, in turn, decrease the statistical power of analyses and introduce bias if the missingness is not random. This approach is suitable only if the data are MCAR and the sample size remains sufficient for statistical power.

Example:

```
# Import necessary libraries
import pandas as pd
# Load the dataset
data = pd.read_csv('example_data/data.csv', index_col=0)
print("Original DataFrame:")
print(data)
# Apply Listwise Deletion (removes any row with missing values)
data_listwise_deleted = data.dropna()
print("\nDataFrame after Listwise Deletion:")
print(data_listwise_deleted)
```

**Pairwise deletion** is used primarily in the context of statistical analyses, particularly in calculating correlations or covariance [34]. In this approach, only the missing values are omitted, and all available data points are used. For instance, when calculating the correlation between two variables, only pairs of values that are complete for both variables are considered. This method allows for more data to be retained compared to listwise deletion, but it can lead to biases and inconsistencies, especially if the pattern of missingness is not random.

Pairwise deletion is generally used when performing calculations like correlations or covariance matrices where only the available pairs are needed, and it might not be directly applicable for other types of analyses. If the goal is to perform different types of statistical analyses using pairwise deletion, customized code should be written based on the type of analysis.

Example:

```
# Import necessary libraries
import pandas as pd
# Load the dataset
data = pd.read_csv('example_data/data.csv', index_col=0)
print("Original DataFrame:")
print(data)
# Calculate pairwise correlations using pairwise deletion
# (excluding NA/null values)
# The `pairwise deletion` is applied by using the 'min_periods=1'
# parameter
# which will calculate correlation for available pairs without
# dropping rows entirely.
correlation_matrix = data.corr(min_periods=1)
print("\nPairwise Correlation Matrix with Pairwise Deletion:")
print(correlation_matrix)
```

**Threshold-based removal** is a technique used to remove columns or rows that contain missing values exceeding a specified threshold. This method is particularly useful in scenarios where some features or samples have too many missing values to be reliable for analysis. The threshold can be defined as a proportion of missing values (e.g., remove columns with more than 20% missing data). This method is ideal when certain features or samples are too incomplete to provide meaningful information. However, setting the threshold requires careful consideration, as overly strict thresholds may lead to excessive data loss, while lenient thresholds might retain too much incomplete data, affecting the robustness of the analysis.

Example:

```
# Import necessary library
import pandas as pd
# Load the dataset
data = pd.read_csv('example_data/data.csv', index_col=0)
print("Original DataFrame:")
print(data)
# Define a threshold for column removal (e.g., remove columns with
# more than 20% missing data)
column_threshold = 0.2
# Remove columns where the proportion of missing values is
# greater than the threshold
df_column_threshold = data.loc[:, data.isnull().mean()
                        <= column_threshold]
print("\nDataFrame after Column Threshold-based Removal"
"(20% missing data threshold):")
print(df_column_threshold)
# Define a threshold for row removal (e.g., remove rows with more
# than 20% missing data)
row_threshold = 0.2
# Remove rows where the proportion of missing values is greater
# than the threshold
df_row_threshold = data[data.isnull().mean(axis=1)
                    <= row_threshold]
print("\nDataFrame after Row Threshold-based Removal"
"(20% missing data threshold):")
print(df_row_threshold)
```

Each of these methods has its advantages and disadvantages. The choice of method depends on factors such as the extent and pattern of missingness in the data, the nature of the study, and the impact of data removal on the results. While removal of incomplete records can simplify the analysis, it is important to consider the implications on the representativeness and bias in the dataset. Therefore, it is often recommended to conduct sensitivity analyses to understand the impact of removing incomplete records on the study [35].

B. *Single Imputation*

**Mean/median/mode imputation**: Replaces missing values with the mean, median, or mode of the observed values for that variable. This method is easy to implement but can underestimate variability and introduce bias, especially if the missing data are not MCAR [36].

Example using Scikit-learn `SimpleImputer`:

```
# Import necessary libraries
import pandas as pd
from sklearn.impute import SimpleImputer
# Load the dataset
data = pd.read_csv('example_data/data2.csv', index_col=0)
print("Original DataFrame:")
print(data)
# Separate numerical and categorical columns
num_col = data.select_dtypes(include=['number']).columns

# Mean Imputation: Replace missing values with the mean of the
# respective column (for numerical columns only)
df_mean_imputed = data.copy()
# Initialize SimpleImputer with mean strategy
mean_imputer = SimpleImputer(strategy='mean')
df_mean_imputed[num_col] = mean_imputer.fit_transform(
df_mean_imputed[num_col])
print("\nDataFrame after Mean Imputation (Numerical Columns):")
print(df_mean_imputed)

# Median Imputation: Replace missing values with the median of
# the respective column (for numerical columns only)
df_median_imputed = data.copy()
# Initialize SimpleImputer with median strategy
median_imputer = SimpleImputer(strategy='median')
df_median_imputed[num_col] = median_imputer.fit_transform(
df_median_imputed[num_col])
print("\nDataFrame after Median Imputation (Numerical Columns):")
print(df_median_imputed)

# Mode Imputation: Replace missing values with the mode of the
# respective column (useful for categorical columns)
df_mode_imputed = data.copy()
# Initialize SimpleImputer with mode strategy
mode_imputer = SimpleImputer(strategy='most_frequent')
df_mode_imputed[data.columns] = mode_imputer.fit_transform(
df_mode_imputed)
print("\nDataFrame after Mode Imputation"
"(Numerical and Categorical Columns):")
print(df_mode_imputed)
```

**Last observation carried forward (LOCF)**: Typically used in longitudinal studies, this method fills in missing values with the last observed value [37]. LOCF assumes that the value remains constant over time, which may not hold true in many contexts, leading to biased results:

```
# Import necessary library
import pandas as pd
# Load the dataset
data = pd.read_csv('example_data/data.csv', index_col=0)
print("Original DataFrame:")
print(data)
# Apply LOCF (Last Observation Carried Forward) imputation
df_locf = data.copy()  # Create a copy of the original dataframe
# Forward fill method to propagate last valid observation forward
df_locf = df_locf.ffill()
print("\nDataFrame after LOCF Imputation:")
print(df_locf)
```

**Hot deck imputation**: Replaces missing values with observed values from similar cases (i.e., cases with similar characteristics) [38]. While more robust than mean imputation, the quality of hot deck imputation depends on how well the matching process is defined.

Example:

```
# Import necessary libraries
import pandas as pd
import numpy as np
# Load the dataset
data = pd.read_csv('example_data/data2.csv', index_col=0)
print("Original DataFrame:")
print(data)
# Hot deck imputation function using grouping and random sampling
def hot_deck_imputation(df, group_column):
    """
    Perform hot deck imputation by filling missing values within
    groups using random sampling.
    Parameters:
    df (pd.DataFrame): Dataframe with missing values
    group_column (str): The column used to group similar records
    (e.g., 'Gender')
    Returns:
    pd.DataFrame: Dataframe with missing values imputed
    """
    # Create a copy of the dataframe to avoid modifying the
    # original
    df_imp = df.copy()
    # Loop through each column in the dataframe that has missing
    # values
    for col in df.columns:
        # Skip the group column itself
        if col == group_column:
            continue
        # Apply imputation within each group defined by the
        # group_column
        for group in df[group_column].unique():
            group_mask = (df[group_column] == group)
            miss_mask = df[col].isnull() & group_mask
```

```
            # Get available values within the group (non-missing
            # values)
            available_values = df.loc[
            group_mask & df[col].notnull(), col]
            # If there are available values to sample from, fill
            missing values with random sampling
            if not available_values.empty:
                df_imp.loc[miss_mask, col] = np.random.choice(
                available_values, size=miss_mask.sum(),
                replace=True)
    return df_imputed

# Perform hot deck imputation using 'groups' as the grouping
# column
df_hot_deck_imputed = hot_deck_imputation(
data, group_column='groups')
print("\nDataFrame after Hot Deck Imputation
(Grouped by 'groups'):")
print(df_hot_deck_imputed)
```

Hot deck imputation can be customized further by choosing more sophisticated matching criteria (e.g., matching on multiple variables) or by using more complex algorithms like nearest neighbors. This implementation uses random sampling for simplicity. In practice, more advanced methods, such as k-nearest neighbors (KNN) or distance-based matching, can be used to find similar records for more accurate imputations.

C. *Model-Based Imputation*

**Regression imputation**: Regression imputation is a method where missing values in a dataset are predicted using regression models [39]. For each variable with missing values, a regression model is built using the other variables as predictors. The predicted values are then used to fill in the missing data. While it preserves relationships between variables, it can artificially inflate correlations and lead to biased estimates if the regression assumptions are violated.

Example:

```
# Import necessary libraries
import pandas as pd
import numpy as np
from sklearn.linear_model import LinearRegression
from sklearn.model_selection import train_test_split
# Load the dataset
data = pd.read_csv('example_data/data.csv', index_col=0)
print("Original DataFrame:")
print(data)
# Function for regression imputation
def regression_imputation(df, target_col):
    """
    Perform regression imputation on a column with missing
    values.
```

```
    Parameters:
    df (pd.DataFrame): The input dataframe with missing values.
    target_col (str): The name of the column to perform regres-
    sion imputation on.
    Returns:
    pd.Series: Column with missing values filled using regression
    imputation.
    """
    # Separate rows with missing values and rows with observed
    # values in the target column
    observed_data = df[df[target_col].notnull()]
    missing_data = df[df[target_col].isnull()]
    # Define the features (all columns except the target column)
    X_train = observed_data.drop(columns=[target_col])
    y_train = observed_data[target_col]
    # Create a Linear Regression model
    regressor = LinearRegression()
    # Train the model using observed data
    regressor.fit(X_train.fillna(0), y_train)
    # Use the trained model to predict missing values
    X_missing = missing_data.drop(columns=[target_col])
    pred_val = regressor.predict(X_missing.fillna(0))
    # Fill missing values with the predicted values
    df.loc[df[target_col].isnull(), target_col] = pred_val
    return df[target_col]

# Apply regression imputation
df = data.copy()
df['feature1'] = regression_imputation(df, 'feature1')
df['feature3'] = regression_imputation(df, 'feature3')
df['feature4'] = regression_imputation(df, 'feature4')
df['feature5'] = regression_imputation(df, 'feature5')
print("\nDataFrame after Regression Imputation:")
print(df)
```



**Stochastic regression imputation**: Stochastic regression imputation is a variation of regression imputation that introduces a random error term (residual) to the predicted values to better reflect the variability of the data [40]. This approach helps prevent the underestimation of variance that is common with standard regression imputation and provides a more realistic representation of the data distribution. However, like standard regression imputation, this method assumes linear relationships between the target column and predictor variables.

Example:

```
# Import necessary libraries
import pandas as pd
import numpy as np
from sklearn.linear_model import LinearRegression
# Load the dataset
data = pd.read_csv('example_data/data.csv', index_col=0)
print("Original DataFrame:")
print(data)
```

```
# Function for stochastic regression imputation
def stochastic_regression_imputation(df, target_col):
    """
    Perform stochastic regression imputation on a column with
    missing values.
    Parameters:
    df (pd.DataFrame): The input dataframe with missing values.
    target_column (str): The name of the column to perform
    stochastic regression imputation on.
    Returns:
    pd.Series: Column with missing values filled using stochastic
    regression imputation.
    """
    # Separate rows with missing values and rows with observed
    # values in the target column
    observed_data = df[df[target_col].notnull()]
    missing_data = df[df[target_col].isnull()]
    # Define the features (all columns except the target column)
    X_train = observed_data.drop(columns=[target_col])
    y_train = observed_data[target_col]
    # Create a Linear Regression model
    regressor = LinearRegression()
    # Train the model using observed data
    regressor.fit(X_train.fillna(0), y_train)
    # Use the trained model to predict missing values
    X_missing = missing_data.drop(columns=[target_col])
    pred_val = regressor.predict(X_missing.fillna(0))
    # Calculate the residuals (difference between observed and
    # predicted values)
    residuals = y_train - regressor.predict(X_train.fillna(0))
    # Calculate the standard deviation of the residuals
    residuals_std = np.std(residuals)
    # Add random noise to the predicted values based on the
    # residuals' standard deviation
    st_pred = pred_val + np.random.normal(
    0, residuals_std, size=pred_val.shape)
    # Fill missing values with the stochastic predictions
    df.loc[df[target_col].isnull(), target_col] = st_pred
    return df[target_col]


# Apply stochastic regression imputation
df = data.copy()
df['feature1'] = stochastic_regression_imputation(df, 'feature1')
df['feature3'] = stochastic_regression_imputation(df, 'feature3')
df['feature4'] = stochastic_regression_imputation(df, 'feature4')
df['feature5'] = stochastic_regression_imputation(df, 'feature5')
print("\nDataFrame after Stochastic Regression Imputation:")
print(df)
```

D. *Multiple Imputation*

Multiple imputation (MI) involves generating several datasets with different imputed values and then combining the results to account for the uncertainty of missing data [7]. It consists of three steps: (1) imputing missing values multiple times to create several complete datasets, (2) analyzing each dataset separately, and (3) pooling the results to produce final estimates. MI is considered a gold-standard technique for handling missing data as it provides valid statistical inferences under MAR conditions and accommodates complex data structures. However, it can be computationally intensive and requires careful implementation to ensure convergence and stability of results [11].

Example using Scikit-learn's `IterativeImputer` [41]:



```
# Import necessary libraries
import pandas as pd
from sklearn.experimental import enable_iterative_imputer
from sklearn.impute import IterativeImputer
# Load the dataset
data = pd.read_csv('example_data/data.csv', index_col=0)
print("Original DataFrame:")
print(data)
# Initialize imputer
imputer = IterativeImputer()
imputed = imputer.fit_transform(data)
df = pd.DataFrame(imputed, index=data.index, columns=data.columns)
print("\nDataFrame after Iterative Imputation:")
print(df)
```

By default, the IterativeImputer uses the BayesianRidge estimator, but it can be changed to use different estimators depending on the nature of your data. It is important to monitor convergence in the iterative imputation process. If the imputer does not converge within the specified number of iterations, increasing max_iter or changing the imputation model might be necessary. The IterativeImputer can introduce some randomness in the imputation process. Setting a random_state ensures reproducibility. This method assumes that the missingness pattern is at random. If the data is missing not at random (MNAR), the imputations might be biased.

Multiple Imputation by Chained Equations (MICE) [42] implementation from Statsmodels:

```
# Import necessary libraries
import pandas as pd
from statsmodels.imputation import mice
import statsmodels.api as sm
# Load the dataset
data = pd.read_csv('example_data/data.csv', index_col=0)
print("Original DataFrame:")
print(data)
# Initialize imputer
mice_data = mice.MICEData(data)
# Prepare MICE model formulas dynamically
cols_with_missing = data.columns[data.isnull().any()].tolist()
# Create a formula and perform MICE only for columns with missing data
for column in cols_with_missing:
# All columns except the current one
other_columns = list(data.columns.drop(column))
```

```
        formula = f"{column} ~ " + " + ".join(other_columns)
        mi_model = mice.MICE(formula, sm.OLS, mice_data)
        # Using 10 imputations with 10 iterations each
        mi_results = mi_model.fit(10, 10)
        print(mi_results.summary())
    imputed_data = mice_data.data
    print("\nDataFrame after MICE Imputation:")
    print(imputed_data)
```

E. *Maximum Likelihood Estimation (MLE)*

MLE estimates model parameters using all available data without imputing missing values, under the assumption that the missing data mechanism is either MCAR or MAR. This method leverages all observed data points, providing efficient estimates [43]. While MLE is robust and widely used in structural equation modeling (SEM), mixed-effects models, and other statistical frameworks, it can be challenging to implement for complex models or non-normally distributed data. One popular library for MLE-based analysis with missing data is `statsmodels` using the expectation-maximization (EM) algorithm, or specialized libraries like `pymc3` or `tensorflow_probability` for Bayesian inference.

F. *Machine Learning Approaches*

**K-nearest neighbors (KNN) imputation**: Replaces missing values based on the values of its k-nearest neighbors. KNN considers the similarity between observations, making it suitable for nonlinear relationships [44]. However, it can be computationally intensive and sensitive to the choice of k and distance metrics.

Example:

```
# Import necessary libraries
import pandas as pd
from sklearn.impute import KNNImputer
# Load the dataset
data = pd.read_csv('example_data/data.csv', index_col=0)
print("Original DataFrame:")
print(data)
# Initialize imputer
imputer = KNNImputer()
imputed = imputer.fit_transform(data)
df = pd.DataFrame(imputed, index=data.index, columns=data.columns)
print("\nDataFrame after KNN Imputation:")
print(df)
```

**Random forest imputation**: Utilizes an ensemble of decision trees to predict missing values [43]. This method captures nonlinear relationships and interactions between variables, providing robust imputation results [45, 46]. However, it requires large sample sizes and computational resources. Random forest imputation can be implemented using the `IterativeImputer` class.

Example:

```
# Import necessary libraries
import pandas as pd
from sklearn.ensemble import RandomForestRegressor
from sklearn.experimental import enable_iterative_imputer
from sklearn.impute import IterativeImputer
# Load the dataset
data = pd.read_csv('example_data/data.csv', index_col=0)
print("Original DataFrame:")
print(data)
# Initialize imputer
imputer = IterativeImputer(estimator=RandomForestRegressor())
imputed = imputer.fit_transform(data)
df = pd.DataFrame(imputed, index=data.index, columns=data.columns)
print("\nDataFrame after Random Forest Imputation:")
print(df)
```

**Deep learning-based imputation**: Recent advancements in deep learning, such as autoencoders and generative adversarial networks (GANs), have been employed for handling complex missing data patterns. These methods require expertise and computational resources to implement effectively [47].

G. *Sensitivity Analysis*

Regardless of the imputation method chosen, conducting a sensitivity analysis is needed to evaluate how different methods impact the study's results [35]. This involves comparing results obtained under various imputation strategies to assess the robustness of the findings, ensuring that conclusions are not unduly influenced by the choice of missing data handling technique.

H. *Considerations During Imputation*

When considering imputation strategies, it is vital to align the chosen method with the data type of the column containing missing values, distinguishing between numerical and categorical types. For instance, mean imputation can be influenced by outliers, making median imputation a more suitable alternative in such scenarios. Moreover, if a considerable portion of data is missing within a feature, it may be more judicious to eliminate that feature altogether, rather than attempting imputation. The nature of the missing data also plays a role; understanding whether the data is missing at random or if there is a discernible pattern can impact the choice of the imputation method. Ultimately, imputation serves to transform incomplete datasets into complete ones by estimating missing values, thereby facilitating more effective and accurate analysis. Therefore, the selection of an imputation technique must be carefully made, taking into account the specific characteristics of the dataset and the objectives of the analysis.

### 4.3.4 Impact of Missing Data on Biomedical Analysis

The impact of missing data on biomedical analysis is multifaceted. Missing data can influence numerous aspects of a study, ranging from reducing its statistical power to compromising the generalizability and validity of its results [48, 49]. ◘ Figure 4.3 describes a brief overview of the various effects that missing data can have on biomedical research.

**Reduction in statistical power**: Statistical power refers to the probability of detecting an effect when one truly exists [49]. Missing data effectively reduces the sample size, leading to diminished statistical power [50]. This reduction makes it harder to identify effects or associations, thereby increasing the likelihood of false-negative results. In clinical trials, for example, lower statistical power can mask the true efficacy or adverse effects of a treatment, affecting patient outcomes.

**Introduction of bias**: Bias is introduced when the missingness of data is not random—such as when missingness is dependent on the value of the missing data itself or on other variables within the dataset (MNAR). For instance, if patients with more severe symptoms are less likely to participate in follow-up visits, the resulting analyses may underestimate the true severity of the condition. Such biases can lead to erroneous conclusions, affecting both clinical decision-making and policy formulation.

**Compromising data integrity**: The integrity of the dataset can be compromised if missing data is substantial or systematic. Large amounts of missing data can lead to skewed data distributions and affect the assumptions of statistical tests and models. This compromises the reliability of any subsequent analysis and reduces confidence in the results obtained.

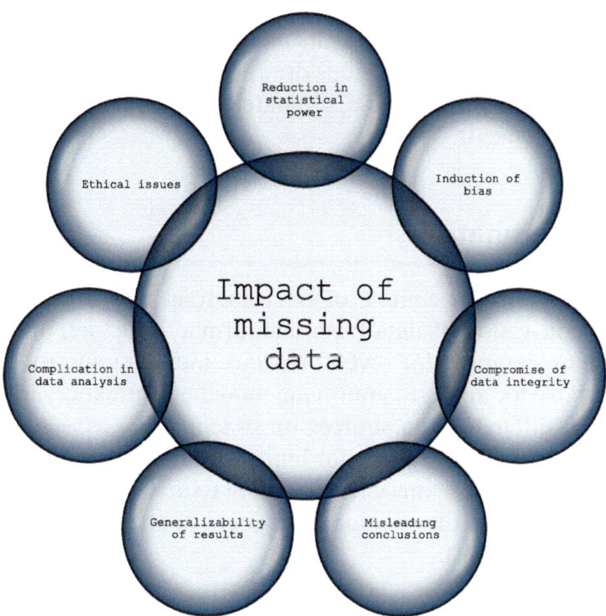


◘ **Fig. 4.3** Impact of missing values on data analysis. The presence of missing values in a dataset can introduce various issues that affect downstream data analysis

**Misleading conclusions**: Due to reduced statistical power and increased risk of bias, missing data can lead to misleading conclusions. In the biomedical field, such erroneous conclusions can have severe implications, leading to the recommendation of ineffective or harmful interventions. For example, a biased estimate of a drug's effectiveness could result in its inappropriate approval or rejection.

**Generalizability of results**: Missing data can affect the generalizability of a study's findings, especially if the missingness leads to a nonrepresentative sample. This is an important issue in biomedical research, where the objective is often to apply results to a broader population. If missing data is not handled correctly, it can limit the applicability of the study's conclusions to the general population, making the results less useful for informing clinical practice or policy.

**Complications in data analysis**: Handling missing data adds complexity to the data analysis process. Deciding on an appropriate imputation method or statistical approach requires a thorough understanding of the nature and pattern of missingness. Implementing advanced methods such as MICE or MLE necessitates additional expertise and computational resources. This increases the burden on researchers, requiring careful planning and validation to ensure accurate results.

**Ethical issues**: In clinical research, handling missing data also has ethical implications. Inaccurate or biased results due to poorly managed missing data can lead to incorrect conclusions about the safety or efficacy of a treatment, directly affecting patient health and safety. Such ethical concerns show the responsibility of researchers to use robust methods and clearly report how missing data was addressed in their analyses.

Missing data in biomedical analysis poses substantial challenges that can undermine the reliability, accuracy, and validity of study results. These issues can lead to flawed conclusions, biased results, and reduced applicability of research to the broader population. Consequently, it is required to employ robust statistical methods for handling missing data and to assess its impact on the analysis. Properly addressing missing data ensures that conclusions drawn from biomedical research are sound, trustworthy, and ethically responsible, ultimately contributing to better patient care and scientific advancements.

## 4.4 Data Transformation

Data transformation, in the context of data analysis and processing, refers to the modification or conversion of data from one format, structure, or value to another [51]. This process is required for several reasons, including improving data quality, making data suitable for analysis, improving model performance, and ensuring compatibility between different data sources or systems. Key aspects and methods of data transformation include, but are not limited to, normalization and standardization, encoding of categorical variables, handling data distribution, feature creation, binning and discretization, dimensionality reduction, and time-series and frequency transformations (◘ Fig. 4.4).

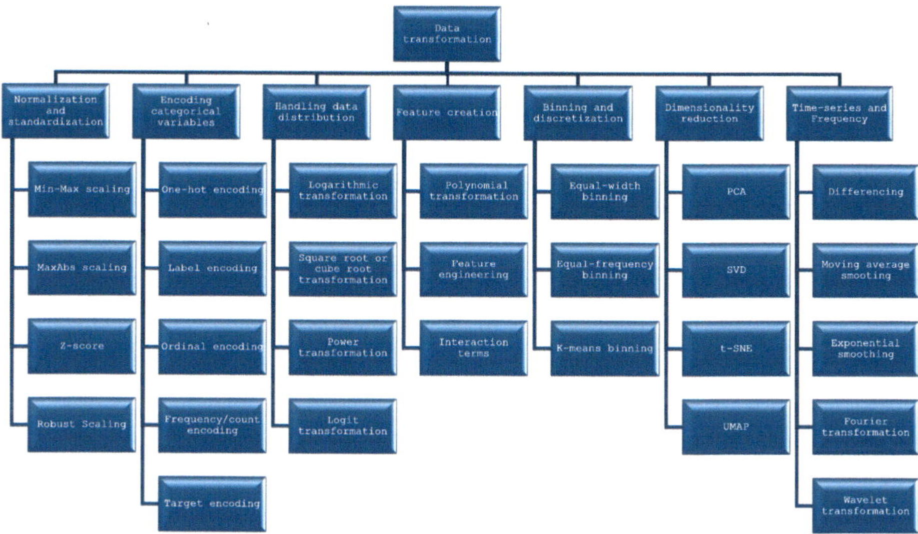


**Fig. 4.4** Overview of data transformation techniques. Various data transformation techniques can be organized into different categories. These transformations are applied based on the data requirements and analysis goals to improve data quality, compatibility, and model performance

## 4.4.1 Normalization and Standardization

**Notebook: Section 4.4.1. Normalization and Standardization**

This notebook provides code examples demonstrating normalization and standardization methods.

Link to the GitHub repository:

► https://github.com/sn-code-inside/BioPy

Go to: Chap. 4—► Sect. 4.4.1

Normalization and standardization are widely used in data preprocessing for biomedical research and machine learning applications [52]. These techniques modify the scale of numerical features to make them comparable, thereby improving the performance and accuracy of models [53]. The goal is to ensure that each feature (e.g., gene expression levels, blood pressure readings, or metabolite concentrations) contributes equally to the analysis, without any one feature dominating due to differences in units, magnitude, or range.

In biomedical datasets, variables often vary in scale. For example, gene expression data may range from 0 to several thousand, while clinical features such as age or cholesterol levels may vary between tens and hundreds. When these features are combined in a machine learning model, the differences in scale can negatively impact performance, making it difficult for the model to properly weigh each feature. Normalization and standardization solve this issue by transforming the data so that the numerical values are on a similar scale.

Common normalization techniques include Min-Max scaling and MaxAbs scaling.

***Min-Max scaling*** method transforms the data to a fixed range, usually [0, 1]. It is considered a normalization technique because it rescales the features without changing their distribution [54, 55]. Min-Max scaling is frequently used when the goal is to bring all the feature values to a uniform scale, particularly for models like neural networks that perform better with normalized inputs:



$$\text{Scaled value} = a + \frac{(\text{Original value} - \text{Minimum value}) \times (b - a)}{\text{Maximum value} - \text{Minimum value}}$$

$$\text{where} : a = \text{lower range and } b = \text{upper range}$$

For a range between 0 and 1:

$$\text{Scaled value} = \frac{(\text{Original value} - \text{Minimum value})}{\text{Maximum value} - \text{Minimum value}}$$

Example code using `MinMaxScaler`:

```
import pandas as pd
from sklearn.preprocessing import MinMaxScaler
# Sample data
data = {
    'Age': [25, 45, 30, 60, 35, 75],
    'Income': [50000, 120000, 80000, 100000, 60000, 150000]
}
df = pd.DataFrame(data)
print("Original DataFrame:")
print(df)
# Initialize MinMaxScaler
scaler = MinMaxScaler(feature_range=(0, 1))
# Fit and transform the data
normalized_data = scaler.fit_transform(df .to_numpy())
# Creating a new DataFrame for the normalized data
normalized_df = pd.DataFrame(normalized_data, columns=df.columns)
print("\nMin-Max scaling:")
print(normalized_df)
```

***MaxAbs scaling*** method scales each feature based on its maximum absolute value, so that the values fall within the range of −1 to 1 [56]. Similar to Min-Max scaling, it maintains the relative relationship of the values but within a fixed range, making it a normalization technique. MaxAbs scaling is typically used for data that has both positive and negative values and when we want to preserve sparsity (e.g., data with zero entries):

$$\text{Scaled value} = \frac{\text{Original value}}{\text{Maximum absolute value}}$$

Example code using `MaxAbsScaler`:

```
import pandas as pd
from sklearn.preprocessing import MaxAbsScaler
# Sample data
data = {
    'Prameter 1': [-5, 0, 7, 4, 1, 4],
    'Prameter 2': [4, -12, 6, 3, 2, -5]
}
df = pd.DataFrame(data)
print("Original DataFrame:")
print(df)
# Initialize MinMaxScaler
scaler = MaxAbsScaler()
# Fit and transform the data
normalized_data = scaler.fit_transform(df .to_numpy())
# Creating a new DataFrame for the normalized data
normalized_df = pd.DataFrame(normalized_data, columns=df.columns)
print("\nMinAbs scaling:")
print(normalized_df)
```

*Z*-score and Robust scaling techniques are widely used standardization methods.

***Z-score*** is a commonly used standardization technique that serves as a preprocessing step in many statistical analyses and machine learning algorithms [54, 57]. It transforms the data such that it has a mean of zero and a standard deviation of one. This process is achieved by subtracting the mean and dividing by the standard deviation for each data point:

$$\text{Scaled value} = \frac{\text{Original value} - \text{Mean value}}{\text{Standard deviation}}$$

*Z*-score brings all variables to the same scale, allowing for fair comparison across variables. Many machine learning algorithms, especially those involving distance calculations like k-nearest neighbors (KNN) and support vector machines (SVM), perform better on standardized data [58]. Furthermore, algorithms that assume normal distribution, such as linear regression, logistic regression, and principal component analysis (PCA), benefit from standardization, as it brings the data closer to a normal distribution. In neural networks and algorithms using gradient descent, standardization helps in faster convergence.

Example code using StandardScaler:

```
import numpy as np
from sklearn.preprocessing import StandardScaler
# Example data
data = np.array([[10, 2.7, 3.6],
                 [15, 3.0, 9.0],
                 [16, 2.5, 5.5],
                 [18, 3.2, 6.8]])
# Create a StandardScaler object
scaler = StandardScaler()
```

```
# Fit and transform the data
standardized_data = scaler.fit_transform(data)
print("Standardized Data:\n", standardized_data)
```

**Robust scaling** method scales the data using statistics like the median and the interquartile range (IQR), making it less sensitive to outliers. It is also considered a standardization technique because it shifts and scales the features based on their robust statistics, rather than their global mean and variance [59]. Robust scaling is particularly useful for datasets with outliers or skewed distributions, where standardization might be heavily influenced by extreme values:



```
import numpy as np
from sklearn.preprocessing import RobustScaler
# Example data
data = np.array([[10, 2.7, 3.6],
                 [15, 3.0, 9.0],
                 [16, 2.5, 5.5],
                 [18, 3.2, 6.8]])
# Create a StandardScaler object
scaler = RobustScaler()
# Fit and transform the data
standardized_data = scaler.fit_transform(data)
print("Standardized Data:\n", standardized_data)
```

### 4.4.2 Encoding Categorical Variables

**Notebook: Section 4.4.2. Encoding Categorical Variables**

This notebook provides code examples demonstrating encoding categorical variables.

Link to the GitHub repository:

▸ https://github.com/sn-code-inside/BioPy

Go to: Chap. 4—▸ Sect. 4.4.2

Categorical variables represent distinct categories or groups, such as color (`red`, `blue`, `green`) or size (`small`, `medium`, `large`). These categories need to be transformed into numerical values because most machine learning algorithms require numerical input for effective model training and evaluation [60]. There are several techniques for encoding categorical variables, each with specific use cases and implications. Here, we discuss common methods along with examples:

***One-hot encoding*** creates new binary columns for each unique category value, assigning 1 or 0 to indicate the presence or absence of a category in each row [61].

This method is particularly useful for nominal categorical variables, where no natural order exists between categories.

We consider a categorical feature `Color` with three categories: `Red`, `Blue`, and `Green`. Using one-hot encoding, the feature would be transformed into three binary columns:

| Color | Red | Blue | Green |
|---|---|---|---|
| Red | 1 | 0 | 0 |
| Blue | 0 | 1 | 0 |
| Green | 0 | 0 | 1 |

Example code:

```
import pandas as pd
from sklearn.preprocessing import OneHotEncoder
# Example dataset
data = pd.DataFrame({
    'Color': ['Red', 'Blue', 'Green']
})
# One-hot encoding
one_hot_encoder = OneHotEncoder()
one_hot_encoded = one_hot_encoder.fit_transform(data).toarray()
# Creating a DataFrame with proper column names
one_hot_encoded_df = pd.DataFrame(one_hot_encoded,
columns=one_hot_encoder.get_feature_names_out(['Color']))
# Display the one-hot encoded DataFrame
print("One-Hot Encoded Data:\n", one_hot_encoded_df)
```

***Label encoding*** assigns a unique integer to each category, making it suitable for ordinal categorical variables where there is a meaningful order among the categories [62].

Let's consider the feature `Country` with categories: `Denmark`, `Germany`, and `Sweden`. Label encoding would convert these categories to numerical values:

| Country | Encoded value |
|---|---|
| Denmark | 0 |
| Germany | 1 |
| Sweden | 2 |

Example codes:

```
import pandas as pd
from sklearn.preprocessing import LabelEncoder
# Example dataset
data = pd.DataFrame({
    'Country': ['Denmark', 'Germany', 'Sweden']
})
# Label encoding
label_encoder = LabelEncoder()
data['Encoded value'] = label_encoder.fit_transform(data['Country'])
# Display the label encoded DataFrame
print("Label Encoded Data:\n", data)
```

***Ordinal encoding*** is similar to label encoding but explicitly designed for ordinal categories where the categories have a specific order [63]. It assigns each category a unique numerical value that reflects this order.

We may consider a feature `Education level` with categories: `High school`, `Bachelor`, `Master's`, and `PhD`. Ordinal encoding would assign values as follows:

| Educational level | Encoded value |
|---|---|
| High school | 0 |
| Bachelor | 1 |
| Master's | 2 |
| PhD | 3 |

Example code:

```
import pandas as pd
from sklearn.preprocessing import OrdinalEncoder
# Example dataset
data = pd.DataFrame({
    'Educational level': ['High school', 'Bachelor', 'Master's',
'PhD']
})
# Defining the order of categories for ordinal encoding
education_order = [['High school', 'Bachelor', 'Master's', 'PhD']]
# Applying Ordinal Encoding
ordinal_encoder = OrdinalEncoder(categories=education_order)
data['Encoded value'] = ordinal_encoder.fit_transform(
data[['Educational level']])
# Display the ordinal encoded DataFrame
print("Ordinal Encoded Data:\n", data)
```

***Frequency/count encoding*** assigns the frequency or count of each category value in the dataset as its numerical value. This method captures information about the prevalence of each category and is useful when the occurrence frequency of categories is significant [64].

For example, we have a categorical feature `Cancer type` with values `Breast`, `Bladder`, and `Ovary` appearing 50, 30, and 20 times, respectively. Frequency encoding would transform these categories as:

| Cancer type | Frequency value |
|---|---|
| Breast | 50 |
| Bladder | 30 |
| Ovary | 20 |

Example code:

```
import pandas as pd
# Example dataset
data = pd.DataFrame({
    'Cancer type': ['Breast', 'Bladder', 'Ovary', 'Breast',
    'Ovary', 'Breast', 'Bladder', 'Breast', 'Ovary', 'Bladder']
})
# Frequency encoding: Calculate frequency of each category in the
# 'Cancer type' column
frequency_encoding = data['Cancer type'].value_counts().to_dict()
# Map the frequency values back to the 'Cancer type' column
data['Frequency Value'] =
data['Cancer type'].map(
frequency_encoding)
# Display the frequency encoded DataFrame
print("Frequency Encoded Data:\n", data)
```

***Target encoding*** (also known as mean encoding) replaces categorical values with the mean of the target variable for each category. This method is particularly effective for categorical features in supervised learning, where the relationship between a categorical feature and the target variable is important.

We consider a categorical feature `Cancer type` and a target variable `Treatment cost`. If the average treatment cost per year for `Breast cancer` is `$75,000`, for `Bladder cancer` is `$90,000`, and for Ovarian cancer is $86,000, then target encoding would replace these categories with the corresponding mean values:

| Cancer type | Mean treatment costs |
|---|---|
| Breast cancer | $75,000 |
| Bladder cancer | $90,000 |
| Ovarian cancer | $86,000 |

Example code:

```
import pandas as pd
# Example dataset
data = pd.DataFrame({
    'Cancer type > Type': ['Breast', 'Bladder', 'Ovary',
    'Breast', 'Bladder', 'Ovary', 'Breast'],
    'Treatment cost > Cost': [65000, 85000, 78000, 75000, 95000,
    94000, 85000]
})
# Calculate the mean Treatment cost for each Cancer type
target_encoding = data.groupby('Type')['Cost'].mean().to_dict()
# Map the mean values back to the 'Cancer type' column
data['Target Encoded Value'] = data['Type'].map(target_encoding)
# Display the target encoded DataFrame
print("Target Encoded Data:\n", data)
```



### 4.4.3 Handling Data Distribution

**Notebook: Section 4.4.3. Handling Data Distribution**

This notebook provides code examples demonstrating handling data distribution.

Link to the GitHub repository:

▶ https://github.com/sn-code-inside/BioPy

Go to: Chap. 4—▶ Sect. 4.4.3

Many statistical techniques assume that the data follows a normal (Gaussian) distribution with constant variance [65]. However, real-world data often deviate from normality due to skewness, kurtosis, or the presence of outliers. Transforming data is a common approach to mitigate these issues, making the data more suitable for analysis. This section explores various transformation techniques used to handle data distribution effectively.

***Log transformation*** is one of the most widely used methods for stabilizing variance and normalizing data that exhibit exponential growth or multiplicative effects [66]. It is particularly effective for right-skewed distributions:

$$\text{Transformed value} = \log\left(\text{Original value}\right)$$

The logarithmic transformation compresses the scale of large values, reducing the impact of outliers and making the distribution more symmetric. It helps in achieving homoscedasticity (constant variance) across different levels of an independent variable. Furthermore, it transforms exponential relationships into linear ones, facilitating linear regression modeling. However, the logarithm is undefined for zero and negative numbers. If the dataset contains zeros or negative values, a constant can be added to shift the data into the positive range. After transformation, differences in the log scale correspond to ratios in the original scale, affecting the interpretation of coefficients in regression models.

Example code:

```
import numpy as np
import pandas as pd
# Generate synthetic gene expression data
np.random.seed(42)  # For reproducibility
generated_gene_expression = np.random.exponential(scale=1000,
size=500)
# Create a DataFrame
df = pd.DataFrame({'Generated_Gene_Expression':
generated_gene_expression})
# Apply logarithmic transformation
df['Log_Expression'] = np.log2(df['Generated_Gene_Expression'])
# Display first few rows of the DataFrame
print(df.head())
```

Figure 4.5 displays $\log_2$ transformed data.

***Square root or cube root transformations*** are useful for reducing moderate skewness and stabilizing variance. Square root transformation may be effective for data that represent counts of occurrences, such as the number of defects or occurrences of an event, but it will provide complex numbers with negative values. Cube root transformation can handle negative values since the cube root of a negative number is negative. Both transformations reduce right skewness by compressing larger values more than smaller ones and can handle zeros. Transformation is less aggressive than logarithmic transformation and may not be sufficient for highly skewed data. Such types of transformations are well suited for data with a limited range and moderate skewness.

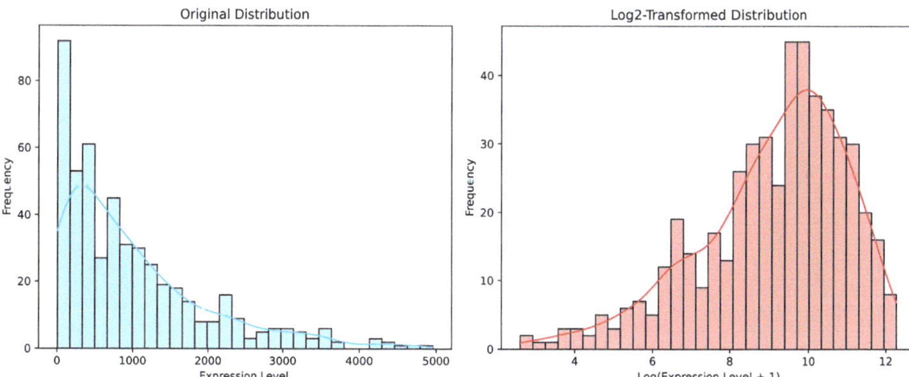


**Fig. 4.5** Example of log transformation of data distribution. The left panel shows the original data distribution, while the right panel shows the $\log_2$-transformed distribution

Example code:

```
import numpy as np
import pandas as pd
# Generate synthetic bacterial colony count data
np.random.seed(42)  # For reproducibility
colony_counts = np.random.poisson(lam=25, size=500)
# Introduce some higher counts to simulate overgrowth
extra_counts = np.random.poisson(lam=100, size=50)
colony_counts = np.concatenate([colony_counts, extra_counts])
# Create a DataFrame
df = pd.DataFrame({'Colony_Counts': colony_counts})
# Apply square root transformation
df['Sqrt_Counts'] = np.sqrt(df['Colony_Counts'])
# Display first few rows of the DataFrame
print(df.head())
```

4

Figure 4.6 shows the distributions of original and square root-transformed data.

***Power transformations*** involve raising data to a specific exponent (power) to achieve normality or stabilize variance. The Box-Cox transformation is a popular method that identifies the optimal power parameter for a given dataset:

$$\text{General power transformation, } y = x^{\lambda}$$

$$\text{Box} - \text{Cox transformation, } y(\lambda) = \frac{x^{\lambda} - 1}{\lambda}, \text{if } \lambda \neq 0 \text{ or } y(\lambda) = \ln(x), \text{if } \lambda = 0$$

Power transformation is highly flexible; by varying the power parameter $\lambda$, different transformations (including logarithmic, square root) can be applied. Furthermore, the Box-Cox method identifies the $\lambda$ that maximizes the likelihood of the data being normally distributed [67]. These transformations help in achieving homoscedasticity

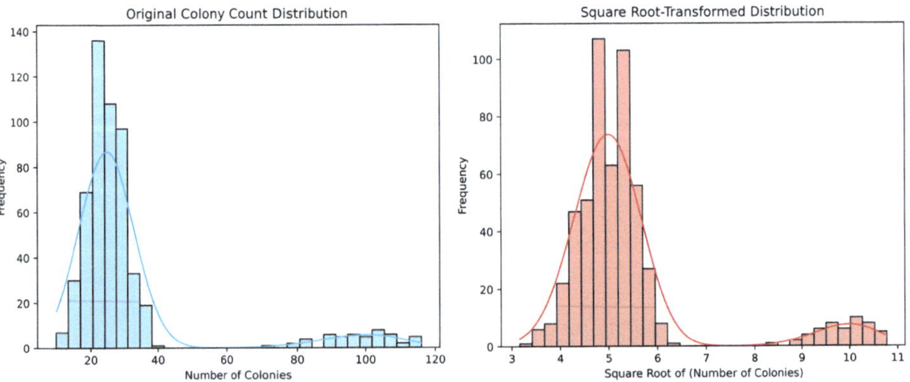


**Fig. 4.6** Example of square root transformation of data distribution. The left panel shows the original data distribution, while the right panel shows the square root-transformed distribution

across the range of data. Limitations of this method include the computational complexity of determining $\lambda$ and the transformed data may be less intuitive to interpret, especially with non-integer $\lambda$.

Example code for Box-Cox transformation:

```
import numpy as np
import pandas as pd
from scipy import stats
# Generate synthetic systolic blood pressure data
np.random.seed(42)  # For reproducibility
# Simulate skewed blood pressure data
blood_pressure = np.random.gamma(shape=2.0, scale=15.0, size=500) + 100
# Create a DataFrame
df = pd.DataFrame({'Systolic_BP': blood_pressure})
# Apply Box-Cox transformation
# The Box-Cox transformation requires positive data
fitted_data, fitted_lambda = stats.boxcox(df['Systolic_BP'])
# Store the transformed data in the DataFrame
df['BoxCox_BP'] = fitted_data
# Display first few rows of the DataFrame
print(df.head())
# Print the lambda value used in the transformation
print(f"Optimal Lambda for Box-Cox Transformation: {fitted_lambda:.4f}")
```

Figure 4.7 displays an example of Box-Cox transformation.

***Logit transformation*** is specifically designed for data representing proportions or probabilities bounded between 0 and 1 [68]. It maps these values onto the entire real number line, facilitating linear modeling:

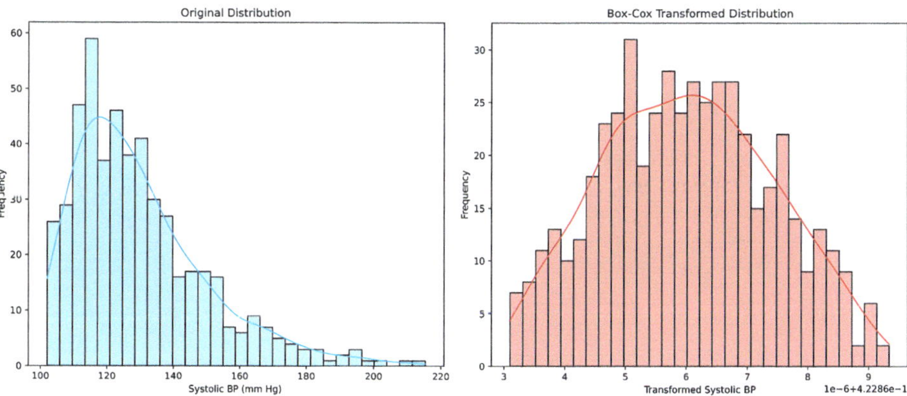


**Fig. 4.7** Example of Box-Cox transformation of data distribution. The left panel shows the original data distribution, while the right panel shows the Box-Cox-transformed distribution

$$\text{Logit transformed value, } y = \ln\left(\frac{x}{1-x}\right)$$

Logit transformation is ideal for transforming proportions, such as success rates or percentages. It is useful for linearizing S-shaped curves as it converts sigmoid relationships into linear ones, aiding in logistic regression analysis. Furthermore, logit transformation stabilizes variance addressing heteroscedasticity in binomial data where variance depends on the mean. The transformation is undefined for values exactly equal to 0 or 1. Adjustments (e.g., adding a small constant) may be necessary. Finally, the transformed values represent log-odds, which may be less intuitive to interpret than probabilities.

Example code:

```
import numpy as np
import pandas as pd
from scipy.special import expit, logit
# Generate synthetic dose-response data
np.random.seed(42)  # For reproducibility
# Simulate dosage levels
dosage = np.linspace(0, 100, 100)
# True parameters for the sigmoid function
beta_0 = -5   # Intercept (logit scale)
beta_1 = 0.1  # Slope (logit scale)
# Calculate true probabilities using the logistic function
prob_response = expit(beta_0 + beta_1 * dosage)
# Simulate observed responses with binomial variability
n_patients = 100  # Number of patients at each dosage level
responses = np.random.binomial(n=n_patients, p=prob_response)
# Calculate observed proportions
obs_prop = responses / n_patients
# Create a DataFrame
df = pd.DataFrame({
    'Dosage': dosage,
    'Responses': responses,
    'Total': n_patients,
    'Obs_Prop': obs_prop
})
# Apply logit transformation to the observed proportions
# Adjust proportions to avoid logit of 0 or 1
epsilon = 1e-5
df['Adj_Prop']= df['Obs_Prop'].clip(epsilon,1 - epsilon)
df['Logit_Proportion'] = logit(df['Adj_Prop'])
# Display first few rows of the DataFrame
print(df.head())
```

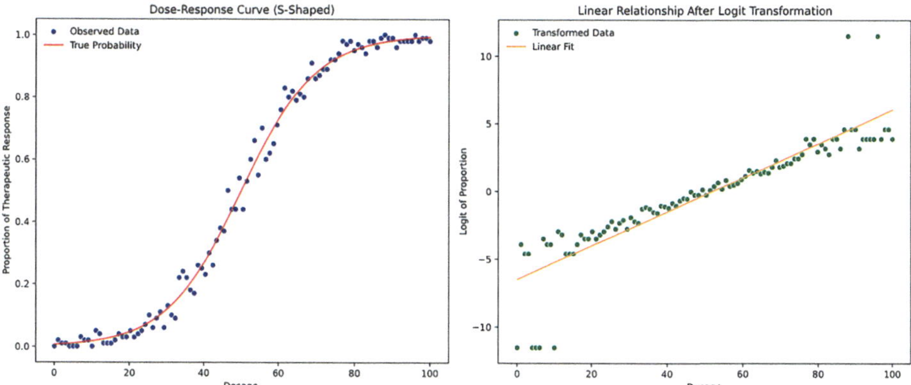


**Fig. 4.8** Example of logit transformation of S-shaped data. The left panel shows the original S-shaped data, while the right panel shows the logit-transformed data

Figure 4.8 shows linearization of S-shaped data by logit transformation.

**Notebook: Section 4.4.4. Feature Creation**

This notebook provides code examples demonstrating feature creation methods.

Link to the GitHub repository:

▶ https://github.com/sn-code-inside/BioPy

Go to: Chap. 4—example data—▶ Sect. 4.4.4

### 4.4.4 Feature Creation

In data science and machine learning, feature creation involves generating new features from existing data to improve the performance of predictive models. By transforming and combining variables, we can capture complex relationships, improve model expressiveness, and ultimately achieve better predictive accuracy. We explore various techniques for feature creation, including polynomial transformations, feature engineering strategies, and the incorporation of interaction terms.

***Polynomial transformation*** involves augmenting the dataset with polynomial features of the original variables [69]. By introducing nonlinear combinations, such as squares or higher-degree terms, we enable linear models to capture nonlinear relationships between variables and the target outcome [70]. Given a feature $x$, polynomial transformation creates new features $x^2$, $x^3$, …, $x^n$, where $n$ is the degree of the polynomial. Such type of transformation can enhance the ability of linear models to fit nonlinear data by adding polynomial terms. It increases the dimensionality of the data, allowing models to learn more complex patterns. However, polynomial transformation introduces additional complexity and therefore, careful selection of the polynomial degree is necessary to balance model complexity and avoid overfitting.

Example code:

```
import pandas as pd
from sklearn.preprocessing import PolynomialFeatures
# Original feature
df = pd.read_csv('example_data/data4.csv', header=0, index_col=0)
print("Original feature:")
print(df['feature1'])
# Create polynomial features up to degree 2
poly = PolynomialFeatures(degree=3, include_bias=False)
df_poly = pd.DataFrame(poly.fit_transform(df[['feature1']]))
print("\nFeatures after polynomial transformation:")
print(df_poly)
```

***Feature engineering*** is the process of using domain knowledge to extract features from raw data, increasing the predictive power of machine learning models [71]. It encompasses creating new features, transforming existing ones, and selecting relevant variables overlapping several other categories and can be slightly repetitive. We will skip codes for feature engineering here as some of them were provided earlier and we will discuss separately some of the others:

- *Creation of new features* entails developing additional attributes derived from the existing dataset. It is not merely about adding more data but about extracting meaningful and informative characteristics that could help in making better predictions or understanding patterns in the data. For example, from a date feature, one might extract day of the week, month, or year as separate features, as these might have different relationships with the target variable.
- *Modification of existing features* involves altering current features to make them more suitable for machine learning algorithms. This could be as simple as normalizing or scaling a feature to ensure that all features contribute equally to the model's predictions. Alternatively, it could involve more complex transformations like applying logarithmic or exponential functions to change the distribution of a feature.
- *Feature combination* can be useful when individual features might not have a strong predictive power. For instance, in a real estate dataset, combining latitude and longitude into a single feature representing geographical location could be more useful than considering them separately.
- *Decomposition of features* is the opposite of feature combination, where a single feature is broken down into multiple features. This is often useful when dealing with complex, multifaceted data. For instance, an address feature could be decomposed into street name, city, and postal code.
- *Feature encoding* is particularly important for handling categorical data. Techniques such as one-hot encoding or label encoding transform categorical variables into a form that can be provided to ML algorithms (discussed above).
- *Feature selection* is a vital part of feature engineering. It involves identifying which features are most important for a model and removing redundant or irrel-

evant features. This simplification can improve model performance and reduce overfitting.
- *Temporal features* in time-series data, such as engineered features to capture trends, seasonality, and cycles, can improve model performance. Examples of such feature engineering include rolling averages, time lags, and exponential smoothing.
- *Domain-specific features*, which leverage domain knowledge to create new features, can be particularly useful. For example, in drug discovery, features representing the chemical properties of compounds or genetic information relevant to a disease can be useful.

***Interaction terms*** represent the product of two or more variables and are used to capture the effect of variables acting together on the response. They allow models to consider that the relationship between a predictor and the outcome may depend on the value of another predictor.

An interaction term between features $x_1$ and $x_2$ can be created by multiplying them:

$$\text{Interaction term} = x_1 \times x_2$$

Interaction terms can be used in linear regression models to capture interactions between predictors [72]. It helps in identifying synergistic effects where the combined influence of variables differs from their individual effects. Interactions can be between continuous variables, categorical variables, or a mix of both.

**Notebook: Section 4.4.5. Binning and Discretization**
This notebook provides code examples demonstrating binning and discretization methods.

Link to the GitHub repository:
▶ https://github.com/sn-code-inside/BioPy
Go to: Chap. 4—▶ Sect. 4.4.5

### 4.4.5 Binning and Discretization

Binning and discretization are techniques used to transform continuous variables into discrete ones by dividing the range of the variable into intervals, or "bins." This process can simplify data analysis, reduce the impact of noise, and help capture nonlinear relationships in statistical models. Discretization is particularly useful when dealing with algorithms that require categorical input or when the exact numerical values are less important than the ranges they fall into. Various methods of binning include but not limited to equal-width binning, equal-frequency binning, quantile-based binning, etc.

***Equal-width binning*** divides the range of the continuous variable into intervals of the same width. The entire range from the minimum to the maximum value is partitioned into $k$ bins, each of equal size:

$$\text{Bin width}, w = \frac{\text{max value} - \text{min value}}{k}$$



Example code:

```
import numpy as np
import pandas as pd
# Sample data
np.random.seed(0)
glucose_levels = np.random.normal(loc=120, scale=20, size=100)
# Ensure values are within 50-200
glucose_levels = np.clip(glucose_levels, 50, 200)
df = pd.DataFrame({'Glucose': glucose_levels})
# Define bin edges
bin_edges = np.linspace(50, 200, num=6)  # 5 bins
# Assign bins
df['Glucose_Bin'] = pd.cut(df['Glucose'], bins=bin_edges, labels=False,
include_lowest=True)
print(df.head())
```

***Equal-frequency binning, also known as quantile binning, divides the data*** into bins such that each bin contains approximately the same number of data points.
Example code:

```
# Assign bins using quantiles
df['Glucose_Bin_EqualFreq'] = pd.qcut(df['Glucose'], q=5, labels=False)

print(df.head())
```

***K-means binning***, also known as K-Means discretization, is a method that uses the K-Means clustering algorithm to partition continuous numerical data into discrete bins. Unlike traditional binning methods like equal-width or equal-frequency binning, K-Means binning groups data points based on similarity, aiming to minimize the variance within each bin.

Example code:

```
import numpy as np
import pandas as pd
from sklearn.preprocessing import KBinsDiscretizer
# Generate synthetic data
np.random.seed(42)
```

```
data_size = 500
# Create a skewed distribution (e.g., exponential distribution)
X = np.random.exponential(scale=2, size=data_size)
df = pd.DataFrame({'Value': X})
# Apply K-Means binning
k = 5  # Number of bins/clusters
est = KBinsDiscretizer(n_bins=k, encode='ordinal', strategy='kmeans',
                 random_state=42, subsample=None)
# Reshape data for the transformer
X_reshaped = df[['Value']]
# Fit and transform the data
df['KMeans_Bin'] = est.fit_transform(X_reshaped).astype(int)
# Display the first few rows
print(df.head())
# Print the cluster centers (bin edges)
print("Cluster Centers (Bin Edges):")
print(est.bin_edges_[0])
```

**Notebook: Section 4.4.6. Dimensionality Reduction**
This notebook provides code examples demonstrating dimensionality reduction methods.

Link to the GitHub repository:
▶ https://github.com/sn-code-inside/BioPy
Go to: Chap. 4—▶ Sect. 4.4.6

### 4.4.6 Dimensionality Reduction

In the era of big data, datasets often contain a vast number of features, many of which may be redundant, irrelevant, or highly correlated. High-dimensional data can pose challenges for data analysis and machine learning algorithms, including increased computational complexity, risk of overfitting, and difficulty in visualizing the data. Dimensionality reduction is a set of techniques aimed at reducing the number of input variables in a dataset while preserving as much relevant information as possible.

Dimensionality reduction methods, including principal component analysis (PCA), singular value decomposition (SVD), t-distributed stochastic neighbor embedding (t-SNE), and uniform manifold approximation and projection (UMAP), help simplify data analysis, enhance visualization, and improve the performance of machine learning models [73].

***Principal component analysis (PCA)*** is a widely used linear dimensionality reduction technique that transforms original variables into uncorrelated new sets of variables referred to as principal components [74]. Transformed variables are ordered so that the first few retain most of the variation present in the original dataset. PCA operates by projecting the data onto directions that maximize variance. The principal components are orthogonal, meaning they are uncorrelated with each other, ensuring that each captures unique information in the data. By selecting the top $k$ principal components, we can reduce the dimensionality of the dataset while retaining most of its variance.

To perform PCA, the data is first standardized by centering and scaling it to have a mean of zero and unit variance. This step ensures that variables with larger scales do not dominate the principal components. Next, the covariance matrix is calculated to understand the relationships between variables. Eigenvalues and eigenvectors of this covariance matrix are then computed through a process called eigendecomposition. The eigenvectors represent the directions of maximum variance (the principal components), and the eigenvalues indicate the magnitude of variance explained by each component. Finally, the original data is projected onto the selected principal components to obtain the reduced dataset.

PCA has several applications across different fields. It is used for data compression by reducing storage requirements through representing data with fewer variables. In noise reduction, PCA eliminates components associated with low variance, which often correspond to noise in the data. For visualization purposes, PCA allows high-dimensional data to be projected onto two or three dimensions, making it easier to interpret and analyze patterns or clusters. Furthermore, PCA serves as a preprocessing step in machine learning pipelines, simplifying datasets before applying algorithms to improve computational efficiency and model performance.

Example code:

```
import numpy as np
import pandas as pd
from sklearn.decomposition import PCA
# Generate synthetic gene expression data
np.random.seed(42)
n_samples = 100
n_genes = 500
gene_expression = np.random.normal(size=(n_samples, n_genes))
# Create a DataFrame
df = pd.DataFrame(gene_expression,
columns=[f'Gene_{i}' for i in range(n_genes)])
# Standardize the data
from sklearn.preprocessing import StandardScaler
scaler = StandardScaler()
X_scaled = scaler.fit_transform(df)
# Apply PCA
pca = PCA(n_components=2)
principal_components = pca.fit_transform(X_scaled)
# Create a DataFrame with principal components
pc_df = pd.DataFrame(data=principal_components, columns=['PC1', 'PC2'])
```

***Singular value decomposition (SVD)*** is a mathematical technique that factorizes a matrix into three component matrices [75]. Given a data matrix $X$, SVD decomposes it into three components, left singular vector ($U$), diagonal matrix of singular value ($\Sigma$), and transposed right singular vectors ($V^T$):

$$X = U\Sigma V^{\mathrm{T}}$$

SVD is closely related to PCA and is widely used in dimensionality reduction, data compression, and noise reduction. In the context of PCA, the principal components are related to the singular vectors obtained from the SVD of the data matrix. Specifically, the eigenvectors in PCA correspond to the singular vectors in SVD, and the squares of the singular values in $\Sigma$ represent the variance explained by each principal component. This relationship allows SVD to capture the patterns in the data by identifying directions (components) that account for the most variance.

Applications of SVD span various fields. In recommender systems, SVD is employed in collaborative filtering techniques to predict user preferences by decomposing user-item interaction matrices [76]. This helps in making personalized recommendations based on patterns extracted from large datasets. In image compression, SVD reduces the size of image files by approximating them with fewer singular values [77], effectively capturing important features while discarding insignificant details (▫ Fig. 4.9). This results in compressed images that retain most of the original quality. In natural language processing, SVD is used in latent semantic analysis (LSA) to analyze relationships between terms and documents. Decomposing term-document matrices, SVD helps in extracting semantic structures from large text corpora, enabling tasks like topic modeling and information retrieval.

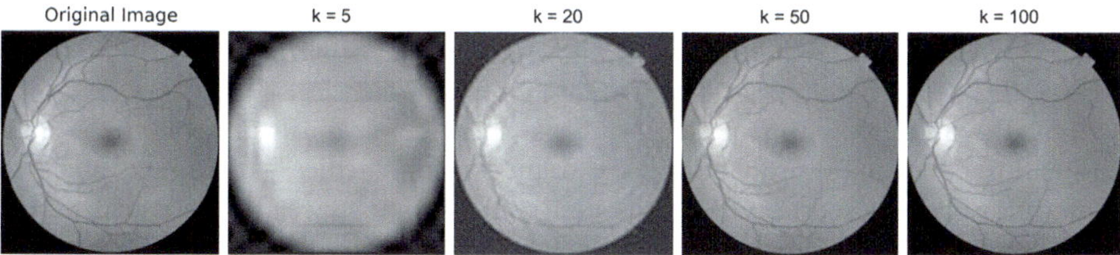


▫ **Fig. 4.9** SVD to reconstruct images. The left to right images show the original image and reconstructed images using different subsets of singular values

Example code:

```
import numpy as np
import matplotlib.pyplot as plt
from skimage import data, color
# Load and preprocess the image
original_image = color.rgb2gray(data.retina())
U, S, Vt = np.linalg.svd(original_image, full_matrices=False)
# Reconstruct the image with different numbers of singular values
singular_values = [5, 20, 50, 100]
fig, axes = plt.subplots(1, len(singular_values)+1, figsize=(15, 5))
# Original Image
axes[0].imshow(original_image, cmap='gray')
axes[0].set_title('Original Image')
axes[0].axis('off')
# Compressed Images
for i, k in enumerate(singular_values):
    # Reconstruct image using k singular values
    S_k = np.diag(S[:k])
    U_k = U[:, :k]
    Vt_k = Vt[:k, :]
    compressed_image = U_k @ S_k @ Vt_k
    axes[i+1].imshow(compressed_image, cmap='gray')
    axes[i+1].set_title(f'k = {k}')
    axes[i+1].axis('off')
plt.show()
```

***t-distributed stochastic neighbor embedding (t-SNE)*** is a nonlinear dimensionality reduction method well suited for embedding high-dimensional data into a low-dimensional space, typically two or three dimensions, for visualization purposes. It focuses on preserving local structures in the data, allowing for the visualization of complex patterns and clusters that may not be apparent in higher dimensions [78].

t-SNE works by converting high-dimensional Euclidean distances between data points into conditional probabilities that represent similarities. It computes pairwise similarities between data points in the high-dimensional space and transforms these similarities into probabilities using a Gaussian distribution [79]. In the low-dimensional mapping, it initializes points and computes pairwise similarities using a Student's t-distribution, which has heavier tails than the Gaussian distribution. This choice helps to maintain the local structure while preventing the "crowding problem," where too many points are projected into a small area. The technique then minimizes the divergence between the joint probabilities of the high-dimensional data and those of the low-dimensional embedding. This is achieved by minimizing the Kullback-Leibler divergence between the two distributions using gradient descent optimization. The cost function focuses on preserving the local relationships among data points, capturing complex, nonlinear relationships that linear methods might miss.

t-SNE has several applications across different fields. In data visualization, it is used to reveal clusters and patterns in high-dimensional datasets, making it easier to interpret and analyze the underlying structures (◘ Fig. 4.10). For anomaly detection, t-SNE helps identify outliers or unusual data points that deviate from normal patterns. In gene expression analysis, it facilitates the visualization of complex rela-

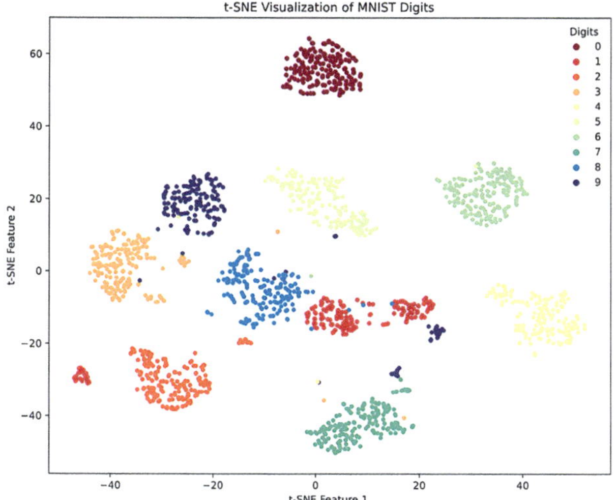


**Fig. 4.10** t-SNE on MNIST dataset. Handwritten digits from the MNIST dataset were clustered using t-SNE

tionships in genomic data, such as differentiating cell types in single-cell RNA sequencing studies.

Example code:

```
import matplotlib.pyplot as plt
from sklearn.datasets import load_digits
from sklearn.manifold import TSNE
# Load the MNIST digits dataset
digits = load_digits()
X = digits.data
y = digits.target
# Apply t-SNE
tsne = TSNE(n_components=2, random_state=42)
X_embedded = tsne.fit_transform(X)
# Plot the results
plt.figure(figsize=(10, 8))
scatter = plt.scatter(X_embedded[:, 0], X_embedded[:, 1], c=y,
cmap='Spectral', s=15)
plt.legend(*scatter.legend_elements(), title="Digits")
plt.title('t-SNE Visualization of MNIST Digits')
plt.xlabel('t-SNE Feature 1')
plt.ylabel('t-SNE Feature 2')
plt.show()
```

***Uniform manifold approximation and projection (UMAP)*** is another nonlinear dimensionality reduction technique utilized for visualization and capturing complex relationships within data. Similar to t-SNE, UMAP projects high-dimensional data into a lower-dimensional space, typically two or three dimensions, to make it more accessible for analysis and interpretation. However, UMAP distinguishes itself by being based on manifold learning, aiming to preserve both local and global data

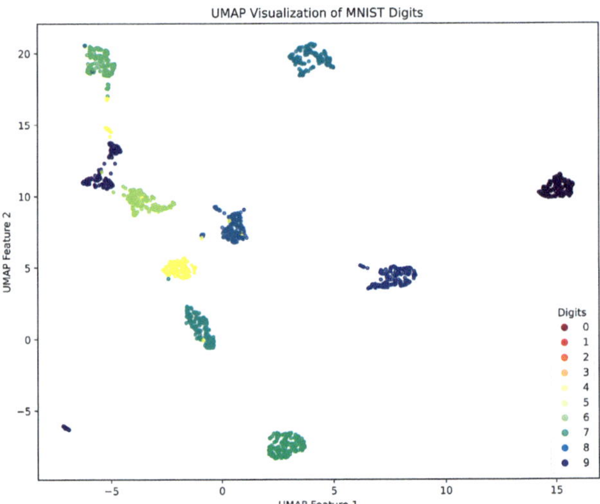


Fig. 4.11 UMAP on MNIST dataset. Handwritten digits from the MNIST dataset were clustered using UMAP

structures. This approach allows UMAP to maintain the overall topology of the data more effectively than some other methods.

UMAP operates under the assumption that the data lies on a manifold—a continuous, smooth surface embedded within the high-dimensional space [80]. It seeks a low-dimensional representation of this manifold by leveraging mathematical foundations from topology and metric space theory. These foundations enable UMAP to better capture the intrinsic structure of the data. The technique begins by constructing a fuzzy topological representation, building a graph that represents the manifold structure of the high-dimensional data. Once the graph is constructed, UMAP employs optimization techniques to find a low-dimensional embedding. Specifically, it uses stochastic gradient descent to optimize the layout of the graph in the low-dimensional space while preserving the data's topological structure. This optimization process is efficient, contributing to UMAP's speed and scalability. As a result, UMAP is generally faster than t-SNE and can handle larger datasets effectively.

UMAP has several key applications. In data visualization, it is used to reveal patterns and clusters within complex datasets, often preserving more of the global structure compared to other methods like t-SNE. This makes it useful for exploratory data analysis, where understanding both local and global relationships is important (Fig. 4.11). In clustering tasks, UMAP can uncover groups in the data that might be less apparent with other dimensionality reduction techniques. Its ability to process large-scale data efficiently makes it suitable for analyzing extensive datasets in fields such as bioinformatics, genomics, and big data analytics.

Example code:

```
import matplotlib.pyplot as plt
from sklearn.datasets import load_digits
import umap
```

```
# Load the MNIST digits dataset
digits = load_digits()
X = digits.data
y = digits.target
# Apply UMAP
reducer = umap.UMAP(n_neighbors=15, min_dist=0.1)
embedding = reducer.fit_transform(X)
# Plot the results
plt.figure(figsize=(10, 8))
plt.scatter(embedding[:, 0], embedding[:, 1], s=10, alpha=0.7, c=y)
plt.legend(*scatter.legend_elements(), title="Digits")
plt.title('UMAP Visualization of MNIST Digits')
plt.xlabel('UMAP Feature 1')
plt.ylabel('UMAP Feature 2')
plt.show()
```

### 4.4.7 Time-Series and Frequency Transformation

Time-series data consist of observations collected sequentially over time, often showing patterns such as trends, cycles, and seasonal variations. Biomedical time-series data—such as patient health metrics over time, gene expression levels, or drug response curves—often present unique challenges, including trends, seasonality, irregular sampling intervals, and noise. The goal of time-series transformation in this field is to preprocess this data to make it more amenable for analysis, particularly for time-series forecasting models. Analyzing such data requires specialized techniques to uncover underlying structures and make accurate forecasts. Time-series and frequency transformations are useful tools in this process, and Python has several rich libraries to handle time-series and frequency data. We briefly describe several basic methods below:

***Differencing*** is a technique used to stabilize the mean of a time-series by removing changes in the level of the series, thereby eliminating trends and seasonality. In the field of biomedicine, this method is particularly useful for analyzing longitudinal clinical data or physiological signals that show trends over time, such as heart rate, blood pressure, or hormone levels. By transforming a nonstationary time-series into a stationary one, differencing allows for the application of time-series modeling methods like ARIMA,[1] which require stationarity for accurate forecasting and analysis [81].

---

1 ARIMA (autoregressive integrated moving average) is a popular statistical modeling technique used for analyzing and forecasting time-series data. It combines three components: the autoregressive (AR) part models the variable of interest using a linear combination of its own past values; the integrated (I) part involves differencing the time-series data to make it stationary, which is essential for accurate modeling; and the moving average (MA) part models the relationship between the variable and the residual errors from a moving average model applied to lagged observations. In biomedicine, ARIMA models are employed to predict future trends in various physiological measurements or disease incidence rates. For example, they can forecast patient vital signs in intensive care units, anticipate outbreaks in epidemiology, or monitor long-term changes in biomarkers. The ability of ARIMA models to handle different types of patterns in time-series data makes them a valuable tool for clinicians and researchers aiming to make data-driven decisions.

The first-order difference of a time-series $y_t$ is calculated as:

$$\nabla y_t = y_t - y_{t-1}$$

For higher-order differencing, the process is repeated on the differenced series:

$$\nabla^2 y_t = \nabla y_t - \nabla y_{t-1}$$

Differencing eliminates linear trends in biomedical data, making the series stationary. This is useful when analyzing time-dependent variables like enzyme activities or biomarker levels, where underlying trends could obscure true physiological changes. Furthermore, seasonal differencing removes periodic patterns by subtracting the value from the same period in the previous cycle. This is especially useful in epidemiology for adjusting data that show seasonal effects, such as influenza cases peaking during winter months.

***Moving average smoothing*** is a technique used to reduce short-term fluctuations in time-series data, thereby displaying longer-term trends or cycles. In biomedicine, this method is particularly important for analyzing physiological signals or clinical measurements that show variability due to noise or transient events [82]. By calculating the average of different subsets of the complete dataset, moving average smoothing helps in visualizing underlying patterns that may be obscured by random fluctuations [83].

The simple moving average (SMA) over $k$ periods is calculated as:

$$\text{SMA}_t = \frac{1}{k}\sum_{i=0}^{k-1} y_{t-i},$$

where,

$\text{SMA}_t$ is the moving average on day $t$, $t$ represent the current time point,

$y_{t-i}$ represents the data point at time $t - i$,

$i$ is an index variable that helps to move backward in time.

While SMA assigns equal weights to all observations within the window, another type of moving average such as weighted moving average (WMA) assigns different weights to observations, typically giving more importance to recent data points.

***Exponential smoothing*** is a time-series forecasting technique that assigns exponentially decreasing weights to past observations, giving more significance to recent data points. In biomedicine, this method is particularly useful for predicting future trends in clinical measurements or physiological signals, such as patient vital signs, lab test results, or epidemiological data. By emphasizing recent observations, exponential smoothing adapts quickly to changes, making it suitable for monitoring patient conditions where timely detection of trends is relevant.

The single exponential smoothing is expressed by:

$$S_t = \alpha y_t + (1-\alpha) S_{t-1}$$

where:

$S_t$ is smoothed value at time $t$,

$y_t$ is the actual observed value at time $t$,

$S_{t-1}$ is the smoothed value at the previous time point,

$\alpha$ is the smoothing constant between 0 and 1.

and, $S_{t-1} = \alpha y_{t-1} + (1 - \alpha) S_{t-2}$

Therefore, the smoothed value $S_t$ can be expanded as:

$$S_t = \alpha y_t + \alpha(1-\alpha) y_{t-1} + \alpha(1-\alpha)^2 y_{t-2} + \alpha(1-\alpha)^3 y_{t-3} + \dots$$

Here, the weights $\alpha(1 - \alpha)^k$ for $k = 0, 1, 2, \dots$ form a geometric series that decreases exponentially as $k$ increases.

Besides the single exponential smoothing, different variants of exponential smoothing are used depending on the characteristics of the data. Single exponential smoothing is applied when the data show no trend or seasonality. As described above, it smooths the data by averaging past observations with exponentially decreasing weights. Double exponential smoothing (Holt's linear method) extends the single method by incorporating a trend component, making it suitable for data with trends but no seasonality. Triple exponential smoothing (Holt-Winters method) adds a seasonal component to the double smoothing method, allowing it to handle data with both trends and seasonal patterns.

***Fourier transform*** is a mathematical technique that decomposes a time-series signal into its constituent frequencies, allowing for the analysis of the frequency components present in the data. In biomedicine, this method is particularly useful for analyzing physiological signals that show periodic or oscillatory behavior, such as electrocardiograms (ECG), electroencephalograms (EEG), and other biomedical signals. By transforming data from the time domain to the frequency domain, clinicians and researchers can identify patterns and anomalies that are not easily detectable in the raw time-series data [84].

The discrete Fourier transform (DFT) for a time-series $x_n$ of length $N$ is defined as:

$$X_k = \sum_{n=0}^{N-1} x_n e^{-\frac{i2\pi kn}{N}}$$

where:

$X_k$ represents the amplitude and phase of the frequency component at frequency $k$,

$x_n$ is the $n$-th sample of the time-domain signal,
$i$ is the imaginary unit,
$n$ is the time index,
$k$ is the frequency index ranging from 0 to $N - 1$

In biomedicine, Fourier transform has several applications:

- *Analysis of electrocardiograms (ECG)*: ECG signals represent the electrical activity of the heart and contain important frequency components related to different cardiac events. By applying Fourier transform, clinicians can isolate specific frequency bands associated with arrhythmias or ischemic events. For instance, high-frequency noise can be filtered out to improve the detection of the QRS complex, aiding in the diagnosis of cardiac conditions.
- *Electroencephalograms (EEG) interpretation*: EEG signals capture the electrical activity of the brain and are inherently complex due to the superposition of multiple neural sources. Fourier transform helps decompose EEG signals into delta, theta, alpha, beta, and gamma frequency bands. Analyzing these bands enables the identification of neurological disorders such as epilepsy, sleep disorders, and encephalopathies by detecting abnormal rhythmic activities.

- *Medical imaging reconstruction*: In magnetic resonance imaging (MRI), the raw data collected are in the frequency domain (k-space). Fourier transform is used to reconstruct spatial images from these frequency components. This transformation helps generate high-resolution images that allow for the detailed examination of internal anatomical structures, facilitating accurate diagnoses and treatment planning.
- *Heart rate variability (HRV) analysis*: HRV is a measure of the variation in time between consecutive heartbeats and is an important indicator of autonomic nervous system function. By applying Fourier transform to HRV data, researchers can assess the balance between sympathetic and parasympathetic activity by analyzing low-frequency and high-frequency components. This analysis is required in stress research, cardiology, and monitoring the effects of various interventions.
- *Vibration analysis in biomechanics*: Fourier transform is employed to study the mechanical properties of biological tissues and prosthetic devices. For example, analyzing the frequency response of bone vibrations can provide information into bone density and strength, which is important in osteoporosis research.



***Wavelet transform*** is a mathematical technique that provides a time-frequency representation of a time-series signal, allowing for the analysis of signals whose frequency content changes over time [85]. In biomedicine, this method is particularly useful for analyzing nonstationary physiological signals, such as electroencephalograms (EEG), electrocardiograms (ECG), and other biomedical data that show transient events or localized features [86]. By decomposing a signal into wavelets—small wave-like oscillations that are localized in both time and frequency—the wavelet transform enables the detection and characterization of patterns that are not easily discernible using traditional Fourier analysis.

Unlike the Fourier transform, which represents a signal as a sum of infinite-duration sine and cosine functions, the wavelet transform uses finite-duration wavelets that can be stretched and shifted to match various features of the signal. This flexibility allows for multi-resolution analysis, where the signal is examined at different scales or resolutions, capturing both high-frequency, short-duration events and low-frequency, long-duration trends.

Applications in biomedicine include the following:

- *Analysis of EEG signals*: EEG recordings capture the electrical activity of the brain and are often complex and nonstationary [87]. Wavelet transform allows for the detection and localization of transient events such as epileptic spikes, sleep spindles, or event-related potentials. Analyzing the signal at multiple scales, clinicians can identify abnormal brain activities associated with neurological disorders like epilepsy, sleep disorders, and cognitive impairments.
- *ECG signal processing*: In ECG analysis, wavelet transform aids in the detection of cardiac events by isolating specific components of the heart's electrical activity. It can improve the identification of QRS complexes, P-waves, and T-waves [88], which are required for diagnosing arrhythmias, myocardial infarctions, and other cardiac conditions. The ability to filter out noise and artifacts improves the accuracy of automated ECG interpretation systems.
- *Medical image analysis*: Wavelet transform is utilized in processing medical images such as MRI, CT scans, and ultrasound images [89–91]. It facilitates image

compression by decomposing the image into wavelet coefficients, allowing for efficient storage and transmission without loss of diagnostic information. Furthermore, it improves image denoising, edge detection, and feature extraction, which are useful for accurate image interpretation and computer-aided diagnosis.

- *Heart rate variability (HRV) analysis*: HRV is an important indicator of autonomic nervous system function and cardiovascular health. Wavelet transform enables the analysis of HRV signals by decomposing them into time-frequency components [92]. This provides information about sympathetic and parasympathetic nervous system activities, aiding in the assessment of stress, fatigue, and susceptibility to cardiac events.
- *Gene expression data analysis*: In genomics, wavelet transform can analyze gene expression time-series data to identify patterns of gene activation and repression over time [93]. This helps in understanding regulatory mechanisms, cellular responses to stimuli, and the progression of diseases at the molecular level.

## 4.5 Dealing with Outliers

Outliers are data points that deviate significantly from the rest of the dataset and can arise due to various reasons such as measurement errors, data entry mistakes, or natural variability in biological processes [94]. Proper handling of outliers is required because they can affect the results of data analysis and statistical modeling, leading to incorrect conclusions or misguided clinical decisions.

### 4.5.1 Outlier Detection Methods

The selection of an appropriate outlier detection method depends on the nature of the data and the specific context of the analysis [95, 96]. Commonly used outlier detection methods can be categorized into several groups using the method of detection (◘ Fig. 4.12).

### 4.5.2 Statistical Methods for Outlier Detection

**Notebook: Section 4.5.2. Statistical Methods for Outlier Detection**
This notebook provides code examples demonstrating statistical methods for outlier detection.

Link to the GitHub repository:
▶ https://github.com/sn-code-inside/BioPy
Go to: Chap. 4—▶ Sect. 4.5.2

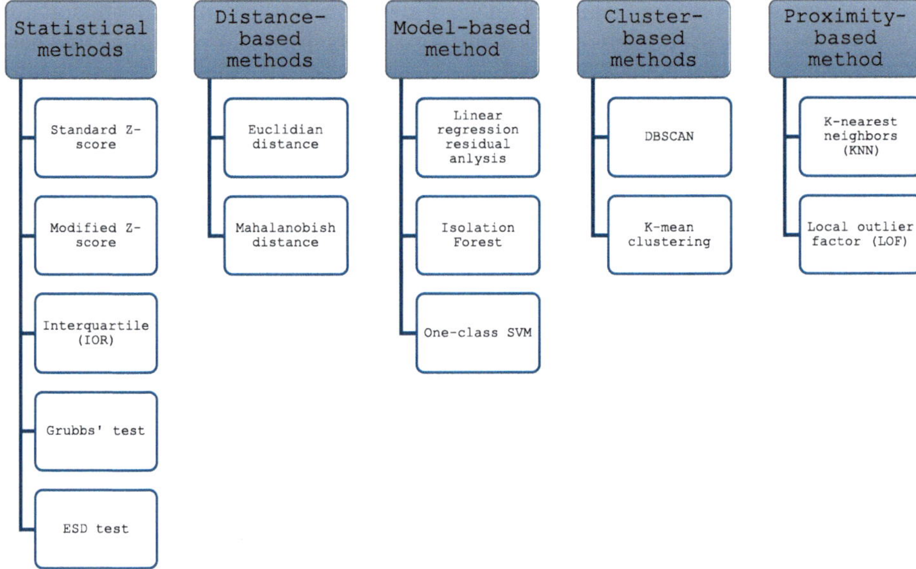


**Fig. 4.12** Common methods for outlier detection. The classification was done using underlying methods used for outlier detection without considering whether methods are supervised or unsupervised and whether univariate methods (dealing with single-variable data) and multivariate methods (dealing with multivariable data)

Statistical methods are grounded in probability theory and statistical measures, making them particularly effective when the distributional properties of the data are well understood.

***Z-score method***: The *Z*-score method evaluates how many standard deviations a data point is from the mean of the dataset. It is commonly used when the data follows a normal distribution. The *Z*-score for a data point *x* is calculated as:

$$Z = \frac{\text{Original value} - \text{Mean value}}{\text{Standard deviation}}$$

Data points with *Z*-scores greater than a threshold (e.g., $|Z| > 3$) are considered outliers. This method is simple and effective but sensitive to extreme values, which can distort the mean and standard deviation.

Example code:

```
import numpy as np
from scipy import stats
# Define a function for Z-score-based outlier detection
def z_score_method(data, threshold=3):
    # Calculate the Z-scores for each data point
    z_scores = np.abs(stats.zscore(data))
    # Get indices of data points where the Z-score is above the
    # threshold
    outlier_indices = np.where(z_scores > threshold)
    # Get the actual outlier values
    outlier_values = data[outlier_indices]
```

```
        return outlier_indices, outlier_values
    # Generate sample data with some outliers
    # Generate 100 points from a normal distribution
    data = np.random.normal(0, 1, 100)
    data = np.concatenate([data, [3, -3, 3.5]])  # Adding outliers
    # Detect outliers using the Z-score method
    outlier_indices, outlier_values = z_score_method(data)
    # Print out the results
    print("Indices of outliers:", outlier_indices)
    print("Outlier values:", outlier_values)
    print("Original data:", data)
```

***Modified Z-score method***: The modified $Z$-score uses the median and median absolute deviation (MAD) instead of the mean and standard deviation, making it more robust to existing outliers [97]. It is calculated as:

$$\text{Modified } Z = \frac{\text{Original value} - \text{Median value}}{1.4826 \times \text{MAD}}$$

For a Gaussian distribution standard deviation term is replaced by MAD multiplied by a constant 1.4826 [98].

Example code:

```
import numpy as np
# Define a function for the Modified Z-Score method
def modified_z_score_method(data, threshold=3.5):
    # Convert input data to a numpy array if not already in that format
    data = np.asarray(data)
    # Calculate the median of the data
    median = np.median(data)
    # Calculate the Median Absolute Deviation (MAD)
    mad = np.median(np.abs(data - median))
    # Calculate the modified Z-scores
    modified_z_scores = (data - median) / (1.4826 * mad)
    # Get indices and values of data points
    outlier_indices = np.where(np.abs(modified_z_scores) > threshold)
    outlier_values = data[outlier_indices]
    return outlier_indices, outlier_values
# Example usage with sample data including some outliers
# Generate 100 points from a normal distribution
data = np.random.normal(0, 1, 100)
data = np.concatenate([data, [10, -10, 15]])  # Adding outliers
# Detect outliers using the Modified Z-Score method
outlier_indices, outlier_values = modified_z_score_method(data)
# Print out the results
print("Indices of outliers:", outlier_indices)
print("Outlier values:", outlier_values)
print("Original data:", data)
```

***Interquartile range (IQR) method***: The IQR method is a statistical technique used to detect outliers in a univariate dataset. It is a nonparametric method, meaning it does not make assumptions about the underlying distribution of the data, making it suitable for datasets with unknown or non-normal distributions. The method uses the spread of the middle 50% of the data to identify data points that fall below or above this central range [99].

The IQR method for detecting outliers is based on the calculation of quartiles, which divide the data into four equal parts. The first quartile ($Q1$) represents the value below which 25% of the data falls, while the third quartile ($Q3$) is the value below which 75% of the data falls. The IQR is then computed as the difference between $Q3$ and $Q1$, representing the spread of the middle 50% of the data.

To identify outliers, we use the following criteria:

*Lower bound*: Any data point less than $Q1 - 1.5 \times \text{IQR}$ is considered as an outlier.

*Upper bound*: Any data point greater than $Q3 + 1.5 \times \text{IQR}$ is also flagged as an outlier.

These limits are known as "fences," and points outside these fences are considered outliers.

Furthermore, extreme outliers can be identified using stricter criteria:

Lower extreme bound: $Q1 - 3 \times \text{IQR}$

Upper extreme bound: $Q3 + 3 \times \text{IQR}$

Points falling beyond these extreme bounds are classified as extreme outliers.

The IQR method has several advantages. It is robust because it uses the median and quartiles, making it unaffected by extreme values in the dataset. It is simple to compute and interpret, which makes it useful for exploratory data analysis. Moreover, the method does not assume normality, allowing it to work well with datasets that do not follow a normal distribution. However, the IQR method has some limitations. It may not perform well with small datasets, as quartiles and the IQR can become unstable in these cases. Furthermore, in multivariate data, it treats each variable independently, overlooking outliers that only become apparent when the relationships between variables are considered together.

Example code:

```
import numpy as np
# Define a function for outlier detection using the IQR method
def iqr_method(data):
    # Calculate Q1 (25th percentile) and Q3 (75th percentile)
    Q1 = np.percentile(data, 25)
    Q3 = np.percentile(data, 75)
    # Compute the IQR
    IQR = Q3 - Q1
    # Determine the lower and upper bounds for outliers
    lower_bound = Q1 - 1.5 * IQR
    upper_bound = Q3 + 1.5 * IQR
    # Identify outliers
    outliers = data[(data < lower_bound) | (data > upper_bound)]
    return outliers, lower_bound, upper_bound
# Example data with potential outliers
data = np.array([10, 12, 14, 15, 15, 16, 16, 18, 19, 20, 25, 30, 100])
# Outlier at 100
```

```
# Detect outliers using the IQR method
outliers, lower_bound, upper_bound = iqr_method(data)
# Output the result
print(f"Outliers: {outliers}")
print(f"Lower Bound: {lower_bound}, Upper Bound: {upper_bound}")
```

***Grubbs' test***: Grubbs' test is used to detect a single outlier in a normally distributed univariate dataset [100]. The test computes a test statistic, $G$, which is the maximum absolute deviation from the mean normalized by the standard deviation:

$$G = \max \frac{|x_i - \overline{x}|}{\sigma}$$

where $x_i$ is the suspected outlier, $\overline{x}$ is the mean, and $\sigma$ is the standard deviation. If $G$ is greater than a critical value, the point is considered an outlier [101]. Grubbs' test is effective for single outliers but needs to be used iteratively for multiple outliers.

Example code:

```
import numpy as np
from scipy import stats
# Function to calculate Grubbs' Test statistic and detect outlier
def grubbs_test(data, alpha=0.05):
    # Compute mean and standard deviation
    mean_y = np.mean(data)
    std_y = np.std(data, ddof=1)
    # Find the data point furthest from the mean (potential outlier)
    deviations = np.abs(data - mean_y)
    max_deviation_index = np.argmax(deviations)
    G = deviations[max_deviation_index] / std_y
    n = len(data)
    # Compute the critical value for Grubbs' test
    t_crit = stats.t.ppf(1 - alpha / (2 * n), n - 2)
    G_crit = ((n - 1) / np.sqrt(n)) * np.sqrt(t_crit**2 / (n - 2 +
    t_crit**2))
    # Return the test statistic, critical value, and the potential
    # outlier
    return G, G_crit, data[max_deviation_index]
# Example data with a potential outlier
data = np.array([10, 12, 12, 13, 12, 12, 11, 25])
# '25' might be an outlier
# Perform Grubbs' Test
G_stat, G_crit, outlier = grubbs_test(data)
print(f"Grubbs' Test Statistic: {G_stat}")
print(f"Grubbs' Critical Value: {G_crit}")
if G_stat > G_crit:
    print(f"Outlier detected: {outlier}")
else:
    print("No outliers detected.")
```

***ESD test***: The Extreme Studentized Deviate (ESD) test is a statistical test used to detect multiple outliers in a dataset, particularly when the number of outliers is unknown. It is an extension of Grubbs' test and is commonly used for datasets that are assumed to be normally distributed [102]. In the Extreme Studentized Deviate (ESD) test, the data must follow an approximately normal distribution. The null hypothesis ($H_0$) assumes there are no outliers in the dataset, while the alternative hypothesis ($H_1$) assumes there is at least one outlier. The test iteratively calculates the test statistic $R_i$ for each observation and removes the most extreme value in each iteration.



The formula for the test statistic is:

$$R_i = \frac{\max\left|x_i - \overline{x}\right|}{\sigma}$$

where $\overline{x}$ is the mean and $\sigma$ is the standard deviation. The ESD test proceeds by removing the value with the largest test statistic in each iteration, and the procedure is repeated until no more outliers are detected or the maximum number of outliers is found. The critical value is calculated using the t-distribution. If the test statistic $R_i$ exceeds the critical value, the corresponding data point is considered an outlier.

Example code:

```
import numpy as np
from scipy import stats
def esd_test(data, max_outliers, alpha=0.05):
    n = len(data)
    outliers = []
    for i in range(1, max_outliers + 1):
        mean_data = np.mean(data)
        std_data = np.std(data, ddof=1)
        # Find the value with the largest deviation from the mean
        abs_deviation = np.abs(data - mean_data)
        max_deviation_index = np.argmax(abs_deviation)
        max_deviation_value = data[max_deviation_index]
        # Calculate the test statistic
        test_statistic = abs_deviation[
        max_deviation_index] / std_data
        # Compute the critical value using the t-distribution
        p = 1 - alpha / (2 * (n - i + 1))
        t_crit = stats.t.ppf(p, df=n - i - 1)
        critical_value = ((n - i) * t_crit) / np.sqrt
                         ((n - i - 1 + t_crit**2) * (n - i + 1))
        # Compare the test statistic with the critical value
        if test_statistic > critical_value:
            outliers.append(max_deviation_value)
            # Remove the detected outlier from the data for the
            # next iteration
            data = np.delete(data, max_deviation_index)
        else:
            break
    return outliers, data
# Example usage
```

```
data = np.array([10, 12, 12, 13, 12, 12, 11, 25, 100])
# 25 and 100 are potential outliers
# Perform ESD Test for up to 2 outliers
outliers, data = esd_test(data, max_outliers=2)
print(f"Outliers: {outliers}")
print(f"Data without outliers: {data}")
```

### 4.5.3 Distance-Based Methods for Outlier Detection

Distance-based methods are commonly used where data points are spatially distributed or when the relationships between points are more important than the actual values [103]. These methods operate under the assumption that normal data points are located in dense regions, while outliers are situated far from other points in a dataset. The key idea is that an outlier is a data point that is significantly distant from most other points.

***Euclidean distance***: Euclidean distance measures the straight-line distance between two points in a multidimensional space. In the context of outlier detection, Euclidean distance helps identify data points that are far from the bulk of the data, suggesting they could be outliers.

For two points $A$ and $B$, in a 2D space with coordinates $A(x_1, y_1)$ and $B(x_2, y_2)$, the Euclidean distance is given by:

$$d(A,B) = \sqrt{(x_2 - x_1)^2 - (y_2 - y_1)^2}$$

For higher dimensions, the equation can be extended as:

$$d(A,B) = \sqrt{\sum_{i=1}^{n} (B_i - A_i)^2}$$

where $n$ is the number of dimensions and $A_i$ and $B_i$ are the values of the coordinates for points $A$ and $B$ in the $i$-th dimension.

In outlier detection, Euclidean distance is used to measure how far each data point is from other points or from a central point like the mean or median of the data [104]. Points with large Euclidean distances from the rest of the data are likely to be outliers. For example, in the k-nearest neighbors (K-NN) approach, the Euclidean distance to the nearest neighbors is computed for each point. If the average or maximum distance to its nearest neighbors is larger than that of other points, the point is flagged as an outlier [105].

Another approach calculates the Euclidean distance of each point from the centroid (mean) of the dataset. Points farthest from the centroid are outliers. In a dataset of health parameters such as heart rate, blood pressure, and cholesterol levels, a patient whose health measurements lie far from the average population might be flagged as an outlier, possibly indicating an abnormal health condition.

Euclidean distance has several advantages for outlier detection. It is simple to compute and interpret, especially in low-dimensional spaces, and works well for detecting outliers when the dataset is low-dimensional and has a simple structure.

However, it has some limitations. In high-dimensional spaces, its usefulness decreases because as the number of dimensions increases, distances between points tend to become similar, reducing its effectiveness. Euclidean distance is sensitive to the scale of the variables. If one feature has a larger range than others, it can dominate the distance calculation unless the data is normalized or standardized beforehand.

***Mahalanobis distance***: Unlike Euclidean distance, which simply measures the straight-line distance between points, Mahalanobis distance takes into account the correlations between variables, making it particularly effective for detecting outliers in data where different variables are interrelated [106].



Mahalanobis distance measures the distance between a point and the mean of a distribution while considering the variance and covariance of the variables. In simpler terms, it identifies how far a point is from the center of the data, but it adjusts for the shape of the data distribution. This adjustment makes it more sensitive to the relationships between different variables.

For a multivariate data point $X$, the Mahalanobis distance from the mean vector $\mu$ of the dataset is calculated as:

$$D = \sqrt{(X - \mu)^{\mathrm{T}} S^{-1} (X - \mu)}$$

where:

$X$ is the vector representing the data point.

$\mu$ is the mean vector of the dataset.

$S$ is the covariance matrix of the dataset and $S^{-1}$ is the inverse of this matrix.

$(X - \mu)^{\mathrm{T}}$ denotes the transpose of the difference vector.

Mahalanobis distance adjusts for correlations between variables. For example, if two variables are highly correlated, an outlier that deviates only along one dimension will be flagged, whereas Euclidean distance might miss it. Since Mahalanobis distance accounts for the variance of the data, points in low-variance areas will need to be closer to the mean to avoid being flagged as outliers, while points in high-variance areas can be farther away.

Mahalanobis distance is particularly useful in detecting outliers in high-dimensional or multivariate data where relationships between variables are important. One of the advantages of Mahalanobis distance is that it handles correlations between variables, making it more robust for multivariate outlier detection than simpler distance metrics like Euclidean distance. Furthermore, it is effectively normalized by the variance-covariance structure of the data, ensuring that it handles data on different scales properly. However, Mahalanobis distance comes with certain limitations. It requires the computation of the inverse covariance matrix, which can be computationally expensive for large datasets. If the covariance matrix is singular (non-invertible), Mahalanobis distance cannot be computed directly. Furthermore, it works best when the data follows a multivariate normal distribution. If the data deviates from normality, Mahalanobis distance may not perform well.

### 4.5.4 Model-Based Methods for Outlier Detection

**Notebook: Section 4.5.4. Model-Based Methods for Outlier Detection**
This notebook provides code examples demonstrating model-based methods for outlier detection.

Link to the GitHub repository:
▶ https://github.com/sn-code-inside/BioPy
Go to: Chap. 4—▶ Sect. 4.5.4

Model-based methods are a category of outlier detection techniques that involve building models to describe the underlying structure of the data and identifying outliers as points that do not fit well within this model. These methods are particularly useful in supervised or semi-supervised learning environments, but they can also be adapted for unsupervised tasks. Below are some commonly used model-based methods for outlier detection [99]:

**Linear regression residual analysis**: In linear regression models, outliers can be detected by analyzing the residuals, which are the differences between the observed values and the values predicted by the model [107]. Residuals provide insight into how well the model fits each data point. If a residual is larger or smaller than expected, the corresponding data point may be an outlier, indicating that it deviates substantially from the general pattern captured by the model.

To detect outliers, the first step is to calculate the residuals for each data point after fitting the regression model. Residuals are simply the vertical distances between the actual data points and the regression line, which represents the model's predicted values. Large residuals indicate points where the model's predictions are far off from the observed values. Typically, these large residuals can be standardized (or normalized) to ensure comparability across different datasets or variables. Standardized residuals greater than $|3|$ are often flagged as outliers, as this indicates that the data point is more than three standard deviations away from the model's prediction.

Example code:

```
import numpy as np
import pandas as pd
import statsmodels.api as sm
# Example dataset: Height (in inches) vs Weight (in pounds)
data = {'Height': [60, 62, 64, 66, 68, 70, 72, 74, 76, 80, 85],
        'Weight': [50, 52, 55, 62, 65, 60, 130, 80, 85, 95, 105]}
# Convert the dataset into a DataFrame
```

```
df = pd.DataFrame(data)
# Define independent (X) and dependent (Y) variables
X = df['Height']
Y = df['Weight']
# Add a constant term to the independent variable matrix for the intercept
X = sm.add_constant(X)
# Fit the linear regression model
model = sm.OLS(Y, X).fit()
# Predict Y values and calculate residuals
predictions = model.predict(X)
residuals = Y - predictions
# Standardize the residuals
standardized_residuals = (residuals -
np.mean(residuals)) / np.std(residuals)
# Identify outliers where standardized residuals are greater than |3|
outliers = df[np.abs(standardized_residuals) > 3]
# Output the results
print("Standardized Residuals:\n", standardized_residuals)
print("Outliers Detected:\n", outliers)
```



**Isolation Forest**: Isolation Forest is a machine learning algorithm designed specifically for anomaly detection [108]. It isolates data points by recursively partitioning the dataset, and points that are isolated quickly (requiring fewer partitions) are considered outliers. The algorithm randomly selects a feature and then chooses a random split value within the range of that feature to divide the data. Outliers, which lie far from the normal data points, tend to be isolated faster because they fall into partitions that can be separated easily, resulting in shorter paths in the tree structure. Points that require fewer partitions to isolate are assigned higher anomaly scores, indicating they are more likely to be outliers. Isolation Forest is particularly effective in high-dimensional datasets, making it well suited for detecting anomalies in areas like cybersecurity, fraud detection, and bioinformatics.

Example code:

```
import numpy as np
import pandas as pd
from sklearn.ensemble import IsolationForest
# Generate sample data with outliers
np.random.seed(42)
# 100 normal data points
normal_data = np.random.normal(0, 1, (100, 2))
# 5 outliers
outliers = np.random.uniform(low=-6, high=6, size=(5, 2))

# Combine the data into a DataFrame
data = np.vstack([normal_data, outliers])
df = pd.DataFrame(data, columns=['Feature1', 'Feature2'])
# Initialize the Isolation Forest model
iso_forest = IsolationForest(contamination=0.05, random_state=42)
# Fit the model
```

```
iso_forest.fit(df)
# Predict outliers (-1 indicates outliers, 1 indicates normal points)
df['Outlier'] = iso_forest.predict(df)
# Print the detected outliers
outliers_detected = df[df['Outlier'] == -1]
print("Detected Outliers:\n", outliers_detected)
```

**One-class SVM**: One-class support vector machine (SVM) is a model-based method used for outlier detection, particularly in semi-supervised and unsupervised learning environments. It works by creating a decision boundary around the normal data points to separate them from anomalies [109]. The algorithm is trained on a dataset that primarily consists of normal data points and attempts to build a boundary that encompasses most of these points. Data points that fall outside this boundary are classified as outliers. One-class SVM can use different kernel functions, such as linear or radial basis function (RBF), to capture both linear and nonlinear relationships within the data.

One-class SVM is commonly applied in fields like intrusion detection systems, fraud detection in financial transactions, and fault detection in industrial systems. Model-based methods like one-class SVM are flexible, supporting supervised, semi-supervised, and unsupervised learning setups. They are effective in handling complex, high-dimensional datasets and can model both linear and nonlinear relationships. However, these methods require careful parameter tuning, such as selecting the appropriate kernel function, and models like Isolation Forest can be computationally expensive for large datasets.

Example code:

```
import numpy as np
from sklearn import svm
from sklearn.datasets import make_blobs
# Generate sample data with normal points and outliers
# Generating normal data (two clusters)
X, _ = make_blobs(n_samples=300, centers=2, cluster_std=0.60,
random_state=42)
# Adding some outliers
X_outliers = np.random.uniform(low=-6, high=6, size=(20, 2))
X = np.vstack([X, X_outliers])
# Fit the One-Class SVM model
# nu is an upper bound on the fraction of training errors and a
# lower bound of the fraction of support vectors
# kernel defines the type of decision boundary (we use 'rbf' for
# non-linear boundary)
model = svm.OneClassSVM(nu=0.05, kernel="rbf", gamma=0.1)
model.fit(X)
# Predict the inliers (1) and outliers (-1)
predictions = model.predict(X)
# Separate the inliers and outliers for plotting
inliers = X[predictions == 1]
outliers = X[predictions == -1]
# Print outlier data points
print("Detected Outliers:\n", outliers)
```

### 4.5.5 Cluster-Based Methods for Outlier Detection

**Notebook: Section 4.5.5. Cluster-Based Methods for Outlier Detection**
This notebook provides code examples demonstrating cluster-based methods for outlier detection.

Link to the GitHub repository:
► https://github.com/sn-code-inside/BioPy
Go to: Chap. 4—► Sect. 4.5.5



Cluster-based methods are commonly used in outlier detection by identifying points that do not belong to any dense cluster or are located far from the centroids of clusters. These methods work on the principle that normal data points tend to form dense clusters, while outliers are either isolated or belong to sparse regions.

**DBSCAN**: Density-based spatial clustering of applications with noise (DBSCAN) is a popular density-based clustering algorithm that detects outliers by identifying points that do not belong to any cluster [110]. DBSCAN works by grouping data points into clusters based on their density. Points in dense regions are clustered together, and points that do not have enough neighbors (determined by parameters like eps, the maximum distance between points, and min_samples, the minimum number of points needed to form a cluster) are considered outliers. DBSCAN is particularly effective at identifying clusters of varying shapes and sizes and is robust to noise. It works well in scenarios where the data is not uniformly distributed and can detect outliers as part of its natural operation.

Example code:

```
import numpy as np
import pandas as pd
from sklearn.cluster import DBSCAN
# Generate sample data with outliers
np.random.seed(42)
# Generate 100 normal data points
X = np.random.normal(0, 1, (100, 2))
# Add some outliers
outliers = np.random.uniform(low=-6, high=6, size=(10, 2))
X = np.vstack([X, outliers])
# Fit DBSCAN model
# eps: maximum distance between two samples for them to be
# considered as in the same neighborhood
# min_samples: minimum number of samples in a neighborhood to
# form a cluster
db = DBSCAN(eps=0.5, min_samples=5).fit(X)
# Identify core points, border points, and noise points
```

```
labels = db.labels_
# Outliers are labeled as -1
outliers = X[labels == -1]
clusters = X[labels != -1]
# Print the detected outliers
print("Detected Outliers:\n", outliers)
```

**K-means clustering**: K-means clustering is another widely used cluster-based method for outlier detection [111, 112]. K-means partitions the dataset into K clusters by minimizing the distance between data points and the cluster centroids. Once the clusters are formed, the distance of each point to its nearest centroid is calculated. Data points that are farthest from their cluster centroids can be considered outliers because they deviate from the "center" of their respective clusters. However, K-means is more sensitive to the initial placement of centroids and assumes clusters are spherical, which may limit its effectiveness in detecting outliers in more complex data distributions. Despite these limitations, K-means is commonly used in various applications due to its simplicity and speed.

Example code:

```
import numpy as np
import pandas as pd
from sklearn.cluster import KMeans
# Generate sample data with outliers
np.random.seed(42)
# Generate 100 normal data points
X = np.random.normal(0, 1, (100, 2))
# Add some outliers
outliers = np.random.uniform(low=-6, high=6, size=(5, 2))
X = np.vstack([X, outliers])
# Apply KMeans clustering
kmeans = KMeans(n_clusters=3, random_state=42)
kmeans.fit(X)
# Get the cluster centroids and labels
centroids = kmeans.cluster_centers_
labels = kmeans.labels_
# Calculate the distance of each point to its assigned cluster centroid
distances = np.linalg.norm(X - centroids[labels], axis=1)
# Identify outliers as points that are farthest from their
# cluster centroids
threshold = np.percentile(distances, 95)  # For example, we
# consider the top 5% farthest points as outliers
outliers = X[distances > threshold]
# Print the detected outliers
print("Detected Outliers:\n", outliers)
```

### 4.5.6 Proximity-Based Methods for Outlier Detection

**Notebook: Section 4.5.6. Proximity-Based Methods for Outlier Detection**
This notebook provides code examples demonstrating proximity-based methods for outlier detection.

Link to the GitHub repository:
▸ https://github.com/sn-code-inside/BioPy
Go to: Chap. 4—▸ Sect. 4.5.6



Proximity-based methods for outlier detection use similarity and distances to identify outliers. Widely used proximity-based methods include the k-nearest neighbors (KNN) algorithm and local outlier factor (LOF).

**K-nearest neighbors (KNN)**: KNN identifies outliers by examining the distance between data points and their nearest neighbors. The core idea is that normal data points are close to their neighbors, while outliers are isolated and far from their nearest neighbors [105].

In KNN-based outlier detection, for each point in the dataset, the algorithm calculates the distances to its k-nearest neighbors. If the average or maximum distance to the k-nearest neighbors is larger compared to other points in the dataset, that point can be considered an outlier. In other words, points with unusually large distances to their nearest neighbors are flagged as outliers.

KNN is effective because it can work with both low-dimensional and high-dimensional data. However, it can be computationally expensive for very large datasets because the distances between all points need to be calculated.

Example code:

```
import numpy as np
import pandas as pd
from sklearn.neighbors import NearestNeighbors
# Generate sample data with outliers
np.random.seed(42)
# Generate 100 normal data points
X = np.random.normal(0, 1, (100, 2))
# Add some outliers
outliers = np.random.uniform(low=-6, high=6, size=(5, 2))
X = np.vstack([X, outliers])
# Fit the KNN model
k = 5  # Number of neighbors
knn = NearestNeighbors(n_neighbors=k)
knn.fit(X)
# Calculate distances to the k-nearest neighbors
distances, _ = knn.kneighbors(X)
# The outlier score can be the distance to the k-th nearest
# neighbor (max distance)
outlier_scores = distances[:, -1]
```

```
# Set a threshold to identify outliers (e.g., top 5% farthest points)
threshold = np.percentile(outlier_scores, 95)
outliers = X[outlier_scores > threshold]
# Print detected outliers
print("Detected Outliers:\n", outliers)
```

**Local outlier factor (LOF)**: LOF is a widely used proximity-based method that compares the local density of a point to the densities of its neighbors. The idea is that outliers typically reside in low-density regions, far from the denser clusters of normal points [113]. LOF computes the "local density" of a point by looking at its nearest neighbors. It then compares this density to the local densities of neighboring points. A point is considered an outlier if its local density is lower than that of its neighbors. LOF is effective for detecting outliers in datasets with varying densities, as it takes the relative density of points into account.

Example code:

```
import numpy as np
import pandas as pd
from sklearn.neighbors import LocalOutlierFactor
# Generate sample data with outliers
np.random.seed(42)
# Generate 100 normal data points
X = np.random.normal(0, 1, (100, 2))
# Add some outliers
outliers = np.random.uniform(low=-6, high=6, size=(5, 2))
X = np.vstack([X, outliers])
# Fit the LOF model
# n_neighbors: Number of neighbors to use in calculating the LOF score
lof = LocalOutlierFactor(n_neighbors=20, contamination=0.05)
y_pred = lof.fit_predict(X)
# LOF assigns -1 to outliers and 1 to inliers
outliers_detected = X[y_pred == -1]
inliers_detected = X[y_pred == 1]
# Print the detected outliers
print("Detected Outliers:\n", outliers_detected)
```

### 4.5.7 Strategies to Manage Outliers

Outliers are data points that deviate from the rest of a dataset [114]. These values can arise due to measurement errors, variability in the population, or unusual events that may or may not be relevant to the analysis. Managing outliers is an important step in data preprocessing because their presence can skew results and lead to inaccurate conclusions. However, the treatment of outliers requires careful consideration. Depending on the context, outliers can either provide useful information or distort the interpretation of the data [115]. We will briefly discuss strategies for managing outliers, including exclusion, transformation of data, and robust scaling. It also shows the importance of understanding the underlying reasons for outliers and using context-specific methods to handle them. Properly managing outliers ensures that the analysis remains accurate and representative, without compromising the integrity of the data.

**Exclusion of outliers**: Outliers should be excluded when they represent clear errors or are not representative of the population being studied [116]. Common causes of such outliers include data entry errors, instrument malfunction, or rare events that are unrelated to the research question. For instance, in a study on human height, a recorded height of 300 cm would likely be considered a data entry error and should be excluded from analysis. While excluding outliers can prevent skewed results due to incorrect data points, it also carries risks. Excluding important data points may result in the loss of important information that could provide information about rare but useful events or patterns. Furthermore, the exclusion of outliers can introduce bias, especially if the decision is subjective or made without a clear and objective rationale. This could result in misleading conclusions, especially if the outliers reflect real-world phenomena that deviate from the expected pattern but are still valid.

**Transformation of data**: Data transformation is widely used to manage skewed data. Applying a logarithmic transformation to the data is a common method to handle skewed data. It reduces the impact of extreme values by compressing the scale of the data, particularly useful for managing datasets where there are large variations between values [117]. For example, gene expression datasets often benefit from log transformation due to the large differences between low and high values.

Winsorization is another technique used to limit the influence of extreme values by replacing outliers with a less extreme value that lies at a predefined percentile, for example, replacing values above the 95th percentile with the value at the 95th percentile [118]. This helps maintain the overall structure of the dataset while minimizing the distorting effect of extreme outliers.

Methods like robust scaling, which uses the median and interquartile range (IQR), are less affected by outliers compared to standard normalization methods that use the mean and standard deviation [54, 119]. This technique ensures that extreme values do not disproportionately influence the data's scale. Robust scaling is particularly useful when the dataset contains a small number of outliers that affect the mean and standard deviation.

**Considerations**: Before applying any method to handle outliers, it is important to understand the nature of the data and the reasons for the presence of outliers [120]. In some cases, outliers can be meaningful and reveal important overviews about rare events, anomalies, or other useful phenomena. For example, outliers might indicate important biological variation in clinical studies, which could lead to a deeper understanding of certain conditions or treatments [95, 120].

The choice of method for managing outliers should be context-specific and driven by the research goals. For example, in highly regulated environments like clinical trials, the exclusion of data points may be restricted or require detailed justification [121, 122]. In contrast, other fields may permit more flexibility, allowing for transformations or exclusions based on the specific characteristics of the dataset. Therefore, it is useful to document any decisions made regarding the handling of outliers. Researchers should clearly record the rationale for excluding or transforming outliers and the method used. This transparency ensures that the research remains reproducible and that others can understand how outliers were managed, maintaining the integrity of the research [121]. Proper documentation is also important for addressing any biases that may arise from decisions made during data preprocessing.

### 4.5.8 Consequences of Not Handling Outliers

Outliers, if not handled properly, can have serious consequences in data analysis and modeling. One of the key issues is that outliers can skew statistical measures such as the mean and standard deviation, resulting in an inaccurate representation of the data [123]. This can lead to a biased analysis where the central tendency and variability of the data are not properly captured.

In machine learning, outliers can negatively impact model performance by causing the model to learn from noise instead of meaningful patterns in the data [115, 124]. This results in models that perform poorly when applied to new, unseen data. Specifically, outliers can lead to overfitting, where the model becomes too closely fitted to the outliers, limiting its ability to generalize. In regression analysis, outliers can distort the regression line, affecting the slope and intercept and potentially leading to incorrect conclusions about relationships between variables.

To summarize, properly detecting and handling outliers in data preprocessing may ensure reliable analysis and modeling. The choice of strategy, whether exclusion, transformation, or robust scaling, should be driven by the context of the data and the objectives of the study. It is also useful to document and justify the handling of outliers to maintain transparency and reproducibility.

## 4.6 Data Integration

Data integration in the context of biomedical research involves the process of combining datasets from diverse sources, including but not limited to genomic, transcriptomic, proteomic, metabolomic, clinical, and pharmacological data [125]. This integration is useful for achieving a systems-level understanding of complex biological processes and diseases [126]. It plays a role in enabling advanced biomedical research by allowing researchers to uncover novel insights that would be difficult to detect from any single data type alone [127].

One of the primary objectives of data integration is to bridge the gap between different layers of biological information, enabling the development of more accurate models of biological systems [128]. This holistic approach supports the identification of new biomarkers, drug targets, and mechanistic overview on diseases, which are useful in advancing drug discovery, target discovery, and drug sensitivity prediction [129].

The integration process faces numerous challenges, such as differences in data formats, experimental protocols, and the inherent heterogeneity of biological systems [130]. Addressing these challenges requires the use of sophisticated computational techniques, including machine learning algorithms, network-based approaches, and statistical models [128, 131]. The development of standardized formats and ontologies also helps facilitate the interoperability and integration of data across different platforms [132].

Furthermore, the integration of clinical data, such as electronic health records (EHRs), with molecular datasets provides a link between laboratory research and patient outcomes [133, 134]. For example, integrating genomic and pharmacological data allows for the identification of drug response patterns and the prediction of drug sensitivity in specific patient populations, accelerating personalized medicine initiatives [129, 135].

### 4.6.1 Importance



Biomedical data is heterogeneous, stemming from a wide variety of experimental methods and biological levels. For example, data from genomics, transcriptomics, proteomics, metabolomics, and epigenomics are generated from different high-throughput technologies, each providing a unique view of the biological system [136, 137]. In addition to molecular-level data, clinical data, such as patient demographics, clinical outcomes, medical histories, and treatment responses, add another dimension of complexity. Integrating these diverse data types is used to gain a holistic understanding of disease mechanisms and to translate observations into clinical applications [138, 139].

The integration of multi-omics data with clinical information provides a picture of the underlying biological pathways and processes that drive disease progression. This perspective is useful for identifying novel drug targets, as it allows researchers to uncover molecular interactions and pathways that would remain hidden when examining a single type of data in isolation [140–142]. For example, combining proteomic data (which reflects the functional protein levels) with transcriptomic data (which provides gene expression information) can reveal posttranscriptional regulatory mechanisms, giving a deeper understanding of gene function and its relevance to disease [143].

Furthermore, data integration is useful for improving drug sensitivity prediction and tailoring treatments to individual patients. Combining pharmacogenomic data with other molecular profiles, researchers can better predict how different patients will respond to specific drugs, thereby advancing personalized medicine [143]. This enables more precise therapeutic interventions, minimizing adverse drug reactions and improving patient outcomes.

Moreover, integrating data across different biological layers helps identify biomarkers that are predictive of disease progression, treatment efficacy, or patient prognosis. These biomarkers are instrumental in developing diagnostic tools, monitoring disease states, and guiding therapeutic decisions. For instance, integrating genomic data with clinical outcomes has led to the identification of genetic mutations that predict response to targeted therapies in cancer treatment [144, 145].

The importance of data integration lies in its ability to provide a systems-level understanding of complex biological processes and disease states. Unifying data from various sources, researchers can uncover novel perspectives, drive drug discovery, and create more personalized treatment strategies that ultimately improve patient care.

### 4.6.2 Challenges

The integration of biomedical data presents a range of challenges that stem from the heterogeneity of the data, differences in quality, and the need for compliance with ethical and legal standards [1, 146, 150]. Addressing these challenges is needed to ensure that integrated datasets can be effectively analyzed and used to advance biomedical research and clinical applications.

**Data heterogeneity**: Biomedical datasets are often collected using different technologies, measurement techniques, and scales, which creates inconsistencies that must be reconciled before integration [148, 149]. For example, genomic data may be in the form of sequencing reads, while proteomic data is typically presented as mass spectrometry intensities, and clinical data may be recorded in electronic health records (EHRs) using standardized or unstandardized terminologies.

Addressing this heterogeneity requires extensive preprocessing steps, such as data cleaning, normalization, and transformation [150–152]. Standardization efforts must ensure that data from diverse sources can be aligned and analyzed together, which often involves developing mapping strategies or using ontologies and common data models. Without effective preprocessing, integrating disparate datasets can lead to inaccuracies, misinterpretations, and flawed conclusions [151].

**Data quality and compatibility**: Another key challenge is the variability in data quality and compatibility across datasets. Differences in experimental design, data collection protocols, and technical variations between platforms can introduce noise and biases into the data [1, 153]. For instance, clinical datasets may be incomplete or contain errors, while molecular datasets may suffer from batch effects or technical artifacts.

Ensuring data quality and consistency often requires robust quality control (QC) processes to assess the accuracy, completeness, and reliability of the data before integration [153, 154]. Data compatibility also involves harmonizing units, measurement scales, and variable definitions across datasets. Automated and semiautomated tools, such as imputation techniques and machine learning algorithms, can help mitigate missing data and variability, but careful curation and domain expertise are often necessary to handle complex integration scenarios [155, 156].

**Ethical and legal considerations**: When integrating biomedical data, particularly data involving human subjects, ethical and legal considerations must be prioritized. The integration of clinical and molecular data often involves sensitive information related to patient health, genomic data, and treatment responses, raising data privacy and patient consent concerns [157, 158].

Regulations such as the HIPAA in the United States and the GDPR in the European Union provide frameworks for safeguarding patient data [158, 159]. Researchers must ensure that data integration efforts comply with these regulations by de-identifying patient information where necessary and obtaining informed consent for data use. Failure to comply with these regulations can result in legal consequences and erode public trust in biomedical research.

Furthermore, ethical considerations extend beyond privacy to issues such as data ownership, access rights, and bias in integrated datasets. For example, biases in data collection or integration may disproportionately affect underrepresented populations, leading to skewed research findings or inequitable health outcomes [160–162, 163]. Ethical frameworks and governance structures, including data access committees, must be established to ensure the responsible use of integrated data.

**Technical and computational challenges**: The sheer volume and complexity of biomedical datasets require advanced computational infrastructures and algorithms for efficient data integration [163]. High-dimensional omics data, for example, can be computationally expensive to process and analyze, especially when combined with large-scale clinical data [133, 164].

Efficient storage, retrieval, and processing of large integrated datasets demand scalable computing resources and optimized algorithms for parallel processing and data mining [1, 165]. Moreover, the need for specialized software tools capable of handling diverse data formats, such as genomic variant files, mass spectrometry outputs, and clinical records, adds another layer of complexity to the data integration process [125, 151]. Developing and maintaining these tools and infrastructures can require time and resources.



### 4.6.3 Methods

Integrating biomedical data from various sources requires a combination of preprocessing, alignment, and harmonization techniques. These methods ensure that datasets, which often differ in structure, format, and scale, can be meaningfully combined to perform analysis [166]. Below are some of the most commonly employed methods in data integration:

#### ▪ Data Preprocessing and Normalization

Before integrating data, preprocessing steps are useful to standardize and prepare the datasets for analysis. This includes normalization, transformation, and handling missing values which we discussed above. Below we briefly recap in this context.

*Normalization*: such as z-score normalization, which rescales data to have a mean of zero and a standard deviation of one, or min-max scaling, which rescales data to a specific range (e.g., 0 to 1), are widely used to ensure that datasets with different measurement units or ranges are comparable [52].

*Transformation*: Data transformation, such as log transformation or Box-Cox transformation, is sometimes necessary to normalize skewed distributions or stabilize variance, ensuring more accurate downstream analyses [66, 67].

Preprocessing also involves handling missing data (using imputation techniques) and correcting for batch effects, which are technical variations introduced by different experimental conditions or sequencing runs. These steps are needed to avoid introducing biases and ensure that integrated datasets can be directly compared.

#### ▪ Statistical and Computational Approaches

Numerous statistical and computational approaches are employed to facilitate the integration of diverse data types. Some of those methods we have discussed and will briefly name below:

*Machine learning algorithms*: Machine learning techniques, such as supervised learning (e.g., random forests, support vector machines) and unsupervised learning (e.g., clustering algorithms), are frequently used for pattern recognition, prediction, and classification in integrated biomedical data [167, 168].

*Dimensionality reduction*: Techniques such as PCA, t-SNE, and UMAP are employed to reduce the dimensionality of large, high-dimensional datasets while preserving important data patterns. These methods make it easier to visualize and analyze integrated datasets [169, 170].

*Network-based approaches*: In biological systems, network-based methods, such as gene regulatory networks, protein-protein interaction networks, or drug-target interaction networks, are used for modeling the complex relationships between integrated datasets [171, 172]. These approaches help to identify key hubs or nodes in the networks that may serve as drug targets or biomarkers.

#### ▪ Ontologies and Data Standards

Ontologies and standardized data formats play a role in ensuring compatibility across datasets. Providing a common framework for representing biomedical data, ontologies enable the integration of information from different sources:

*Ontologies*: Resources like the Gene Ontology (GO) for biological processes, the Human Phenotype Ontology (HPO) for disease phenotypes, and the Disease Ontology (DO) provide standardized vocabularies for describing biomedical entities [173]. These ontologies allow for more consistent annotation and interpretation of data across different datasets.

*Standardized data formats*: The use of standardized formats, such as FASTA for genomic sequences, DICOM (Digital Imaging and Communications in Medicine) for medical imaging data, and HL7 for clinical data, ensures that data from different platforms can be integrated and interpreted in a cohesive manner.

#### ▪ Alignment Techniques

Alignment is critical for ensuring that data from different sources or time points can be meaningfully compared:

*Temporal alignment*: For time-series data, such as longitudinal clinical data or real-time monitoring data, temporal alignment ensures that measurements from different datasets correspond to the same time intervals. This is particularly important in studies tracking disease progression or treatment responses over time [179]. Techniques like interpolation or resampling may be used to synchronize data points recorded at different frequencies [174].

*Spatial alignment*: In imaging data, spatial alignment ensures that data from different imaging modalities (e.g., MRI, CT, PET scans) are registered into a common coordinate system. Image registration techniques use transformations (such as scaling, rotation, and translation) to overlay images from different sources, allowing for accurate comparison and analysis. For instance, in cancer studies, aligning tumor images from different scans enables more precise diagnosis and treatment planning [175].

#### ▪ Formatting for Analysis

Once data have been aligned, they must be formatted appropriately for analysis. This involves:

*Structuring data*: Organizing data into rows and columns with appropriate variable names, units, and labels is required for ensuring compatibility with statistical or machine learning algorithms. Data should be labeled consistently across datasets to avoid errors during analysis.

*Handling missing data*: As discussed above missing data can introduce biases into integrated datasets. Techniques such as multiple imputation or interpolation are often used to estimate missing values based on the existing data.

*Addressing outliers*: Outliers or anomalies in the data can skew results and should be identified and handled appropriately (discussed above). Depending on the context, outliers may be removed, corrected, or retained for further analysis.

#### ▪ Data Harmonization

Data harmonization is the process of aligning and standardizing datasets from heterogeneous sources to facilitate integration and analysis. This involves:

*Data harmonization in health systems*: In healthcare, harmonization ensures that datasets from various sources, such as hospital information systems, patient records, or laboratory data, can be combined coherently [176, 177]. This allows for the pooling of data from multiple studies or institutions, enabling large-scale comparative analyses across populations.

*Use of unique identifiers*: Unique patient or sample identifiers are often employed in data harmonization to link datasets from different sources without revealing personally identifiable information. This allows for secure and compliant integration of sensitive data while preserving the ability to track data across studies [176, 178].

*Harmonization of qualitative data*: While most data harmonization efforts focus on quantitative data, there is growing interest in harmonizing qualitative data, such as patient narratives or clinician notes, across healthcare settings. Techniques for harmonizing qualitative data remain less well developed than those for quantitative data, presenting an area for future research [179].

Applying these methods, researchers can ensure that integrated biomedical datasets are standardized, aligned, and ready for meaningful analysis. Data harmonization, preprocessing, and advanced statistical techniques are useful to overcoming the challenges posed by heterogeneous biomedical data and studying disease mechanisms, drug response, and patient care.

### 4.6.4 Addressing Inconsistencies and Duplicate Data

In biomedical data integration, ensuring the consistency and uniqueness of data is useful for maintaining dataset integrity and improving the accuracy of downstream analyses. Inconsistencies and duplicate data are common challenges, particularly when merging datasets from multiple sources, each with its own collection protocols, standards, and formats.

#### ▪ Identifying and Resolving Inconsistencies

Inconsistencies in datasets often arise from variations in data collection, differing units of measurement, or errors in data entry. For example, clinical data might record a patient's weight in kilograms in one system and pounds in another, or genomic data might use different reference genomes across studies.

To identify these inconsistencies, data profiling tools can be used to systematically analyze the dataset and detect irregularities such as conflicting entries, out-of-range values, or nonstandard formats [180, 181]. Once inconsistencies are identified, they can be addressed through data standardization, which involves converting data into common formats or units, for instance, ensuring that all weight measurements are expressed in kilograms or converting all concentration measurements to either molar or milligrams per milliliter (mg/ml).

In addition to standardization, data validation rules can be implemented to prevent future inconsistencies. These rules ensure that data entries conform to predefined formats, ranges, and logical constraints. For example, setting validation rules can ensure that patient ages fall within biologically reasonable limits, or that concentration values cannot be negative. These checks help maintain data integrity during both data entry and the integration process [182, 183].

#### ■ Handling Duplicate Records

Duplicate data entries are common when integrating datasets from multiple sources, particularly when records are collected for the same subject, sample, or event. Duplicates can introduce bias, inflate sample sizes, and distort results, making it important to identify and address them during the integration process.

Data deduplication tools use algorithms to detect duplicate records based on matching identifiers, such as patient IDs, sample labels, or event timestamps. However, detecting duplicates is not always straightforward, as duplicates might not be exact matches. In such cases, more sophisticated techniques such as fuzzy matching can be employed to identify similar but not identical records (e.g., due to typographical errors or slight variations in recorded data) [184].

Once flagged, duplicates should be reviewed carefully to ensure they are indeed erroneous duplicates rather than valid but similar records. This requires domain-specific knowledge and context-specific analysis to determine the uniqueness of each record. For example, two records might appear similar but could correspond to different individuals, time points, or experimental conditions. Automated tools can assist in the deduplication process, but manual review is often necessary for complex cases where domain expertise is required [185].

#### ■ Consistency Checks Post-Integration

After the integration of data from multiple sources, performing consistency checks is vital to ensure that the process has not introduced errors or inconsistencies into the dataset. These checks help validate the quality and reliability of the integrated data, ensuring that it is fit for further analysis or research [186, 187].

Consistency checks can involve the following:

*Benchmark validation*: Comparing integrated data against known standards or reference datasets to ensure accuracy [188, 189]. For example, if integrating genomic data, the final dataset can be validated against a reference genome to check for alignment or annotation errors.

*Logical consistency*: Ensuring that the data maintains logical coherence, such as avoiding temporal inconsistencies in time-series data (e.g., a patient's treatment recorded after their discharge). Logical checks also include verifying that clinical events follow a plausible order (e.g., diagnosis before treatment) and that no duplicated or missing events appear [190, 191].

*Relational constraints*: For relational databases, it is important to verify that foreign key constraints and relationships between tables (such as between patient records and treatment data) are preserved after integration. This ensures that no data relationships are lost or corrupted during the merging process.

These consistency checks, when combined with validation and deduplication strategies, help ensure that the final integrated dataset is both reliable and ready for analysis. Consistent, clean data allows researchers to confidently draw conclusions from integrated datasets and supports reproducibility in biomedical research.

### 4.6.5 Metadata Management



Metadata or data about data, provides context and descriptive information about the dataset, including details about its source, format, collection methods, and any transformations or processing steps it has undergone. Effective management of this metadata is required for ensuring data integrity, usability, and compliance with scientific and regulatory standards.

#### ▪ Role of Metadata in Data Integration

Metadata plays multiple roles during the data integration process, helping to ensure that integrated datasets are properly understood, traceable, and usable across different research contexts. Below are the key roles that metadata serves:

*Traceability*: Metadata provides a historical record of the data, including its origin (e.g., institution or experiment), collection methods, and any transformations or processing steps it has undergone [192]. This traceability can be useful for verifying the authenticity of the data, tracking any changes made over time, and understanding its journey from raw data collection to final analysis. This is especially important in biomedical research, where the validity of data is paramount.

*Reproducibility*: Metadata facilitates reproducibility by providing detailed descriptions of how data was collected, processed, and transformed [193, 194]. Ensuring that other researchers have access to this contextual information, metadata helps them replicate studies accurately and validate the original findings.

*Data understanding*: Metadata offers information about the limitations, applicability, and context of the data [195]. For instance, understanding how a dataset was generated, whether through next-generation sequencing, mass spectrometry, or patient-reported surveys, allows researchers to assess its reliability, accuracy, and relevance to their particular study. This ensures that data is applied appropriately in different biomedical research contexts, such as drug discovery or epidemiology.

*Data quality assessment*: Metadata can also include information about the quality of the data, such as completeness, accuracy, consistency, and any known limitations or biases in the data collection process. This allows researchers to assess whether a particular dataset is suitable for a given analysis and to identify areas where data quality improvements might be necessary. For example, metadata may note if there were any missing values, technical errors, or variations in the experimental protocol that could impact the data's reliability.

#### ▪ Effective Metadata Management Practices

To ensure that metadata serves its intended purpose in data integration, it is important to adopt effective management practices. These practices help maintain metadata's accuracy, accessibility, and usability, improving its value in the research process [196, 197].

*Standardize metadata*: Standardizing metadata across datasets is useful for ensuring consistency and interoperability. Adopting established metadata standards and frameworks, such as the Minimum Information About a Microarray Experiment

(MIAME) or the Minimum Information About a Proteomics Experiment (MIAPE), makes it easier for researchers to interpret and integrate data across different platforms and studies [198]. Standardized metadata enables easier sharing and collaboration within the scientific community and ensures that data can be reused for future analyses.

*Integrate metadata with data*: Metadata should be linked to the data it describes and made accessible within the same management systems. This can be achieved through integrated data management platforms that store both data and metadata together in a cohesive structure, allowing users to retrieve relevant metadata at the point of data access. For instance, in clinical research, electronic health record (EHR) systems often incorporate metadata about patient demographics, treatments, and outcomes, providing context for the clinical data [199].

*Update metadata*: As data undergoes transformations and updates or is used in new research contexts, the associated metadata should be updated to reflect these changes. This ensures that metadata remains accurate, relevant, and informative throughout the data lifecycle [200, 201]. For example, if data is transformed through normalization or merged with other datasets, the metadata should record these changes to maintain transparency and traceability.

*Ensure accessibility*: Metadata should be made readily accessible to all relevant stakeholders, including researchers, clinicians, and data managers, with appropriate security measures in place to protect sensitive information. For sensitive biomedical data, such as patient information, it is useful to strike a balance between accessibility and security, ensuring that authorized users can access metadata while complying with regulations like the HIPAA and GDPR [202].

## 4.7 Ensuring Data Privacy and Security

Ensuring data privacy and security is a paramount concern in data management. It is not only a legal obligation but also an ethical imperative. Proper data privacy and security are important for maintaining public trust, upholding research integrity, and safeguarding the rights and welfare of individuals whose data are used in research. Failure to adequately address these concerns can result in legal repercussions, loss of credibility, and harm to individuals and communities. Below, we explore the key considerations and methods for ensuring privacy and security in biomedical data integration.

### 4.7.1 Data Privacy Considerations

Data privacy revolves around the responsible handling of sensitive information to protect individuals' rights and ensure confidentiality. In biomedical research, patient data typically contains personal health information (PHI) protected by laws such as the HIPAA in the United States and GDPR in the European Union. Ensuring data privacy involves several key practices:

**De-identification or anonymization**: De-identification involves removing or encrypting personal identifiers (e.g., names, social security numbers, patient IDs) from datasets to protect individual identities. Careful attention is required, as even indirect identifiers (such as birthdates or rare conditions) can sometimes be used to reidentify individuals. Anonymization methods ensure that individuals cannot be reidentified from data, even in combination with other datasets [203, 204].

**Consent and ethical approval**: Obtaining informed consent is important for collecting and using personal data in biomedical research. Ethical approval from institutional review boards (IRBs) or ethics committees ensures that data use aligns with ethical guidelines and complies with legal frameworks [205]. Consent forms must clearly describe how data will be collected, used, and protected [206].

**Access controls**: Limiting access to sensitive data is needed for privacy protection. Role-based access control (RBAC) mechanisms restrict access to authorized personnel, ensuring that only those with a legitimate need can view sensitive data. Logging and monitoring access activity helps track usage and identify any unauthorized access [207, 208].

### 4.7.2 Data Security Measures

Data security is concerned with protecting data from unauthorized access, breaches, and loss. In biomedical research, maintaining the confidentiality, integrity, and availability of data is needed, especially given the sensitive nature of health information [209]. Key data security measures include:

**Encryption**: Encrypting data, both at rest and in transit, is one of the most effective ways to prevent unauthorized access. Advanced encryption methods ensure that data cannot be read or tampered with if intercepted by malicious actors [210]. Common encryption protocols include Advanced Encryption Standard (AES) for data at rest and Transport Layer Security (TLS) for data in transit.

**Regular audits and monitoring**: Conducting regular security audits helps identify vulnerabilities in the system. Monitoring tools can track access attempts, identify unusual activities, and detect breaches in real time, enabling quick responses to threats [211].

**Data backup and recovery plans**: Having robust backup systems and disaster recovery plans is required to prevent data loss from system failures, cyberattacks, or other unforeseen events. Backup solutions should be secure and follow the 3-2-1 rule (three copies of data, two different storage media, one copy off-site) to ensure data can be restored.

**Training and awareness**: Educating researchers and data handlers about the importance of data security is key to preventing accidental breaches or mishandling of sensitive information. Regular training on best practices, including recognizing phishing attacks and securely managing data, reduces human error, which is often a cause of data breaches.

### 4.7.3 Anonymization Techniques

Data anonymization is a process for protecting patient privacy while allowing researchers to derive information from health data. Anonymization techniques vary in their level of protection and utility, with some providing simple de-identification and others advanced privacy guarantees.

Data anonymization involves modifying or removing personal information in a dataset to prevent the identification of individuals. This is especially important in biomedical research, where patient data can contain sensitive health information. Effective anonymization is often a legal requirement under privacy laws like HIPAA and GDPR, as it protects individuals from risks such as discrimination, stigmatization, or identity theft [203, 212].

**De-identification methods**: De-identification involves two primary steps:

*Removal of direct identifiers*: Stripping data of directly identifiable information such as names, social security numbers, or patient IDs.

*Handling of indirect identifiers*: Ensuring that combinations of attributes (such as age, gender, and rare diseases) do not inadvertently reveal identities. This may involve generalization (e.g., replacing exact birthdates with age ranges) or suppression (removing specific attributes that pose identification risks).

*Data masking*: Data masking is a technique where sensitive data is replaced with fictitious, yet realistic data. For example, patient names may be replaced with pseudonyms or random identifiers. This allows data to be used for testing or analysis without exposing real identities [213]. Data masking is commonly used in environments where data integrity must be maintained for research but privacy must be protected.

*K-anonymity*: K-anonymity is a technique that ensures each record in a dataset is indistinguishable from at least $k - 1$ other records in terms of key identifying attributes. This approach prevents re-identification by ensuring that groups of records are indistinguishable from one another [214, 215]. For example, modifying a dataset so that any combination of characteristics (e.g., age, gender) appears in at least five records provides k-anonymity with $k = 5$.

*Differential privacy*: Differential privacy is a more advanced and mathematically rigorous technique that provides strong privacy guarantees. It involves adding controlled noise to either the data or the outputs of data queries, ensuring that individual-specific information cannot be inferred from the dataset [216, 217]. This method allows researchers to perform analysis while minimizing the risk of reidentifying individuals. Differential privacy has become a gold standard in data anonymization, especially when datasets are shared publicly or used for large-scale analysis.

## 4.8 Exercises and Questions

We include exercises and questions to complement the material covered in this chapter.

### Exercises

The following exercises will enhance your understanding of the materials which we covered in this chapter.

1. Clean a provided clinical dataset: Remove duplicate entries and correct typographical errors in patient names and ID numbers.
2. Normalize gene expression levels: Use a provided script to normalize gene expression data obtained from different batches in a laboratory experiment.

3. Standardize measurements: Convert heights from centimeters to meters and weights from pounds to kilograms in a patient dataset, ensuring consistency across records.
4. Impute missing BMI values: Use the mean BMI of the sample to fill in missing BMI values in a dataset of diabetic patients.
5. Identify and remove outliers: Using the interquartile range method, identify and remove outliers in a dataset of cholesterol levels.
6. Encode categorical variables: Convert blood type data from text to numeric codes in a blood donor dataset.
7. Log transform high-range data: Apply log transformation to normalize highly skewed liver enzyme test results in a liver disease study.
8. Feature engineering on time-to-event data: Create a new feature that represents the duration from diagnosis to treatment start date in a cancer patient dataset.
9. Discretize age data: Group ages into categories ("Young," "Adult," "Senior") in a dataset concerning age-related diseases.
10. Transform time-series data: Apply a moving average to smooth out short-term fluctuations in heart rate monitor data from a cardiovascular study.



## Questions to Be Answered

1. What are common sources of errors in biomedical datasets, and how can they be mitigated?
2. How do privacy concerns affect the preprocessing steps in biomedical data handling?
3. What is the impact of outliers on biomedical data analysis, and what methods can be used to detect them?
4. Discuss the advantages and limitations of missing data imputation techniques in clinical trials.
5. How does normalization differ from standardization, and when should each be used in biomedical data?
6. What is the role of log transformation in biomedical data analysis?
7. Why is feature engineering important in predictive modeling with biomedical data?
8. How can categorical data be encoded effectively in genetic studies?
9. Describe the process and significance of data discretization in handling continuous variables.
10. What challenges arise in integrating data from different biomedical studies, and how can they be addressed?

**Acknowledgement** The language of the human-generated text was corrected with the assistance of artificial intelligence (AI) tools [GPT-3.5 and GTP-4 from OpenAI]. GitHub Co-Pilot was used to check the correctness of the codes. The text underwent subsequent human revision to ensure its accuracy.

## References

1. Martinez-Garcia M, Hernandez-Lemus E. Data integration challenges for machine learning in precision medicine. Front Med (Lausanne). 2022;8:784455. https://doi.org/10.3389/fmed.2021.784455.
2. Pendergrass SA, Crawford DC. Using electronic health records to generate phenotypes for research. Curr Protoc Hum Genet. 2019;100(1):e80.
3. Tanjo T, Kawai Y, Tokunaga K, Ogasawara O, Nagasaki M. Practical guide for managing large-scale human genome data in research. J Hum Genet. 2021;66(1):39–52.
4. Florkow MC, Willemsen K, Mascarenhas VV, Oei EHG, van Stralen M, Seevinck PR. Magnetic resonance imaging versus computed tomography for three-dimensional bone imaging of musculoskeletal pathologies: a review. J Magn Reson Imaging. 2022;56(1):11–34.
5. Marino M, Lucas J, Latour E, Heintzman JD. Missing data in primary care research: importance, implications and approaches. Fam Pract. 2021;38(2):200–3.
6. Hogan JW, Roy J, Korkontzelou C. Handling drop-out in longitudinal studies. Stat Med. 2004;23(9):1455–97. https://doi.org/10.1002/sim.1728.
7. Li P, Stuart EA, Allison DB. Multiple imputation: a flexible tool for handling missing data. JAMA. 2015;314(18):1966–7. https://doi.org/10.1001/jama.2015.15281.
8. Younus S, Rönnstrand L, Kazi JU. Xputer: bridging data gaps with NMF, XGBoost, and a streamlined GUI experience. Front Artif Intell. 2024;7:1345179. https://doi.org/10.3389/frai.2024.1345179.
9. Kim MO, Coiera E, Magrabi F. Problems with health information technology and their effects on care delivery and patient outcomes: a systematic review. J Am Med Inform Assoc. 2017;24(2):246–250. https://doi.org/10.1093/jamia/ocw154.
10. Zhan C, Hicks RW, Blanchette CM, Keyes MA, Cousins DD. Potential benefits and problems with computerized prescriber order entry: analysis of a voluntary medication error-reporting database. Am J Health Syst Pharm. 2006;63(4):353–8. https://doi.org/10.2146/ajhp050379.
11. Hughes RA, Heron J, Sterne JAC, Tilling K. Accounting for missing data in statistical analyses: multiple imputation is not always the answer. Int J Epidemiol. 2019;48(4):1294–1304. https://doi.org/10.1093/ije/dyz032.
12. Carpenter JR, Smuk M. Missing data: A statistical framework for practice. Biom J. 2021 Jun;63(5):915–947. https://doi.org/10.1002/bimj.202000196.
13. Graham JW. Missing data analysis: making it work in the real world. Annu Rev Psychol. 2009;60:549–76. https://doi.org/10.1146/annurev.psych.58.110405.085530.
14. Salgado CM, Azevedo C, Proença H, Vieira SM. Noise Versus Outliers. 2016 Sep 10. In: MIT Critical Data, editor. Secondary Analysis of Electronic Health Records [Internet]. Cham (CH): Springer; 2016. Chapter 14. https://doi.org/10.1007/978-3-319-43742-2_14. PMID: 31314268.
15. Yan F, Robert M, Li Y. Statistical methods and common problems in medical or biomedical science research. Int J Physiol Pathophysiol Pharmacol. 2017;9(5):157–63. PMID: 29209453
16. Badrick T. Biological variation: Understanding why it is so important? Pract Lab Med. 2021;23:e00199. https://doi.org/10.1016/j.plabm.2020.e00199.
17. Cook CN, Freeman AR, Liao JC, Mangiamele LA. The philosophy of outliers: reintegrating rare events into biological science. Integr Comp Biol. 2022;61(6):2191–98. https://doi.org/10.1093/icb/icab166.
18. Villerc S. Common data elements repository. Med Ref Serv Q. 2024;43(2):182–90.
19. Matsuzaki K, Kitayama M, Yamamoto K, Aida R, Imai T, Ishida M, Katafuchi R, Kawamura T, Yokoo T, Narita I, Suzuki Y. A pragmatic method to integrate data from preexisting cohort studies using the clinical data interchange standards consortium (cdisc) study data tabulation model: case study. JMIR Med Inform. 2023;11:e46725. https://doi.org/10.2196/46725.
20. Hettinger BJ, Brazile RP. Health Level Seven (HL7): standard for healthcare electronic data transmissions. Comput Nurs. 1994;12(1):13–6. PMID: 8149297.
21. Kasthurirathne SN, Mamlin B, Kumara H, Grieve G, Biondich P. Enabling better interoper ability for healthcare: lessons in developing a standards based application programing interface for electronic medical record systems. J Med Syst. 2015;39(11):182. https://doi.org/10.1007/s10916-015-0356-6.

22. Loshin D. 11 – Data quality assessment. In: Loshin D, editor. The practitioner's guide to data quality improvement. Boston: Morgan Kaufmann; 2011. p. 191–206.
23. Derakhshan P, Azadmanjir Z, Naghdi K, Habibi Arejan R, Safdarian M, Zarei MR, et al. The impact of data quality assurance and control solutions on the completeness, accuracy, and consistency of data in a national spinal cord injury registry of Iran (NSCIR-IR). Spinal Cord Ser Cases. 2021;7(1):51.
24. Schmidt CO, Struckmann S, Enzenbach C, Reineke A, Stausberg J, Damerow S, et al. Facilitating harmonized data quality assessments. A data quality framework for observational health research data collections with software implementations in R. BMC Med Res Methodol. 2021;21(1):63.
25. Ghalavand H, Shirshahi S, Rahimi A, Zarrinabadi Z, Amani F. Common data quality elements for health information systems: a systematic review. BMC Med Inform Decis Mak. 2024;24(1):243. https://doi.org/10.1186/s12911-024-02644-7.
26. Bosu MF, MacDonell SG. Data quality in empirical software engineering: a targeted review. In: Proceedings of the 17th international conference on evaluation and assessment in software engineering. Porto de Galinhas, Brazil: Association for Computing Machinery; 2013. p. 171–6.
27. Hassenstein MJ, Vanella P. Data quality—concepts and problems. Encyclopedia. 2022;2(1):498–510.
28. Sebastian-Coleman L. Chapter 4 – Data quality and measurement. In: Sebastian-Coleman L, editor. Measuring data quality for ongoing improvement. Boston: Morgan Kaufmann; 2013. p. 39–53.
29. Little RJA. A test of missing completely at random for multivariate data with missing values. J Am Stat Assoc. 1988;83:1198–202.
30. Mousafi Alasal L, Hammarlund EU, Pienta KJ, Rönnstrand L, Kazi JU. XeroGraph: enhancing data integrity in the presence of missing values with statistical and predictive analysis. Bioinform Adv. 2025;5(1):vbaf035. https://doi.org/10.1093/bioadv/vbaf035.
31. Lee KJ, Carlin JB, Simpson JA, Moreno-Betancur M. Assumptions and analysis planning in studies with missing data in multiple variables: moving beyond the MCAR/MAR/MNAR classification. Int J Epidemiol. 2023;52(4):1268–75. https://doi.org/10.1093/ije/dyad008.
32. Little RJ. Regression with missing X's: a review. J Am Stat Assoc. 1992;87(420):1227–37
33. Ross RK, Breskin A, Westreich D. When is a complete-case approach to missing data valid? The importance of effect-measure modification. Am J Epidemiol. 2020;189(12):1583–89. https://doi.org/10.1093/aje/kwaa124.
34. Kim JO, Curry J. The treatment of missing data in multivariate analysis. Sociol Methods Res. 1977;6:215–241
35. Mathur MB. The M-value: a simple sensitivity analysis for bias due to missing data in treatment effect estimates. Am J Epidemiol. 2023;192(4):612–20. https://doi.org/10.1093/aje/kwac207.
36. Zhang Z. Missing data imputation: focusing on single imputation. Ann Transl Med. 2016;4(1):9. https://doi.org/10.3978/j.issn.2305-5839.2015.12.38.
37. Lachin JM. Fallacies of last observation carried forward analyses. Clin Trials. 2016;13(2):161–8. https://doi.org/10.1177/1740774515602688.
38. Andridge RR, Little RJ. A review of hot deck imputation for survey non-response. Int Stat Rev. 2010;78(1):40–64. https://doi.org/10.1111/j.1751-5823.2010.00103.x.
39. Solomon N, Lokhnygina Y, Halabi S. Comparison of regression imputation methods of baseline covariates that predict survival outcomes. J Clin Transl Sci. 2020;5(1):e40. https://doi.org/10.1017/cts.2020.533.
40. Wallace ML, Anderson SJ, Mazumdar S. A stochastic multiple imputation algorithm for missing covariate data in tree-structured survival analysis. Stat Med. 2010;29(29):3004–16. https://doi.org/10.1002/sim.4079.
41. Pedregosa F, Varoquaux G, Gramfort A, Michel V, Thirion B, Grisel O, et al. Scikit-learn: machine learning in python. J Mach Learn Res. 2011;12:2825–30.
42. van Buuren S, Groothuis-Oudshoorn K. mice: multivariate imputation by chained equations in R. J Stat Softw. 2011;45(3):1–67.
43. Baker SG. Maximum likelihood estimation with missing outcomes: from simplicity to complexity. Stat Med. 2019;38(22):4453–4474. https://doi.org/10.1002/sim.8319.
44. Murti DMP, Pujianto U, Wibawa AP, Akbar MI, K-Nearest Neighbor (K-NN) based Missing Data Imputation, 2019 5th International Conference on Science in Information Technology (ICSITech), Yogyakarta, Indonesia, 2019, pp. 83–88, https://doi.org/10.1109/ICSITech46713.2019.8987530.

45. Tang F, Ishwaran H. Random forest missing data algorithms. Stat Anal Data Min. 2017;10(6):363–77. https://doi.org/10.1002/sam.11348.
46. Shah AD, Bartlett JW, Carpenter J, Nicholas O, Hemingway H. Comparison of random forest and parametric imputation models for imputing missing data using mice: a caliber study. Am J Epidemiol. 2014;179(6):764–74. https://doi.org/10.1093/aje/kwt312.
47. Liu M, Li S, Yuan H, Ong MEH, Ning Y, Xie F, Saffari SE, Shang Y, Volovici V, Chakraborty B, Liu N. Handling missing values in healthcare data: a systematic review of deep learning-based imputation techniques. Artif Intell Med. 2023;142:102587. https://doi.org/10.1016/j.artmed.2023.102587.
48. Groenwold RHH, Dekkers OM. Missing data: the impact of what is not there. Eur J Endocrinol. 2020;183(4):E7–9.
49. Ayilara OF, Zhang L, Sajobi TT, Sawatzky R, Bohm E, Lix LM. Impact of missing data on bias and precision when estimating change in patient-reported outcomes from a clinical registry. Health Qual Life Outcomes. 2019;17(1):106.
50. Huang L, Wang C, Rosenberg NA. The relationship between imputation error and statistical power in genetic association studies in diverse populations. Am J Hum Genet. 2009;85(5):692–8. https://doi.org/10.1016/j.ajhg.2009.09.017.
51. Nisbet R, Miner G, Yale K. Chapter 4 - Data understanding and preparation, handbook of statistical analysis and data mining applications. 2nd ed. Academic Press; 2018. p. 55–82, https://doi.org/10.1016/B978-0-12-416632-5.00004-9.
52. Federico A, Serra A, Ha MK, Kohonen P, Choi JS, Liampa I, et al. Transcriptomics in toxicogenomics, part II: preprocessing and differential expression analysis for high quality data. Nanomaterials (Basel). 2020;10(5):903.
53. Wongoutong C. The impact of neglecting feature scaling in k-means clustering. PLoS One. 2024;19(12):e0310839. https://doi.org/10.1371/journal.pone.0310839.
54. Cao XH, Stojkovic I, Obradovic Z. A robust data scaling algorithm to improve classification accuracies in biomedical data. BMC Bioinform. 2016;17(1):359. https://doi.org/10.1186/s12859-016-1236-x.
55. SciKit-Learn. “Sklearn.preprocessing.MinMaxScaler — Scikit-Learn 0.22.1 Documentation.” Scikit-Learn.org, 2019, scikit-learn.org/stable/modules/generated/sklearn.preprocessing.MinMaxScaler.html.
56. Scikit-Learn. “Sklearn.preprocessing.MaxAbsScaler.” scikit-learn.org/stable/modules/generated/sklearn.preprocessing.MaxAbsScaler.html.
57. Scikit-learn. “StandardScaler.” Scikit-Learn.org, 2019, scikit-learn.org/stable/modules/generated/sklearn.preprocessing.StandardScaler.html.
58. Raju VNG, Lakshmi KP, Jain VM, Kalidindi A, Padma V. Study the influence of normalization/transformation process on the accuracy of supervised classification. 2020 Third International conference on Smart Systems and Inventive Technology (ICSSIT), Tirunelveli, India; 2020. p. 729–35. https://doi.org/10.1109/ICSSIT48917.2020.9214160.
59. “Sklearn.preprocessing.RobustScaler — Scikit-Learn 0.24.2 Documentation.” Scikit-Learn.org, scikit-learn.org/stable/modules/generated/sklearn.preprocessing.RobustScaler.html.
60. Hermes M, Wolters T, Schult N, Hein A. Pre-processing of categorical features within medical analysis systems. Stud Health Technol Inform. 2024;316:741–45. https://doi.org/10.3233/SHTI240520.
61. Scikit-learn. “Sklearn.preprocessing.OneHotEncoder — Scikit-Learn 0.22 Documentation.” Scikit-Learn.org, 2019, scikit-learn.org/stable/modules/generated/sklearn.preprocessing.OneHotEncoder.html
62. Scikit-learn. “Sklearn.preprocessing.LabelEncoder — Scikit-Learn 0.22.1 Documentation.” Scikit-Learn.org, 2019, scikit-learn.org/stable/modules/generated/sklearn.preprocessing.LabelEncoder.html
63. Scikit-learn. “Sklearn.preprocessing.OrdinalEncoder.” scikit-learn.org/stable/modules/generated/sklearn.preprocessing.OrdinalEncoder.html
64. Pargent, F., Pfisterer, F., Thomas, J. et al. Regularized target encoding outperforms traditional methods in supervised machine learning with high cardinality features. Comput Stat. 2022;37:2671–92. https://doi.org/10.1007/s00180-022-01207-6.
65. Altman DG, Bland JM. Statistics notes: the normal distribution. BMJ. 1995;310(6975):298. https://doi.org/10.1136/bmj.310.6975.298.

66. Bland JM, Altman DG. Statistics notes. Logarithms. BMJ. 1996;312(7032):700. https://doi.org/10.1136/bmj.312.7032.700.
67. Blum L, Elgendi M, Menon C. Impact of box-cox transformation on machine-learning algorithms. Front Artif Intell. 2022;5:877569. https://doi.org/10.3389/frai.2022.877569.
68. Seiffert S, Weber S, Sack U, Keller T. Use of logit transformation within statistical analyses of experimental results obtained as proportions: example of method validation experiments and EQA in flow cytometry. Front Mol Biosci. 2024;11:1335174. https://doi.org/10.3389/fmolb.2024.1335174.
69. Shanableh T, Assaleh K., Feature modeling using polynomial classifiers and stepwise regression, Neurocomputing. 2010;73(10–12):1752–1759, https://doi.org/10.1016/j.neucom.2009.11.045.
70. Scikit-learn, "Sklearn.preprocessing.PolynomialFeatures — Scikit-Learn 0.23.2 Documentation." Scikit-Learn.org, scikit-learn.org/stable/modules/generated/sklearn.preprocessing.PolynomialFeatures.html.
71. Huang HN, Chen HM, Lin WW, Huang CJ, Chen YC, Wang YH, Yang CT. Employing feature engineering strategies to improve the performance of machine learning algorithms on echocardiogram dataset. Digit Health. 2023;9:20552076231207589. https://doi.org/10.1177/20552076231207589.
72. Karaca-Mandic P, Norton EC, Dowd B. Interaction terms in nonlinear models. Health Serv Res. 2012;47(1 Pt 1):255–74. https://doi.org/10.1111/j.1475-6773.2011.01314.x.
73. Huang H, Wang Y, Rudin C, Browne EP. Towards a comprehensive evaluation of dimension reduction methods for transcriptomic data visualization. Commun Biol. 2022;5(1):719. https://doi.org/10.1038/s42003-022-03628-x.
74. Greenacre M, Groenen PJF., Hastie T. et al. Principal component analysis. Nat Rev Methods Primers. 2022;2:100. https://doi.org/10.1038/s43586-022-00184-w
75. Klema V. and Laub A. The singular value decomposition: its computation and some applications. IEEE Trans Automat Contr. 1980;25(2):164–76. https://doi.org/10.1109/TAC.1980.1102314.
76. Mehta R and Rana K. A review on matrix factorization techniques in recommender systems. In: 2nd International Conference on Communication Systems, Computing and IT Applications (CSCITA), Mumbai, India; 2017. p. 269–74, https://doi.org/10.1109/CSCITA.2017.8066567.
77. Li SJ, Pang H, Li PY, Li YN, Liu ZX. Image compression based on SVD algorithm. In: International Conference on Computer Information Science and Artificial Intelligence (CISAI), Kunming, China; 2021. p. 306–9. https://doi.org/10.1109/CISAI54367.2021.00065.
78. Cieslak MC, Castelfranco AM, Roncalli V, Lenz PH, Hartline DK. t-Distributed Stochastic Neighbor Embedding (t-SNE): a tool for eco-physiological transcriptomic analysis. Mar Genomics. 2020;51:100723.
79. van der Maaten L, Hinton G. Visualizing data using t-SNE. J Mach Learn Res. 2008;9(86):2579–605.
80. McInnes L, Healy J, Melville J. UMAP: uniform manifold approximation and projection for dimension reduction. arXiv. 2020;1802.03426.
81. Earnest A, Evans SM, Sampurno F, Millar J. Forecasting annual incidence and mortality rate for prostate cancer in Australia until 2022 using autoregressive integrated moving average (ARIMA) models. BMJ Open. 2019;9(8):e031331.
82. Lukić V, Ignjatović S. Optimizing moving average control procedures for small-volume laboratories: can it be done? Biochem Med (Zagreb). 2019;29(3):030710. https://doi.org/10.11613/BM.2019.030710.
83. Zbezhkhovska U, Chumachenko D. Smoothing techniques for improving COVID-19 time series forecasting across countries. Computation. 2025;13(6):136. https://doi.org/10.3390/computation13060136.
84. Safdar MF, Nowak RM, Palka P. Pre-processing techniques and artificial intelligence algorithms for electrocardiogram (ECG) signals analysis: a comprehensive review. Comput Biol Med. 2024;170:107908.
85. Pukhova V, Gorelova E, Ferrini G, Burnasheva S, Time-frequency representation of signals by wavelet transform. In: 2017 IEEE conference of Russian Young Researchers in Electrical and Electronic Engineering (EIConRus). St. Petersburg and Moscow, Russia; 2017. p. 715–8, https://doi.org/10.1109/EIConRus.2017.7910658.

86. Crowe JA, Gibson NM, Woolfson MS, Somekh MG. Wavelet transform as a potential tool for ECG analysis and compression. J Biomed Eng. 1992;14(3):268–72. https://doi.org/10.1016/0141-5425(92)90063-q.
87. Li N, He F, Ma W, Wang R, Jiang L, Zhang X. The identification of ECG signals using wavelet transform and WOA-PNN. Sensors (Basel). 2022;22(12):4343. https://doi.org/10.3390/s22124343.
88. Li C, Zheng C, Tai C. Detection of ECG characteristic points using wavelet transforms. IEEE Trans Biomed Eng. 1995;42(1):21–8. https://doi.org/10.1109/10.362922.
89. Bullmore E, Fadili J, Maxim V, Sendur L, Whitcher B, Suckling J, Brammer M, Breakspear M. Wavelets and functional magnetic resonance imaging of the human brain. Neuroimage. 2004;23 (Suppl 1):S234–49. https://doi.org/10.1016/j.neuroimage.2004.07.012.
90. Guo X, Liu X, Wang H, Liang Z, Wu W, He Q, Li K, Wang W. Enhanced CT images by the wavelet transform improving diagnostic accuracy of chest nodules. J Digit Imaging. 2011;24(1):44–9. https://doi.org/10.1007/s10278-009-9248-y.
91. Rizi FY, Noubari HA, Setarehdan SK. Wavelet-based ultrasound image denoising: performance analysis and comparison. Annu Int Conf IEEE Eng Med Biol Soc. 2011;2011:3917–20. https://doi.org/10.1109/IEMBS.2011.6090973.
92. Pichot V, Gaspoz JM, Molliex S, Antoniadis A, Busso T, Roche F, Costes F, Quintin L, Lacour JR, Barthélémy JC. Wavelet transform to quantify heart rate variability and to assess its instantaneous changes. J Appl Physiol (1985). 1999;86(3):1081–91. https://doi.org/10.1152/jappl.1999.86.3.1081.
93. Hu X, Yoo I, Zhang X, Nanavati P, Debjit Das. Wavelet transformation and cluster ensemble for gene expression analysis. Int J Bioinform Res Appl. 2005;1(4):447–60. https://doi.org/10.1504/IJBRA.2005.008447.
94. McCue C. Chapter 5 - Data, data mining and predictive analysis. 2nd ed. Butterworth-Heinemann; 2015. p. 75–106. https://doi.org/10.1016/B978-0-12-800229-2.00005-5.
95. Gerke S, Minssen T, Cohen G. Ethical and legal challenges of artificial intelligence-driven healthcare. Artificial Intelligence in Healthcare. 2020:295–336. https://doi.org/10.1016/B978-0-12-818438-7.00012-5.
96. Muhr D, Affenzeller M. Little data is often enough for distance-based outlier detection. Procedia Comput Sci. 2022;200:984–92. https://doi.org/10.1016/j.procs.2022.01.297.
97. Smiti A. A critical overview of outlier detection methods. Comput Sci Rev. 2020;38:100306. https://doi.org/10.1016/j.cosrev.2020.100306.
98. Wang H, Bah MJ, Hammad M. Progress in outlier detection techniques: a survey. IEEE Access. 2019;7:107964–8000. https://doi.org/10.1109/ACCESS.2019.2932769.
99. Yaro AS, Maly F, Prazak P and Malý K. Outlier detection performance of a modified Z-score method in time-series RSS observation with hybrid scale estimators. IEEE Access. 2024;12:12785–96. https://doi.org/10.1109/ACCESS.2024.3356731.
100. Rousseeuw PJ, Croux C. Alternatives to the median absolute deviation. J Am Stat Assoc. 1993;88(424):1273–83.
101. Knorr EM, Ng TR. Algorithms for mining distance-based outliers in large datasets. In: Very large data bases conference; 1998. p. 392–403.
102. Dash CSK, Behera AK, Dehuri S, Ghosh A. An outliers detection and elimination framework in classification task of data mining. Decis Anal J. 2023;6:100164. https://doi.org/10.1016/j.dajour.2023.100164.
103. Aslam M. Introducing Grubbs's test for detecting outliers under neutrosophic statistics – an application to medical data. J King Saud Univ Sci. 2020;32(6):2696–700. https://doi.org/10.1016/j.jksus.2020.06.003.
104. Grubbs FE. Procedures for detecting outlying observations in samples. Technometrics. 1969;11(1):1–21.
105. Jain RB. A recursive version of Grubbs' test for detecting multiple outliers in environmental and chemical data. Clin Biochem. 2010;43(12):1030–33, https://doi.org/10.1016/j.clinbiochem.2010.04.071.
106. Yang J, Tan X, Rahardja S. Outlier detection: how to select k for k-nearest-neighbors-based outlier detectors. Pattern Recognit Lett. 2023;174:112–17. https://doi.org/10.1016/j.patrec.2023.08.020.

107. Astivia OLO. A method to simulate multivariate outliers with known mahalanobis distances for normal and non-normal data. Methods Psychol. 2024;11:100157. https://doi.org/10.1016/j.metip.2024.100157.
108. Kim HY. Statistical notes for clinical researchers: simple linear regression 3 - residual analysis. Restor Dent Endod. 2019;44(1):e11. https://doi.org/10.5395/rde.2019.44.e11. PMID: 30834233.
109. Yepmo V, Smits G, Lesot MJ, Pivert O. Leveraging an isolation forest to anomaly detection and data clustering, Data Knowl Eng. 2024;151:102302. https://doi.org/10.1016/j.datak.2024.102302.
110. Yang J, Deng T, Sui R. An adaptive weighted one-class SVM for robust outlier detection. In: Jia Y, Du J, Li H, Zhang W, editors. Proceedings of the 2015 Chinese intelligent systems conference. Lecture Notes in Electrical Engineering. Berlin, Heidelberg: Springer; 2016. https://doi.org/10.1007/978-3-662-48386-2_49.
111. Thang TM. Kim J. The anomaly detection by using DBSCAN clustering with multiple parameters. In: International conference on information science and applications. Jeju, Korea (South); 2011. p. 1–5. https://doi.org/10.1109/ICISA.2011.5772437.
112. Gan G. A k-means algorithm with automatic outlier detection. Electronics. 2025;14(9):1723. https://doi.org/10.3390/electronics14091723.
113. Shrifan NHMM, Akbar MF, Isa NAM. An adaptive outlier removal aided k-means clustering algorithm, J King Saud Univ Comput Inform Sci. 2022;34(8), Part B:6365–76. https://doi.org/10.1016/j.jksuci.2021.07.003.
114. Alghushairy O, Alsini R, Soule T, Ma X. A review of local outlier factor algorithms for outlier detection in big data streams. Big Data Cogn Comput. 2021; 5(1):1. https://doi.org/10.3390/bdcc5010001.
115. Aggarwal CC. An introduction to outlier analysis. In: Outlier analysis. Cham: Springer; 2017. https://doi.org/10.1007/978-3-319-47578-3_1.
116. Hodge V, Austin J. A survey of outlier detection methodologies. Artif Intell Rev. 2004;22:85–126. https://doi.org/10.1023/B:AIRE.0000045502.10941.a9.
117. Aguinis H, Gottfredson R K, Joo H. Best-Practice Recommendations for Defining, Identifying, and Handling Outliers. Organ Res Methods. 2013;16(2):270–301. https://doi.org/10.1177/1094428112470848.
118. West RM. Best practice in statistics: the use of log transformation. Ann Clin Biochem. 2022;59(3):162–5. https://doi.org/10.1177/00045632211050531.
119. Riffenburgh RH., Chapter 28 - Methods you might meet, but not every day, Statistics in Medicine. 3rd ed. Academic Press; 2012. p. 581–91. https://doi.org/10.1016/B978-0-12-384864-2.00028-7.
120. de Amorim LVB, Cavalcanti GDC, Cruz RMO. The choice of scaling technique matters for classification performance, Appl Soft Comput. 2023;133:109924. https://doi.org/10.1016/j.asoc.2022.109924.
121. Janoudi G, Uzun Rada M, Fell DB, Ray JG, Foster AM, Giffen R, Clifford T, Walker MC. Outlier analysis for accelerating clinical discovery: an augmented intelligence framework and a systematic review. PLOS Digit Health. 2024;3(5):e0000515. https://doi.org/10.1371/journal.pdig.0000515.
122. Wu Y, Curhan S, Rosner B et al. Analytical method for detecting outlier evaluators. BMC Med Res Methodol. 2023;23:177. https://doi.org/10.1186/s12874-023-01988-4.
123. Mogoş B. Exploratory data analysis for outlier detection in bioequivalence studies. Biocybern Biomed Eng. 2013;33(3):164–70. https://doi.org/10.1016/j.bbe.2013.07.005.
124. Osborne JW, Overbay A. The power of outliers (and why researchers should ALWAYS check for them). Pract Assess Res Eval. 2004;9(1):6. https://doi.org/10.7275/qf69-7k43.
125. Aggarwal CC. Applications of outlier analysis. In: Outlier analysis. Cham: Springer; 2017. https://doi.org/10.1007/978-3-319-47578-3_13.
126. Hasin Y, Seldin M, Lusis A. Multi-omics approaches to disease. Genome Biol. 2017;18:83. https://doi.org/10.1186/s13059-017-1215-1
127. Manzoni C, Kia DA, Vandrovcova J, Hardy J, Wood NW, Lewis PA, Ferrari R. Genome, transcriptome and proteome: the rise of omics data and their integration in biomedical sciences. Brief Bioinform. 2018;19(2):286–302. https://doi.org/10.1093/bib/bbw114.
128. Kang M, Ko E, Mersha TB. A roadmap for multi-omics data integration using deep learning. Brief Bioinform. 2022;23(1):bbab454. https://doi.org/10.1093/bib/bbab454.

129. Ivanisevic T, Sewduth RN. Multi-omics integration for the design of novel therapies and the identification of novel biomarkers. Proteomes. 2023;11(4):34. https://doi.org/10.3390/proteomes11040034.
130. Fillinger S, de la Garza L, Peltzer A, Kohlbacher O, Nahnsen S. Challenges of big data integration in the life sciences. Anal Bioanal Chem. 2019;411(26):6791–800. https://doi.org/10.1007/s00216-019-02074-9.
131. Hernández-Lemus E, Ochoa S. Methods for multi-omic data integration in cancer research. Front Genet. 2024;15:1425456. https://doi.org/10.3389/fgene.2024.1425456.
132. Zhang H, Guo Y, Li Q, et al. An ontology-guided semantic data integration framework to support integrative data analysis of cancer survival. BMC Med Inform Decis Mak. 2018; 18(Suppl 2):41. https://doi.org/10.1186/s12911-018-0636-4.
133. Phan JH, Quo CF, Cheng C, Wang MD. Multiscale integration of -omic, imaging, and clinical data in biomedical informatics. IEEE Rev Biomed Eng. 2012;5:74–87. https://doi.org/10.1109/RBME.2012.2212427.
134. Zhang H, Lyu T, Yin P, Bost S, He X, Guo Y, Prosperi M, Hogan WR, Bian J. A scoping review of semantic integration of health data and information. Int J Med Inform. 2022;165:104834. https://doi.org/10.1016/j.ijmedinf.2022.104834.
135. Pammi M, Aghaeepour N, Neu J. Multiomics, artificial intelligence, and precision medicine in perinatology. Pediatr Res. 2023;93(2):308–15. https://doi.org/10.1038/s41390-022-02181-x.
136. Alemu R, Sharew NT, Arsano YY, Ahmed M, Tekola-Ayele F, Mersha TB, Amare AT. Multi-omics approaches for understanding gene-environment interactions in noncommunicable diseases: techniques, translation, and equity issues. Hum Genomics. 2025;19(1):8. https://doi.org/10.1186/s40246-025-00718-9.
137. Vangimalla RR, Sreevalsan-Nair J. HCNM: heterogeneous correlation network model for multi-level integrative study of multi-omics data for cancer subtype prediction. Annu Int Conf IEEE Eng Med Biol Soc. 2021;2021:1880–6. https://doi.org/10.1109/EMBC46164.2021.9630781.
138. Ahmed Z. Precision medicine with multi-omics strategies, deep phenotyping, and predictive analysis. Prog Mol Biol Transl Sci. 2022;190(1):101–25. https://doi.org/10.1016/bs.pmbts.2022.02.002.
139. Gliozzo J, Mesiti M, Notaro M, Petrini A, Patak A, Puertas-Gallardo A, Paccanaro A, Valentini G, Casiraghi E. Heterogeneous data integration methods for patient similarity networks. Brief Bioinform. 2022;23(4):bbac207. https://doi.org/10.1093/bib/bbac207.
140. Yu M, Xu J, Dutta R, Trapp B, Pieper AA, Cheng F. Network medicine informed multiomics integration identifies drug targets and repurposable medicines for Amyotrophic Lateral Sclerosis. NPJ Syst Biol Appl. 2024;10(1):128. https://doi.org/10.1038/s41540-024-00449-y.
141. Zhang Z, Li S, Dai X, Li C, Sun P, Qu J, Jiang H, Pan B. Multi-omic data integration reveals drug targets of skin fibrosis. Curr Med Chem. 2025. https://doi.org/10.2174/0109298673379521250630140948.
142. Nam Y, Kim J, Jung SH, Woerner J, Suh EH, Lee DG, Shivakumar M, Lee ME, Kim D. Harnessing artificial intelligence in multimodal omics data integration: paving the path for the next frontier in precision medicine. Annu Rev Biomed Data Sci. 2024;7(1):225–50. https://doi.org/10.1146/annurev-biodatasci-102523-103801.
143. Zhang Z, Zhai M, Bao S, Sun X, Chen R, Wang B, Yang F, Yang L, Zhou M. Integrative multi-omics profiling deciphers tumor microenvironment heterogeneity and immunotherapy vulnerabilities in lung neuroendocrine carcinomas. J Adv Res. 2025:S2090-1232(25)00427-8. https://doi.org/10.1016/j.jare.2025.06.017.
144. Wang ZZ, Li XH, Wen XL, Wang N, Guo Y, Zhu X, Fu SH, Xiong FF, Bai J, Gao XL, Wang HJ. Integration of multi-omics data reveals a novel hybrid breast cancer subtype and its biomarkers. Front Oncol. 2023;13:1130092. https://doi.org/10.3389/fonc.2023.1130092.
145. Kong L, Liu P, Zheng M, Xue B, Liang K, Tan X. Multi-omics analysis based on integrated genomics, epigenomics and transcriptomics in pancreatic cancer. Epigenomics. 2020;12(6):507–24. https://doi.org/10.2217/epi-2019-0374.
146. Chang Q, Yan Z, Zhou M, et al. Mining multi-center heterogeneous medical data with distributed synthetic learning. Nat Commun. 2023;14:5510. https://doi.org/10.1038/s41467-023-40687-y.
147. Tayefi M, Ngo P, Chomutare T, et al. Challenges and opportunities beyond structured data in analysis of electronic health records. WIREs Comput Stat. 2021;13:e1549. https://doi.org/10.1002/wics.1549.

148. Knickerbocker JU, et al. Heterogeneous integration technology demonstrations for future healthcare, IoT, and AI computing solutions. In: IEEE 68th Electronic Components and Technology Conference (ECTC). San Diego, CA, USA; 2018. p. 1519-28, https://doi.org/10.1109/ECTC.2018.00231.
149. Torab-Miandoab A, Samad-Soltani T, Jodati A, Rezaei-Hachesu P. Interoperability of heterogeneous health information systems: a systematic literature review. BMC Med Inform Decis Mak. 2023;23(1):18. https://doi.org/10.1186/s12911-023-02115-5.
150. Torres-Martos Á, Bustos-Aibar M, Ramírez-Mena A, Cámara-Sánchez S, Anguita-Ruiz A, Alcalá R, Aguilera CM, Alcalá-Fdez J. Omics data preprocessing for machine learning: a case study in childhood obesity. Genes. 2023;14(2):248. https://doi.org/10.3390/genes14020248.
151. Chicco D, Cumbo F, Angione C. Ten quick tips for avoiding pitfalls in multi-omics data integration analyses. PLoS Comput Biol. 2023;19(7):e1011224. https://doi.org/10.1371/journal.pcbi.1011224.
152. Athieniti E, Spyrou GM. A guide to multi-omics data collection and integration for translational medicine. Comput Struct Biotechnol J. 2022;21:134–49. https://doi.org/10.1016/j.csbj.2022.11.050.
153. Tripathi A, Waqas A, Venkatesan K, Yilmaz Y, Rasool G. Building flexible, scalable, and machine learning-ready multimodal oncology datasets. Sensors. 2024;24(5):1634. https://doi.org/10.3390/s24051634.
154. Hui HWH, Kong W, Goh WWB. Thinking points for effective batch correction on biomedical data. Brief Bioinform. 2024;25(6):bbae515. https://doi.org/10.1093/bib/bbae515.
155. Čuklina J, Lee CH, Williams EG, Sajic T, Collins BC, Rodríguez Martínez M, Sharma VS, Wendt F, Goetze S, Keele GR, Wollscheid B, Aebersold R, Pedrioli PGA. Diagnostics and correction of batch effects in large-scale proteomic studies: a tutorial. Mol Syst Biol. 2021;17(8):e10240. https://doi.org/10.15252/msb.202110240.
156. Baião AR, Cai Z, Poulos RC, Robinson PJ, Reddel RR, Zhong Q, Vinga S, Gonçalves E. A technical review of multi-omics data integration methods: from classical statistical to deep generative approaches. Brief Bioinform. 2025;26(4):bbaf355. https://doi.org/10.1093/bib/bbaf355.
157. Murdoch, B. Privacy and artificial intelligence: challenges for protecting health information in a new era. BMC Med Ethics. 2022;22:122. https://doi.org/10.1186/s12910-021-00687-3
158. Pham T. Ethical and legal considerations in healthcare AI: innovation and policy for safe and fair use. R Soc Open Sci. 2025;12(5):241873. https://doi.org/10.1098/rsos.241873.
159. Ueda D, Kakinuma T, Fujita S, et al. Fairness of artificial intelligence in healthcare: review and recommendations. Jpn J Radiol. 2024;42:3–15. https://doi.org/10.1007/s11604-023-01474-3.
160. Lvovs D, Creason AL, Levine SS, Noble M, Mahurkar A, White O, Fertig EJ. Balancing ethical data sharing and open science for reproducible research in biomedical data science. Cell Rep Med. 2025;6(4):102080. https://doi.org/10.1016/j.xcrm.2025.102080.
161. Norori N, Hu Q, Aellen FM, Faraci FD, Tzovara A. Addressing bias in big data and AI for health care: A call for open science. Patterns (N Y). 2021;2(10):100347. https://doi.org/10.1016/j.patter.2021.100347.
162. Bellazzi R. Big data and biomedical informatics: a challenging opportunity. Yearb Med Inform. 2014;9(1):8–13. https://doi.org/10.15265/IY-2014-0024.
163. Rahnenführer J, De Bin R, Benner A, et al. Statistical analysis of high-dimensional biomedical data: a gentle introduction to analytical goals, common approaches and challenges. BMC Med. 2023;21:182. https://doi.org/10.1186/s12916-023-02858-y.
164. Zou Q, Mrozek D, Ma Q, Xu Y. Scalable Data Mining Algorithms in Computational Biology and Biomedicine. Biomed Res Int. 2017;2017:5652041. https://doi.org/10.1155/2017/5652041.
165. Gomez-Cabrero D, Abugessaisa I, Maier D, Teschendorff A, Merkenschlager M, Gisel A, et al. Data integration in the era of omics: current and future challenges. BMC Syst Biol. 2014;8(2):I1. https://doi.org/10.1186/1752-0509-8-S2-I1.
166. Schmidt B-M, Colvin CJ, Hohlfeld A, Leon N. Defining and conceptualising data harmonisation: a scoping review protocol. Syst Rev. 2018;7(1):226.
167. Zitnik M, Nguyen F, Wang B, Leskovec J, Goldenberg A, Hoffman MM. Machine learning for integrating data in biology and medicine: principles, practice, and opportunities. Inf Fusion. 2019;50:71–91. https://doi.org/10.1016/j.inffus.2018.09.012.
168. Mirza B, Wang W, Wang J, Choi H, Chung NC, Ping P. Machine learning and integrative analysis of biomedical big data. Genes (Basel). 2019;10(2):87. https://doi.org/10.3390/genes10020087.

169. Garma LD, Osório NS. Demystifying dimensionality reduction techniques in the 'omics' era: A practical approach for biological science students. Biochem Mol Biol Educ. 2024;52(2):165–78. https://doi.org/10.1002/bmb.21800.
170. Sakaue S, Hirata J, Kanai M, Suzuki K, Akiyama M, Lai Too C, Arayssi T, Hammoudeh M, Al Emadi S, Masri BK, Halabi H, Badsha H, Uthman IW, Saxena R, Padyukov L, Hirata M, Matsuda K, Murakami Y, Kamatani Y, Okada Y. Dimensionality reduction reveals fine-scale structure in the Japanese population with consequences for polygenic risk prediction. Nat Commun. 2020;11(1):1569. https://doi.org/10.1038/s41467-020-15194-z.
171. Agamah FE, Bayjanov JR, Niehues A, Njoku KF, Skelton M, Mazandu GK, Ederveen THA, Mulder N, Chimusa ER, 't Hoen PAC. Computational approaches for network-based integrative multi-omics analysis. Front Mol Biosci. 2022;9:967205. https://doi.org/10.3389/fmolb.2022.967205.
172. Tomazou M, Bourdakou MM, Minadakis G, Zachariou M, Oulas A, Karatzas E, Loizidou EM, Kakouri AC, Christodoulou CC, Savva K, Zanti M, Onisiforou A, Afxenti S, Richter J, Christodoulou CG, Kyprianou T, Kolios G, Dietis N, Spyrou GM. Multi-omics data integration and network-based analysis drives a multiplex drug repurposing approach to a shortlist of candidate drugs against COVID-19. Brief Bioinform. 2021;22(6):bbab114. https://doi.org/10.1093/bib/bbab114.
173. Schriml LM, Mitraka E The Disease Ontology: fostering interoperability between biological and clinical human disease-related data. Mamm Genome. 2015;26:584–89. https://doi.org/10.1007/s00335-015-9576-9.
174. Ren Y, Li Y, Loftus TJ, et al. Identifying acute illness phenotypes via deep temporal interpolation and clustering network on physiologic signatures. Sci Rep. 2024;14:8442. https://doi.org/10.1038/s41598-024-59047-x.
175. Sushentsev N, Hamm G, Flint L, et al. Metabolic imaging across scales reveals distinct prostate cancer phenotypes. Nat Commun. 2024;15:5980. https://doi.org/10.1038/s41467-024-50362-5.
176. Schmidt BM, Colvin CJ, Hohlfeld A, Leon N. Definitions, components and processes of data harmonisation in healthcare: a scoping review. BMC Med Inform Decis Mak. 2020;20(1):222. https://doi.org/10.1186/s12911-020-01218-7.
177. Nan Y, Ser JD, Walsh S, Schönlieb C, Roberts M, Selby I, Howard K, Owen J, Neville J, Guiot J, Ernst B, Pastor A, Alberich-Bayarri A, Menzel MI, Walsh S, Vos W, Flerin N, Charbonnier JP, van Rikxoort E, Chatterjee A, Woodruff H, Lambin P, Cerdá-Alberich L, Martí-Bonmatí L, Herrera F, Yang G. Data harmonisation for information fusion in digital healthcare: a state-of-the-art systematic review, meta-analysis and future research directions. Inf Fusion. 2022;82:99–122. https://doi.org/10.1016/j.inffus.2022.01.001.
178. Lee JS, Kibbe WA, Grossman RL. Data harmonization for a molecularly driven health system. Cell. 2018;174(5):1045–48. https://doi.org/10.1016/j.cell.2018.08.012.
179. Higashi RT, et al. Harmonizing qualitative data across multiple health systems to identify quality improvement interventions: a methodological framework using PROSPR II cervical research center data as exemplar. Int J Qual Methods. 2023;22. https://doi.org/10.1177/16094069231157345.
180. Gordon B, Fennessy C, Varma S, Barrett J, McCondochie E, Heritage T, Duroe O, Jeffery R, Rajamani V, Earlam K, Banda V, Sebire N. Evaluation of freely available data profiling tools for health data research application: a functional evaluation review. BMJ Open. 2022;12(5):e054186. https://doi.org/10.1136/bmjopen-2021-054186.
181. Wang P, Pullen D, Garza M, Walden A, Zozus M. Data Profiling in Support of Entity Resolution of Multi-Institutional EHR Data. Stud Health Technol Inform. 2019;257:479–83.
182. Gill IS, Griffiths EJ, Dooley D, Cameron R, Savić Kallesøe S, John NS, Sehar A, Gosal G, Alexander D, Chapel M, Croxen MA, Delisle B, Di Tullio R, Gaston D, Duggan A, Guthrie JL, Horsman M, Joshi E, Kearny L, Knox N, Lau L, LeBlanc JJ, Li V, Lyons P, MacKenzie K, McArthur AG, Panousis EM, Palmer J, Prystajecky N, Smith KN, Tanner J, Townend C, Tyler A, Van Domselaar G, Hsiao WWL. The DataHarmonizer: a tool for faster data harmonization, validation, aggregation and analysis of pathogen genomics contextual information. Microb Genom. 2023;9(1):mgen000908. https://doi.org/10.1099/mgen.0.000908.
183. Johnson SB, Farach FJ, Pelphrey K, Rozenblit L. Data management in clinical research: synthesizing stakeholder perspectives. J Biomed Inform. 2016;60:286–93. https://doi.org/10.1016/j.jbi.2016.02.014.

184. Hammer B, Virgili E, Bilotta F. Evidence-based literature review: de-duplication a cornerstone for quality. World J Methodol. 2023;13(5):390–8. https://doi.org/10.5662/wjm.v13.i5.390.
185. Hair K, Bahor Z, Macleod M, Liao J, Sena ES. The Automated Systematic Search Deduplicator (ASySD): a rapid, open-source, interoperable tool to remove duplicate citations in biomedical systematic reviews. BMC Biol. 2023;21(1):189. https://doi.org/10.1186/s12915-023-01686-z.
186. Cooper DR, Grabowski M, Zimmerman MD, Porebski PJ, Shabalin IG, Woinska M, Domagalski MJ, Zheng H, Sroka P, Cymborowski M, Czub MP, Niedzialkowska E, Venkataramany BS, Osinski T, Fratczak Z, Bajor J, Gonera J, MacLean E, Wojciechowska K, Konina K, Wajerowicz W, Chruszcz M, Minor W. State-of-the-art data management: improving the reproducibility, consistency, and traceability of structural biology and in vitro biochemical experiments. Methods Mol Biol. 2021;2199:209–36.
187. Kamdje Wabo G, Moorthy P, Siegel F, Seuchter SA, Ganslandt T. Evaluating and enhancing the fitness-for-purpose of electronic health record data: qualitative study on current practices and pathway to an automated approach within the medical informatics for research and care in university medicine consortium. JMIR Med Inform. 2024;12:e57153. https://doi.org/10.2196/57153.
188. Majidian S, Agustinho DP, Chin CS, Sedlazeck FJ, Mahmoud M. Genomic variant benchmark: if you cannot measure it, you cannot improve it. Genome Biol. 2023;24(1):221. https://doi.org/10.1186/s13059-023-03061-1.
189. Zook JM, McDaniel J, Olson ND, Wagner J, Parikh H, Heaton H, Irvine SA, Trigg L, Truty R, McLean CY, De La Vega FM, Xiao C, Sherry S, Salit M. An open resource for accurately benchmarking small variant and reference calls. Nat Biotechnol. 2019;37(5):561–6. https://doi.org/10.1038/s41587-019-0074-6.
190. Hao X, Abeysinghe R, Roberts K, Cui L. Logical definition-based identification of potential missing concepts in SNOMED CT. BMC Med Inform Decis Mak. 2023;23(Suppl 1):87. https://doi.org/10.1186/s12911-023-02183-7.
191. Schwabe D, Becker K, Seyferth M, Klaß A, Schaeffter T. The METRIC-framework for assessing data quality for trustworthy AI in medicine: a systematic review. NPJ Digit Med. 2024;7(1):203. https://doi.org/10.1038/s41746-024-01196-4.
192. Hume S, Sarnikar S, Noteboom C. Enhancing Traceability in Clinical Research Data through a Metadata Framework. Methods Inf Med. 2020;59(2-03):75–85. https://doi.org/10.1055/s-0040-1714393.
193. Sahoo SS, Valdez J, Rueschman M. Scientific reproducibility in biomedical research: provenance metadata ontology for semantic annotation of study description. AMIA Annu Symp Proc. 2017;2016:1070–79.
194. Leipzig J, Nüst D, Hoyt CT, Ram K, Greenberg J. The role of metadata in reproducible computational research. Patterns (N Y). 2021;2(9):100322. https://doi.org/10.1016/j.patter.2021.100322.
195. Ulrich H, Kock-Schoppenhauer AK, Deppenwiese N, Gött R, Kern J, Lablans M, Majeed RW, Stöhr MR, Stausberg J, Varghese J, Dugas M, Ingenerf J. Understanding the nature of metadata: systematic review. J Med Internet Res. 2022;24(1):e25440. https://doi.org/10.2196/25440.
196. Wang F, Vergara-Niedermayr C, Liu P. Metadata based management and sharing of distributed biomedical data. Int J Metadata Semant Ontol. 2014;9(1):42–57. https://doi.org/10.1504/IJMSO.2014.059126.
197. Korenblum D, Rubin D, Napel S, Rodriguez C, Beaulieu C. Managing biomedical image metadata for search and retrieval of similar images. J Digit Imaging. 2011;24(4):739–48. https://doi.org/10.1007/s10278-010-9328-z.
198. Rayner TF, Rocca-Serra P, Spellman PT, et al. A simple spreadsheet-based, MIAME-supportive format for microarray data: MAGE-TAB. BMC Bioinform. 2006;7:489. https://doi.org/10.1186/1471-2105-7-489
199. Daniel C, Sinaci A, Ouagne D, Sadou E, Declerck G, Kalra D, Charlet J, Forsberg K, Bain L, Mead C, Hussain S, Laleci Erturkmen GB. Standard-based EHR-enabled applications for clinical research and patient safety: CDISC – IHE QRPH – EHR4CR & SALUS collaboration. AMIA Jt Summits Transl Sci Proc. 2014;2014:19–25
200. Griffin PC, Khadake J, LeMay KS, Lewis SE, Orchard S, Pask A, Pope B, Roessner U, Russell K, Seemann T, Treloar A, Tyagi S, Christiansen JH, Dayalan S, Gladman S, Hangartner SB, Hayden HL, Ho WWH, Keeble-Gagnère G, Korhonen PK, Neish P, Prestes PR, Richardson MF, Watson-Haigh NS, Wyres KL, Young ND, Schneider MV. Best practice data life cycle approaches for the life sciences. F1000Res. 2017;6:1618. https://doi.org/10.12688/f1000research.12344.2.

201. Seep L, Grein S, Splichalova I, Ran D, Mikhael M, Hildebrand S, Lauterbach M, Hiller K, Ribeiro DJS, Sieckmann K, Kardinal R, Huang H, Yu J, Kallabis S, Behrens J, Till A, Peeva V, Strohmeyer A, Bruder J, Blum T, Soriano-Arroquia A, Tischer D, Kuellmer K, Li Y, Beyer M, Gellner AK, Fromme T, Wackerhage H, Klingenspor M, Fenske WK, Scheja L, Meissner F, Schlitzer A, Mass E, Wachten D, Latz E, Pfeifer A, Hasenauer J. From planning stage towards FAIR data: a practical metadatasheet for biomedical scientists. Sci Data. 2024;11(1):524. https://doi.org/10.1038/s41597-024-03349-2.
202. McGraw D, Mandl KD. Privacy protections to encourage use of health-relevant digital data in a learning health system. NPJ Digit Med. 2021;4(1):2. https://doi.org/10.1038/s41746-020-00362-8.
203. Chevrier R, Foufi V, Gaudet-Blavignac C, Robert A, Lovis C. use and understanding of anonymization and de-identification in the biomedical literature: scoping review. J Med Internet Res. 2019;21(5):e13484. https://doi.org/10.2196/13484.
204. Prasser F, Kohlmayer F, Kuhn KA. The importance of context: risk-based de-identification of biomedical data. Methods Inf Med. 2016;55(4):347–55. https://doi.org/10.3414/ME16-01-0012.
205. King CST, Bivens KM, Pumroy E, Rauch S, Koerber A. IRB problems and solutions in health communication research. Health Commun. 2018;33(7):907–16. https://doi.org/10.1080/10410236.2017.1321164.
206. Rebers S, Aaronson NK, van Leeuwen FE, Schmidt MK. Exceptions to the rule of informed consent for research with an intervention. BMC Med Ethics. 2016;17:9. https://doi.org/10.1186/s12910-016-0092-6.
207. Fernández-Alemán JL, Señor IC, Lozoya PÁ, Toval A. Security and privacy in electronic health records: a systematic literature review. J Biomed Inform. 2013;46(3):541–62. https://doi.org/10.1016/j.jbi.2012.12.003.
208. Choe J, Yoo SK. Web-based secure access from multiple patient repositories. Int J Med Inform. 2008;77(4):242–8. https://doi.org/10.1016/j.ijmedinf.2007.06.001.
209. Filkins BL, Kim JY, Roberts B, Armstrong W, Miller MA, Hultner ML, Castillo AP, Ducom JC, Topol EJ, Steinhubl SR. Privacy and security in the era of digital health: what should translational researchers know and do about it? Am J Transl Res. 2016;8(3):1560–80
210. Mehrtak M, SeyedAlinaghi S, MohsseniPour M, Noori T, Karimi A, Shamsabadi A, Heydari M, Barzegary A, Mirzapour P, Soleymanzadeh M, Vahedi F, Mehraeen E, Dadras O. Security challenges and solutions using healthcare cloud computing. J Med Life. 2021;14(4):448–61. https://doi.org/10.25122/jml-2021-0100.
211. Paul M, Maglaras L, Ferrag MA, Almomani I. Digitization of healthcare sector: a study on privacy and security concerns, ICT Express. 2023;9(4):571–88. https://doi.org/10.1016/j.icte.2023.02.007.
212. Zuo Z, Watson M, Budgen D, Hall R, Kennelly C, Al Moubayed N. Data anonymization for pervasive health care: systematic literature mapping study. JMIR Med Inform. 2021;9(10):e29871. https://doi.org/10.2196/29871.
213. Motiwalla L, Li XB. Developing privacy solutions for sharing and analyzing healthcare data. Int J Bus Inf Syst. 2013;13(2):10.1504/IJBIS.2013.054335
214. Karagiannis S, Ntantogian C, Magkos E, et al. Mastering data privacy: leveraging K-anonymity for robust health data sharing. Int. J. Inf. Secur. 2024;23:2189–201. https://doi.org/10.1007/s10207-024-00838-8.
215. El Emam K, Dankar FK, Issa R, Jonker E, Amyot D, Cogo E, Corriveau JP, Walker M, Chowdhury S, Vaillancourt R, Roffey T, Bottomley J. A globally optimal k-anonymity method for the de-identification of health data. J Am Med Inform Assoc. 2009;16(5):670–82. https://doi.org/10.1197/jamia.M3144.
216. Dyda A, Purcell M, Curtis S, Field E, Pillai P, Ricardo K, Weng H, Moore JC, Hewett M, Williams G, Lau CL. Differential privacy for public health data: an innovative tool to optimize information sharing while protecting data confidentiality. Patterns (N Y). 2021;2(12):100366. https://doi.org/10.1016/j.patter.2021.100366.
217. Liu W, Zhang Y, Yang H, Meng Q. A survey on differential privacy for medical data analysis. Ann Data Sci. 2023:1–15. https://doi.org/10.1007/s40745-023-00475-3.

# Basic Biomedical Data Exploration Techniques

Contents




J. U. Kazi, *Python Essentials for Biomedical Data Analysis: An Introductory Textbook*,
https://doi.org/10.1007/978-3-031-85600-6_5

Biomedical datasets range from genetic sequences to clinical trial data and medical imaging, each possessing unique characteristics, formats, and challenges. The process of data exploration is the first step in any data analysis pipeline. It serves as the initial means of understanding and interpreting the large, complex datasets. During this exploratory phase, researchers examine the structure and quality of the data, typically by summarizing the dataset with descriptive statistics, visualizing distributions and relationships with graphs and charts, and assessing the quality of the data by identifying missing values, outliers, or inconsistencies. This systematic exploration also helps to identify patterns and anomalies and ensures that the data are suitable for further, more advanced analysis.

**Learning Goals**

The learning goals for this chapter include developing a basic understanding of the processes involved in data exploration within biomedical research. Students will learn to identify different types of data, such as quantitative and qualitative, and to recognize key properties including shape, size, and dimensionality. They will also learn the purpose of inspecting and cleaning data, including handling missing values and correcting anomalies. Furthermore, this chapter will briefly discuss the challenges associated with categorical data, particularly in contexts of drug discovery and sensitivity prediction.

## 5.1 Data Exploration

Data exploration is the first step in any data analysis process, providing a basic understanding of the data [1]. This step contributes to making informed decisions and guiding subsequent analyses. The process involves examining the dataset to assess its structure, characteristics, and patterns [2]. Techniques such as statistical summaries, visualizations, and preliminary assessments are used in data exploration, allowing analysts to form an early understanding of data distribution, variability, and to identify anomalies [3].

More than just identifying what is present in the data, data exploration helps recognize what might be missing or problematic. This process enables analysts to choose the most suitable analytical techniques, identify areas that need deeper investigation, and generate hypotheses for more complex analysis. The exploratory phase can be compared to a detective's preliminary search for clues, laying the groundwork for solving the larger puzzle that the data presents.

### 5.1.1 Data Exploration in Biomedical Research

Data exploration involves the investigation and analysis of datasets to understand their basic characteristics, structure, and underlying patterns [4]. This phase includes multiple tasks such as examining, cleaning, transforming, and visualizing the data, with the goal of generating preliminary understanding, identifying issues, and forming hypotheses for more detailed analysis (◻ Fig. 5.1). Firstly, data exploration serves as the basis for all subsequent analyses. Without a deep understanding of the

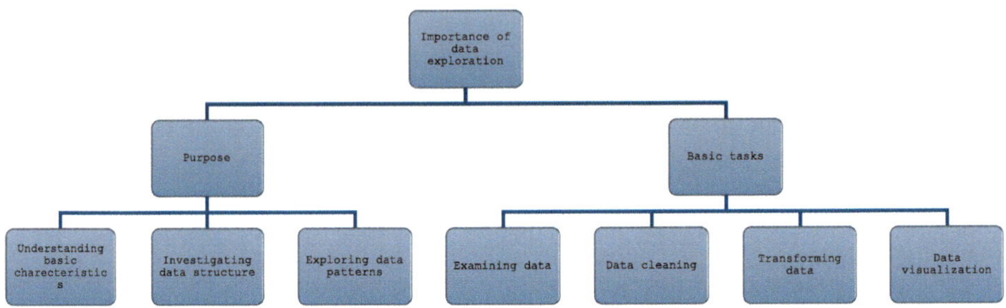


**Fig. 5.1** Importance of data exploration. The purpose of data exploration lies in understanding the basic characteristics of data, investigating data structures, and exploring data patterns through multiple tasks, including but not limited to examining, cleaning, transforming, and visualizing the data



characteristics of the dataset, advanced statistical analyses or machine learning models may be built on incorrect assumptions, which can result in misleading or invalid results. Thorough exploration ensures that researchers are aware of the subtleties of data and can make informed decisions in their analyses. Secondly, biomedical datasets often contain errors, missing values, or anomalies [5]. Early detection of these data quality issues can help to ensure the accuracy and reliability of research output. Addressing data quality issues at the exploratory stage, it may be possible to prevent complications in later analyses. Furthermore, data exploration helps guide research questions. Examining interesting patterns or relationships within the data, researchers can shape the formulation of research questions or hypotheses. This exploratory step can be beneficial as overlooking it may result in missing important aspects of data and opportunities for new discoveries. Furthermore, understanding the nature of the data determines how to process and analyze it. Decisions regarding how to handle missing values, which transformations to apply, or which statistical tests to use are often guided by observations made during initial exploration. An informed approach to data processing ensures that the analytical methods are appropriate and that the results are valid. Finally, datasets often include sensitive health information. Initial data exploration provides an opportunity to address privacy concerns and ensure compliance with ethical standards, such as protecting patient confidentiality. Considering ethical considerations early in the research process, researchers can uphold ethical standards and maintain the trust of participants and the public. Therefore, data exploration in biomedical research is useful for laying the groundwork for accurate and reliable analyses, identifying and addressing data quality issues, guiding the formulation of research questions, informing data processing choices, and ensuring ethical standards are met. Skipping or undervaluing this step can lead to flawed conclusions and ethical breaches.

### 5.1.2 The Data Exploration Process

The data exploration process involves several key steps, each designed to provide a clear overview of the dataset and to prepare it for more advanced analyses. These steps ensure that researchers thoroughly understand the data before applying sophisticated analytical techniques (Fig. 5.2).

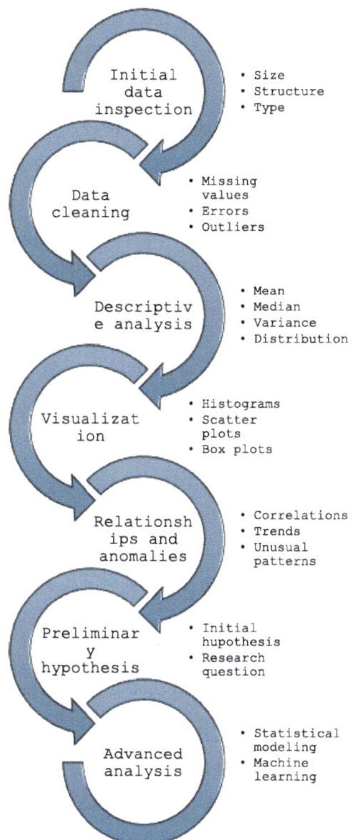


**Fig. 5.2** Steps involved in the data exploration process. The data exploration process involves several steps, including initial data inspection, data cleaning, statistical summaries and descriptive analysis, visualization, identification of relationships and anomalies, preliminary hypothesis formation, and preparation for advanced analysis

The initial data inspection provides researchers with a sense of the size and structure of the dataset and the types of data it contains, such as numerical, categorical, or time-series data. This preliminary review helps establish a general understanding of the dataset's scope and identify any immediate issues that may need attention. In the very next step, data cleaning is conducted to address inconsistencies like missing values, errors, or outliers [6]. This step involves identifying and rectifying these issues to ensure that the dataset is accurate and reliable for further analysis. Data cleaning step maintains the quality and validity of subsequent analysis. Next, researchers may perform statistical summaries and descriptive analysis using descriptive statistics such as mean, median, and variance [7]. These statistics summarize the key features of the data, helping researchers understand central tendencies and data distribution. This knowledge is needed for informing more detailed analyses. Following this, visualization techniques are employed. Visual representations like histograms, scatter plots, and box plots allow researchers to observe trends, patterns, and relationships that may not be immediately apparent from raw data [8]. Visualizations help identify outliers, correlations, and possible relationships between variables. The process continues with the identification of relationships and anomalies. During this step, researchers look for correlations, trends, and unusual patterns within the data. Identifying these elements can show important relationships between variables or pinpoint unexpected anomalies that warrant further investigation. Based on the information gained, researchers engage in preliminary hypothesis formation. They formulate initial hypotheses or research questions to be tested in more detailed analyses. This stage serves as a bridge between basic exploration and advanced investigations, guiding the direction of future research efforts. Finally, the preparation for advanced analysis is undertaken. Information extracted from the data exploration guides the transition to more advanced techniques such as statistical modeling, machine learning, or other detailed analytical methods. A thorough understanding of the dataset ensures that these methods are applied appropriately, maximizing the chance for meaningful discoveries. Collectively, the data exploration process includes initial inspection, data cleaning, descriptive statistics, visualization, identification of relationships and anomalies, preliminary hypothesis formation, and preparation for advanced analysis. Each step builds upon the previous one, ensuring that researchers have a deep and accurate understanding of the data before proceeding to complex analyses.



## 5.2 Understanding the Dataset

A thorough understanding of the dataset starts with identifying the nature of data, distinguishing between quantitative and qualitative variables. It also includes assessing basic properties such as shape, size, and dimensionality, and identifying the specific characteristics of the dataset at hand. Collectively these information help ensure that the data are handled, analyzed, and interpreted in a manner that is both accurate and appropriate for the research goals.

### 5.2.1 Identifying the Type of Data: Quantitative Versus Qualitative

Understanding the type of data determines the appropriate methods for analysis and interpretation. Data can be broadly categorized into quantitative and qualitative types (◘ Fig. 5.3).

**Quantitative data** are numerical and can be either measured or counted, making them suitable for statistical analysis and mathematical modeling [9]. These type of data are common in biomedical research, as numerical precision often drives clinical decisions. For example, patient age is measured in years, blood pressure readings measured in millimeters of mercury (mmHg), cholesterol levels measured in milligrams per deciliter (mg/dL), and gene expression levels in genomic studies indicating how actively specific genes are being transcribed. Quantitative data can be further categorized into discrete and continuous data. Discrete data represent countable quantities and are typically integers such as the number of patients in a study or the number of times a cell divides. Continuous data can take any value within a defined range and are often measurements like enzyme activity levels in a biochemical assay or body temperature measured to a decimal point.

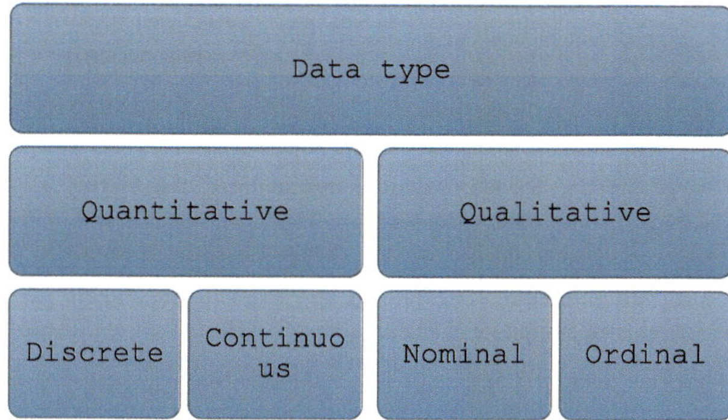


**Fig. 5.3** Types of data. Broadly, data can be subdivided into two groups: quantitative and qualitative data, which are further divided into discrete, continuous, nominal, and ordinal data types

**Qualitative data** are non-numerical and describe characteristics or qualities rather than measurements [10]. These data types often provide categorical or descriptive details about the subjects being studied. In biomedical research, qualitative data are often encountered in areas such as patient symptoms described in terms of type and severity, disease classifications indicating different categories or types of diseases, and treatment outcomes categorized as "improved," "unchanged," or "worsened." Qualitative data can be classified into nominal and ordinal data. Nominal data consist of non-ordered categories; examples include blood types (e.g., A, B, AB, O), gender (e.g., male, female, nonbinary), and disease presence (e.g., positive, negative). Ordinal data involve categories that have a meaningful order or progression. Examples include cancer stages (e.g., Stage I, Stage II, Stage III, Stage IV), where each stage signifies increasing severity; pain scales (e.g., mild, moderate, severe); and patient satisfaction ratings (e.g., very dissatisfied, dissatisfied, neutral, satisfied, very satisfied).

### 5.2.2 Basic Dataset Properties: Shape, Size, and Dimensionality

The shape and size of a dataset are important factors in understanding its structure. The shape typically refers to the layout of the data, described in terms of rows and columns (rows, columns). Rows represent individual samples or observations, such as patients, while columns represent variables or features, such as clinical test results. The size of the dataset is determined by the total number of data points, calculated by multiplying the number of rows by the number of columns (rows x columns). Dimensionality, on the other hand, is often defined as the number of features (columns) in the dataset (Fig. 5.4). Dimensionality also reflects the number of axes needed to index the data. For example, a 1-dimensional dataset might be a simple list, a 2-dimensional dataset would be a table with rows and columns, and a 3-dimensional dataset could be represented as a tensor or array with three axes in its data structure.

Understanding the shape and size of a dataset is important for several reasons. It aids in managing data complexity, as large datasets require appropriate tools for storage, handling, and processing. It also influences the choice of analytical methods, since the complexity and size of the dataset affect the selection of statistical and

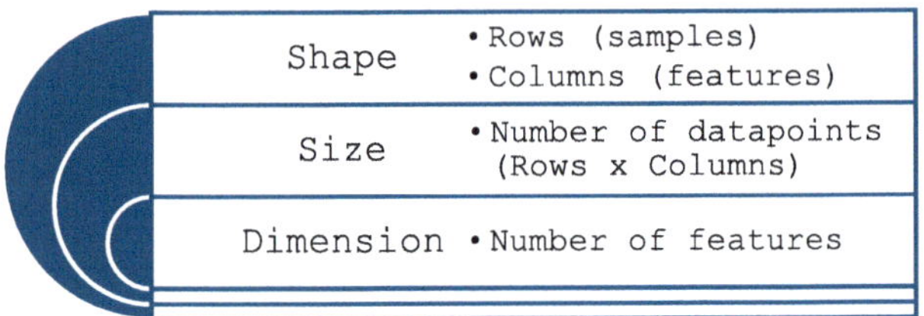


**Fig. 5.4** Shape, size, and dimension. The data shape represents the number of rows and columns, corresponding to the number of samples and features, respectively. Data size is the total number of data points, and data dimensionality refers to the number of variables or features



computational techniques. Furthermore, knowing the dataset's shape and size provides context for interpreting results, helping to understand the breadth and depth of the information it contains.

High-dimensional datasets, which are common in fields like genomics and proteomics, can include thousands of features, each representing different biological or clinical measurements. While high-dimensional data can offer detailed information, it also presents challenges. In machine learning, models trained on high-dimensional data can often overfit, learning noise instead of generalizable patterns. Moreover, high-dimensional datasets demand substantial computational resources, making analysis more time-consuming and complex.

### 5.2.3 Recognizing Different Types of Biomedical Datasets

Since data analysis methods are often tailored to specific data types, and biomedical datasets display a wide variety, it is needed to first identify the data type before proceeding with analysis. The major data types include clinical data, which contain patient-related information such as demographics, medical history, laboratory test results, and treatments administered; these datasets are often used in epidemiology, clinical trials, and patient care management, to identify disease patterns, treatment efficacy, and patient recovery, forming the basis of much biomedical research. Genomic data capture information about an individual's DNA sequences, gene expression profiles, and genetic variations and are central to personalized medicine where treatments are tailored to a patient's genetic makeup; genomic datasets help researchers understand the genetic basis of diseases, identify predispositions to certain conditions, and develop targeted therapies [11]. Transcriptomic data involve the study of RNA sequences, providing a dynamic view of gene expression under various conditions; these data are key to understanding how genes are regulated and how they respond to environmental factors, and is used extensively in cancer research to identify biomarkers and in pharmacogenomics to inform personalized drug responses [12]. Imaging data include various types of medical images such as X-rays, MRIs, and CT scans that provide visual representations of the body's internal structures; imaging data are used for accurate diagnostics, treatment planning, and research fields like radiomics, which involves extracting quantitative features from medical images to improve diagnostic accuracy and personalize treatment approaches [13].

Proteomic data focus on the protein composition of biological samples that include protein expression, modification, and interaction, are vital for understanding cellular processes [14]. Metabolomic data, on the other hand, examine small molecules and biochemical processes, providing information about the metabolic pathways and organism responses to environmental changes [15]. Finally, electronic health records (EHR) datasets combine clinical and administrative data, providing an overview of patient health histories, treatments, and healthcare utilization; they are useful for both individual patient care and system-wide healthcare analysis [16].

## 5.3 Data Inspection and Cleaning

Data inspection and cleaning are early steps in any data analysis workflow, especially in biomedical research, where data quality directly affects the validity of conclusions. This process involves a systematic examination of the data to assess its structure and quality, identify missing values, correct errors, and ensure proper data type conversions (◘ Fig. 5.5). These steps help create a reliable dataset, laying the groundwork for advanced analyses that can lead to accurate and insightful results.

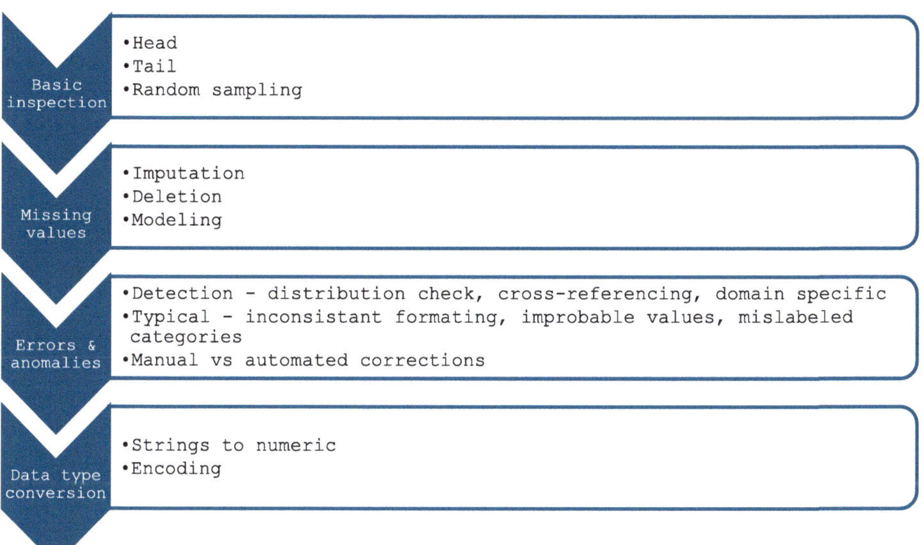


◘ **Fig. 5.5** Basic data inspection and cleaning. Basic data inspection includes checking the data's head, tail, and random samples. This helps to understand the content of the data and to determine the necessary downstream processing steps, such as handling missing values, detecting errors and anomalies, and performing data type conversions

### 5.3.1 Basic Data Inspection: Head, Tail, and Random Samples

**Notebook: Section 5.3.1. Basic Data Inspection: Head, Tail, and Random Samples**
This notebook provides code examples demonstrating basic data inspection methods such as head, tail, and random samples.

Link to the GitHub repository:
▶ https://github.com/sn-code-inside/BioPy
Go to: Chap. 5—example data—▶ Sect. 5.3.1.



A common starting point in data inspection is to view the "head" and "tail" of the dataset, which refer to the first and last few rows, respectively. This provides a quick snapshot of the dataset's general structure, allowing researchers to verify whether the data types, values, and organization are consistent [17]. Head inspection includes a review of the first 5–10 rows, which provides an overview of key variables and their data types, as well as the general nature of the values in the dataset. Tail inspection examines the last few rows to ensure that the data maintain consistency throughout and to help detect any irregularities or missing information that may have been appended incorrectly at the end. Furthermore, random sampling involves selecting a subset of rows from the dataset at random for closer inspection. This is especially useful for large datasets, where examining the head and tail alone might not be representative. By randomly checking rows, researchers can detect irregularities that may not be visible in a sequential view of the data. This technique is particularly helpful for identifying outliers, inconsistencies, or hidden patterns that could affect the analysis.

Example code:

```
import pandas as pd
# Load dataset
df = pd.read_csv('example_data/data.csv')
# Display the first few rows
print('Display the first few rows:\n', df.head())
# Display the last few rows
print('Display the last few rows:\n', df.tail())
# Random sample
print('Display random rows:\n', df.sample(5))
```

### 5.3.2 Identifying and Handling Missing Values

**Notebook: Section 5.3.2. Identifying and Handling Missing Values**
This notebook provides code examples demonstrating methods for identifying and handling missing values.
Link to the GitHub repository:
▶ https://github.com/sn-code-inside/BioPy
Go to: Chap. 5—example data—▶ Sect. 5.3.3.

Missing values can occur due to errors in data entry, incomplete data collection, or technical issues during data acquisition. Properly managing missing values is needed to maintain data integrity and ensuring the validity of research conclusions [18]. Detection of missing values is one of the initial steps and can be done through various methods such as summary statistics or data visualization techniques (also discussed in chapter 4) that show gaps in the data. Python libraries like Pandas offer straightforward functions (e.g., `isnull()` or `info()`) to identify missing data. After detection, several strategies can be employed depending on the nature of the missing data and the analysis goals [19]. For example, imputation can be used to replace missing values with estimates such as the mean or median, or by using more sophisticated techniques like regression models to predict the missing values, examples include filling missing patient height values based on other patient characteristics. Deletion is another method; in cases where missing values are few, it might be appropriate to simply remove those rows or columns. However, this approach should be used with caution, as it reduces the dataset size and can lead to biased results. Furthermore, modeling with missing data using advanced statistical models, particularly in machine learning, can handle missing data by incorporating it directly into the model's structure. Techniques like multiple imputation or maximum likelihood estimation can deal with incomplete data without loss of information. Choosing the right method for handling missing values is important for ensuring that the results of the analysis are both accurate and reliable [20].

### 5.3.3 Detecting and Correcting Errors or Anomalies in the Data

Data errors or anomalies can skew analysis results if not properly addressed. Detecting and correcting these issues is a vital step in ensuring data quality and preventing misleading conclusions. Errors in the data can be identified using several methods, including distribution checks, cross-referencing, and domain expertise. Distribution checks involve analyzing data distributions to reveal unexpected patterns, such as skewness or outliers, which may indicate data entry errors [21, 22]. Cross-referencing entails comparing the data with external, trusted sources (e.g., previous datasets or industry standards) to detect discrepancies or mistakes. Incorporating domain expertise allows experts to identify errors that may not be statistically apparent, such as biologically impossible values like negative ages or unfeasibly high blood pressure readings.

Common errors in biomedical datasets include inconsistent formatting, impossible or improbable values, and mislabeled categories. Inconsistent formatting refers to variations in data entry, such as inconsistent use of units (e.g., different scales for blood pressure readings) or varying date formats. Impossible or improbable values are data that fall outside biologically plausible ranges, such as negative values for a patient's weight or age. Mislabeled categories involve incorrect or inconsistent classification labels, which can lead to improper grouping or analysis of categorical data.



For small datasets, manual review and correction might be feasible and allow for nuanced adjustments. However, it is labor-intensive and unsuitable for large datasets. For larger datasets, automated correction algorithms can efficiently handle common errors. For instance, setting rules to adjust values outside realistic ranges (e.g., capping outliers in cholesterol levels) or using text-matching algorithms to correct variations in spelling for drug names. Automated tools can be built using Python libraries such as Pandas, which allow researchers to standardize, clean, and reformat data efficiently.

### 5.3.4 Data Type Conversions

Ensuring the correct data types for each variable in a dataset is required for smooth and accurate data analysis. Data type mismatches can cause errors during analysis or lead to incorrect conclusions. When data types are not properly assigned, statistical computations might fail or produce misleading results. Therefore, data type conversion is an important step in the data cleaning process.

One common issue is the need to convert strings to numeric values. Quantitative analysis requires numeric data types, but in some cases, numerical values may be stored as text due to the way data was collected or imported. For example, a dataset may record patient ages as strings instead of integers because of formatting issues or data entry methods. In such cases, it is necessary to convert these strings into numeric types to perform mathematical and statistical operations. Failing to do so can prevent the use of functions such as calculating averages, standard deviations, or performing regression analyses.

Another important aspect is categorical encoding. Categorical variables, such as responses recorded as "yes" or "no," must be converted into a numerical format for many analytical techniques, particularly for machine learning models that require numerical input. One common approach is one-hot encoding, which converts each category into a binary column. For example, the categorical variable "disease status" with categories like "healthy" and "sick" would be transformed into two binary columns: one column indicating "healthy" (1 if the patient is healthy, 0 otherwise) and another indicating "sick" (1 if the patient is sick, 0 otherwise). This encoding allows models to interpret categorical data effectively, enabling them to detect patterns and make predictions based on categorical features.

Utilizing Python libraries can greatly facilitate the process of data type conversions. Python libraries like Pandas offer a range of functions for handling data type conversions and categorical encoding. Methods such as `astype()` can be used to convert columns to the correct data types—for instance, converting a column of strings representing numbers into integers or floats. Furthermore, Pandas' `get_dummies()` function provides an easy way to perform one-hot encoding on categorical variables. These tools simplifies the data type conversion process, ensuring that the dataset is formatted correctly for further analysis. Data type conversions help to avoid errors and ensure that their data is properly prepared for analysis.

## 5.4 Dealing with Categorical Data

Categorical data, which are qualitative data consisting of nominal and ordinal values, are a frequent component of biomedical research and analysis. Effectively handling categorical data requires a solid understanding of its nature, tools for summarizing and analyzing it, such as frequency and contingency tables, and appropriate visualization techniques like bar charts and pie charts. These methods are used for interpreting categorical data in areas ranging from clinical trials to genetic studies. Python libraries like Pandas, Matplotlib, and Seaborn provide robust tools for efficiently analyzing and visualizing categorical data.

### 5.4.1 Understanding Categorical Data

Categorical data refer to information grouped by distinct categories rather than numerical values. In biomedical research, many variables are inherently qualitative, such as disease types, patient symptoms, or genetic markers. Unlike quantitative data, categorical data group information based on characteristics or attributes without implying any mathematical relationships between the groups. As discussed above, nominal data represent categories that have no natural order or ranking and are used for labeling or classifying variables. Ordinal data are also categorical but differ from nominal data in that they have a meaningful order or ranking among the categories. The order of these categories holds importance. Nominal data are important for identifying and categorizing distinct biological or clinical features, helping researchers analyze disease types, genetic diversity, and classification of various biological samples. Ordinal data are often used for tracking disease progression, treatment responses, and patient outcomes. Their ability to reflect hierarchy enables researchers and healthcare professionals to make informed decisions based on severity or progression, making them useful in biomedical analysis.

### 5.4.2 Tools for Summarizing and Analyzing Categorical Data

Frequency and contingency tables are widely used tools for analyzing categorical data. Frequency tables display how often each category occurs within a dataset, providing a summary of the distribution of data across various groups. Frequency tables are used to understand disease prevalence by counting how many individuals fall into different disease categories and to analyze drug usage trends by summarizing the frequency of different medications prescribed. These tables may also include relative frequencies (percentages) and cumulative frequencies, helping researchers assess the distribution of categories in more detail. Contingency tables, also known as cross-tabulation tables, examine the relationship between two or more categorical variables. They are indispensable for epidemiological studies, where they might be used to compare different population groups (e.g., exposed vs. unexposed) against an outcome variable (e.g., disease presence), and for genetic studies, where they help explore associations between genetic variants and traits. Contingency tables lay the foundation for statistical tests, such as the chi-square test, used to evaluate whether there is a statistical association between the variables.



### 5.4.3 Importance of Categorical Data in Biomedical Context

Categorical data are useful in biomedical research due to their ability to represent nonnumerical biological and clinical variables. Some of their key roles include the representation of biological variables, such as genotypes, phenotypes, and disease classifications, which are qualitative in nature and can only be represented categorically. They facilitate classification and diagnosis; for example, staging cancers for determining treatment strategies and prognosis. In epidemiological studies, categorical data help classify populations based on exposure to risk factors, symptoms, or health outcomes, enabling epidemiologists to analyze trends and identify public health risks. Furthermore, in personalized medicine, categorical data are frequently used to group patients based on genetic, environmental, and lifestyle factors, which helps tailor treatments to individual needs.

### 5.4.4 Challenges in Handling Categorical Data

While categorical data are widely used in biomedical research, they present several challenges. Data sparsity is one such challenge [23]; in large datasets, some categories may have very few observations, creating sparsity that can hinder the effectiveness of statistical models, as they may struggle to generalize from underrepresented categories. Encoding complexity is another issue [24]; converting categorical data into numerical formats (e.g., one-hot encoding for machine learning models) can increase the dimensionality of a dataset, complicating the modeling process. Encoding decisions must balance informativeness with manageability. There is also a risk of loss of information; when encoding or recoding ordinal data, important information about the natural order of categories may be lost if not handled carefully. Lastly, bias and misclassification can occur [25, 26]; incorrect or biased categorization can lead to misclassification of data, which may compromise the accuracy of clinical decisions and predictions.

### 5.4.5 Significance in Drug Discovery and Sensitivity Prediction

Categorical data are often used in drug discovery and drug sensitivity prediction [27, 28]. In target identification, categorical data help researchers classify biological targets, such as proteins or genes, based on their functions or involvement in disease pathways; this categorization is vital for identifying viable drug targets. In drug classification, drugs are often categorized by their mechanism of action, and understanding these categories aids in predicting drug interactions, side effects, and effectiveness. In predictive modeling, categorical data help identify which drug categories are more likely to be effective based on a patient's genetic or molecular profile; for instance, predictive models often use categorical biomarkers to assess drug efficacy for specific cancer types [29, 30]. Furthermore, in clinical trial design, categorical data are used to stratify patients based on biomarkers or other categorical variables, ensuring that clinical trials are more targeted and efficient.

## 5.5 Initial Data Exploration in Practice

Initial data exploration is the first step in analyzing biomedical datasets. Taking a systematic approach allows researchers to understand the details of the data, identify key challenges, and lay the foundation for subsequent advanced analyses. This phase not only helps in understanding the structure and key characteristics of the dataset but also serves as a vital checkpoint to ensure that the analysis is conducted on a clean, reliable, and well-understood dataset. The knowledge developed during initial exploration directly informs the direction of further research and analysis.

### 5.5.1 Systematic Approach to Data Exploration

The first step in data exploration is familiarizing oneself with the dataset. Researchers should begin by understanding the source of the data, the variables involved, and the context in which the data was collected [31]. In biomedical research, this might include reviewing clinical parameters, patient demographics, treatment regimens, or specific health conditions. Understanding the background and structure of the dataset is useful for correctly interpreting the results and avoiding misinterpretations [32]. Key considerations during data familiarization include understanding variable types, whether quantitative or categorical, and examining the data sources, such as clinical trials, observational studies, or lab experiments. Knowing the provenance of the data helps assess its reliability and relevance. Once researchers are familiar with the dataset, the next step is data cleaning and preprocessing [21]. This involves identifying missing values, outliers, and errors that could skew the analysis. Careful handling of missing data is important, as incomplete or inaccurate data can lead to incorrect conclusions, particularly when patient health is concerned. Preprocessing tasks include missing data imputation, choosing the appropriate strategy to handle missing values (e.g., using mean imputation, removing incomplete records, or employing advanced imputation techniques); outlier detection, identifying data points that fall far outside the normal range and deciding whether they are biologically plausible or errors; and error correction, correcting data entry mistakes (e.g., negative values for age or height) before proceeding with analysis [33, 34]. Descriptive statistics and visualization provide a summary of the data's main features, while visualizations help reveal patterns and relationships that are not immediately apparent (discussed in Chap. 7). Tools such as mean, median, mode, and standard deviation help summarize the central tendencies and variability within the dataset. Common visual tools include histograms, used to visualize the distribution of numerical variables, such as the distribution of patient ages or lab test results; box plots, effective for identifying outliers and understanding the spread of the data across different groups (e.g., patients in different clinical stages); and scatter plots, useful for examining relationships between two variables, such as the correlation between cholesterol levels and heart disease risk.



### 5.5.2 Common Pitfalls

A common pitfall during initial data exploration is neglecting data quality issues, such as missing values or outliers. Failing to address these issues can distort subsequent analyses. To avoid this, it is best practice to thoroughly check for and clean up missing or erroneous data to ensure accuracy and completeness before moving on to more complex analyses. Bias is another major concern, particularly in biomedical research, where biased interpretations can have substantial consequences. Researchers should be vigilant in preventing preconceived notions or expectations from influencing their interpretation of the data. Approaching the data objectively and considering multiple perspectives is recommended for drawing accurate conclusions. Furthermore, biomedical data do not exist in a vacuum; understanding the clinical, demographic, and experimental context is needed. Researchers must consider the broader clinical or biological context of the data, including how certain clinical measurements might vary across populations or how treatment protocols might influence patient outcomes. Without this contextual understanding, there is a risk of misinterpreting results or drawing erroneous conclusions.

### 5.5.3 Transitioning to Advanced Analysis

After completing initial exploration, researchers should have a clear understanding of the data's structure, quality, and basic relationships. This knowledge forms the basis for transitioning to advanced analyses, such as hypothesis testing, where researchers can formally test any hypotheses that arose during initial exploration; predictive modeling, where the cleaned and well-understood dataset is ready for machine learning models or other predictive techniques that may provide useful information on patient outcomes or treatment efficacy; and complex statistical analysis, where researchers may employ regression analysis, survival analysis, or other advanced statistical techniques to explore relationships between variables in greater depth. The knowledge gained from initial exploration informs which variables require careful handling and which relationships might be most worth investigating in further detail.

### 5.5.4 Role of Initial Exploration in Guiding Research

Initial data exploration can be important in shaping the entire research process. Conducting a thorough and systematic exploration of the dataset, researchers can formulate hypotheses, as initial observations often reveal patterns or relationships that lead to new hypotheses worth investigating; select appropriate statistical techniques, as knowledge from data exploration helps determine which statistical methods or models are best suited for analyzing the data; and identify limitations, as early exploration also shows any limitations in the data, such as biases, inconsistencies, or missing values that might impact the reliability of the final results. In biomedical research, where datasets are often complex and heterogeneous, this phase is required to extract useful information that can lead to novel discoveries or advancements in medical science.

## 5.6 Preparing for Advanced Analysis

This section discusses how initial data exploration can inform and support the application of more complex analytical techniques that follow. It uses information gained during basic exploration and sets the stage for deeper investigation, hypothesis formulation, and the application of advanced statistical and machine learning methods. Effective use of this stage may help researchers both conceptually and technically to extract meaningful conclusions from biomedical datasets.

### 5.6.1 Summarizing Insights from Basic Exploration

The first step in preparing data for advanced analysis is to summarize the main observations from the basic exploration phase. This involves compiling key findings about the dataset, including:

Data distribution: Are the variables normally distributed, or do they show skewness?

Key variables: Which variables appear to be the most important or informative?

Possible relationships: Were any correlations or associations between variables observed?

Anomalies and patterns: Were there any outliers, unexpected trends, or missing values that could impact further analysis?

The purpose of this summary is to create a clear overview of the dataset's main characteristics. It serves as a reference point to ensure that any advanced analyses are grounded in a deep understanding of the data. Summarizing these early findings, researchers can better guide the selection of appropriate advanced techniques and reduce the risk of errors or misinterpretation later in the analysis.

### 5.6.2 Identifying Areas for Deeper Analysis

After summarizing the initial exploration, the next step is to identify specific areas within the dataset that warrant deeper investigation. This involves pinpointing major variables, such as biomarkers that show variability across patient subgroups or variables that appear to influence clinical outcomes, and focusing on relationships, exploring correlations or interactions between variables that suggest a link between treatments and patient outcomes or disease progression. Identifying these key areas helps researchers focus their resources and analytical efforts on the most promising aspects of the data.



### 5.6.3 Formulating Hypotheses

Based on what has been learned during the initial exploration and the areas identified for deeper analysis, the next step is to formulate clear, testable hypotheses. These should be informed by both data and relevant biomedical knowledge; for example, a hypothesis might explore the relationship between a specific genetic marker and disease susceptibility. They should direct the focus of advanced analysis, serving as guiding questions for subsequent statistical or machine learning techniques to ensure the analysis is purposeful and focused on answering specific research questions. Moreover, they should be relevant to the broader research goals, in agreement with the overall objectives of the biomedical research, whether identifying new therapeutic targets, understanding disease mechanisms, or improving patient outcomes. Clear hypotheses help frame the research questions and provide structure for the advanced analyses to follow.

### 5.6.4 Transition to Advanced Statistical Methods and Machine Learning

With a solid foundation established during initial exploration, researchers can now transition to advanced analytical methods. This phase involves applying more sophisticated tools to study the complexities of the data. Advanced statistical methods include multivariate analysis, where techniques such as multiple regression or principal component analysis (PCA) helps in understanding the relationships between multiple variables simultaneously; survival analysis, which is common in clinical research and investigates time-to-event data, such as patient survival rates or time until disease progression; and time-series analysis, which, for longitudinal datasets, helps identify trends and patterns over time, particularly in patient monitoring or disease progression. Machine learning techniques include supervised learning, with methods such as decision trees, random forests, or support vector machines that can classify patient outcomes or predict disease risk based on a set of variables, and unsupervised learning, where techniques like clustering (e.g., K-means or hierarchical clustering) are used to group similar patients or samples based on their characteristics, often without predefined labels. The choice of advanced technique should be guided by the research questions, the structure of the data, and the formulated hypotheses, ensuring that the chosen methods are well-suited to the complexity and nature of the dataset.

### 5.6.5 Conceptual and Technical Preparation

Preparing for advanced analysis involves more than just technical skills; it also requires conceptual readiness. Researchers need to be prepared to address complexities and challenges; for example, advanced machine learning models are susceptible to overfitting, where a model performs well on training data but poorly on new, unseen data, and researchers must apply strategies like cross-validation to mitigate this risk. They also need to interpret multivariate relationships, as in biomedical datasets, relationships between variables can be complex, and understanding these relationships and interpreting results from complex models require a strong foundation in both statistics and biomedical knowledge. This phase demands not only technical skills but also critical thinking and problem-solving abilities to ensure that the advanced analysis is both meaningful and accurate.

### 5.6.6 Embracing Opportunities in Advanced Data Analysis

Advanced data analysis creates new possibilities for progress in biomedical research. With sophisticated analytical methods, researchers can develop novel understandings in disease mechanisms; for example, advanced models might reveal previously unknown interactions between genes and environmental factors that contribute to disease susceptibility. They can also identify novel therapeutic targets, as machine learning models can help predict which biological pathways are most likely to be effective targets for new drugs, and personalize medicine by identifying patient subgroups more likely to respond to specific treatments, paving the way for personalized therapies that improve patient outcomes. As researchers transition to advanced analysis, they should remain open to the possibilities that new methods and technologies offer, ensuring that research remains flexible and adaptive to the evolving nature of biomedical science.

## 5.7 Exercises and Questions

This section provides exercises and questions to complement the material covered in this chapter.

**Exercises**

The following exercises will help deepen your understanding of the key concepts discussed in this chapter:

1. Explore basic properties of a dataset: Load a dataset of patient demographics and use statistical software to analyze its shape, size, and dimensionality.

2. Identify data types: Distinguish between quantitative and qualitative data in a dataset containing clinical trial results. Highlight the variables and categorize them accordingly.
3. Perform data inspection: Write a script that displays the head, tail, and random samples from a dataset of blood test results. Use this to understand the structure and identify any preliminary data quality issues.
4. Handle missing values: Practice identifying and imputing missing values in a dataset from a longitudinal study on diabetes management. Try different imputation methods (e.g., mean, median) and evaluate their impact.
5. Correct anomalies: Identify and correct unrealistic or incorrect entries (e.g., negative age values) in a dataset of patient records.
6. Analyze categorical data: Explore the distribution of blood types within a dataset. Discuss the significance of this categorical data in biomedical research, especially in relation to disease risk or treatment compatibility.
7. Visualize dataset distribution: Use histograms and box plots to visualize the distribution of cholesterol levels in a cardiovascular health study. Interpret the results to identify trends or outliers.
8. Summarize dataset insights: After performing an initial exploration of a dataset detailing cancer patient recovery rates, write a summary of the key findings, focusing on patterns and anomalies observed.
9. Formulate hypotheses: Based on exploratory data analysis of smoking habits and lung function test results, formulate hypotheses regarding possible correlations between these variables.
10. Transition to advanced analysis: Outline the steps required to take a dataset from initial exploration to advanced statistical analysis. Use a dataset on medication adherence as an example, detailing how you would prepare for regression analysis or machine learning modeling.



### Questions to Be Answered

These questions encourage reflection on the key concepts and methods covered in this chapter:

1. Why is it important to understand the type of data (quantitative vs. qualitative) in a biomedical dataset?
2. How does the size and shape of a dataset impact the types of analysis that can be performed?
3. What are some common indicators of poor data quality in biomedical datasets, and how can they affect research outcomes?
4. Describe how missing values can affect the outcome of a biomedical research study. What are some methods for handling missing data?
5. What are the implications of improperly handling categorical data in drug sensitivity prediction models?
6. What steps would you take to inspect a new biomedical dataset for the first time, and why are these steps required?
7. How can initial data exploration guide the formulation of research hypotheses, particularly in the biomedical context?

8. Discuss the role of visual tools (e.g., histograms, box plots) in the data exploration process for biomedical datasets. How do they aid in interpretation?
9. What are some best practices for avoiding common pitfalls during data exploration? Provide examples related to biomedical datasets.
10. How does preparing a dataset for advanced analysis differ from basic data exploration, particularly when transitioning to machine learning or advanced statistical methods?

**Acknowledgement** The language of the human-generated text was corrected with the assistance of artificial intelligence (AI) tools [GPT-3.5 and GTP-4 from OpenAI]. GitHub Co-Pilot was used to check the correctness of the codes. The text underwent subsequent human revision to ensure its accuracy.

## References

1. Komorowski M, Marshall DC, Salciccioli JD, Crutain Y. Exploratory data analysis. In: Secondary analysis of electronic health records. Cham: Springer; 2016. p. 185–203.
2. Shreffler J, Huecker MR. Exploratory data analysis: frequencies, descriptive statistics, histograms, and boxplots. [Updated 2023 Nov 3]. In: StatPearls [Internet]. Treasure Island (FL): StatPearls Publishing; 2025 Jan-. Available from: https://www.ncbi.nlm.nih.gov/books/NBK557570/.
3. Daele SV, Janssenswillen G. Identifying the steps in an exploratory data analysis: a process-Oriented Approach. In: Montali M, Senderovich A, Weidlich M, editors. Process mining workshops. ICPM 2022. Lecture Notes in Business Information Processing, (2023) vol 468. Springer, Cham. https://doi.org/10.1007/978-3-031-27815-0_38.
4. Konopka BM, Lwow F, Owczarz M, Laczmanski L. Exploratory data analysis of a clinical study group: development of a procedure for exploring multidimensional data. PLoS One. 2018;13(8):e0201950.
5. Bellazzi R. Big data and biomedical informatics: a challenging opportunity. Yearb Med Inform. 2014;9(1):8–13.
6. Guo M, Wang Y, Yang Q, Li R, Zhao Y, Li C, Zhu M, Cui Y, Jiang X, Sheng S, Li Q, Gao R. Normal workflow and key strategies for data cleaning toward real-world data: viewpoint. Interact J Med Res. 2023 Sep 21;12:e44310. https://doi.org/10.2196/44310.
7. Cooksey RW. Descriptive statistics for summarising data. illustrating statistical procedures: finding meaning in quantitative data. Springer; 2020. p. 61–139 https://doi.org/10.1007/978-981-15-2537-7_5.
8. Li Q. Overview of data visualization. embodying data. Springer; 2020. p. 17–47 https://doi.org/10.1007/978-981-15-5069-0_2.
9. “Quantitative Data.” NNLM, www.nnlm.gov/guides/data-glossary/quantitative-data. Accessed 29 July 2025.
10. “Qualitative Data.” NNLM, www.nnlm.gov/guides/data-glossary/qualitative-data. Accessed 29 July 2025.
11. Robertson AJ, Tan NB, Spurdle AB, Metke-Jimenez A, Sullivan C, Waddell N. Re-analysis of genomic data: An overview of the mechanisms and complexities of clinical adoption. Genet Med. 2022;24(4):798–810. https://doi.org/10.1016/j.gim.2021.12.011.
12. Cockrum C, Kaneshiro KR, Rechtsteiner A, Tabuchi TM, Strome S. A primer for generating and using transcriptome data and gene sets. Development. 2020;147(24):dev193854. https://doi.org/10.1242/dev.193854.
13. Willemink MJ, Koszek WA, Hardell C, Wu J, Fleischmann D, Harvey H, Folio LR, Summers RM, Rubin DL, Lungren MP. Preparing medical imaging data for machine learning. Radiology. 2020;295(1):4–15. https://doi.org/10.1148/radiol.2020192224.
14. Al-Amrani S, Al-Jabri Z, Al-Zaabi A, Alshekaili J, Al-Khabori M. Proteomics: Concepts and applications in human medicine. World J Biol Chem. 2021;12(5):57–69. https://doi.org/10.4331/wjbc.v12.i5.57.

15. Chen Y, Li EM, Xu LY. Guide to metabolomics analysis: a bioinformatics workflow. Metabolites. 2022;12(4):357. https://doi.org/10.3390/metabo12040357.
16. Wang W, Ferrari D, Haddon-Hill G, et al. Electronic health records as source of research data. 2023. In: Colliot O, editor. Machine learning for brain disorders [Internet]. New York: Humana; 2023. Chapter 11. https://doi.org/10.1007/978-1-0716-3195-9_11.
17. GeeksforGeeks. "Difference between Pandas Head, Tail and Sample." GeeksforGeeks, 23 Jul. 2025, www.geeksforgeeks.org/python/difference-between-pandas-head-tail-and-sample/.
18. Ayilara OF, Zhang L, Sajobi TT, Sawatzky R, Bohm E, Lix LM. Impact of missing data on bias and precision when estimating change in patient-reported outcomes from a clinical registry. Health Qual Life Outcomes. 2019;17(1):106.
19. Mousafi Alasal L, Hammarlund EU, Pienta KJ, Rönnstrand L, Kazi JU. XeroGraph: enhancing data integrity in the presence of missing values with statistical and predictive analysis. Bioinform Adv. 2025;5(1):vbaf035. https://doi.org/10.1093/bioadv/vbaf035.
20. Groenwold RHH, Dekkers OM. Missing data: the impact of what is not there. Eur J Endocrinol. 2020;183(4):E7–9.
21. Van den Broeck J, Cunningham SA, Eeckels R, Herbst K. Data cleaning: detecting, diagnosing, and editing data abnormalities. PLoS Med. 2005;2(10):e267. https://doi.org/10.1371/journal.
22. Samariya D, Ma J, Aryal S, Zhao X. Detection and explanation of anomalies in healthcare data. Health Inf Sci Syst. 2023;11(1):20.
23. Dousti Mousavi N, Aldirawi H, Yang J. Categorical data analysis for high-dimensional sparse gene expression data. BioTech. 2023;12:52. https://doi.org/10.3390/biotech12030052.
24. Bolikulov F, Nasimov R, Rashidov A, Akhmedov F, Cho Y-I. Effective methods of categorical data encoding for artificial intelligence algorithms. Mathematics, 2024;12:2553. https://doi.org/10.3390/math12162553.
25. Poon WY, Wang HB. Analysis of ordinal categorical data with misclassification. Br J Math Stat Psychol. 2010;63(Pt 1):17–42. https://doi.org/10.1348/000711008X401314.
26. Johnson CY, Howards PP, Strickland MJ, Waller DK, Flanders WD. National birth defects prevention study. Multiple bias analysis using logistic regression: an example from the National birth defects prevention study. Ann Epidemiol. 2018;28(8):510–514. https://doi.org/10.1016/j.annepidem.2018.05.009.
27. Nasimian A, Ahmed M, Hedenfalk I, Kazi JU. A deep tabular data learning model predicting cisplatin sensitivity identifies BCL2L1 dependency in cancer. Comput Struct Biotechnol J. 2023;21:956–64. https://doi.org/10.1016/j.csbj.2023.01.020.
28. Shah K, Nasimian A, Ahmed M, Al Ashiri L, Denison L, Sime W, Bendak K, Kolosenko I, Siino V, Levander F, Palm-Apergi C, Massoumi R, Lock RB, Kazi JU. PLK1 as a cooperating partner for BCL2-mediated antiapoptotic program in leukemia. Blood Cancer J. 2023;13(1):139. https://doi.org/10.1038/s41408-023-00914-7.
29. Nasimian A, Al Ashiri L, Ahmed M, Duan H, Zhang X, Rönnstrand L, Kazi JU. A receptor tyrosine kinase inhibitor sensitivity prediction model identifies AXL dependency in leukemia. Int J Mol Sci. 2023;24(4):3830. https://doi.org/10.3390/ijms24043830.
30. Shah K, Ahmed M, Kazi JU. The Aurora kinase/β-catenin axis contributes to dexamethasone resistance in leukemia. NPJ Precis Oncol. 2021;5(1):13. https://doi.org/10.1038/s41698-021-00148-5.
31. Chatfield, C. The initial examination of data J R Stat Soc Series A. 1985;148:214–53.
32. Harris M, Nordheim R, Batzli J. Chapter 2: Examining and understanding your data. In: Process of science companion: data analysis, statistics and experimental design 2019. University of Wisconsin-Madison Biocore.
33. Kashina M, Lenivtceva ID, Kopanitsa GD. Preprocessing of unstructured medical data: the impact of each preprocessing stage on classification, Procedia Comput Sci. 2020;178;284–90
34. Gonzalez Zelaya CV. Towards explaining the effects of data preprocessing on machine learning. In: 2019 IEEE 35th International conference on data engineering (ICDE), Macao, China, 2019. p. 2086–90. https://doi.org/10.1109/ICDE.2019.00245.

# Data Visualization in Biomedicine

## Contents




J. U. Kazi, *Python Essentials for Biomedical Data Analysis: An Introductory Textbook*,
https://doi.org/10.1007/978-3-031-85600-6_6

This chapter discusses the role of data visualization in biomedicine. As the volume and complexity of biological information continue to expand, effective visualization techniques have become valuable tools for simplifying the presentation of complex datasets and identifying patterns relevant to biological processes, drug discovery, and disease diagnosis [1]. The chapter presents a range of common visualization methods, and discusses their possible use in biomedical data presentation and analysis.

**Learning Goals**

The learning goals include understanding the role of data visualization in communicating complex biomedical information. Students will learn various data visualization techniques, including histograms, scatter plots, box plots, line graphs, correlation matrices, heatmaps, clustering, bar graphs, and pie plots. They will also learn to identify specific challenges associated with visualizing biomedical data. Additional goals include effective data visualization for communication with non-experts.



## 6.1 The Role of Data Visualization in Biomedicine

Data visualization lies at the intersection of graphic design, statistics, and information technology, and it has become an integral part of modern biomedicine. In an era characterized by large and complex datasets, the ability to transform data into understandable visual formats is not just helpful; it is key to advancing biomedical research and improving patient care [2]. The significance of data visualization in biomedicine is briefly depicted in ◘ Fig. 6.1.

*Enhancing data comprehension:* Biomedical data often involve complex, multidimensional relationships that can be challenging to interpret using traditional tabular or textual formats [3]. Visualization techniques help to distill these complexities, allowing researchers, clinicians, and healthcare professionals to interpret data more intuitively. This leads to better understanding of biological processes, disease mechanisms, and treatment outcomes, which deepens our knowledge of health and disease.

*Enabling rapid analysis for decision-making:* In clinical settings, timely and evidence-based decisions are paramount. Visualization tools help healthcare professionals to quickly access and interpret data by presenting information in an intuitive format that supports rapid, accurate analysis. Reducing the cognitive load associated with complex data interpretation, visualization facilitates faster decision-making and supports improved patient care outcomes [4].

*Supporting hypothesis generation and validation:* Visualization allows researchers to explore deeper into datasets, often uncovering patterns, correlations, or anomalies that may be overlooked through numerical analysis alone [5]. These observations can lead to new hypotheses, while visual tools further aid in validating these hypotheses by presenting evidence in an interpretable, accessible format that either supports or refutes scientific claims. This process drives innovation and discovery within biomedical research.

*Enhancing communication and collaboration:* Data visualization serves as a bridge across disciplines in biomedical research and healthcare, promoting effective com-

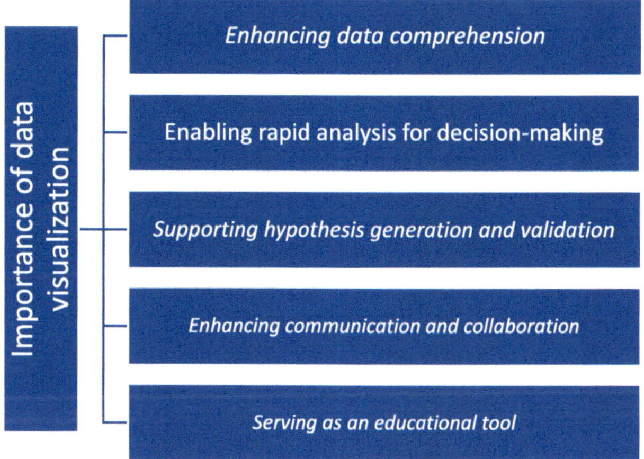


**Fig. 6.1** The importance of data visualization in biomedicine. Key elements that represent its role in biomedical research and healthcare, include enhancing data comprehension, enabling rapid analysis for clinical decision-making, supporting hypothesis generation and validation, facilitating interdisciplinary communication and collaboration, and serving as an educational tool for students and professionals in the field

munication and collaboration. Converting complex datasets into accessible visuals enables diverse teams, such as scientists, clinicians, and policymakers, to share knowledge and promote interdisciplinary understanding. Visual representation facilitates the exchange of ideas and findings, increasing the overall impact of research and clinical efforts [6].

*Serving as an educational tool:* For students and trainees in biomedicine, data visualizations provide an intuitive method for understanding complex biological systems, disease processes, and pharmacological interactions. Visual aids transform abstract scientific concepts into comprehensible forms, making learning materials more engaging and accessible, ultimately enriching the educational experience and supporting knowledge retention.

## 6.2 Basic Data Visualization Tools in Python

Visualization techniques are instrumental in the analysis, interpretation, and communication of complex datasets [7]. Python libraries such as Matplotlib, Seaborn, Plotly, and Bokeh can help researchers to create a diverse array of visualizations, from basic histograms to interactive, web-based plots, that can help to identify patterns, relationships, and trends hidden within raw data. These visualizations facilitate not only the identification of trends and anomalies but also help to gain deeper insights that might be missed in purely statistical analyses.

Converting data into intuitive visual representations aids in identifying trends, detecting outliers, and appreciating subtle nuances that might otherwise go unnoticed. This visual approach strengthens the comprehension of complex data, supports hypothesis generation, and informs further statistical exploration. Visual data exploration is useful not only for scientific audiences but also for clinicians, policymakers, and stakeholders who rely on clear and accessible visual communication of biomedical information.

### 6.2.1 Matplotlib

Matplotlib is a core plotting library in Python that provides a wide array of plotting functions. It allows for the creation of static, publication-quality figures in various formats. Basic plotting functions include, but are not limited to, line plots, scatter plots, bar graphs, histograms, and more. Matplotlib enables quick and straightforward visualization of data. Moreover, it offers extensive customization options, allowing users to modify colors, labels, scales, and other plot attributes to tailor visuals to specific needs. Matplotlib can be easily installed from PyPI using the command `pip install matplotlib`. Comprehensive and up-to-date user guides, tutorials, and examples are available at ▶ matplotlib.org.

6

### 6.2.2 Seaborn

Seaborn is built on top of Matplotlib and specializes in statistical data visualization [8]. It simplifies complex plotting with high-level interfaces and attractive default styles. Seaborn provides functions for visualizing both univariate and multivariate data, emphasizing statistical relationships and patterns. With built-in themes and color palettes, Seaborn enhances the aesthetic appeal of plots, making them more engaging and easier to interpret. Like Matplotlib, Seaborn can be installed via PyPI using the command `pip install seaborn`. Detailed application descriptions, tutorials, and examples are available at ▶ seaborn.pydata.org.

### 6.2.3 Plotly and Bokeh

Plotly and Bokeh are libraries designed for creating interactive visualizations that can be rendered directly in web browsers. These libraries support interactive features such as zooming, panning, and tooltips, allowing users to explore data dynamically. Both libraries enable the integration of plots into web applications, facilitating the sharing of interactive visualizations online. The open-source Plotly library is available for free at ▶ plotly.com/python, with a commercial version marketed as Dash apps. Similarly, Bokeh is a community-developed library available at ▶ bokeh.org. Both Plotly and Bokeh can be installed from PyPI using the respective commands `pip install plotly` and `pip install bokeh`. They also provide extensive user documentation to help users navigate their features effectively.

## 6.3 Creating Basic Plots

We will discuss a range of basic visualization techniques like line plots, histograms, scatter plots, box plots, bar graphs, and pie plots. Each of these visualization techniques serves as a useful tool for data exploration and interpretation in biomedicine.

### 6.3.1 Line Graphs

**Notebook: Section 6.3.1. Line Graphs**
This notebook provides code examples demonstrating line graphs.
Link to the GitHub repository:
▶ https://github.com/sn-code-inside/BioPy
Go to: Chap. 6—▶ Sect. 6.3.1.

Line graphs are particularly effective for displaying trends over time. They are well-suited for time-series data analysis, where the primary objective is to understand how specific variables change over a given period. In a line graph, each point represents a data observation, with the x-axis typically representing time and the y-axis showing the variable of interest. A line connects these points to illustrate the trend or pattern over time.

In biomedical research, line graphs are instrumental for visualizing trends such as disease progression, patient recovery, or changes in biological markers in response to treatment. These graphs provide a clear way to observe changes over time, helping researchers and clinicians identify important patterns such as cyclic behaviors, abrupt changes, or long-term trends.

Below is an example using Matplotlib to create a line graph with multiple series of random data:

```
import matplotlib.pyplot as plt
import numpy as np
days = range(1, 11) # Sample time series data for multiple lines
values1 = [2, 3, 4, 3.5, 4, 4.5, 5, 6, 7, 8]
values2 = np.random.rand(10) * 8
values3 = np.random.rand(10) * 8
# Creating multiple line graphs
plt.plot(days, values1, marker='o', label='Series 1')
plt.plot(days, values2, marker='s', label='Series 2')
plt.plot(days, values3, marker='^', label='Series 3')
# Adding titles and labels
plt.title('Trend Over Time')
plt.xlabel('Days')
plt.ylabel('Value')
plt.legend() # Adding a legend
plt.show() # Show the plot
```

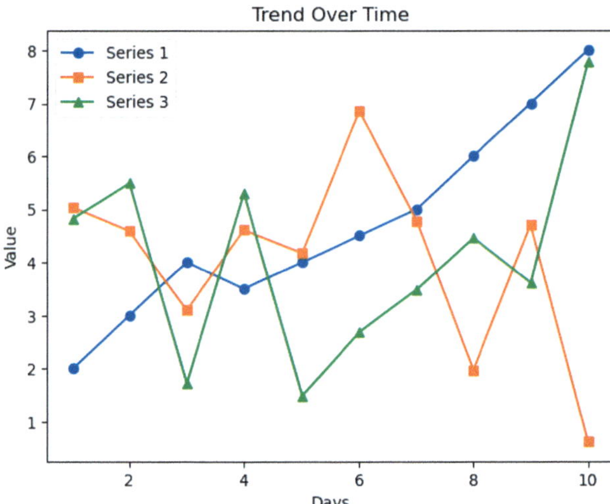


**Fig. 6.2** Comparative line graphs of three data series over time. This figure presents the trends of three distinct data series across a span of 10 days. Each line graph corresponds to a series with unique markers: Series 1 is denoted by blue circles, Series 2 by orange squares, and Series 3 by green triangles



In this example, we have a simple time-series dataset representing random values (such as a biological measurement) over a period of 10 days. The `plt.plot` function is used to create the line graph, where the "days" array is mapped on the x-axis, and the "values" array on the y-axis (Fig. 6.2).

Line graphs are especially useful for visualizing changes in a patient's health metrics over time, such as tumor marker levels, which are often monitored to assess disease progression or treatment response. Below is an example of how to visualize tumor marker levels using a line graph:

```
import matplotlib.pyplot as plt
# Sample data: patient's tumor marker levels over time
time = np.arange(0, 12, 1) # Months
tumor_markers = np.random.normal(30, 5, 12) # Mock data
plt.plot(time, tumor_markers, marker='o', color='green')
plt.title('Tumor Marker Levels Over Time')
plt.xlabel('Time (Months)')
plt.ylabel('Marker Level')
plt.show()
```

The code snippet is designed to plot the variation of a patient's tumor marker levels over a period of 12 months. It employs the matplotlib.pyplot library to generate a line graph with data points marked by circles. The time variable represents the months from 0 to 11, and tumor_markers simulates the patient's tumor marker levels, generated as random values from a normal distribution centered around 30 with a standard deviation of 5 (Fig. 6.3).

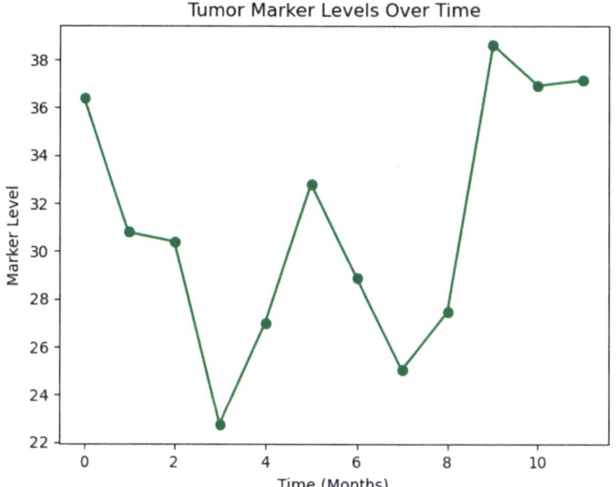


**Fig. 6.3** Tumor marker levels over 12 months. The graph depicts the fluctuation of tumor marker levels in a patient over a year, with each point representing the level recorded each month

### 6.3.2 Histograms

**Notebook: Section 6.3.2. Histograms**

This notebook provides code examples demonstrating histograms.

Link to the GitHub repository:

▶ https://github.com/sn-code-inside/BioPy

Go to: Chap. 6—▶ Sect. 6.3.2.

Histograms are graphical representations of the distribution of numerical data often used for visualizing and understanding the spread of a single numerical variable. In biomedical research, histograms are particularly useful for illustrating biomarker distributions, patient age groups, gene expression levels, and other key variables. They work by dividing the data range into bins, with each bin representing a specific range of values. The height of each bar in the histogram corresponds to the frequency of data points within each bin, providing a clear visual of how the data is distributed.

Histograms are useful for identifying the overall shape of a distribution, such as whether it is symmetric or skewed, detecting multiple modes (peaks), and detecting outliers or gaps in the data. They also provide an estimate of the probability distribution of a continuous variable, aiding in understanding the underlying data structure.

Below is a Python example using Matplotlib to generate a histogram of 1000 data points drawn from a normal distribution:

```
import matplotlib.pyplot as plt
import numpy as np
# Example data: 1000 data points from a normal distribution
data = np.random.normal(0, 1, 1000)
# Creating the histogram
plt.hist(data, bins=30, color='tab:blue',
              edgecolor='white', alpha=0.7)
plt.title('Histogram of Example Data')
plt.xlabel('Value')
plt.ylabel('Frequency')
plt.show()
```



In this example, `np.random.normal` is used to generate a sample dataset of 1000 data points with a mean of 0 and a standard deviation of 1, which are then plotted as a histogram using `plt.hist()`. The `bins` parameter determines how many bins the data range is divided into—in this case, 30. Adjusting the number of bins can change the level of granularity in the histogram, allowing for more detailed or more generalized views of the distribution (◘ Fig. 6.4). The `edgecolor` parameter is used to define the border color of each bin for better visual distinction. Analyzing the histogram, one can infer whether the data is normally distributed, skewed, or if there are any anomalies such as outliers or multiple peaks. This kind of visual analysis may help in understanding your data before proceeding to more complex analyses.

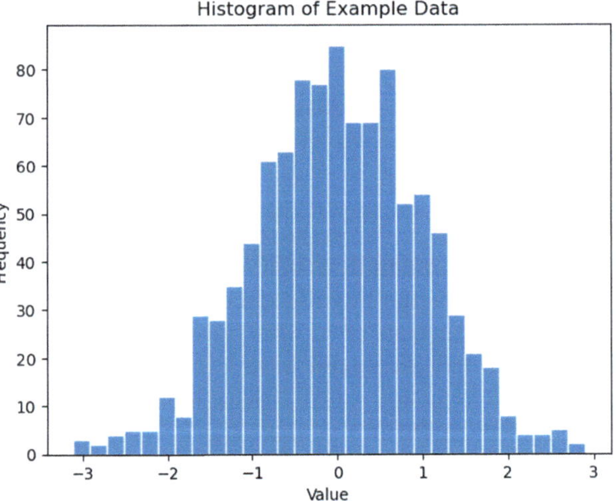


◘ **Fig. 6.4** Histogram of normally distributed data. This histogram represents the frequency distribution of 1000 randomly generated data points following a normal distribution with a mean of 0 and a standard deviation of 1. The x-axis (Value) indicates the data values, while the y-axis (Frequency) shows the number of occurrences for each data bin. The data is divided into 30 bins, providing a detailed view of the distribution's shape. The nearly symmetrical, bell-shaped distribution visible here is characteristic of a normal distribution

In biomedical research, histograms are commonly used to display distributions of variables like gene expression levels. Below is a second example demonstrating how to generate a histogram to visualize simulated gene expression data:

```
import matplotlib.pyplot as plt
import numpy as np
# Sample data: Simulated gene expression levels
# Mock gene expression data
gene_expression = np.random.normal(50, 15, 200)
# Creating the histogram
plt.hist(gene_expression, bins=20, color='tab:blue',
              edgecolor='white', alpha=0.7)
plt.title('Gene Expression Distribution')
plt.xlabel('Expression Level')
plt.ylabel('Frequency')
plt.show()
```

In this example, `random.normal` was used to generate an array of 200 random numbers. These numbers are drawn from a normal (Gaussian) distribution with a mean (loc) of 50 and a standard deviation (scale) of 15. The generated numbers are intended to simulate gene expression levels in a biological dataset (◘ Fig. 6.5).

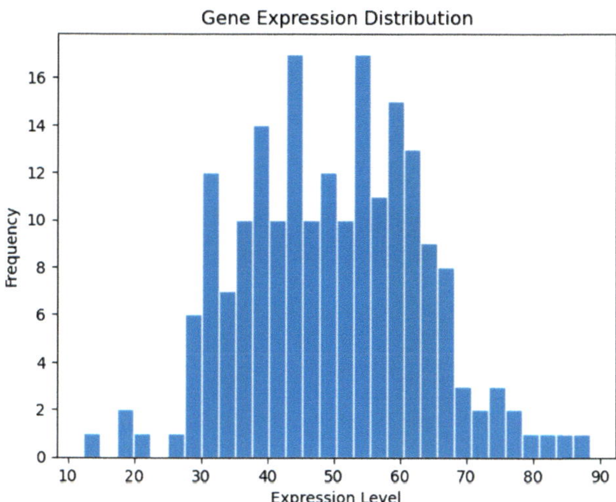


◘ **Fig. 6.5** Histogram of simulated gene expression levels. This histogram displays the distribution of gene expression levels for a sample of 200 data points, which have been generated to follow a normal distribution with a mean of 50 and a standard deviation of 15. The x-axis, labeled "Expression Level," represents the range of expression levels measured, while the y-axis, labeled "Frequency," shows the number of occurrences within each of the 20 bins into which the data range has been divided

### 6.3.3 Scatter Plots

**Notebook: Section 6.3.3. Scatter Plots**
This notebook provides code examples demonstrating scatter plots.
Link to the GitHub repository:
▶ https://github.com/sn-code-inside/BioPy
Go to: Chap. 6—▶ Sect. 6.3.3.



Scatter plots (also known as pair plots) are widely used for exploring relationships between two numerical variables. In a scatter plot, each point represents an observation, plotted according to its values on the x-axis and y-axis. This type of plot is particularly used for identifying patterns, such as correlations or trends, between variables and for detecting outliers that deviate from the general trend.

Scatter plots are frequently used in biomedical research to visualize relationships between different physiological or clinical measurements, such as plotting cholesterol levels against blood pressure readings. They help to reveal whether there is a linear, exponential, or no apparent relationship between the variables. Furthermore, scatter plots can help to identify clusters or outliers, providing information about variability within the dataset.

Here is an example of how to create a scatter plot in Python using Matplotlib and NumPy:

```
import matplotlib.pyplot as plt
import numpy as np
# Generating sample data
x = np.random.rand(50)
y = x + np.random.normal(0, 0.1, 50) # Adding some noise
# Creating the scatter plot
plt.scatter(x, y)
plt.title('Scatter Plot of Sample Data')
plt.xlabel('X Value')
plt.ylabel('Y Value')
plt.show()
```

In this example, `np.random.rand` and `np.random.normal` are used to generate a set of random sample data, representing two variables. The `plt.scatter` function then plots these data points on a scatter plot. Here, x and y are arrays of the same length, where each pair (`x[i]`, `y[i]`) represents a point on the plot (◘ Fig. 6.6). Examining the scatter plot, one can infer the nature of the relationship between the two variables. For instance, if the points roughly form a straight line, it suggests a linear relationship. A cloud of points without any discernible pattern might indicate

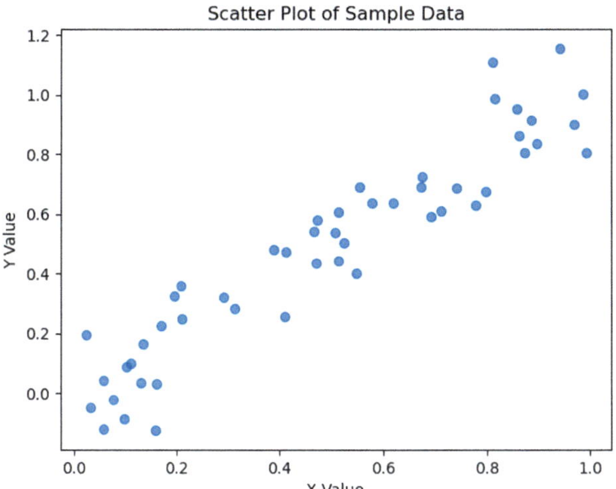


**Fig. 6.6** Scatter plot of correlated sample data. This scatter plot illustrates the relationship between two variables, represented along the x-axis and y-axis, within a dataset of paired values. The distribution of points suggests a positive correlation, as the values of the y-variable tend to increase with increasing values of the x-variable. The visualization is useful for identifying trends, relationships, and possible clusters within the data

no relationship. Any points that significantly deviate from the general pattern can be investigated as possible outliers. Scatter plots are thus a fundamental tool in the exploratory data analysis phase, helping to guide further statistical testing or modeling based on the observed relationships.

Scatter plots are often used in biomedical research to visualize relationships between different biological variables, such as gene expression levels. Below is an example of how to visualize the correlation between two simulated gene expression datasets:

```
import matplotlib.pyplot as plt
# Sample data: two different gene expressions
gene_exp1 = np.random.normal(60, 10, 100) # Mock data
gene_exp2 = gene_exp1 + np.random.normal(0, 5, 100) # Mock data
plt.scatter(gene_exp1, gene_exp2, alpha=0.5)
plt.title('Gene Expression Correlation')
plt.xlabel('Gene 1 Expression Level')
plt.ylabel('Gene 2 Expression Level')
plt.show()
```

The code generates a set of 100 mock data points for `gene_exp1` using a normal distribution centered at 60 with a standard deviation of 10 (Fig. 6.7). Then, it creates a related dataset, `gene_exp2`, by adding a normally distributed random noise with a mean of 0 and a standard deviation of 5 to `gene_exp1`.

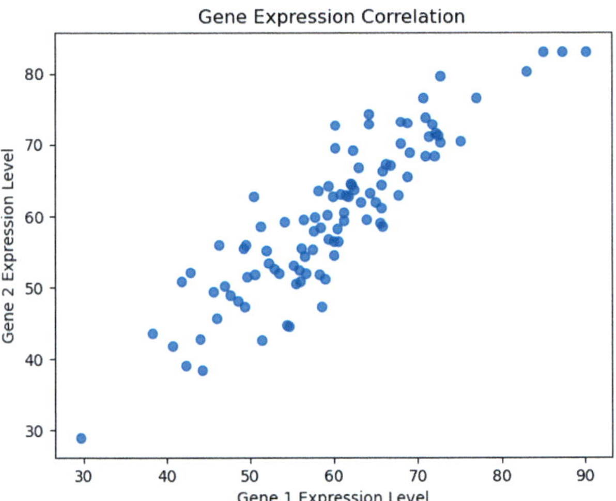




**Fig. 6.7** Scatter plot of gene expression levels correlation. This plot represents the relationship between the expression levels of two genes within a sample of 100 data points. Each point denotes an individual measurement with the expression level of Gene 1 on the x-axis and Gene 2 on the y-axis. The distribution of points suggests a positive correlation, indicating that as the expression level of Gene 1 increases, the expression level of Gene 2 tends to increase as well

The Seaborn library in Python offers advanced functionality for visualizing scatter plots, especially for multivariate data. Below is an example using Seaborn's `pairplot()` function, which creates scatter plots for all variable pairs in a dataset:

```
import seaborn as sns
import matplotlib.pyplot as plt
import pandas as pd
import numpy as np
# Generating sample data
np.random.seed(0)
data = pd.DataFrame(data=np.random.rand(50, 4), columns=['A',
'B', 'C', 'D'])
# Creating the pair plot
sns.pairplot(data)
plt.show()
```

In the provided example, we start by creating a `DataFrame` named `data` using Pandas, consisting of 50 samples and four variables labeled A, B, C, and D. These variables are filled with random numbers. To visualize the relationships between these variables, we use the `sns.pairplot` function from the Seaborn library. This function efficiently generates a grid of axes, arranging each variable of the data along the y-axes of a single row and along the x-axes of a single column. On the diagonal axes, rather than showing scatter plots, the function plots the univariate distribution of the data for each variable. The outcome of this process is a matrix of scatter plots, one for each pair of variables (Fig. 6.8).

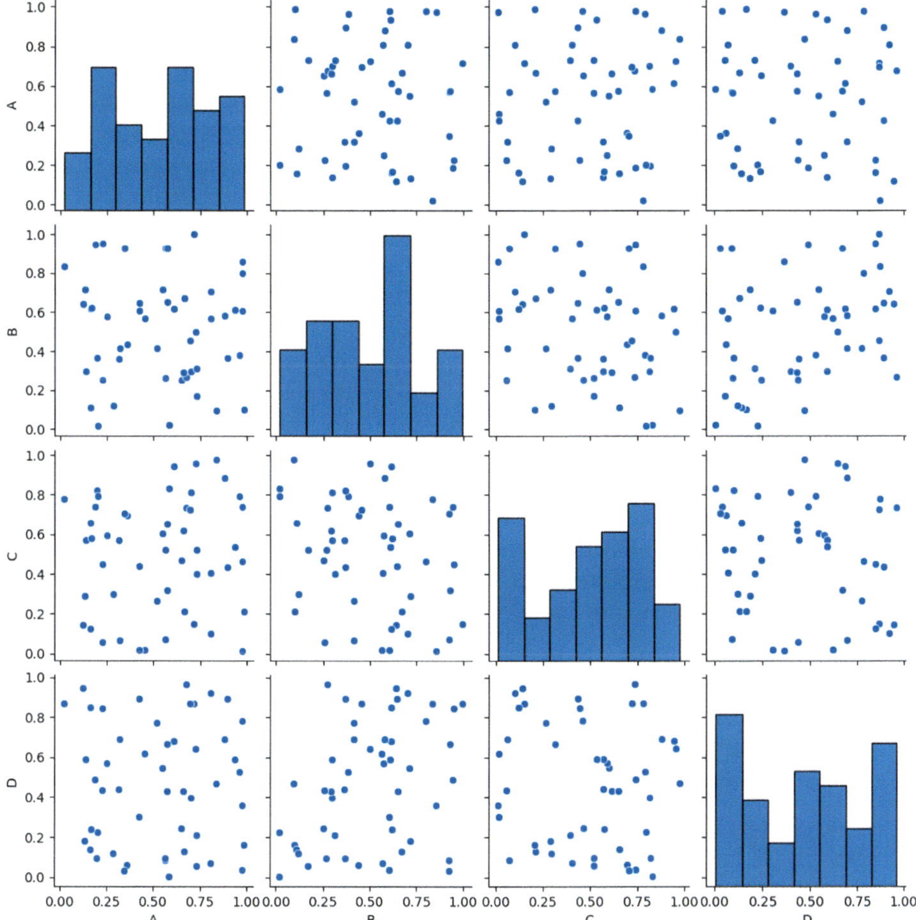


**Fig. 6.8** Pairwise relationships and distributions for multivariate data. This pair plot illustrates the pairwise relationships between four variables (A, B, C, and D) across a dataset of 50 observations. Scatter plots on the off-diagonal panels reveal correlations and patterns between pairs of variables, while histograms on the diagonal show the univariate distribution of each variable

## 6.3.4 Box Plots

**Notebook: Section 6.3.4. Box Plots**

This notebook provides code examples demonstrating box plots.

Link to the GitHub repository:

▶ https://github.com/sn-code-inside/BioPy

Go to: Chap. 6—▶ Sect. 6.3.4.

Box plots, also known as box-and-whisker plots, are highly informative statistical graphs used to display the distribution of datasets. They are particularly effective for visualizing central tendency, variability, and detecting outliers within a dataset. Box plots are commonly used in biomedical research to compare distributions across different groups or categories, making them useful tools for analyzing variables like patient biomarker levels, treatment responses, or physiological measurements.

A box plot visually summarizes a dataset by dividing it into quartiles:

- The box represents the interquartile range (IQR), which contains the middle 50% of the data.
- The median (second quartile) is displayed as a line inside the box.
- The whiskers extend from the box to show the range of the data, excluding outliers.
- Any data points outside the whiskers are considered outliers and are often plotted as individual points.



Box plots provide a concise summary of data distribution, showing important characteristics such as the spread, skewness, and presence of outliers. They are particularly useful when comparing multiple groups, allowing for quick visual comparison of differences between categories.

Below is an example using Matplotlib and NumPy to create a box plot for multiple groups:

```
import matplotlib.pyplot as plt
import numpy as np
# Set a random seed for reproducibility
np.random.seed(0)
# Generating sample data for 15 groups arranged randomly
data = [np.random.normal(np.random.randint(70, 130),
 np.random.randint(5, 30), 200) for _ in range(15)]
# Creating the box plot with 15 groups
plt.boxplot(data, labels=[f'Group {i+1}' for i in range(15)])
# Rotate the x-axis labels for better readability
plt.xticks(rotation=45)
plt.title('Box Plot of Sample Data')
plt.ylabel('Values')
plt.show()
```

In this example, we generate sample data for 15 different groups, each group containing 200 data points. The `plt.boxplot` function is then used to create the box plot. Each box in the plot corresponds to one group of data, showing the median, interquartile range, and any outliers (◘ Fig. 6.9).

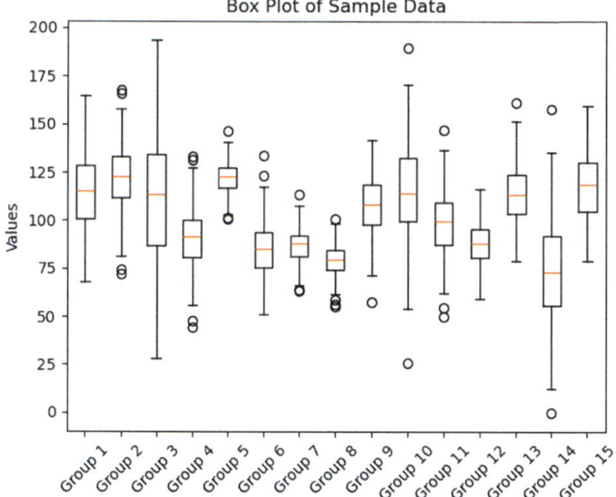


**Fig. 6.9** Box plot comparing 15 random groups. This figure displays a box plot for 15 distinct groups with values ranging approximately from 25 to 200. Each box represents the interquartile range (IQR) of the data for that group, with the horizontal line within the box indicating the median value

### 6.3.5 Correlation Matrices, Heatmaps, and Clustering

**Notebook: Section 6.3.5. Correlation Matrices, Heatmaps, and Clustering**
This notebook provides code examples demonstrating correlation matrices, heatmaps, and clustering.
Link to the GitHub repository:
▶ https://github.com/sn-code-inside/BioPy
Go to: Chap. 6—▶ Sect. 6.3.5.

Heatmaps transform complex data matrices into a color-coded format, making it easier to perceive patterns and relationships within the data. They are particularly effective in visualizing correlation matrices, where the strength and direction of relationships between multiple variables can be assessed quickly through color variations. In biomedical research, heatmaps are widely used to represent data such as gene expression levels across different conditions, biomarker correlations, or clinical measurements across patient groups.

Heatmaps provide an intuitive way to visualize correlations and detect patterns or clusters in the data. For example, in a heatmap of a gene expression dataset, rows might represent different genes while columns represent experimental conditions or patient samples. Color intensity indicates the magnitude of expression levels, allowing researchers to identify genes that behave similarly across conditions.

Here is an example of creating a heatmap in Python using the Seaborn library, which is built on top of Matplotlib:

```
import seaborn as sns
import numpy as np
import matplotlib.pyplot as plt
# Generating a sample correlation matrix
data = np.random.rand(10, 10)
correlation_matrix = np.corrcoef(data)
# Creating the heatmap
sns.heatmap(correlation_matrix, annot=False, cmap='coolwarm')
plt.title('Heatmap of Correlation Matrix')
plt.show()
```



In this example, `np.random.rand` is used to generate random `data`, and `np.corrcoef` computes the correlation coefficient for each pair of variables, resulting in a correlation matrix. The `sns.heatmap` function from Seaborn is then used to create the heatmap. The `cmap = 'coolwarm'` specifies the color palette used to represent these values (Fig. 6.10). Heatmaps are effective in showing patterns,

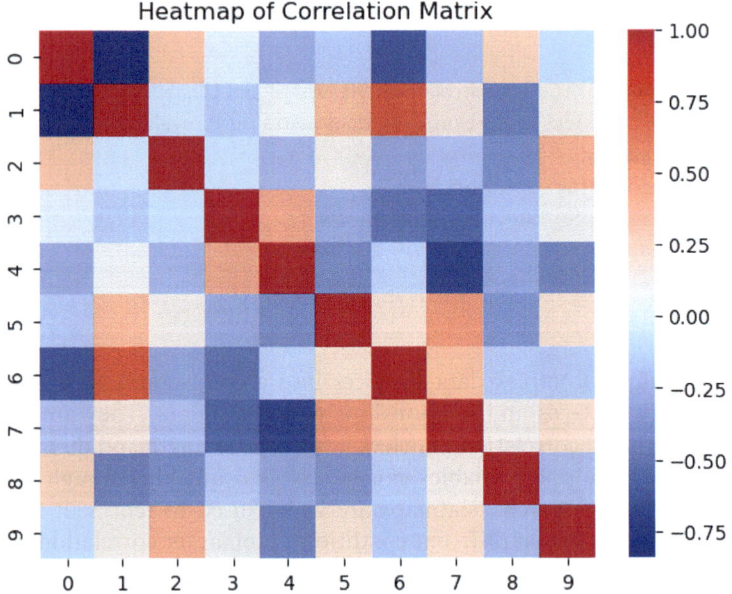


**Fig. 6.10** Heatmap representing a correlation matrix. The heatmap displays the correlation coefficients between pairs of nine variables, ranging from −1 to 1. Each square's color intensity and tone correspond to the strength and nature of the correlation: blue for positive, red for negative, and white for no correlation. The scale on the right provides a reference for interpreting the colors in terms of correlation values

clusters, and outliers within the data. Different colors can indicate strong positive correlations (typically represented by warmer colors like red or orange), strong negative correlations (cooler colors like blue), or lack of correlation (colors close to neutral).

Hierarchical clustering is a technique often combined with heatmaps to uncover groups of variables or samples that behave similarly. In the context of gene expression data, clustering helps group genes with similar expression patterns and experimental conditions that produce comparable gene expression responses. Seaborn's clustermap function performs hierarchical clustering and arranges both rows and columns based on the similarity of their patterns.

Here is an example of how to generate a clustered heatmap:

```
import seaborn as sns; sns.set_theme()
import matplotlib.pyplot as plt
import numpy as np
# Sample data: gene expression data for multiple genes across
# different conditions
data = np.random.normal(0, 1, (10, 10))
# Mock data for 10 genes across 10 conditions
# Generate a heatmap with clustering
cg = sns.clustermap(data, cmap='viridis', method='average',
standard_scale=1)
# Adding axis labels
cg.ax_heatmap.set_xlabel('Conditions')
cg.ax_heatmap.set_ylabel('Genes')
# Rotate the y-axis labels for better readability
plt.setp(cg.ax_heatmap.yaxis.get_majorticklabels(), rotation=0)
# Show the plot
plt.show()
```

In this code, Seaborn's `clustermap` function is used. This function performs hierarchical clustering and arranges the rows and columns of the heatmap based on the similarity of their patterns (◘ Fig. 6.11). It also includes dendrograms at the sides representing the results of this hierarchical clustering.

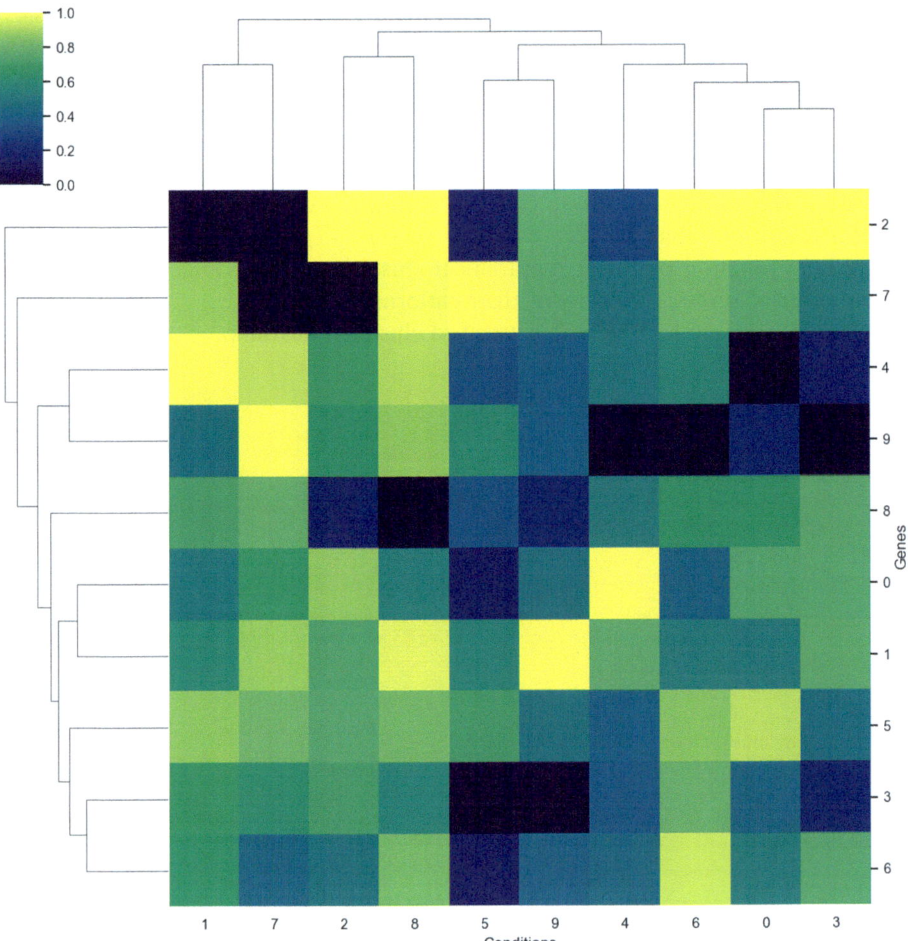


**Fig. 6.11** Clustered heatmap of gene expression data. This clustered heatmap illustrates the standardized expression levels of 10 genes across 10 different conditions, depicted by the color intensity, with violet representing lower and yellow representing higher expression levels. Rows and columns are hierarchically clustered, as shown by the dendrograms on the left and top, grouping together genes with similar expression patterns and conditions with similar overall gene expression profiles, respectively

## 6.3.6 Bar Graphs

**Notebook: Section 6.3.6. Bar Graphs**

This notebook provides code examples demonstrating bar graphs.

Link to the GitHub repository:

▶ https://github.com/sn-code-inside/BioPy

Go to: Chap. 6—▶ Sect. 6.3.6.

Bar graphs are well suited for comparing quantities across different groups or categories. They are primarily used for visualizing categorical data, where each category is represented by a bar, and the height or length of the bar reflects a specific value, such as frequency, mean, or proportion.

In biomedical research, bar graphs are widely used to compare various types of data, including the following:

- Incidence rates of diseases across different demographic groups (e.g., age, gender).
- Average biomarker levels in different patient categories.
- Treatment responses among distinct experimental or clinical groups.

Bar graphs offer a clear way to represent categorical data, making it easy for researchers and healthcare professionals to compare data across groups and draw meaningful conclusions. They are particularly effective for communicating results to both scientific and nonscientific audiences.

Below is an example using Matplotlib to create a bar graph that includes error bars, which represent the standard error of the mean (SEM) for each category:

```
import matplotlib.pyplot as plt
import numpy as np
# Sample data with replicates for each group
data = {
     'Group A': [22, 23, 24],
     'Group B': [43, 45, 47],
     'Group C': [54, 56, 58],
     'Group D': [34, 35, 36],
     'Group E': [58, 60, 62],
     'Group F': [40, 42, 44],
     'Group G': [53, 55, 57],
     'Group H': [46, 48, 50]
}
# Calculating means and SEMs for each group
means = {group: np.mean(values) for group, values in data.items()}
sems = {group: np.std(values, ddof=1) / np.sqrt(len(values)) for group,
values in data.items()}
# Creating the bar cgraph with error bars (SEM)
plt.bar(means.keys(), means.values(), yerr=sems.values(),
color='blue', alpha=0.7, capsize=5)
plt.title('Bar Graph with SEM of Sample Data')
plt.xticks(rotation=45)
plt.xlabel('Category')
plt.ylabel('Value')
plt.show()
```

In this example, we have eight categories and three corresponding values for each. The `plt.bar` function creates the bar graph, with `categories` representing the x-axis and mean values the height of the bars (Fig. 6.12).

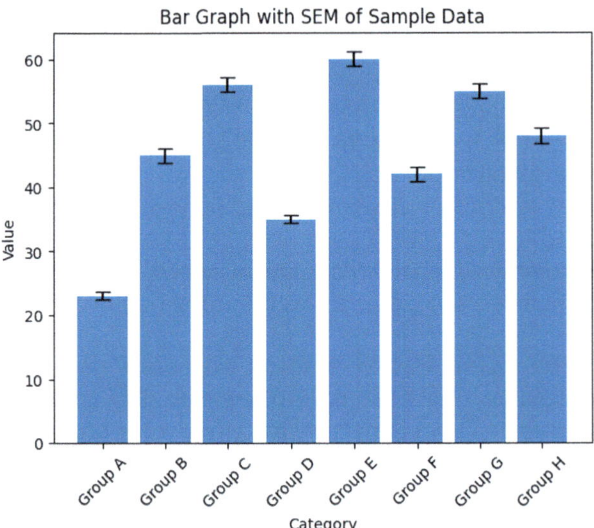


**Fig. 6.12** The bar graph illustrates the mean values for eight distinct groups. The error bars represent the standard error of the mean (SEM), indicating the variability of the data within each group. The consistent use of blue across all bars ensures a clear and cohesive visual comparison of the groups' values



### 6.3.7 Pie Plots

**Notebook: Section 6.3.7. Pie Plots**

This notebook provides code examples demonstrating pie plots.

Link to the GitHub repository:

▸ https://github.com/sn-code-inside/BioPy

Go to: Chap. 6—▸ Sect. 6.3.7.

Pie plots are graphical representations used to illustrate the proportions of different categories within a whole. They display data in a circular graph, where each slice of the pie represents a category, and the size of each slice is proportional to the value or quantity it represents. While pie plots are less commonly used in scientific analysis due to the potential for misinterpretation, especially when comparing multiple groups or dealing with many categories, they can still be useful for showing simple compositions and proportions in a visually appealing format.

In biomedical research, pie plots may be used to visualize data such as:

- The distribution of a population by disease subtype.
- The proportion of patients responding to different treatment protocols.
- The breakdown of a cohort by demographic characteristics, such as age, gender, or ethnicity.

Though they are most effective when dealing with a small number of categories and clear relative differences between the slices, pie plots are a useful tool for presenting data to nontechnical audiences, as they offer a quick, intuitive way to understand proportions.

Below is an example using Matplotlib to create a pie plot in Python:

```
import matplotlib.pyplot as plt
# Sample data
labels = ['Type A', 'Type B', 'Type C', 'Type D']
sizes = [15, 30, 45, 10]
# Creating the pie plot
plt.pie(sizes, labels=labels, autopct='%1.1f%%', startangle=140,
colors=['red', 'green', 'blue', 'orange'])
plt.title('Pie Plot of Sample Data')
# Equal aspect ratio ensures the pie plot is circular.
plt.axis('equal')
plt.show()
```

In this example, `labels` represents the different categories, and `sizes` contains the corresponding values for each category. The `plt.pie` function creates the pie plot, with `autopct` used to display the percentage value of each slice, `startangle` determining the starting angle of the first slice, and colors specifying the color of each slice. Despite their limitations, pie plots can be a useful tool for presenting data in a format that is easy to understand at a glance, particularly for nontechnical audiences (Fig. 6.13). They are best used when the number of categories is small, and the relative size differences between slices are clear.

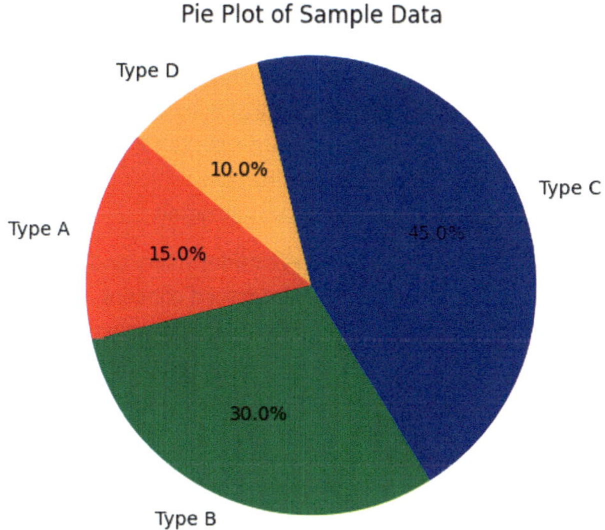


**Fig. 6.13** Pie plot for data distribution. This pie plot depicts the percentage breakdown of data into four categories: Type A, Type B, Type C, and Type D. The chart shows that Type C constitutes the majority with 45.0%, followed by Type B at 30.0%. Type A accounts for 15.0%, while Type D is the smallest at 10.0%. Each category is represented by a unique color for visual differentiation—red for Type A, green for Type B, blue for Type C, and yellow for Type D

## 6.4 Complexities in Visualization of Biomedical Data

Despite its value in interpreting complex biomedical datasets, data visualization in this field faces challenges that can limit its effectiveness. A major issue is handling the sheer volume and complexity of biomedical data, which often includes high-dimensional information like genomic sequences, proteomic profiles, and medical imaging [9]. Visualizing such large datasets requires advanced computational tools and innovative visualization methodologies to ensure that key information is accurately represented. There is also the risk of misinterpretation due to visual artifacts or biases introduced during data processing and visualization. Overcoming these challenges requires not only technological advancements but also a deep understanding of possible pitfalls in data representation and interpretation. Ensuring accuracy, avoiding misinterpretation, and effectively managing large datasets are all important for the successful application of visualization techniques in biomedical research and clinical practice (Fig. 6.14). This involves developing user-friendly visualization interfaces and providing training for researchers and clinicians to interpret visual data correctly. Addressing these issues may enhance transparency of data visualization, thereby advancing biomedical science and improving patient care.



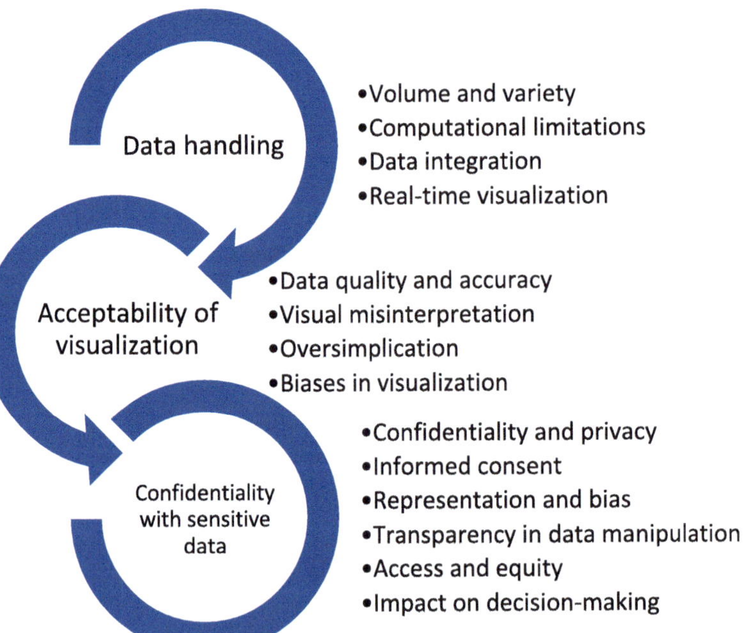


**Fig. 6.14** Data visualization challenges. Data visualization challenges include properly and efficiently handling large datasets, accurately manipulating data while preserving its original integrity, avoiding misinterpretation, and ensuring ethical considerations are met

### 6.4.1 Effective Data Handling

The immense scale and complexity of biomedical datasets pose challenges for effective visualization. As data sources expand in size and variety, including genomics, proteomics, imaging data, and electronic health records, it becomes increasingly difficult to manage and represent these datasets in a meaningful way. Each data type has distinct structures, scales, and characteristics, requiring flexible visualization tools that can adapt to this heterogeneity. Large datasets often exceed the computational capabilities of standard systems, necessitating advanced computing resources with high memory capacity, processing power, and specialized software capable of handling complex data. Integrating data from multiple sources into a unified visual format adds another layer of complexity, as biomedical datasets often differ in scale, format, and quality, making it challenging to create coherent visualizations that effectively combine all available information. Furthermore, in clinical settings, real-time visualization is often required to inform decisions quickly. This demands highly efficient data processing and visualization algorithms that can deliver rapid overviews without sacrificing accuracy.

### 6.4.2 Acceptability of Visualization

Ensuring the accuracy of biomedical data visualizations and preventing misinterpretation are important, given that such visualizations often influence decisions in healthcare and research. The accuracy of visualizations is directly linked to the quality of the underlying data; if the data contains inaccuracies due to collection, processing, or analysis errors, the resulting visualizations can be misleading and lead to incorrect conclusions or decisions. Moreover, the design of a visualization can strongly influence its interpretation. Poor choices in color schemes, scales, or graphical representations can obscure important patterns or mislead viewers, for instance, inappropriate use of color gradients in heatmaps can hide data relationships. While simplifying complex data is one of the goals of visualization, oversimplification can result in the loss of important information. Maintaining a balance between making data comprehensible and retaining its complexity may avoid misrepresenting the data. Moreover, unintentional biases can emerge during data selection, processing, or the visualization itself. Focusing on specific data subsets while excluding others can lead to skewed interpretations, potentially overlooking key trends or outliers.

### 6.4.3 Confidentiality with Sensitive Data

Ethical considerations in biomedical data visualization are applicable, as data visualizations can influence research outcomes, clinical decisions, and public health policies. Ethical data visualization practices address issues such as privacy, informed consent, fairness, and transparency. When visualizing patient data, especially from electronic health records or genetic information, maintaining confidentiality is necessary, and data must be anonymized to prevent unintended disclosure of personal information. It is also important to ensure that informed consent has been obtained for visualizations based on data from human subjects, particularly when the data are displayed publicly or used in research contexts; participants should be aware that their data might be used for visualization purposes.

Careful attention is required to avoid biases that could misrepresent the data, such as ensuring fair representation in demographic visualizations to prevent healthcare disparities like the underrepresentation of minority groups. Transparency in data manipulation is also required; when data is altered for visualization through methods like normalization or scaling, documenting and justifying these changes helps viewers understand how the visualization was created and acknowledges any limitations. Furthermore, ensuring accessibility of visualizations is an ethical responsibility, particularly in public health contexts. Visualizations should be designed to be accessible to a broad audience, including individuals with visual impairments and those without specialized knowledge. Since biomedical visualizations often influence decisions in healthcare and research, ethical practices require that these visualizations be accurate, clear, and structured in a way that supports informed, evidence-based decision-making without misleading the viewer.



## 6.5 Data Visualization in Biomedical Communication

Data visualization can enhance biomedical communication by facilitating better understanding for a wide range of audiences, ranging from non-experts to professionals. It can be used for disseminating research, educating patients, conveying public health messages, and supporting clinical decision-making [10]. The impact of data visualization in biomedical research communication is briefly depicted in ◘ Fig. 6.15.

### 6.5.1 Presenting Data to Nonexperts

Data visualization can be used for translating complex biomedical concepts into accessible formats for nonexperts, including students, patients, policymakers, and the general public [11]. Visual tools simplify complicated subjects such as genetic mutations, disease pathways, and drug mechanisms more understandable by turning abstract ideas into clear, practical information and supporting better comprehension. In educational settings, visualizations play a role in teaching complicated biomedical subjects; interactive models, diagrams, and animations enhance comprehension and retention, particularly for visual learners. For patient communication, visual aids can help in explaining diagnoses, treatment plans, and health risks in a more relatable and less intimidating manner than technical language, enabling patients to make informed decisions about their care. During public health crises, clear and effective visualizations, such as infographics, dashboards, and maps, are useful for conveying important information to the public. These tools help people understand the severity of the situation and the need for preventive measures, facilitating behavioral changes.

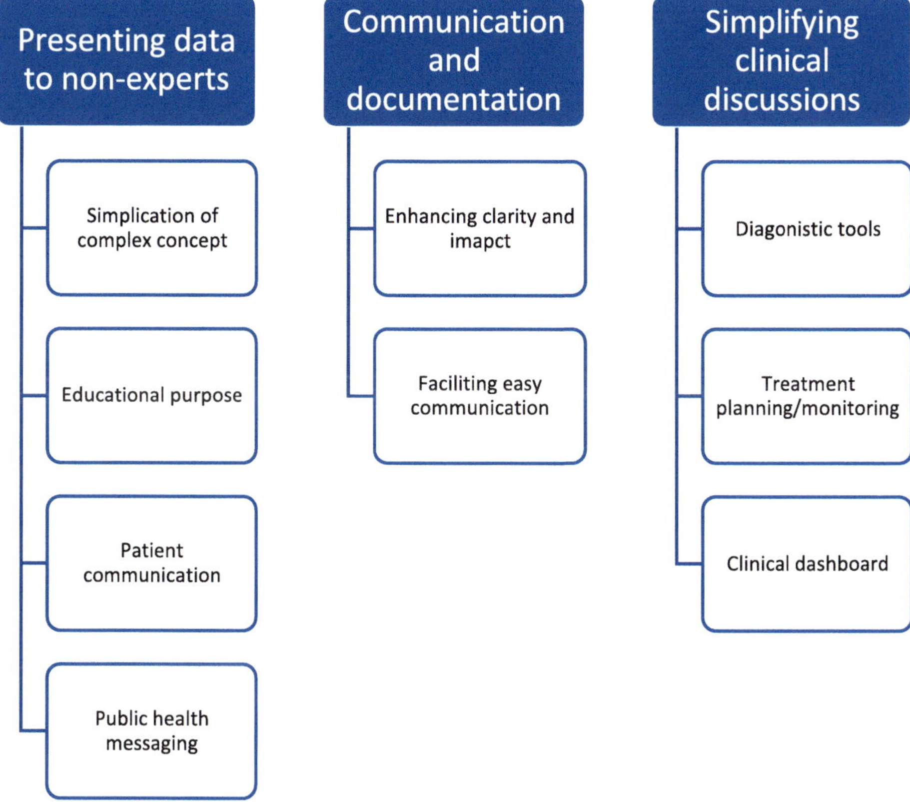


**Fig. 6.15** Data visualization in biomedical communications. Visualization of biomedical data enhances understanding and communication by facilitating easy interpretation and visual explanation to nonexpert audiences, making it easier for decision-making processes in clinical settings

### 6.5.2 Communication and Documentation

In scientific research, the quality and clarity of visualizations can influence the reception and understanding of research findings. Effective visualizations make complex data more digestible and impactful; clear and concise visual elements allow researchers to convey detailed information quickly and efficiently in both publications and presentations, enhancing overall comprehension and engagement. High-quality visualizations also simplify the peer review process by enabling reviewers to quickly assess research methodologies and results. They facilitate collaboration by promoting the exchange of ideas and information, improving communication between researchers from different disciplines. Visualizations also play a role in grant proposals by illustrating the impact, innovation, and feasibility of research projects. Well-designed visuals can help secure funding by clearly communicating the objectives and expected outcomes of the research, thereby emphasizing its significance and viability.

### 6.5.3 Simplifying Clinical Discussions

In clinical practice, visualizations can be used for diagnosis, treatment planning, and patient care, shaping key decision-making processes. Visual representations of imaging data, such as X-rays, MRIs, and CT scans, are needed for disease diagnosis, allowing clinicians to detect abnormalities and identify pathological changes. Moreover, visualizations of laboratory results and other patient data enable clinicians to quickly identify trends or patterns that may influence the diagnostic process. Visual tools can also help with treatment planning; for instance, precise mapping is necessary for radiation therapy in cancer patients to target tumors accurately while minimizing damage to surrounding healthy tissue. These tools aid in monitoring disease progression or treatment response, providing clinicians with a dynamic view of a patient's health over time. Furthermore, clinical dashboards integrate various patient data points into a visual overview, offering clinicians an accessible means for tracking patient health. These dashboards facilitate informed decision-making by presenting relevant data, such as vital signs, lab results, and medication adherence, in an easily interpretable format, thereby increasing the efficiency and effectiveness of patient care.



## 6.6 Exercises and Questions

This section includes exercises and questions designed to reinforce and complement the material covered in this chapter.

**Exercises**

The following exercises are designed to enhance your understanding of the visualization techniques discussed in this chapter:

1. Create a histogram: Use a dataset of patient cholesterol levels to create a histogram that displays the distribution of cholesterol levels among different age groups.
2. Generate scatter plots: Visualize the relationship between body mass index (BMI) and blood pressure in patients from a cardiovascular health study.
3. Develop box plots: Construct box plots to compare the distribution of liver enzyme levels across different patient demographics.
4. Line graphs for trend analysis: Plot the trend of average daily insulin dosage used by diabetic patients over a year.
5. Correlation matrices and heatmaps: Create a heatmap to illustrate the correlation between various blood markers in a metabolic syndrome study.
6. Bar graphs: Design a bar graph to show the prevalence of different types of cancer in a population.
7. Pie plots: Use pie plots to represent the proportion of patients adhering to different types of treatment regimens in a mental health study.
8. Interactive visualization: Create an interactive dashboard that allows clinicians to explore various blood markers and their impact on kidney function.

## Questions to Be Answered

These questions will help you reflect on and apply the concepts covered in this chapter:

1. Why is data visualization particularly important in the field of biomedicine?
2. How can histograms help in understanding the distribution of data in biomedical research?
3. What are the advantages of using scatter plots to analyze relationships between two variables?
4. Describe how box plots can be utilized to detect outliers in clinical data.
5. What insights can correlation matrices and heatmaps provide in a study involving multiple biomarkers?
6. In what situations would a bar graph be more effective than a pie plot in presenting biomedical data?
7. What are some challenges faced when visualizing large and complex biomedical datasets?
8. How do ethical considerations impact the way data should be visualized and presented in biomedicine?

**Acknowledgement** The language of the human-generated text was corrected with the assistance of artificial intelligence (AI) tools [GPT-3.5 and GTP-4 from OpenAI]. GitHub Co-Pilot was used to check the correctness of the codes. The text underwent subsequent human revision to ensure its accuracy.

## References

1. Midway SR. Principles of effective data visualization. Patterns (N Y). 2020;1(9):100141.
2. Aung T, Niyeha D, Heidkamp R. Leveraging data visualization to improve the use of data for global health decision-making. J Glob Health. 2019;9(2):020319.
3. Weissgerber TL, Winham SJ, Heinzen EP, Milin-Lazovic JS, Garcia-Valencia O, Bukumiric Z, et al. Reveal, don't conceal: transforming data visualization to improve transparency. Circulation. 2019;140(18):1506–18.
4. Abudiyab NA, Alanazi AT. Visualization techniques in healthcare applications: a narrative review. Cureus. 2022;14(11):e31355.
5. Li Q. Overview of data visualization. Embodying Data. 2020:17–47. https://doi.org/10.1007/978-981-15-5069-0_2.
6. Schneeweiss S, Glynn RJ. Real-world data analytics fit for regulatory decision-making. Am J Law Med. 2018;44(2–3):197–217.
7. Han S, Kwak IY. Mastering data visualization with Python: practical tips for researchers. J Minim Invasive Surg. 2023;26(4):167–75.
8. Waskom ML. Seaborn: statistical data visualization. J Open Source Softw. 2021;6(60):3021.
9. O'Donoghue SI. Grand challenges in bioinformatics data visualization. Front Bioinform. 2021;1:669186.
10. Organ JM, Taylor AM. Science communication and biomedical visualization: two sides of the same coin. Adv Exp Med Biol. 2023;1421:3–13.
11. Lundkvist A, El-Khatib Z, Kalra N, Pantoja T, Leach-Kemon K, Gapp C, et al. Policy-makers' views on translating burden of disease estimates in health policies: bridging the gap through data visualization. Arch Public Health. 2021;79(1):17.

# Statistical Analysis in Biomedicine

## Contents




J. U. Kazi, *Python Essentials for Biomedical Data Analysis: An Introductory Textbook*,
https://doi.org/10.1007/978-3-031-85600-6_7

This chapter provides an overview of statistical methodologies often used for analyzing biomedical data. It discusses the role of statistical analysis in interpreting biomedical datasets and describes statistical techniques often applied to study drug discovery, disease modeling, and genetic research. The chapter also briefly covers advanced statistical models such as regression analysis, multivariate analysis, and survival analysis.

**Learning Goals**

The learning goals include understanding the characteristics of biomedical data and the role of statistical analysis in biomedical research. Students will learn basic statistical concepts such as descriptive statistics, probability distributions, hypothesis testing, and confidence intervals. They will also learn to apply exploratory data analysis techniques to detect patterns and anomalies and perform preliminary data screenings. Furthermore, this chapter briefly describes advanced statistical methods, including regression analysis, multivariate analysis, and survival analysis.



## 7.1 Statistics in Biomedical Data Analysis

Biomedical data analysis involves the examination and interpretation of various data types derived from biological research and healthcare. It needs a wide range of techniques, from statistical evaluations to advanced computational modeling, with the goal of identifying patterns, associations, and trends within complex biological datasets. This field supports deeper understanding of biological systems, guiding clinical decision-making, and shaping the development of new therapeutic strategies. Consequently, it is integral to drug discovery, personalized medicine, and the broader scope of biomedicine, emphasizing the need for precise, methodologically sound analysis.

### 7.1.1 Overview of Biomedical Data Characteristics

Biomedical data includes a wide array of information types, each with unique characteristics relevant to medical research and clinical practice. Data is originated from sources such as clinical trials, patient records, laboratory experiments, and genomic studies, this data is inherently diverse and complex. Its heterogeneity is evident in the spectrum from structured data like electronic health records to unstructured data such as medical images and genomic sequences [1, 2]. The volume of biomedical data has surged, especially with the frequent use of high-throughput technologies in genomics and proteomics, presenting both opportunities and challenges in data management. Moreover, the multidimensional complexity arises from various layers of information, including genetic, epigenetic, and phenotypic data. Biomedical data is also dynamic, reflecting changes over time like disease progression, treatment responses, and evolutionary patterns in genetics. Given its association with individual health information, it requires strict adherence to data privacy rules and ethical standards [3].

### 7.1.2 Importance of Statistical Analysis in Biomedical Research

Statistical analysis validates scientific claims by determining whether results are statistically significant, ensuring findings are not due to random chance [4]. Due to the natural variability in biological data, statistical methods quantify this variability, allowing for more accurate interpretations. Statistical principles can be applied in experimental design, including determining appropriate sample sizes and minimizing biases, which improve the robustness of research results. In drug discovery and disease modeling, statistical analysis supports predictive modeling that assesses treatment effects and helps to monitor disease progression. Furthermore, statistical methods contribute to personalized medicine by analyzing patient-specific data to develop tailored treatment strategies [5]. The application of statistical analysis ensures the reliability and validity of biomedical research, advancing healthcare and medicine.

### 7.1.3 Challenges in Biomedical Data Analysis

Biomedical data analysis, despite its importance, faces several challenges that can limit the accurate interpretation of data. One major challenge is data quality and integration; issues such as missing data, noise, and bias can affect the reliability of analyses. Integrating heterogeneous data from diverse sources, such as genomic data, electronic health records, and medical imaging, adds layers of complexity due to differences in formats, standards, and terminologies. Another hurdle is high dimensionality, as biomedical datasets often contain more variables (features) than observations. This "curse of dimensionality" complicates statistical analysis and increases the risk of overfitting, where models perform well on training data but poorly on new data [6]. Concerns around privacy and data protection are also prominent, especially when dealing with sensitive patient information. Adherence to legal and ethical requirements, such as the Health Insurance Portability and Accountability Act (HIPAA), demands robust data security measures [7]. The interdisciplinary nature of biomedical research requires effective collaboration among statisticians, biologists, clinicians, and computer scientists, which can be hindered by communication barriers and differing methodological approaches. Moreover, keeping pace with technological advances is demanding, as rapid developments in biotechnology often outpace the creation of new statistical methods, requiring researchers to continuously adapt and learn to keep up with emerging technologies.

## 7.2 Fundamental Statistical Concepts

The basic statistical concepts and tools provide the groundwork for analyzing and interpreting biomedical data [8]. Understanding and applying them correctly is required for drawing accurate and reliable conclusions.

### 7.2.1 Descriptive Statistics

**Notebook: Section 7.2.1. Descriptive Statistics**
This notebook provides code examples demonstrating descriptive statistics.
Link to the GitHub repository:
▶ https://github.com/sn-code-inside/BioPy
Go to: Chap. 7—▶ Sect. 7.2.1.

Descriptive statistics provide a summary of key characteristics of a dataset, including measures of central tendency, variability, skewness, and kurtosis. Python libraries such as Pandas and NumPy can be used for computation of these metrics.



- **Calculating Central Tendency Measures**

Calculating measures of central tendency (*mean, median, and mode*) and variability (range, interquartile range (IQR), standard deviation, and variance) is useful, as they provide a clear view of how data are distributed [9]. Each of these measures reflects different aspects of the dataset. For instance, mean provides a general idea of the data's central value, but it can be influenced by outliers. The median, being the middle value, is less affected by extreme values. The mode shows the most common value, which is particularly useful in categorical data. Range, IQR, standard deviation, and variance are all measures that describe how spread out the data are, with variance and standard deviation providing more detailed information about how values deviate from the mean.

Here is an overview of these measures along with example Python codes for each:

*Mean:* The average value of the dataset is the sum of all values divided by the number of values. It is useful for understanding the typical value of a dataset, but it can be skewed by outliers or non-normal distributions.

$$\text{Mean} = \frac{1}{n}\sum_{i=1}^{n} x_i,$$

where $n$ is the number of values and $x_i$ $i$th value of $x$.

```
import numpy as np
data = np.array([1, 2, 3, 4, 4, 5, 6])
mean = np.mean(data)
print(mean)
```

*Median:* The middle value in the dataset, when it is sorted in ascending order. Median is more robust than the mean in the presence of outliers or skewed distributions. It is particularly relevant in datasets with skewed distributions, common in biomedical data (e.g., survival times).

$$\text{Median} = \begin{cases} x_{\left(\frac{n+1}{2}\right)} & \text{if } n \text{ is odd} \\ \frac{1}{2}\left(x_{\left(\frac{n}{2}\right)} + x_{\left(\frac{n}{2}+1\right)}\right) & \text{if } n \text{ is even} \end{cases}$$

```
import numpy as np
data = np.array([1, 2, 3, 4, 4, 5, 6])
median = np.median(data)
print(median)
```

*Mode:* The most frequently occurring value in the dataset. It is important for identifying the most common category or value, especially in categorical data. There may be one mode, more than one, or none (if all values are unique).

```
from scipy import stats
data = np.array([1, 2, 3, 4, 4, 5, 6])
mode = stats.mode(data)[0]
print(mode)
```

*Range:* The difference between the maximum and minimum values in the dataset. It gives a sense of the spread of the data.

$$\text{Range} = \max(x_i) - \min(x_i)$$

```
import numpy as np
data = np.array([1, 2, 3, 4, 4, 5, 6])
range = np.ptp(data)
print(range)
```

*Interquartile range (IQR):* The range between the first quartile (25th percentile) and the third quartile (75th percentile). It is useful for understanding the spread of the middle 50% of the data and less affected by outliers.

$$\text{IQR} = Q_3 - Q_1$$

```
import numpy as np
data = np.array([1, 2, 3, 4, 4, 5, 6])
Q1 = np.percentile(data, 25)
Q3 = np.percentile(data, 75)
IQR = Q3 - Q1
print(IQR)
```

*Standard deviation:* A measure of the amount of variation or dispersion in a set of values. The standard deviation indicates the average distance of each data point from the mean, showing the spread or variability within the dataset.

$$\mathrm{SD} = \sqrt{\frac{1}{n}\sum_{i=1}^{n}\left(x_i - \mathrm{Mean}\right)^2}$$

```
import numpy as np
data = np.array([1, 2, 3, 4, 4, 5, 6])
sd = np.std(data)
print(sd)
```

*Variance:* Variance is the average of the squared differences from the mean, providing a measure of how spread out the values are. It is the square of the standard deviation.



$$\mathrm{Variance} = \frac{1}{n}\sum_{i=1}^{n}\left(x_i - \mathrm{Mean}\right)^2$$

```
import numpy as np
data = np.array([1, 2, 3, 4, 4, 5, 6])
variance = np.var(data)
print(variance)
```

### ■ Skewness and Kurtosis in Biomedical Data

Skewness and kurtosis are important statistical concepts used to describe the shape and characteristics of the distribution of data in various fields, including biomedical research [10, 11].

*Skewness* refers to the degree of asymmetry observed in a probability distribution. In simpler terms, it indicates whether the data points tend to lean toward the left (negative skew) or right (positive skew) side of the distribution. In a perfectly symmetrical distribution, such as a normal distribution, the skewness is zero. In biomedical data, skewness can reveal important characteristics about the data. For example, the distribution of certain biomarkers or physiological measurements might be skewed due to underlying biological processes or treatment effects. Understanding the skewness in such data can be informative for choosing the right statistical tests and models, as many standard techniques assume a normal distribution.

*Kurtosis*, on the other hand, measures the "tailedness" of the distribution, basically how sharp the peak of the distribution is and how heavy the tails are compared to a normal distribution. High kurtosis (leptokurtic) indicates a distribution with heavy tails and a sharp peak, often signifying the presence of outliers. Low kurtosis (platykurtic) suggests a distribution with lighter tails and a flatter peak. In biomedical data, kurtosis can be indicative of the presence of outliers or unusual data points, which could be useful in clinical research for identifying rare but significant occurrences or reactions in patients.

Both skewness and kurtosis provide deeper understanding beyond the basic measures of central tendency and variability. They allow researchers to understand the underlying structure of the data, make more informed decisions about data processing and analysis, and interpret results more accurately in the context of biomedical research. A clear picture of these aspects of data distribution can be beneficial for hypothesis testing, model building, and interpreting the results in a meaningful way.

```
import pandas as pd
import numpy as np
from scipy.stats import skew, kurtosis
np.random.seed(0) # Creating normally distributed data
sample_data = np.random.normal(0, 1, 1000) # 1000 data points
data = pd.DataFrame(sample_data, columns=['Sample Values'])
# Calculate skewness and kurtosis
data_skewness = skew(data['Sample Values'])
data_kurtosis = kurtosis(data['Sample Values'], fisher=False)
print(f'Skewness: {data_skewness}; Kurtosis: {data_kurtosis}')
```

When this code is run, it yields the skewness and kurtosis of the sample data. In this example, the skewness is approximately 0.034, indicating that the data distribution is fairly symmetrical. The kurtosis is approximately 2.953, which is close to 3, indicating that the data's peakedness is close to that of a normal distribution (◘ Fig. 7.1).

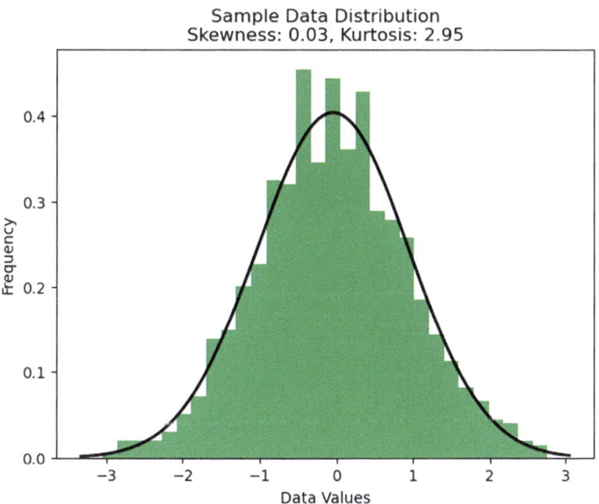


◘ **Fig. 7.1** The frequency distribution of example data. The histogram in green represents the observed frequency of data values, while the smooth curve indicates the expected normal distribution based on the sample's mean and standard deviation. Skewness and kurtosis are annotated in the top right, quantifying the asymmetry (0.03) and tailedness (2.95) of the distribution relative to a normal distribution

### 7.2.2 Probability Distributions

**Notebook: Section 7.2.2. Probability Distributions**
This notebook provides code examples demonstrating probability distributions.
Link to the GitHub repository:
▶ https://github.com/sn-code-inside/BioPy
Go to: Chap. 7—▶ Sect. 7.2.2.

Probability distributions are mathematical functions that describe the probabilities of occurrence of different possible outcomes for a random variable. They are useful for modeling the randomness in various phenomena and processes.

**Discrete distributions** are used when the random variable takes on distinct, countable values. The Binomial Distribution is a classic example of a discrete distribution. It describes the number of successes in a fixed number of independent Bernoulli trials (like coin flips), each with the same probability of success. For instance, it can model the number of patients, out of a total, who respond to a certain treatment.

```
import numpy as np
import matplotlib.pyplot as plt
from scipy.stats import binom, norm
# modeling the number of patients responding to a treatment
n_patients = 100 # Total number of patients
# Probability of a patient responding positively to the treatment
P_success = 0.65
# Generate a binomial distribution
binomial_distribution = binom(n_patients, p_success)
# Plot the probability mass function (PMF)
x = np.arange(0, n_patients+1)
pmf = binomial_distribution.pmf(x)
# Removed the use_line_collection argument
plt.stem(x, pmf, linefmt='b-', markerfmt='bo', basefmt='r-')
plt.title('Binomial Distribution PMF')
plt.xlabel('Number of Patients Responding Positively')
plt.ylabel('Probability')
plt.grid(True)
plt.show()
```

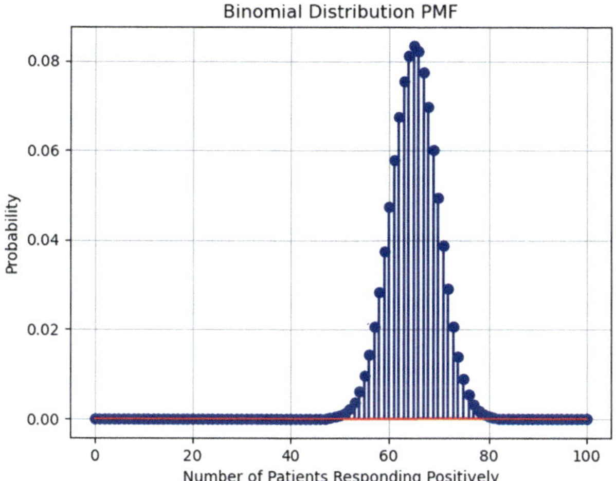


**Fig. 7.2** The probability mass function (PMF) of a binomial distribution. The binomial distribution represents the likelihood of varying numbers of patients responding positively to treatment within a sample of 100 patients, assuming a 65% individual success probability. The vertical stems, with blue markers at each data point, indicate the probability of each outcome, ranging from 0 to 100 patients. The peak around 65 patients reflects the highest probability, with the expected number of positive responses given the success rate, and illustrates the distribution's symmetry around the mean value

This code models the probability of patients responding positively to a treatment using a binomial distribution, given a sample of 100 patients and a 65% success rate (Fig. 7.2). It calculates the probability mass function (PMF) to determine the likelihood of any given number of successful responses. These probabilities are then plotted as a stem plot with blue stems and markers and a red baseline, which visually represents the discrete nature of the binomial distribution's possible outcomes against the number of positive responses.

**Continuous distributions**, on the other hand, apply when the random variable can take on any value in a continuous range. The Normal Distribution, also known as the Gaussian distribution, is the most prominent continuous distribution. It is often used to model natural phenomena like the distribution of blood pressure readings across a population, because many biological measurements tend to cluster around a mean value with a symmetric spread on either side.

```
import numpy as np
import matplotlib.pyplot as plt
from scipy.stats import norm
mean_bp = 120 # Mean systolic blood pressure
std_dev_bp = 15 # Standard deviation of systolic blood pressure
# Generate a normal distribution
normal_distribution = norm(mean_bp, std_dev_bp)
# Plot the probability density function (PDF)
x = np.linspace(mean_bp - 4*std_dev_bp, mean_bp + 4*std_dev_bp,
1000)
pdf = normal_distribution.pdf(x)
plt.plot(x, pdf, 'b-')
plt.title('Normal Distribution PDF')
plt.xlabel('Systolic Blood Pressure Readings')
plt.ylabel('Probability Density')
plt.grid(True)
plt.show()
```



This code generates a plot of the probability density function for a normal distribution with a mean of 120 and a standard deviation of 15, representing systolic blood pressure readings. It uses `numpy` for numerical range creation and `matplotlib` for plotting (◘ Fig. 7.3).

*Let's take another example:*

```
import matplotlib.pyplot as plt
import numpy as np
from scipy.stats import norm, binom, shapiro
# For reproducibility
np.random.seed(0)
# Sample size
n_patients = 100
# Generating synthetic data for different health parameters
blood_sugar = np.random.normal(100, 20, n_patients)
# Normally distributed
blood_pressure = np.random.normal(120, 15, n_patients)
# Normally distributed
diabetes = np.random.binomial(1, 0.2, n_patients)
# Binomially distributed (20% have diabetes)
age = np.random.normal(40, 12, n_patients)  # Normally distributed
pulse = np.random.normal(70, 10, n_patients)  # Normally distributed
wbc_count = np.random.normal(6.5, 1.5, n_patients)
# Normally distributed
hb = np.random.normal(14, 1.5, n_patients)  # Normally distributed
rbc = np.random.normal(5, 0.5, n_patients)  # Normally distributed
# Collecting all parameters in a dictionary for ease of iteration
parameters = {
    'Blood Sugar': blood_sugar,
    'Blood Pressure': blood_pressure,
    'Diabetes': diabetes,
    'Age': age,
    'Pulse': pulse,
    'WBC Count': wbc_count,
```

```
        'Hemoglobin (Hb)': hb,
        'Red Blood Cell (RBC) Count': rb}
# Set up the figure and axes for multiple subplots
fig, axs = plt.subplots(2, 4, figsize=(12, 6))
axs = axs.ravel()
# Iterate through the parameters and plot each distribution
for i, (param_name, values) in enumerate(parameters.items()):
    # Perform Shapiro-Wilk test to assess the distribution
    shapiro_test = shapiro(values)
    # Check if the distribution is binomial (for diabetes)
    if param_name == 'Diabetes':
        # Use bar plot for binomial distribution
        axs[i].bar(['No', 'Yes'], [n_patients - sum(values),
        sum(values)], color='orange')
    else:
        # Use histogram for normal distribution
        axs[i].hist(values, bins=15, density=True, color='skyblue',
        alpha=0.7)
        # Fit a normal distribution curve
        mu, std = norm.fit(values)
        xmin, xmax = axs[i].get_xlim()
        x = np.linspace(xmin, xmax, 100)
        p = norm.pdf(x, mu, std)
        axs[i].plot(x, p, 'k', linewidth=2)
    axs[i].title.set_text(
    f'{param_name}\nShapiro-Wilk p-value: {shapiro_test.pvalue:.2e}')
    # Set common labels
    axs[i].set_xlabel('Value')
    axs[i].set_ylabel('Probability')
# Adjust layout to prevent overlap
plt.tight_layout()
plt.show()
```

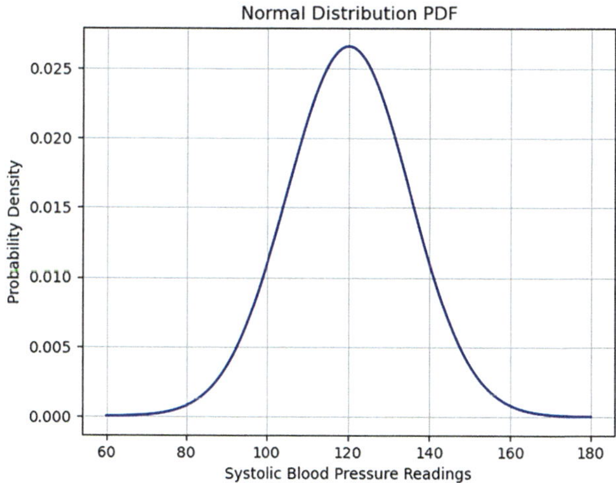


**Fig. 7.3** The probability density function (PDF) for a normal distribution. The PDF for a normal distribution of systolic blood pressure readings. Centered around a mean of 120 with a standard deviation of 15, the curve depicts the likelihood of various blood pressure readings within the population, showing a bell-shaped distribution that is characteristic of normally distributed data

This example code generates a series of subplots for various health parameters of 100 patients, including normally distributed measurements like blood sugar, blood pressure, age, pulse, white blood cell (WBC) count, hemoglobin (Hb), and red blood cell (RBC) count, and a binomially distributed parameter for diabetes (yes/no). Each subplot displays the distribution of the data with a histogram for continuous variables and a bar plot for the binary diabetes variable (◘ Fig. 7.4). Superimposed on the histograms are the probability density functions, fitted assuming a normal distribution. Additionally, the Shapiro-Wilk test is performed for each parameter to assess the normality of the distribution, with the p-values indicated on each subplot. The lower p-values suggest a deviation from the normal distribution, as seen significantly in the diabetes parameter, which follows a binomial distribution.

### 7.2.3 Hypothesis Testing



**Notebook: Section 7.2.3. Hypothesis Testing**

This notebook provides code examples demonstrating hypothesis testing.

Link to the GitHub repository:

▸ https://github.com/sn-code-inside/BioPy

Go to: Chap. 7—▸ Sect. 7.2.3.

Hypothesis testing allows us to draw conclusions about populations based on sample data. The process begins by proposing two competing hypotheses: the null hypothesis ($H_0$), which suggests that there is no effect or difference, and the alternative hypothesis ($H_1$), which suggests that there is an effect or a difference. To determine which hypothesis is supported by the sample data, a significance level ($\alpha$) is chosen, typically set at 0.05, representing a 5% risk of concluding that a difference exists when there is none.

A test statistic is then calculated from the sample data, which measures how compatible the data are with H0. Depending on the nature of the data and the hypothesis, different tests are used, such as t-tests for comparing means or chi-square tests for categorical data. A decision is then made on whether the test statistic falls within the threshold defined by $\alpha$. If it does, $H_0$ is rejected in favor of $H_1$; otherwise, we fail to reject $H_0$.

Here is an example code for a two-sample t-test, which is used to determine if there are statistically significant differences between the means of two independent groups:

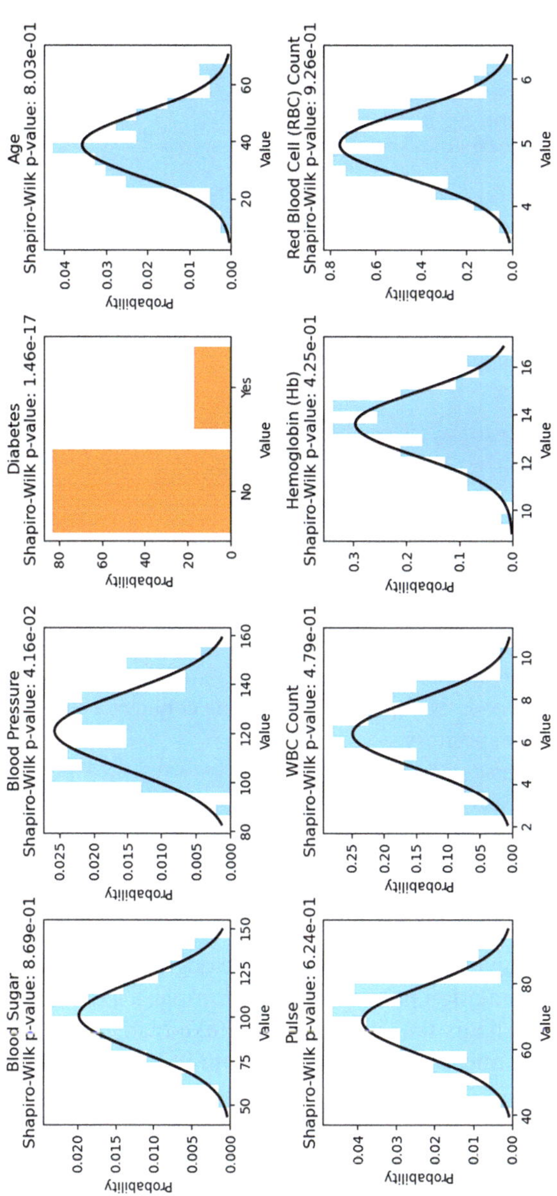


**Fig. 7.4** The probability distributions of various health parameters. The plots represent the distributions of various health parameters. Each plot for continuous variables includes a fitted normal distribution curve for reference. The Shapiro-Wilk test results are provided for each variable, assessing the adherence to a normal distribution

```
import numpy as np
from scipy import stats
# Generating random sample data
# Let's say Group A represents a control group and Group B
# represent a treatment group
np.random.seed(0)
group_A = np.random.normal(100, 10, 50) # Control group
group_B = np.random.normal(102, 10, 50) # Treatment group
# Step 1: Formulate the hypotheses
# H0: μA = μB (the means are equal)
# H1: μA ≠ μB (the means are not equal)
# Step 2: Select a significance level
alpha = 0.05
# Step 3: Calculate the test statistic
t_statistic, p_value = stats.ttest_ind(group_A, group_B)
# Step 4: Make a decision
print(f'T-statistic: {t_statistic}')
print(f'P-value: {p_value}')
if p_value < alpha:
    print("Reject the null hypothesis: There is a significant"
    "difference between the groups.")
else:
    print("Fail to reject the null hypothesis: There is"
    "no significant difference between the groups.")
```



### 7.2.4 Confidence Intervals

**Notebook: Section 7.2.4. Confidence Intervals**

This notebook provides code examples demonstrating confidence intervals.

Link to the GitHub repository:

▶ https://github.com/sn-code-inside/BioPy

Go to: Chap. 7—▶ Sect. 7.2.4.

Confidence intervals (CI) are used to express the reliability of an estimate. They provide a range around a sample estimate to express the degree of uncertainty associated with the estimate of a population parameter. For example, a 95% confidence interval around a sample mean tells us that if we were to take many samples and build an interval estimate for each one, we would expect about 95% of those intervals to contain the population mean.

The width of a confidence interval gives us some idea about how uncertain we are about the estimated population parameter: narrower intervals represent more precise estimates, while wider intervals indicate less precision. Confidence intervals are commonly used in research to demonstrate the reliability of an estimate.

Here is an example using Python to calculate the 95% confidence interval for the mean effect of a drug:import numpy as np

```
import scipy.stats as stats
# Generating a synthetic dataset: effects of a drug on blood
# pressure reduction
np.random.seed(0)
sample_size = 100
# mean effect is 10, with a standard deviation of 3
drug_effect = np.random.normal(10, 3, sample_size)
# Calculate the sample mean and standard error of the mean (SEM)
sample_mean = np.mean(drug_effect)
sem = stats.sem(drug_effect)
# Calculate the 95% confidence interval
confidence_level = 0.95
ci = stats.t.interval(confidence_level, df=sample_size-1,
loc=sample_mean, scale=sem)
print(f"The 95% confidence interval for the mean effect of"
f" the drug is: {ci}")
```

This code will output a 95% confidence interval for the mean effect of a drug based on a sample. It assumes the drug effect follows a normal distribution, which is a common assumption in the calculation of confidence intervals. The `stats.t.interval` function is used to calculate the interval, accounting for the t-distribution that is appropriate for small sample sizes. If the sample size were larger (e.g., greater than 30), the normal distribution could be used instead.

## 7.3 Statistics in Exploratory Data Analysis

Exploratory data analysis (EDA) in biomedicine is used to understand and prepare data for in-depth statistical modeling and hypothesis testing [12]. It involves summarizing the main characteristics of the data, often with visual methods, to identify patterns, detect anomalies, and assess underlying structure. This process can guide the direction of subsequent analysis and supports more informed research decisions.

### 7.3.1 Descriptive Statistics and Data Visualization Techniques in EDA

Descriptive statistics and data visualization are components of exploratory data analysis, used to summarize and interpret complex biomedical datasets effectively. As discussed above as well, calculating measures of central tendency, such as the mean, median, and mode, helps understand the typical values within a dataset, like average blood pressure readings in a patient cohort. Assessing measures of dispersion, including the range, interquartile range, variance, and standard deviation, provides information about the variability of the data, which is required for understanding the spread of biomarkers or genetic expression levels.

Data visualization techniques complement these statistical measures by providing graphical representations. Histograms depict the frequency distribution of a single continuous variable, such as the distribution of patient ages, revealing patterns like skewness or modality. Box plots summarize the distribution of a variable using quartiles and point out outliers, which is useful in comparing clinical measurements across different patient groups. Scatter plots illustrate relationships between two continuous variables, aiding in detecting correlations, for example, between drug dosage and therapeutic response. Heatmaps visualize large-scale data such as gene expression profiles, where color intensity represents the magnitude of measurements across conditions or time points. Line graphs show trends over time, are useful for monitoring disease progression or treatment effects.

### 7.3.2 Identifying Patterns and Anomalies

> **Notebook: Section 7.3.2. Identifying Patterns and Anomalies**
> This notebook provides code examples demonstrating confidence intervals.
> Link to the GitHub repository:
> ► https://github.com/sn-code-inside/BioPy
> Go to: Chap. 7—example data—► Sect. 7.3.2.



Identifying patterns and detecting anomalies are key steps in generating hypotheses and recognizing meaningful results in biomedical research. To facilitate this process, several statistical methods including correlation analysis, cluster analysis, Principal Component Analysis (PCA), and anomaly detection are often used. Correlation analysis quantifies the degree to which two variables are related using statistical measures like Pearson's or Spearman's correlation coefficients, assisting in identifying associations between clinical parameters. Cluster analysis groups data points based on similarity, employing methods like K-means or hierarchical clustering, and is often used in genomics to identify subtypes of diseases based on gene expression patterns. PCA reduces dimensionality by transforming correlated variables into a smaller number of uncorrelated variables called principal components, capturing most of the variance in the data and revealing underlying structures. Anomaly detection, meanwhile, utilizes statistical techniques to identify outliers that deviate significantly from the norm, which could indicate measurement errors or novel biological phenomena requiring further investigation. These statistical methods are integral in exploratory data analysis for pattern recognition and anomaly detection, providing a deeper understanding of the data's structure and relationships.

```
import numpy as np
import pandas as pd
from scipy.stats import zscore
from sklearn.decomposition import PCA
from sklearn.cluster import KMeans
import seaborn as sns
import matplotlib.pyplot as plt

# Example data
data = pd.read_csv('example_data/data.csv')
```

```
# Correlation Analysis
correlation_matrix = data.corr()
sns.heatmap(correlation_matrix, annot=True)
plt.title('Correlation Analysis')
plt.show()

# Cluster Analysis
kmeans = KMeans(n_clusters=3)
clusters = kmeans.fit_predict(data)
data['Cluster'] = clusters
sns.pairplot(data, hue='Cluster')
plt.title('Cluster analysis')
plt.show()

# Principal Component Analysis (PCA)
pca = PCA(n_components=2)
principal_components = pca.fit_transform(data.drop('Cluster',
axis=1))
pca_df = pd.DataFrame(data = principal_components,
                      columns = ['PC1', 'PC2'])
sns.scatterplot(x='PC1', y='PC2', data=pca_df)
plt.title('PCA Result')
plt.show()


# Anomaly Detection using Z-scores
z_scores = np.abs(zscore(data.drop('Cluster', axis=1)))
threshold = 3
outliers = np.where(z_scores > threshold)
outlier_data = data.iloc[outliers[0]]
plt.scatter(data.index, data['PLK1'])
plt.scatter(outlier_data.index, outlier_data['PLK1'], color='r')
plt.title('Anomaly Detection')
plt.show()
```

This code will plot a heatmap for correlation analysis, a pair plot for cluster analysis, a scatter plot for PCA, and a scatter plot identifying outliers for anomaly detection. Each plot helps in visualizing different aspects of pattern and anomaly detection.

### 7.3.3 Preliminary Data Screening Methods

Prior to advanced analyses, preliminary data screening ensures data integrity and suitability for statistical testing. Evaluating the extent and mechanisms of missing data, whether missing completely at random, missing at random, or not missing at random, is important, and determining appropriate imputation methods helps handle these gaps effectively. This process, known as missing data analysis, reduces the risk of biased results.

Data cleaning involves correcting or removing erroneous data points, handling duplicates, and addressing inconsistencies, all of which are vital for maintaining data quality. This step ensures that the dataset accurately reflects the information intended for analysis, reducing the risk of misleading results due to flawed data.

Normalization and transformation apply statistical techniques to adjust data scales, such as using log transformations for skewed data or Z-score normalization. These adjustments ensure that variables are comparable and meet the assumptions of statistical tests, which is needed for accurate modeling and interpretation.

Feature selection uses statistical criteria—such as variance thresholding, correlation coefficients, or p-values from univariate analyses—to identify the most relevant variables. This process improves model performance and interpretability by focusing on the most significant factors within the dataset, thereby improving efficiency and reducing computational complexity.

## 7.4 Advanced Statistical Methods



Advanced statistical methods allow researchers to investigate complex relationships within data, model time-dependent phenomena, and identify factors affecting patient outcomes. Techniques such as multivariate regression, survival analysis, time-series modeling, and machine learning algorithms enable the analysis of high-dimensional and longitudinal data. Each method serves a specific type of research question and data structures, and their proper use requires careful planning and interpretation to ensure reliable conclusions.

### 7.4.1 Regression Analysis

Regression analysis is used to model and analyze the relationship between a dependent (outcome) variable and one or more independent (predictor) variables [13]. In biomedicine, it is widely used for tasks such as predicting patient outcomes, understanding risk factors, and evaluating the impact of treatments or interventions. We will further discuss it in Chap. 8.

*Linear Regression:* Linear regression is employed when the relationship between the dependent and independent variables is assumed to be linear. It estimates the coefficients of the linear equation involving one or more independent variables that best predict the value of the dependent variable. An example includes predicting blood glucose levels based on body mass index (BMI) and age.

*Example code for linear regression:*

```
import statsmodels.api as sm
import numpy as np
import pandas as pd

# Assuming X and y are defined as pandas DataFrame/Series or numpy arrays
# X: Independent variable(s)
# y: Dependent variable

# Add a constant term (intercept) to the predictors
X = sm.add_constant(X)
```

```
# Fit the Ordinary Least Squares (OLS) linear regression model
model = sm.OLS(y, X).fit()

# Generate predictions
predictions = model.predict(X)

# Print the summary of the model
print(model.summary())
```

`model.summary()` provides a summary of the regression results, including coefficients, R-squared value, p-values, and confidence intervals.

*Logistic Regression:* Logistic regression is suitable for modeling binary outcome variables, where the dependent variable is categorical with two possible outcomes (e.g., disease presence: yes/no). It models the probability that a given input point belongs to a particular category. An example is predicting the occurrence of a disease based on various risk factors such as genetic markers, lifestyle factors, and clinical measurements.

*Example code for logistic regression:*

```
from sklearn.linear_model import LogisticRegression
from sklearn.metrics import accuracy_score, classification_report

# Assuming X and y are defined
# X: Independent variable(s)
# y: Dependent binary variable (values 0 or 1)

# Initialize the logistic regression model
model = LogisticRegression(max_iter=1000)
# Increased max_iter for convergence

# Fit the model to the data
model.fit(X, y)

# Make predictions
predictions = model.predict(X)

# Calculate accuracy
accuracy = model.score(X, y)
print(f"Model Accuracy: {accuracy:.2f}")

# Detailed classification report
print(classification_report(y, predictions))

# Access model coefficients
coefficients = model.coef_
intercept = model.intercept_
print(f"Coefficients: {coefficients}")
print(f"Intercept: {intercept}")
```

`classification_report(y, predictions)` generates a detailed report including precision, recall, F1 score, and support. These metrics help quantify different aspects of model accuracy and error distribution. Precision reflects the proportion of true positives among all predicted positives, making it important in contexts where false positives carry a high cost. Recall (or sensitivity) measures the proportion of actual positives correctly identified by the model, which is especially relevant in medical diagnostics where missing a true case can have serious consequences. The F1 score combines precision and recall into a single value via their harmonic mean, offering a balanced view of performance—particularly useful in imbalanced datasets. Support indicates the number of actual instances per class, helping contextualize the other metrics. Beyond performance metrics, `model.coef_` and `model.intercept_` provide the coefficients and intercept of the model, respectively, indicating the influence of each predictor variable. These parameters enable researchers to examine how input variables influence the probability of the target class. To assess whether these effects are statistically meaningful, inferential methods such as the Wald test or likelihood ratio test can be applied to determine the strength of association between predictors and outcome.



### 7.4.2 Multivariate Analysis

Multivariate analysis involves the simultaneous observation and analysis of more than two statistical outcome variables. This approach allows researchers to understand complex relationships and interactions between multiple variables, which is particularly useful in biomedical research where datasets often contain numerous interrelated variables.

*Principal Component Analysis (PCA):* As discussed above as well, PCA is used to reduce the number of variables in a dataset while preserving as much variability as possible. PCA transforms the original variables into a new set of uncorrelated variables called principal components, which are ordered so that the first few retain most of the variation present in all of the original variables. In biomedicine, applications of PCA include reducing thousands of gene expression measurements in genomics and proteomics to a few principal components for visualization and analysis, and compressing high-dimensional imaging data in medical imaging while retaining key features for diagnosis.

*Factor Analysis:* Factor analysis is a statistical method used to identify underlying relationships between measured variables. It models observed variables and their covariance structure in terms of a smaller number of unobserved variables called factors. This technique helps in identifying latent variables that explain patterns of correlations within the observed data. In biomedicine, applications of factor analysis include psychometrics, where it is used to identify underlying psychological factors from survey responses, and symptom clustering, which involves determining common factors contributing to multiple disease symptoms to aid in diagnosis and treatment planning. Key points include Exploratory Factor Analysis (EFA), which is used when the underlying structure is unknown and aims to discover the number and nature of latent factors, and Confirmatory Factor Analysis (CFA), which is used when testing specific hypotheses or theories about the structure of the factors.

Both PCA and Factor Analysis share several important considerations. They both assume linear relationships between variables, meaning that the methods are based on the premise that variables interact in a linear fashion. Reliable results from these techniques typically require a large sample size, as a substantial number of observations increases the validity and stability of the findings. Furthermore, standardization of variables is needed when they are measured on different scales, ensuring that each variable contributes equally to the analysis and preventing variables with larger scales from dominating the results.

When choosing the number of components or factors to retain, different approaches are used for PCA and Factor Analysis. In PCA, one examines the explained variance ratio to decide how many principal components to retain, focusing on the components that capture the most variance in the data. For Factor Analysis, methods like the scree plot, Kaiser criterion, or parallel analysis are employed to determine the number of factors, aiming to identify the underlying latent structures that explain the observed correlations.

There are key differences between PCA and Factor Analysis in terms of their goals and the type of variance they consider. The goal of PCA is to capture the maximum total variance in the data with a few principal components, effectively summarizing the data without necessarily interpreting underlying structures. In contrast, Factor Analysis seeks to model the underlying latent structure that explains the observed correlations among variables, focusing on identifying unobserved factors that influence the measured variables. Regarding variance, PCA is based on the total variance in the data, including both shared and unique variances, whereas Factor Analysis is based solely on the shared variance (covariance) among variables, emphasizing the commonality between them.

### 7.4.3 Survival Analysis

Survival analysis models time-to-event data, which is particularly important in clinical trials and studies focusing on patient survival and other time-dependent events. It allows researchers to estimate survival probabilities over time and assess the impact of various factors on the time until an event occurs. Widely used survival analysis models include Kaplan-Meier estimator and Cox proportional hazards model.

*Kaplan-Meier Estimator:* The Kaplan-Meier estimator is a nonparametric statistic used to estimate the survival probability from lifetime data [14]. In simpler terms, it measures the proportion of subjects who have not experienced a particular event at different points in time. This event can vary depending on the context, for example, it might refer to death in a clinical trial, recurrence of disease, or failure of a mechanical component. The method allows us to track how long subjects remain event-free and to compare these patterns across different groups. One of its strengths is the ability to account for censored data, subjects who have not experienced the event of interest during the study period or are lost to follow-up [15]. At each time an event takes place, the estimator computes the survival probability, considering only those individuals who were still at risk immediately prior to that point. The output is a step function known as the Kaplan–Meier curve, which provides a visual summary of how survival probabilities evolve over time. In clinical trials comparing treatments, such as two approaches for cancer therapy, Kaplan–Meier curves enable researchers to observe differences in patient outcomes over the study duration. These plots make it possible to identify when survival patterns begin to diverge and to apply statistical tests like the log-rank test to assess whether observed differences are likely due to random variation [16]. The resulting plot displays survival probability on the y-axis and time on the x-axis, offering a straightforward view of treatment performance throughout the study period.

*Example code for Kaplan-Meier estimator:*

```
import pandas as pd
from lifelines import KaplanMeierFitter
import matplotlib.pyplot as plt

# Load the dataset
# Assuming 'df' is a pandas DataFrame with 'time' and 'event' columns

# Initialize the Kaplan-Meier fitter
kmf = KaplanMeierFitter()

# Fit the model to the data
kmf.fit(durations=df['time'], event_observed=df['event'])

# Plot the survival function
kmf.plot_survival_function()
plt.title('Kaplan-Meier Survival Curve')
plt.xlabel('Time')
plt.ylabel('Survival Probability')
plt.show()
```



*Cox Proportional Hazards Model:* The Cox proportional hazards model is a semi-parametric regression model that assesses the effect of several variables on survival time [17]. While the Kaplan-Meier estimator describes survival patterns, it does not account for the influence of covariates (predictor variables). The Cox proportional hazards model addresses this limitation by modeling the hazard function, which describes the risk of the event occurring at a given time, while incorporating both baseline hazard and covariates [18]. The model is used for assessing the impact of clinical factors (e.g., age, treatment type, biomarkers) on patient survival rates, identifying risk factors associated with increased or decreased hazard, etc. The model outputs hazard ratios for each covariate, which reflect how the event risk changes with a one-unit increase in that variable; hazard ratios = 1 indicates the covariate has no effect on the hazard of the event, hazard ratios > 1 indicates the covariate is associated with an increased hazard of the event, and hazard ratios < 1 indicates the covariate is associated with a decreased hazard of the event. For instance, a hazard ratio above 1 for "smoking status" suggests that smokers face a higher immediate risk of the event (such as death or disease progression) compared to non-smokers, controlling for other factors. These ratios help pinpoint which variables are independently linked to outcomes. The summary results from the model include estimated coefficients, p-values, and confidence intervals, offering a basis for assessing the strength and reliability of each variable's association with survival time.

*Code for Cox proportional hazards model:*

```
import pandas as pd
from lifelines import CoxPHFitter
import matplotlib.pyplot as plt

# Load the dataset
# The DataFrame 'df' should include 'time', 'event', and
# covariate columns

# Initialize the Cox Proportional Hazards model
cph = CoxPHFitter()

# Fit the model to the dataset
cph.fit(df, duration_col='time', event_col='event')

# Print the summary of the model
cph.print_summary()

# Plot the hazard ratios
cph.plot(hazard_ratios=True)
plt.title('Cox Proportional Hazards Model')
plt.show()
```

## 7.5 Statistical Methods in Genomics and Proteomics

Advanced statistical methods and tools allow researchers to explore the complexities of genetic and protein data, supporting a better understanding of biological mechanisms and the identification of therapeutic targets. The effectiveness of these approaches depends on careful implementation and interpretation of these methods.

### 7.5.1 Analysis of Gene Expression Data

Gene expression analysis involves examining the expression levels of genes across different samples, conditions, or time points. It is used for understanding biological processes and disease mechanisms.

*Differential Expression Analysis:* This involves identifying genes whose expression levels significantly differ between groups. For example, comparing gene expression in cancerous tissue versus normal tissue can reveal genes involved in tumorigenesis.

Example code for differential expression analysis using DESeq2 in R via rpy2 in Python (needs adjustment):

```
import rpy2.robjects as robjects
from rpy2.robjects import pandas2ri
import pandas as pd

# Activate the automatic conversion between pandas and R dataframes
# pandas2ri.activate()

# Assuming countData and colData are defined as pandas DataFrames
# countData: rows are genes, columns are samples
# colData: sample information including condition (e.g., 'treatment'
# or 'control')

# Load your data into pandas DataFrames (replace with your actual data)
# Example:
# countData = pd.read_csv('count_data.csv', index_col=0)
# colData = pd.read_csv('col_data.csv', index_col=0)

# Convert pandas DataFrames to R data frames
robjects.globalenv['count_data_r'] = pandas2ri.py2rpy(countData)
robjects.globalenv['col_data_r'] = pandas2ri.py2rpy(colData)

# R code for differential expression analysis
r_code = '''
library(DESeq2)

# Create DESeq2 dataset
dds <- DESeqDataSetFromMatrix(countData = count_data_r,
                              colData = col_data_r,
                              design = ~ condition)

# Run the differential expression analysis
dds <- DESeq(dds)

# Get the results
res <- results(dds)

# Convert results to a data frame
res_df <- as.data.frame(res)
'''

# Execute R code
robjects.r(r_code)

# Retrieve the results from R to Python
res_df = robjects.r('res_df')

# Convert R data frame to pandas DataFrame
res_df = pandas2ri.rpy2py(res_df)


# Display the results
print(res_df.head())
```

7

In the code example using `rpy2`, any R code that you write within the triple-quoted string assigned to the variable `r_code` will be read and executed by R when you call `robjects.r(r_code)`. This mechanism allows you to embed R code directly within your Python script and execute it in the R environment.

Clustering Analysis: This involves grouping genes with similar expression patterns across samples, which can reveal co-expressed genes and suggest shared regulatory mechanisms.

Example code for Clustering Analysis using K-means:

```
from sklearn.cluster import KMeans
import pandas as pd
from sklearn.preprocessing import StandardScaler
import matplotlib.pyplot as plt
import seaborn as sns

# Assuming gene_expression_data is a pandas DataFrame
# Rows are samples, columns are genes

# Transpose data to have genes as rows if necessary
gene_expression_data_T = gene_expression_data.T

# Standardize the data
scaler = StandardScaler()
scaled_data = scaler.fit_transform(gene_expression_data_T)

# Perform K-Means clustering
kmeans = KMeans(n_clusters=5, random_state=42, n_init=10)
clusters = kmeans.fit_predict(scaled_data)

# Add cluster labels to the DataFrame
gene_expression_data_T['Cluster'] = clusters

# Visualize the clusters using a heatmap
sns.clustermap(gene_expression_data_T.drop('Cluster', axis=1),
      row_cluster=False, col_cluster=False,
      row_colors=gene_expression_data_T['Cluster'].map({i: f'C{i}'
      for i in range(5)}),
      cmap='viridis')
plt.show()

# Print cluster assignments
print(gene_expression_data_T[['Cluster']])
```

### 7.5.2 Proteomic Data Analysis

Proteomic data analysis involves the large-scale study of proteins, focusing on their structures, functions, and interactions which help understand biological processes at the molecular level and for identifying biomarkers and therapeutic targets in various diseases.

Several Python tools have been developed to facilitate proteomic data analysis. One such tool is Pyteomics, a collection of lightweight and user-friendly modules that assist in the processing of proteomics data, particularly mass spectrometry data [19]. Pyteomics provides functionalities for reading and writing various mass spectrometry file formats, parsing protein and peptide sequences, and performing common tasks such as calculating molecular weights and isotopic distributions. Another important tool is PyOpenMS, which serves as a Python interface to the OpenMS library—a C++ framework for mass spectrometry data analysis [20]. PyOpenMS allows researchers to perform advanced data processing tasks, including peak picking, feature detection, and quantification, enabling integration of these processes into Python workflows. Moreover, Biopython offers tools for computational biology and bioinformatics, including functionalities relevant to proteomics. It provides modules for sequence analysis, structural bioinformatics, and accessing biological databases. Biopython can handle protein sequence parsing, modification, and analysis, which are basic tasks in proteomic studies.



### 7.5.3 Bioinformatics Tools for Statistical Analysis

Bioinformatics tools are widely used for statistical analysis in genomics and proteomics, and Python offers a rich ecosystem of such tools. These tools enable researchers to process, analyze, and interpret large-scale biological data efficiently, aiding discoveries in genetics, molecular biology, and disease mechanisms. As discussed above, Biopython provides tools for biological computation, supporting various functionalities required for bioinformatics [21]. It allows for sequence parsing and analysis, enabling researchers to read and write sequence files in multiple formats such as FASTA and GenBank, perform sequence alignments, and compute sequence statistics. Biopython also facilitates access to biological databases, allowing retrieval of data from online resources like NCBI, UniProt, and PDB. Furthermore, it supports phylogenetics by constructing and analyzing phylogenetic trees and offers capabilities in structural bioinformatics for working with protein structures, parsing PDB files, and performing structural analyses.

Beyond core libraries like NumPy and Pandas, which are required for data manipulation and numerical computations in bioinformatics, several specialized packages provide advanced analysis options. For instance, Pyteomics and PyOpenMS offer tools for proteomics data analysis, including mass spectrometry data processing. Libraries like HTSeq and Pysam are useful for genomic data analysis, such as handling sequencing data formats and read counting. For handling very large datasets that exceed memory capacity, Dask provides parallel computing capabilities, enabling big data processing by scaling computations across multiple cores or nodes and allowing for the analysis of large genomic datasets. Statsmodels allows users to explore data, estimate statistical models, and perform statistical tests. It supports statistical modeling by implementing models like linear regression, logistic regression, and time-series analysis and provides advanced statistical tests and model diagnostics. For interactive visualization and application development, libraries like

Plotly and Dash are useful. Plotly enables the creation of interactive, web-based graphs and visualizations, while Dash allows for building interactive web applications for data analysis without requiring extensive web development skills. Additional Python tools include BioPandas [22], which extends Pandas for working with molecular structures in bioinformatics, and Anvio, which is used for microbial genomics and metagenomics data visualization and analysis [23]. There are also next-generation sequencing tools like Poretools for handling sequencing data [24].

These Python tools form a flexible framework for statistical analysis in genomics and proteomics. With Python's broad bioinformatics libraries, researchers can perform complex analyses, integrate various data types, and develop custom workflows tailored to their specific research questions.

## 7.6 Exercises and Questions

In this section, we include exercises and questions to complement the material covered in this chapter.

### Exercises

The following exercises will enhance your understanding of the materials we covered in this chapter.

1. Calculating descriptive statistics: Use a dataset of patient blood pressure readings to calculate mean, median, mode, variance, and standard deviation.
2. Analyzing probability distributions: Examine the distribution of cholesterol levels in a patient cohort to determine if they follow a normal distribution.
3. Performing hypothesis testing: Conduct a t-test to compare the efficacy of two antihypertensive drugs based on systolic blood pressure outcomes.
4. Creating confidence intervals: Calculate the 95% confidence interval for average glucose levels in a diabetic patient group.
5. Visualizing data patterns: Use scatter plots to identify relationships between age and liver enzyme levels in a dataset.
6. Detecting anomalies: Apply outlier detection techniques to a dataset of heart rate readings from patients with cardiovascular disease.
7. Regression analysis: Perform linear regression to explore the relationship between BMI and heart disease risk.
8. Multivariate analysis: Use logistic regression to assess the impact of multiple risk factors on the likelihood of developing type 2 diabetes.
9. Time-series analysis: Analyze monthly data on influenza cases to model trends and seasonal variations.
10. Survival analysis: Conduct a Kaplan-Meier analysis to estimate survival probabilities for patients undergoing a new cancer treatment.

## Questions to Be Answered

1. What are the key characteristics of biomedical data that make statistical analysis challenging?
2. How do descriptive statistics help in understanding biomedical data?
3. Why is it important to understand the type of probability distribution a dataset follows?
4. What is hypothesis testing, and how can it be used to draw conclusions in biomedical research?
5. How are confidence intervals used to infer population parameters in biomedicine?
6. What role does exploratory data analysis play in biomedical research?
7. Describe how regression analysis can be used to understand relationships in biomedical data.
8. Explain the significance of multivariate analysis in the context of biomedicine.
9. What are the challenges and benefits of conducting time-series analysis in biomedicine?
10. How is survival analysis particularly useful in clinical research?



**Acknowledgement** The language of the human-generated text was corrected with the assistance of artificial intelligence (AI) tools [GPT-3.5 and GTP-4 from OpenAI]. GitHub Co-Pilot was used to check the correctness of the codes. The text underwent subsequent human revision to ensure its accuracy.

## References

1. Sedlakova J, Daniore P, Horn Wintsch A, Wolf M, Stanikic M, Haag C, Sieber C, Schneider G, Staub K, Alois Ettlin D, Grübner O, Rinaldi F, von Wyl V; University of Zurich Digital Society Initiative (UZH-DSI) Health Community. Challenges and best practices for digital unstructured data enrichment in health research: a systematic narrative review. PLOS Digit Health. 2023;2(10):e0000347. https://doi.org/10.1371/journal.pdig.0000347.
2. Safarova MS, Kullo IJ. Using the electronic health record for genomics research. Curr Opin Lipidol. 2020;31(2):85–93. https://doi.org/10.1097/MOL.0000000000000662.
3. Yang J, Li Y, Liu Q, Li L, Feng A, Wang T, et al. Brief introduction of medical database and data mining technology in big data era. J Evid Based Med. 2020;13(1):57–69.
4. Robinson R, Haviland JS. Understanding statistical significance and avoiding common pitfalls. Clin Oncol (R Coll Radiol). 2021;33(12):804–6.
5. Collins FS, Varmus H. A new initiative on precision medicine. N Engl J Med. 2015;372(9):793–5.
6. Berisha V, Krantsevich C, Hahn PR, Hahn S, Dasarathy G, Turaga P, Liss J. Digital medicine and the curse of dimensionality. NPJ Digit Med. 2021;4(1):153. https://doi.org/10.1038/s41746-021-00521-5.
7. Rothstein MA. Is deidentification sufficient to protect health privacy in research? Am J Bioeth. 2010;10(9):3–11.
8. Katnic I, Orlandic M. Fundamentals of biomedical statistics. Stud Health Technol Inform. 2020;274:111–21.
9. Mishra P, Pandey CM, Singh U, Gupta A, Sahu C, Keshri A. Descriptive statistics and normality tests for statistical data. Ann Card Anaesth. 2019;22(1):67–72.
10. McDermid R. Statistics in medicine. Anaesth Intensive Care Med. 2024;25(5):361–9.
11. Royston P. Which measures of skewness and kurtosis are best? Stat Med. 1992;11(3):333–43.
12. Skaf Y, Laubenbacher R. Topological data analysis in biomedicine: a review. J Biomed Inform. 2022;130:104082.
13. Sharp DS, Gahlinger PM. Regression analysis in biological research: sample size and statistical power. Med Sci Sports Exerc. 1988;20(6):605–10.

14. Goel MK, Khanna P, Kishore J. Understanding survival analysis: Kaplan-Meier estimate. Int J Ayurveda Res. 2010;1(4):274–8. https://doi.org/10.4103/0974-7788.76794.
15. Rich JT, Neely JG, Paniello RC, Voelker CC, Nussenbaum B, Wang EW. A practical guide to understanding Kaplan-Meier curves. Otolaryngol Head Neck Surg. 2010;143(3):331–6. https://doi.org/10.1016/j.otohns.2010.05.007.
16. Dudley WN, Wickham R, Coombs N. An Introduction to Survival Statistics: Kaplan-Meier Analysis. J Adv Pract Oncol. 2016;7(1):91–100. https://doi.org/10.6004/jadpro.2016.7.1.8.
17. Abd ElHafeez S, D'Arrigo G, Leonardis D, Fusaro M, Tripepi G, Roumeliotis S. Methods to analyze time-to-event data: the Cox regression analysis. Oxid Med Cell Longev. 2021;2021:1302811. https://doi.org/10.1155/2021/1302811.
18. Cox DR. Regression models and life-tables. J R Stat Soc Ser B Methodol. 1972;34(2):187–202. https://www.jstor.org/stable/2985181.
19. Goloborodko AA, Levitsky LI, Ivanov MV, Gorshkov MV. Pyteomics–a Python framework for exploratory data analysis and rapid software prototyping in proteomics. J Am Soc Mass Spectrom. 2013;24(2):301–4.
20. Rost HL, Schmitt U, Aebersold R, Malmstrom L. pyOpenMS: a Python-based interface to the OpenMS mass-spectrometry algorithm library. Proteomics. 2014;14(1):74–7.
21. Cock PJ, Antao T, Chang JT, Chapman BA, Cox CJ, Dalke A, et al. Biopython: freely available Python tools for computational molecular biology and bioinformatics. Bioinformatics. 2009;25(11):1422–3.
22. Raschka S. Biopandas: working with molecular structures in pandas dataframes. J Open Source Softw. 2017;2(14). https://doi.org/10.21105/joss.00279. http://dx.doi.org/10.21105/joss.00279.
23. Eren AM, Kiefl E, Shaiber A, Veseli I, Miller SE, Schechter MS, Fink I, Pan JN, Yousef M, Fogarty EC, Trigodet F, Watson AR, Esen ÖC, Moore RM, Clayssen Q, Lee MD, Kivenson V, Graham ED, Merrill BD, Karkman A, Blankenberg D, Eppley JM, Sjödin A, Scott JJ, Vázquez-Campos X, McKay LJ, McDaniel EA, Stevens SLR, Anderson RE, Fuessel J, Fernandez-Guerra A, Maignien L, Delmont TO, Willis AD. Community-led, integrated, reproducible multi-omics with anvi'o Nat Microbiol. 2021;6(1):3–6. https://doi.org/10.1038/s41564-020-00834-3.
24. Loman NJ, Quinlan AR. Poretools: a toolkit for analyzing nanopore sequence data. Bioinformatics. 2014;30(23):3399–401. https://doi.org/10.1093/bioinformatics/btu555.

# Machine Learning in Biomedicine

Contents




J. U. Kazi, *Python Essentials for Biomedical Data Analysis: An Introductory Textbook*,
https://doi.org/10.1007/978-3-031-85600-6_8

This chapter presents an overview of machine learning concepts and their applications in biomedicine, with a focus on methods and basic understanding. It outlines the main categories of machine learning and describes supervised learning techniques such as linear regression, logistic regression, decision trees, ensemble methods, support vector machines, and deep learning. The chapter also discusses unsupervised learning methods like clustering, dimensionality reduction, and anomaly detection, as well as semi-supervised and reinforcement learning. Techniques for data preprocessing and model optimization are discussed, along with methods for model evaluation. The chapter concludes with example implementations on drug sensitivity prediction using logistic regression, Random Forest, and the AlphaML platform.

### Learning Goals

The learning goals include gaining a basic understanding of the principles and methodologies of machine learning and their applications in biomedical research. Students will learn the differences between supervised, unsupervised, reinforcement, and deep learning techniques, as well as how these approaches can be used in biomedical tasks such as drug sensitivity prediction. This chapter also addresses common issues encountered when applying machine learning models, including how to optimize model parameters and assess their performance.

## 8.1 Basic Concepts of Machine Learning

Machine learning is a subset of artificial intelligence that focuses on developing algorithms capable of learning from data to make predictions, decisions, or identify patterns without explicit programming (◘ Fig. 8.1). Unlike traditional programming, where the computer follows predefined instructions, machine learning relies on statistical methods to recognize relationships in data and improve performance over time [1]. Common tasks include, but are not limited to, classification, regression, clustering, and anomaly detection.

Artificial intelligence, as the broader field, aims to create intelligent systems that mimic human decision-making and problem-solving. Machine learning is distinguished within artificial intelligence by its focus on algorithms and statistical models, enabling machines to learn and adapt from data. Within machine learning, deep learning is a specialized area that uses neural networks with multiple layers (hence "deep") to model complex patterns in large datasets. This approach is widely used in areas such as image analysis and natural language processing [2, 3].

In biomedicine, machine learning examines complex biological and medical data, detect patterns, and generate predictions. Some applications include drug discovery, disease diagnosis, and personalized patient care. This interdisciplinary approach blends biology, medicine, and computer science, making it particularly impactful in areas like drug sensitivity prediction and target identification. The use of machine learning allows researchers to work with large, complex datasets and supports progress in biomedical research and healthcare.

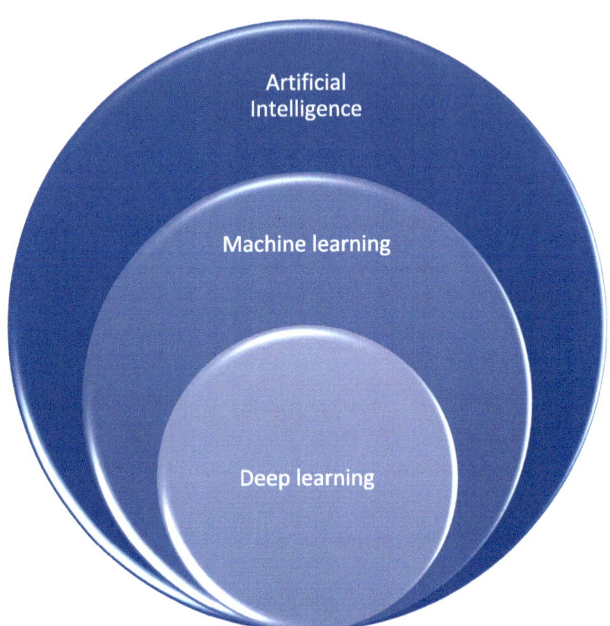




■ **Fig. 8.1** Relationship between artificial intelligence, machine learning, and deep learning. The diagram illustrates the hierarchical relationship between artificial intelligence (AI), machine learning (ML), and deep learning (DL). Artificial intelligence represents the overarching field aimed at developing intelligent systems capable of human-like decision-making and problem-solving. Machine learning is a subset of artificial intelligence that focuses on algorithms and statistical models to learn from data and improve performance on tasks. Deep learning, a specialized branch of machine learning, uses neural networks with multiple layers to model complex patterns and achieve breakthroughs in tasks such as image recognition and natural language processing

### 8.1.1 Machine Learning: Brief Classifications

Machine learning (ML) can be broadly classified into four main types: supervised learning, unsupervised learning, semi-supervised learning, and reinforcement learning (■ Fig. 8.2). Supervised learning relies on labeled data to train algorithms, allowing for tasks such as classification and regression where input-output relationships are known [4]. Unsupervised learning, on the other hand, works with unlabeled data to identify patterns, clusters, or structures using techniques like clustering and dimensionality reduction [5]. Semi-supervised learning combines elements of both supervised and unsupervised approaches, using a small amount of labeled data along with a larger set of unlabeled data to improve learning efficiency, especially in domains where labeled data is scarce [6]. Finally, reinforcement learning trains an agent to make sequential decisions in an environment by maximizing a reward signal, often applied in areas like robotics and game AI [7]. These classifications form the backbone of machine learning applications across diverse fields, including biomedicine.

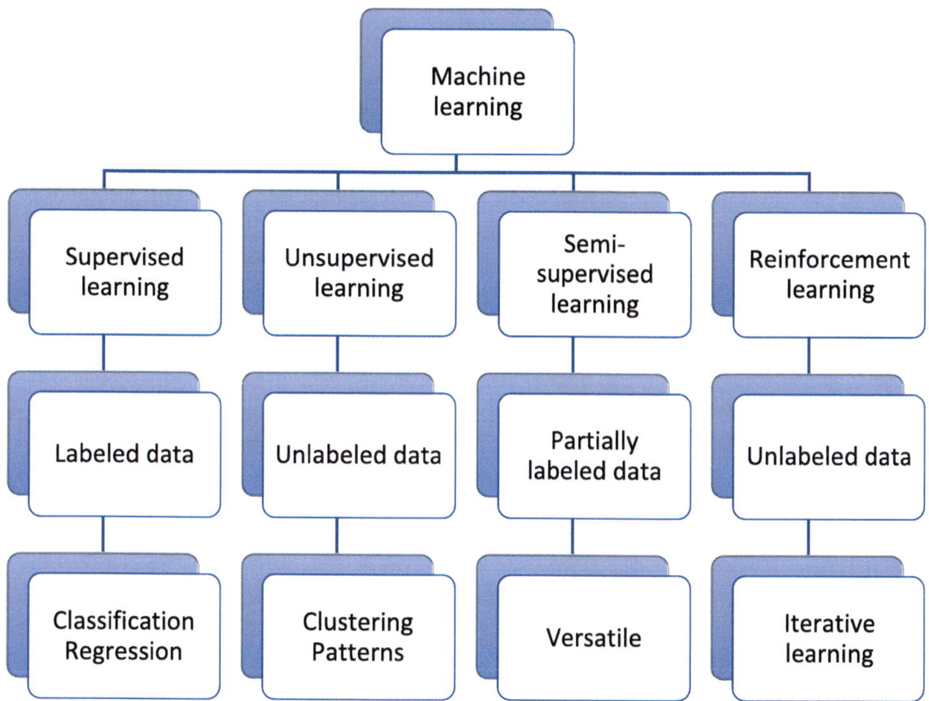


**Fig. 8.2** Taxonomy of machine learning approaches. This diagram provides an overview of the main categories of machine learning approaches: supervised learning, unsupervised learning, semi-supervised learning, and reinforcement learning. Each category is defined by the type of data it utilizes and its core objectives. Supervised learning uses labeled data for tasks like classification and regression. Unsupervised learning relies on unlabeled data to uncover clusters or patterns. Semi-supervised learning combines partially labeled and unlabeled data, showing versatility in scenarios with limited labeled datasets. Reinforcement learning uses unlabeled data in an interactive environment where agents learn to maximize rewards through iterative decision-making processes. This structure suggests the diverse applications and methodologies within machine learning

## 8.2 Supervised Learning

This is a common approach in biomedicine where the algorithm learns from labeled training data to make predictions [4, 8]. For example, supervised learning can be used to predict the efficacy of drug compounds based on known outcomes [9, 10]. These approaches are applied to estimate drug efficacy, toxicity, and possible new applications, which can help facilitate the drug development process by making it faster, more cost-effective, and more targeted toward successful outcomes [11]. However, challenges like data quality, model interpretability, and generalizability must be carefully addressed to fully realize the benefits of supervised learning in this domain. Widely used supervised learning algorithms can be classified into different groups as shown (Fig. 8.3).

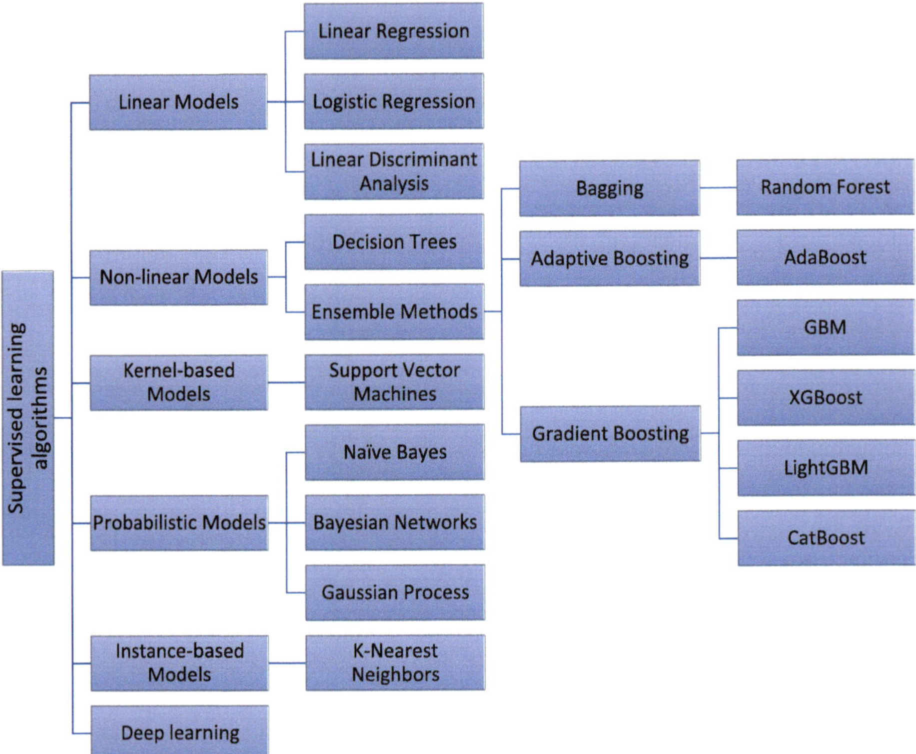


**Fig. 8.3** Supervised learning algorithms. Supervised learning algorithms can be applied to both classification and regression tasks and can be categorized in various ways. This figure provides a model-based high-level classification of algorithms. While some algorithms are task-specific, many are capable of handling both classification and regression tasks with minor modifications



### 8.2.1 Linear Regression

Linear regression is a statistical technique in machine learning, employed to model the relationship between a dependent variable and one or more independent variables [12]. The objective is to predict values of the dependent variable using a linear equation fitted to the input data. A simple linear regression with one independent variable takes the form:

$$Y = \beta_0 + \beta_1 X + \varepsilon,$$

where

$Y$ is the dependent variable.
$X$ is the independent variable.
$\beta_0$ is the intercept term.
$\beta_1$ is the slope coefficient.
$\varepsilon$ is the error term.

Extending this to multiple independent variables, the equation becomes [13]:

$$Y = \beta_0 + \beta_1 X_1 + \beta_2 X_2 + \ldots + \beta_n X_n + \varepsilon.$$

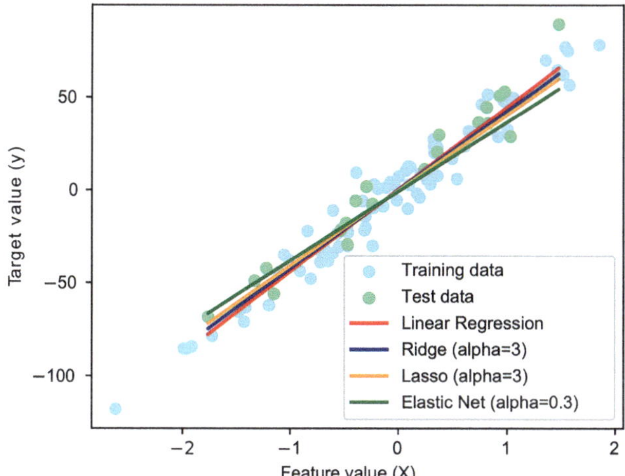


**Fig. 8.4** Visualization of linear regression models. This figure illustrates linear regression and its variations: Ridge, lasso, and elastic net regressions, and shows how each model fits the training and test data. The $x$-axis represents the feature value, while the $y$-axis shows the target value for linear regression. The figure was generated using sklearn.linear_model, where alpha is equevalent to λ

Coefficients are estimated using the least squares method, which minimizes the sum of squared residuals between observed and predicted values [13]. The measure of model fit, $R^2$, indicates the extent to which variability in $Y$ is explained by the model; however, it does not evaluate the model's predictive performance on unseen data.

Linear regression algorithms are highly valued for their simplicity and efficacy, serving as a popular choice within the spectrum of machine learning algorithms. As described so far, the canonical form, least squares regression, minimizes the sum of squared residuals. This sum serves as the cost function, which represents the total error of the model. While efficient with simple datasets, as complexity grows, the algorithm may show overfitting, characterized by low bias but high variance. To mitigate overfitting, techniques such as ridge regression, lasso regression, and elastic net are introduced, which modify the cost function to incorporate a penalty term [14]. Ridge regression employs L2 regularization, adding the squared magnitude of coefficients scaled by a regularization parameter $\lambda$ to the cost function [15]. On the other hand, lasso regression incorporates L1 regularization, adding the absolute values of the coefficients, also scaled by $\lambda$ [16]. Elastic net combines both L1 and L2 regularization terms [17]. The strength of the regularization term is governed by $\lambda$ (*in figure denoted as alpha*), with higher values imposing stricter penalties on the size of the coefficients, thereby increasing bias and reducing variance (Fig. 8.4). When $\lambda$ approaches zero, these methods converge to the least squares regression model.

These approaches help to prevent overfitting and support better performance on new data. Moreover, lasso and elastic net also perform feature selection by shrinking some coefficients to zero, allowing identification of the most relevant features in predictive modeling. Tuning λ, models can be optimized for performance, ensuring robustness and generalizability to new data.

### 8.2.2 Logistic Regression

Logistic regression is predominantly used for classification problems [18]. Unlike linear regression, which predicts continuous outcomes, logistic regression is designed for mainly binary classification tasks, where the dependent variable is categorical, typically taking values 0 or 1. The goal is to model the probability that a given input belongs to a particular category. Logistic regression predicts the probability of the dependent variable being 1 using the logistic function, an S-shaped curve known as the sigmoid function [19]. The logistic function transforms any input into a value between 0 and 1, interpreting it as a probability.

The basic form of a logistic regression model with one independent variable is expressed as:



$$P(Y=1)=\frac{e^{\beta_0+\beta_1 X}}{1+e^{\beta_0+\beta_1 X}},$$

where

$P(Y=1)$ is the probability of the dependent variable equaling 1.

$X$ is the independent variable.

$\beta_0$ and $\beta_1$ are the model coefficients.

Extending logistic regression to multiple independent variables, the model generalizes to:

$$P(Y=1)=\frac{e^{\beta_0+\beta_1 X_1+\beta_2 X_2+\ldots+\beta_n X_n}}{1+e^{\beta_0+\beta_1 X_1+\beta_2 X_2+\ldots+\beta_n X_n}}.$$

Each coefficient $\beta$ represents the change in the log odds of $Y = 1$ for a one-unit increase in the corresponding independent variable. Coefficients are estimated using maximum likelihood estimation (MLE), which seeks to find the coefficient values that maximize the likelihood of observing the sample data. Unlike the least squares method in linear regression, MLE involves an iterative optimization process.

One of the key metrics for evaluating the performance of a logistic regression model is the Receiver Operating Characteristic (ROC) curve and the associated Area Under the Curve (AUC) metric. The ROC curve plots the true positive rate against the false positive rate at various threshold levels [20], while the AUC quantifies the model's ability to discriminate between the two categories, with higher values indicating better performance [21].

Logistic regression is valued for its simplicity, interpretability, and efficiency in binary classification tasks. However, it assumes a linear relationship between the independent variables and the log odds of the dependent variable. It may also suffer from overfitting, particularly in datasets with irrelevant features or large feature sets. To address these limitations, regularization techniques such as L1 (lasso) and L2

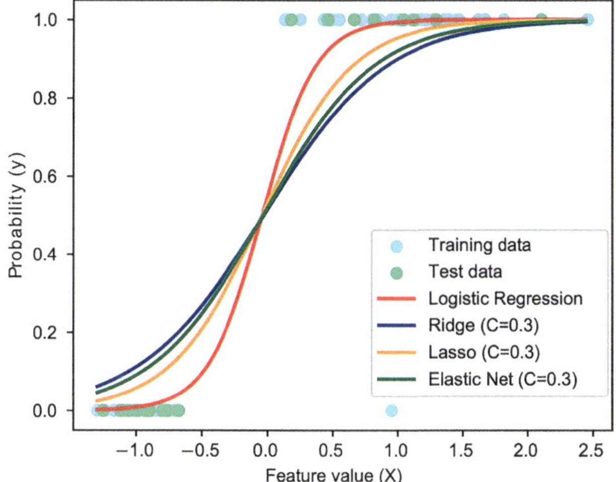


**Fig. 8.5** Visualization of logistic regression models. This figure displays logistic regression along with its regularized versions: Ridge, lasso, and elastic net, depicting the probability curve fit for binary classification. The $x$-axis represents the feature value, while the $y$-axis shows the probability for logistic regression

(ridge) regularization can be applied. L1 regularization encourages sparsity by shrinking some coefficients to zero, aiding feature selection, while L2 regularization prevents any single feature from dominating by uniformly shrinking coefficients. Regularization techniques penalize large coefficients in the model, reducing variance and improving generalization by controlling the bias-variance tradeoff. By tuning the regularization parameter $\lambda$ (in figure denoted as C), logistic regression models can be fine-tuned to ensure robustness and generalizability to new data (Fig. 8.5).

### 8.2.3 Linear Discriminant Analysis (LDA)

Linear discriminant analysis (LDA, we have used LDA in Sect. 8.3.6 for latent Dirichlet allocation) is a supervised learning algorithm widely used for classification and dimensionality reduction tasks [22]. It is particularly effective when the dataset has clear class separability and assumes that the predictor variables (features) follow a multivariate normal distribution. The primary objective of LDA is to find a linear combination of features that best separates the classes in the data while reducing the dimensionality of the feature space.

LDA works by maximizing the ratio of between-class variance to within-class variance, ensuring that the resulting linear boundaries optimally separate the classes. Mathematically, this involves finding a projection matrix $W$ that transforms the input feature space into a lower-dimensional space while preserving the class distinctions [22]. For a dataset with $k$ classes, the algorithm seeks to maximize the Fisher criterion:

$$J(W) = \frac{\left|W^{\mathrm{T}} S_B W\right|}{\left|W^{\mathrm{T}} S_W W\right|},$$

where $S_B$ is the between-class scatter matrix, representing the variance between class means, and $S_W$ is the within-class scatter matrix, representing the variance within each class. The projection matrix $W$ is obtained by solving this optimization problem. The between-class scatter matrix ($S_B$) measures the spread of the class means relative to the overall mean of the data. It quantifies how far apart the different class centers are, reflecting the degree of separation between classes. The within-class scatter matrix ($S_W$) measures the spread of the data points within each class around their respective class means. It reflects how tightly grouped the points are within each class; lower within-class scatter indicates that the points in each class are closer to each other.

LDA makes several key assumptions. For example, the features within each class are normally distributed (multivariate normality), all classes share the same covariance matrix (equal covariance), and classes can be separated by linear boundaries in the projected space (linear decision boundaries) [23]. While these assumptions are often violated in real-world datasets, LDA remains robust and performs well in practice, especially when the violations are minor. In classification, LDA calculates the posterior probability for each class and assigns the data point to the class with the highest probability. Beyond classification, LDA is also used for dimensionality reduction. For a dataset with $k$ classes, LDA reduces the feature space to $k - 1$ dimensions, capturing the most discriminative information for class separability. This is particularly useful for high-dimensional datasets, such as gene expression data or image recognition tasks. Preprocessing plays a key role in LDA's performance. Features should be scaled appropriately, especially when their ranges differ. For datasets with nonlinear class separability, extensions like Quadratic Discriminant Analysis (QDA) or kernel-based methods may be more suitable [23].



### 8.2.4 Decision Trees

Decision trees are interpretable machine learning models used for both categorical and numerical predictive tasks [24]. They organize decision-making into a branched hierarchical structure, making them easy to understand and use. A decision tree consists of internal nodes, branches, and terminal leaves. Internal nodes test predictor variables, branches represent the outcomes of these tests, and leaves provide the predicted responses, whether categorical (classification) or numerical (regression). The path from the root to a leaf represents the decision-making process.

Decision trees work by recursively partitioning the feature space into subsets. At each node, the algorithm selects the feature and threshold that best separates the data based on a chosen criterion. For classification tasks, popular criteria include Gini impurity and information gain. Gini impurity measures the probability of incorrect classification for a randomly chosen instance, while information gain quantifies the reduction in entropy (or uncertainty) after splitting the data (Fig. 8.6). For regression tasks, the algorithm typically uses variance reduction to minimize the error in predicting numerical values within each partition.

A key advantage of decision trees is their simplicity and interpretability. The decision-making process is straightforward, and predictions can be easily explained without requiring advanced statistical knowledge. This makes decision trees particularly useful in applications where transparency is required. However, they also have limitations. Decision trees are prone to overfitting, especially when the tree becomes

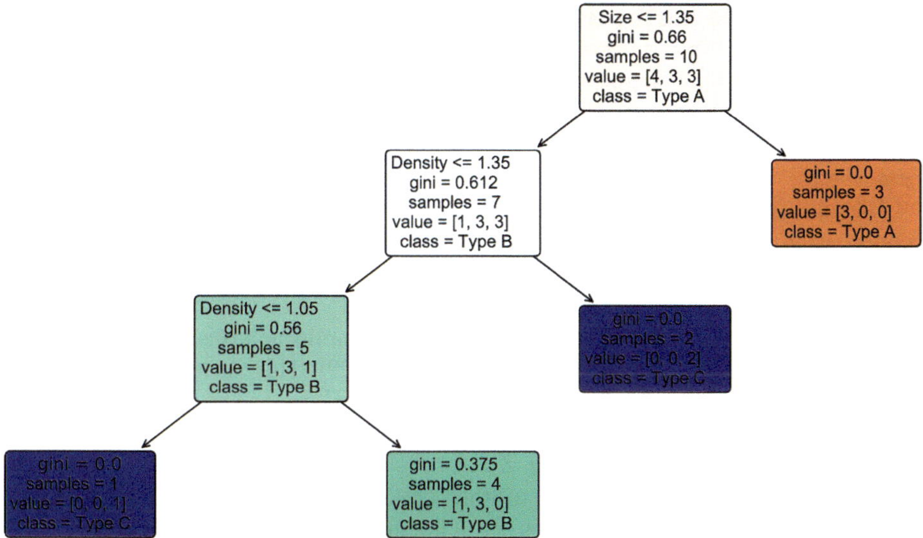


**Fig. 8.6** Example decision tree for classification. This figure illustrates a decision tree developed from a hypothetical dataset. The tree is structured with nodes, branches, and leaves, representing the decision-making process. Each internal node (square) corresponds to a decision point based on one of four features. Branches emanate from these nodes, leading to either subsequent decision nodes or leaf nodes (ovals), which represent the final classification into one of three types (Type A, Type B, and Type C). The decision tree was generated using a sample dataset of ten instances, each described by the aforementioned features. The Gini impurity measure is used at each node to guide the splitting process, aiming to maximize the homogeneity of classes within the branches emerging from a decision node

overly complex and begins to model noise in the data rather than its underlying patterns [25]. Techniques such as pruning (removing less significant branches) or setting a maximum depth can mitigate overfitting.

The Classification and Regression Tree (CART) algorithm is a early method that supports both classification and regression tasks through binary data splits [26]. It inspired subsequent algorithms, including ID3, which uses a greedy approach to maximize information gain during data segmentation [27]. Later advancements such as C4.5 and C5.0 improved upon CART by adding support for continuous variables, handling missing values, and introducing post hoc pruning to prevent overfitting [28]. C5.0, a refinement of C4.5, offers greater computational efficiency and reduced memory requirements.

Despite their advantages, decision trees can be unstable. Small variations in the training data may produce entirely different trees, reducing reliability. This instability is often addressed through ensemble methods like Random Forests and Gradient Boosting, which combine predictions from multiple trees to improve accuracy and stability [29, 30]. Another challenge is that decision trees can be biased toward attributes with many levels. Techniques like binning can mitigate this issue, ensuring a more balanced and accurate model.

Overall, decision trees are widely used in machine learning due to their simplicity, interpretability, and flexibility in handling diverse types of data. They are especially useful in tasks where understanding the decision-making process is as important as achieving accurate predictions.

### 8.2.5 Ensemble Methods

Ensemble methods in machine learning integrate multiple algorithms to improve predictive performance compared to using a single model [31]. The core principle is to use the strengths of diverse models to acheive higher accuracy and greater robustness.

Bagging (Bootstrap Aggregating) is an ensemble approach that builds multiple models, each trained on different subsets of the training dataset generated through bootstrapping. This process introduces diversity in training sets thereby the models and helps reduce the risk of overfitting by mimizing data-specific noise. Random Forest is a well-known example of bagging, where multiple decision trees are constructed, each using a random subset of features [29]. This strategy decreases the correlation between trees and leading to lower overall error. The final prediction is obtained by combining the predictions from individual trees, resulting in a more accurate and stable outcome (◘ Fig. 8.7a).

Boosting algorithms represent another key ensemble strategy. These methods sequentially build models, with each subsequent model correcting the errors of its predecessors. The goal is to convert weak learners into a strong collective learner. AdaBoost is a notable example (◘ Fig. 8.7b), where incorrectly classified instances are assigned higher weights, encouraging the next model to focus on harder-to-classify cases [32]. Gradient Boosting Machines (GBM) extend this approach by iteratively optimizing the model to minimize a specified loss function (◘ Fig. 8.7c) [33]. Modern implementations like XGBoost (eXtreme Gradient Boosting) are designed for speed and scalability. XGBoost incorporates regularization to prevent overfitting and effectively handles sparse data (◘ Fig. 8.7d) [34]. It includes regularized boosting to prevent overfitting and is capable of handling sparse data (◘ Fig. 8.3d). LightGBM further improves computational efficiency by using a histogram-based algorithm, which reduces memory usage and speeds up training on large datasets [35]. CatBoost, another boosting algorithm, specializes in handling categorical data efficiently, reducing the need for extensive preprocessing and offering improved performance on such datasets [36].

Stacking takes a different approach to ensemble learning by combining predictions from multiple models through a meta-classifier or meta-regressor [37]. Base models are trained on the dataset, and their predictions are used as inputs for a higher-level model. This meta-model learns to leverage the distinct capabilities of the base models to construct a more detailed and accurate predictive model.

Each ensemble method has unique characteristics, making them suitable for different types of data and predictive tasks. Bagging methods like Random Forest is effective in reducing variance and overfitting, making them reliable for high-dimensional datasets. Boosting methods like XGBoost and LightGBM are often used for handling complex, large datasets and achieving state-of-the-art performance in competitions. Stacking, with its flexibility, can combine diverse model types to achieve better generalization. The choice of ensemble method depends on specific problem requirements, data characteristics, and desired model performance.

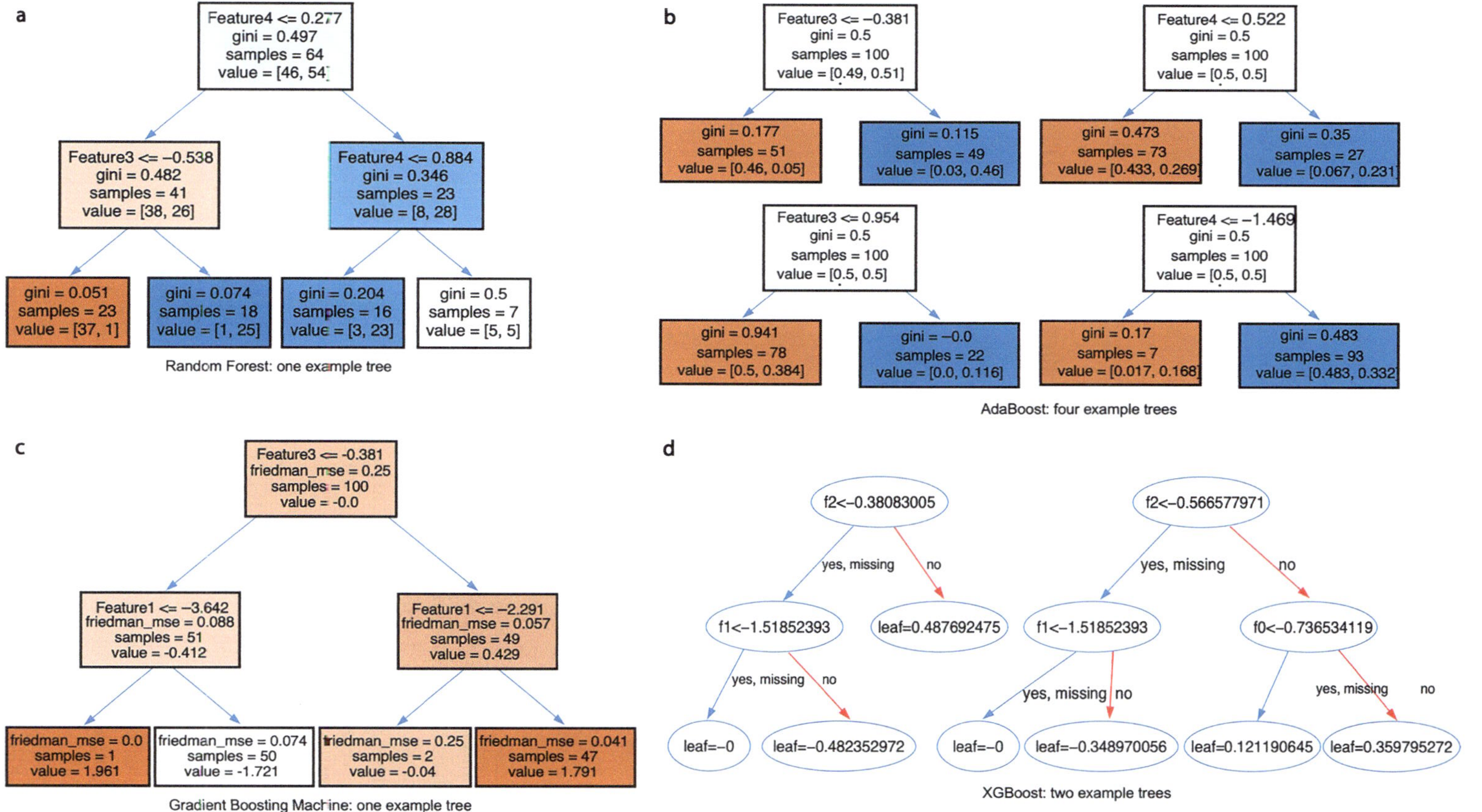


**Fig. 8.7** The figure illustrates various tree structures from ensemble learning models, displaying examples from Random Forest (**a**), AdaBoost (**b**), Gradient Boosting Machine (**c**), and XGBoost (**d**). Each tree shows decision paths based on different features, demonstrating the unique split criteria and leaf nodes. The trees represent the diverse approaches these algorithms take in modeling data, reflecting the complexities and subtleties in predicting outcomes based on various attributes. This comparative visualization helps in understanding how each model processes data and makes predictions

### 8.2.6 Support Vector Machines

Support Vector Machines (SVMs) are a set of supervised learning methods used for classification, regression, and outlier detection [38–40]. The versatility of SVMs in both linear and nonlinear problems makes them particularly useful in a wide range of applications, including drug discovery and bioinformatics.

For classification tasks, the Support Vector Classifier (SVC) is a commonly used SVM variant that constructs a hyperplane to optimally separate the classes in the feature space [41]. If the data is two-dimensional, this hyperplane is a line. Mathematically, a hyperplane can be expressed as $w \cdot x + b = 0$, where $w$ is the weight vector, $x$ is the feature vector, and $b$ is the bias. SVC aims to maximize the margin (minimal distance of a sample to the hyperplane, can be measured by the length of weight vector $w$) between the hyperplane and the nearest data points from each class, known as support vectors [40]. The optimization problem for SVC becomes:

$$\min_{w,b} \frac{1}{2}\|w\|^2, \text{ subject to } y_i\left(w \cdot x_i + b\right) \geq 1 \text{ for each data point } \left(x_i, y_i\right),$$



where $y_i$ is the class label.

In cases where perfect separation is impossible due to noise or overlapping data, soft margin SVC is used. Soft margin SVC introduces slack variables $\xi_i$ to allow for some misclassification:

$$\min_{w,b} \frac{1}{2}\|w\|^2 + C\sum_{i=1}^{n}\xi_i, \text{ subject to } y_i\left(w \cdot x_i + b\right) \geq 1 - \xi_i \text{ and } \xi_i \geq 0,$$

where $C$ is a regularization parameter.

For nonlinear classification, SVC uses the kernel trick to map the input space into a higher-dimensional feature space where a linear separator may exist. The kernel function computes the inner product of two points in the feature space without explicitly mapping the points [42]. Common kernel functions include:

Linear: $K(x_i, x_j) = x_i \cdot x_j$

Polynomial: $K(x_i, x_j) = (\gamma x_i \cdot x_j + r)^d$, where the sacale parameter $\gamma > 0$, $r$ is a constant term and $d$ is the degree

Radial basis function (RBF): $K(x_i, x_j) = \exp(-\gamma\|x_i - x_j\|^2)$, where $\gamma > 0$

Sigmoid: $K(x_i, x_j) = \tanh(\gamma x_i \cdot x_j + r)$, where $\gamma > 0$

Figure 8.8 showcases a comparative analysis of different kernel methods used in Support Vector Machines (SVMs) for classification tasks on a two-dimensional synthetic dataset.

In addition to classification via SVC, SVMs are also adapted for regression tasks using Support Vector Regression (SVR). SVR attempts to fit the error within a specified threshold, called epsilon, rather than minimizing the error directly [43]. This makes SVR particularly suitable for applications requiring robust predictions with tolerance for small deviations. Moreover, in unsupervised learning, techniques like Support Vector Clustering use the concept of decision boundaries to group unlabeled data.

The optimization problem in SVMs is typically solved using quadratic programming methods, which are efficient for small to medium-sized datasets but can become computationally expensive for large-scale problems. Despite these challenges, SVMs is useful in handling high-dimensional data and offer flexibility in modeling diverse

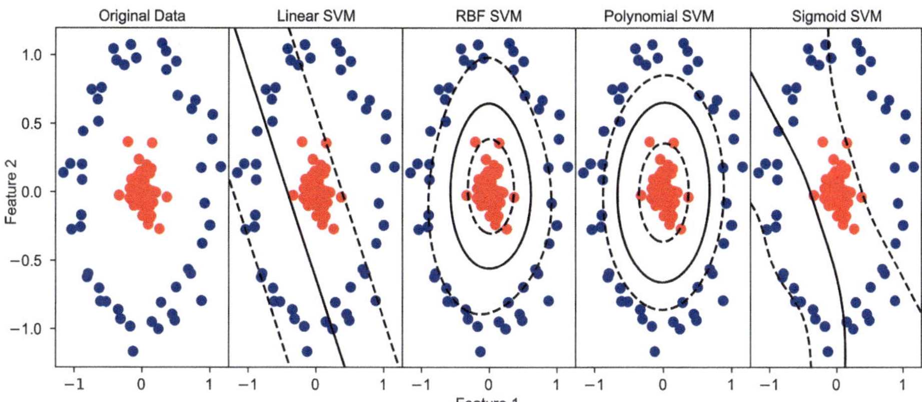


**Fig. 8.8** Comparative analysis of SVC kernel methods on a two-dimensional synthetic dataset. This figure illustrates the classification performance of different SVC kernels on a nonlinearly separable dataset. Subplot 1 displays the original dataset with two distinct classes. Subplot 2 shows the decision boundary of a linear SVC, showing its limitations in handling nonlinear separations. Subplot 3 presents the RBF kernel SVC, demonstrating its effectiveness in capturing complex data patterns. Subplot 4 depicts the polynomial kernel SVC, which offers a balance between simplicity and nonlinear fitting. Subplot 5 showcases the sigmoid kernel SVC, providing a unique approach to nonlinear classification. Each subplot includes the data points colored according to their class and the decision boundaries generated by the respective SVC kernel

datasets through the kernel trick. Their application in drug discovery is remarkable, where SVC can classify compounds, and SVR can predict drug sensitivities with high accuracy. However, the performance of SVMs depends heavily on the choice of kernel and the tuning of hyperparameters like $C$, $\gamma$, and $\epsilon$.

### 8.2.7 Naive Bayes

Naive Bayes is a simple machine learning algorithm based on Bayes' theorem. Bayes' theorem, named after Reverend Thomas Bayes, is a base principle in probability theory [44]. It describes the probability of an event based on prior knowledge of conditions that may influence the event. The formula for Bayes' theorem is:

$$P(Y \mid X) = \frac{P(X \mid Y)P(Y)}{P(X)},$$

where

$P(Y|X)$ is the posterior probability of class (target) given predictor (attribute).

$P(Y)$ is the prior probability of class.

$P(X|Y)$ is the likelihood which is the probability of predictor given class.

$P(X)$ is the prior probability of predictor.

Naive Bayes is "naive," because it assumes that all features (predictor variables) are independent of each other, an assumption rarely true in real-world datasets [45]. Despite this, the algorithm often performs remarkably well, as the independence assumption simplifies computation and reduces the risk of overfitting [46]. In a classification task, Naive Bayes uses this formula to calculate the posterior probability

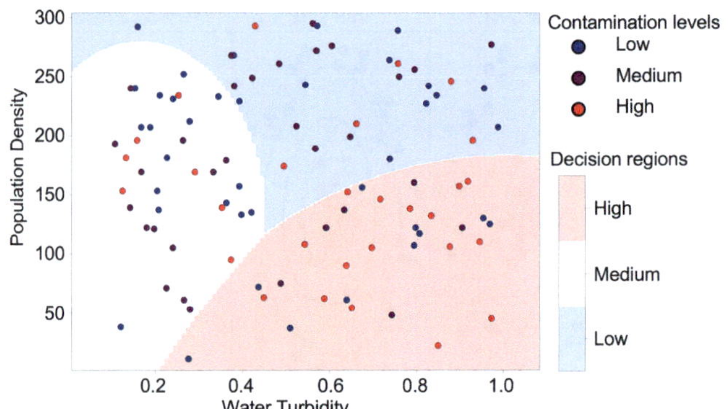


**Fig. 8.9** Illustration of the decision regions for hypothetical contamination prediction using a Gaussian Naive Bayes classifier. The *x*-axis represents "Water Turbidity," while the *y*-axis indicates "Population Density." Different colors in the background represent the decision regions determined by the classifier, each corresponding to a predicted contamination level: low, medium, or high. The distinct colored dots represent the actual data points, with their colors indicating the true contamination levels as per the dataset. Blue, purple, and red dots signify low, medium, and high levels of contamination, respectively. The overlay of the decision regions and actual data points shows model's classification accuracy and areas where misclassifications occur



for each class and selects the class with the highest probability. For example, Naive Bayes could be applied to predict hypothetical contamination levels based on independent features such as turbidity and population density (Fig. 8.9).

Naive Bayes and its varient have been applied for text classification, spam filtering, sentiment analysis, and disease diagnosis [47]. While it is computationally efficient and robust to irrelevant features, it struggles with datasets where features are highly correlated, as the independence assumption does not hold.

Despite its limitations, Naive Bayes remains a popular choice for classification tasks, due to its simplicity, speed, and good performance on a wide range of problems.

## 8.2.8 Bayesian Networks

Bayesian Networks (BNs) are probabilistic graphical models that represent relationships among variables using a directed acyclic graph (DAG) [48]. DAG is a structure made up of nodes connected by arrows that show direction, with the restriction that no path leads back to the starting node, in other words, there are no loops or cycles in a DAG [49]. Bayesian Network provide a flexible approach that can be applied to classification, regression, anomaly detection, and causal inference [50, 51]. For classification tasks, Bayesian Networks predict the most probable class for a given set of features, and more broadly, they support the modeling of complex probabilistic relationships and uncertainty within data.

A Bayesian Network consists of three key components: nodes, which represent variables that can be either discrete or continuous; edges, which are directed connections between nodes representing conditional dependencies; and conditional probability distributions (CPDs), which specify the probability of each variable given the values of its parent nodes in the graph [52].

Bayesian Networks provide a unified framework for classification and regression tasks, supporting probabilistic reasoning, uncertainty quantification, and causal inference [50, 51, 53].

### 8.2.9 Gaussian Process

A Gaussian Process (GP) is a nonparametric approach used for regression, classification, and other machine learning tasks [54–56]. It defines a distribution over functions, allowing for predictions that include a measure of uncertainty. Unlike parametric models, which assume a fixed form for the function mapping inputs to outputs, GPs provide a flexible framework where the function is inferred directly from the data [54, 57]. This makes GPs particularly suitable for tasks involving small datasets, where capturing uncertainty is important.

The core of a Gaussian Process lies in the assumption that any finite set of points in the input space follows a joint Gaussian distribution [54]. GPs are often effective in scenarios where data is scarce or noisy. They provide interpretable models with clear measures of uncertainty [58]. However, GPs have limitations, including high computational complexity for large datasets and the need for careful selection of the kernel function. Despite these challenges, Gaussian Processes are widely applied in fields such as geostatistics (kriging), robotics, and environmental modeling, where careful handling of uncertainty and prediction accuracy are required.

### 8.2.10 k-Nearest Neighbors (k-NN)

Like GPs, k-Nearest Neighbors (k-NN) is also a nonparametric algorithm used for both classification and regression tasks [59]. As a nonparametric method like GPs, it does not make assumptions about the underlying distribution of data, making it particularly suitable for real-world scenarios where such assumptions are often invalid. The central idea behind k-NN is that similar data points exist in close proximity, meaning that data points with similar features tend to have similar outputs.

The term $k$ in k-NN refers to the number of nearest neighbors considered when making predictions [60]. The choice of $k$ can affect the model's performance. A smaller $k$ makes the model sensitive to noise, while a larger $k$ can oversmooth predictions, ignoring important patterns in the data.

k-NN relies on a distance metric to identify the closest neighbors to a given data point. Common distance metrics include:

Euclidean distance:

$$d(x,y) = \sqrt{\sum_{i=1}^{n} (x_i - y_i)^2},$$

where $x$ and $y$ represent two points in the feature space, and n is the number of features. Euclidean distance is suitable for general numerical data [61].

Manhattan distance:

$$d(x,y) = \sum_{i=1}^{n} |x_i - y_i|.$$

Similarly, $x$ and $y$ represent two points in the feature space, and n is the number of features. Manhattan distance is useful for grid-like data [62].

Chebyshev distance:

$$d(x,y) = \max_i |x_i - y_i|.$$

Measures the maximum difference along any coordinate dimension [63].

Minkowski distance:

$$d(x,y) = \left( \sum_{i=1}^{n} |x_i - y_i|^p \right)^{1/p},$$

where $x$ and $y$ represent two points in the feature space, n is the number of features, and p is a constant. When $p = 1$, the Minkowski distance is equivalent to the Manhattan distance, which is also known as L1 norm. When $p = 2$, it is equivalent to the Euclidean distance, L2 norm [64]. As p approaches infinity, the Minkowski distance converges to the Chebyshev distance, in this case, the distance is determined by the greatest of their differences along any coordinate dimension.

Mahalanobis distance:

$$d(x,y) = \sqrt{(x-y)^{\mathrm{T}} S^{-1} (x-y)},$$

where $S$ is the covariance matrix of the data, $x$ and $y$ are the feature vectors. Mahalanobis distance accounts for correlations between features [65].

Cosine similarity:

$$\cos(x,y) = \frac{x.y}{\|x\| \, \|y\|},$$

where $x$ and $y$ are the feature vectors. Cosine similarity is commonly used for high-dimensional data.

In classification tasks, k-NN assigns the most frequent label (mode) among the $k$ nearest neighbors to the data point. For example, if $k = 3$, and the three closest neighbors are classified as 'A', 'A', and 'B', the data point is classified as 'A', the majority label. This approach is effective for categorizing data into discrete classes.

Conversely, in regression tasks, k-NN predicts a continuous output by averaging the values of the $k$ nearest neighbors. For instance, if $k = 3$, and the nearest neighbors have values 10, 15, and 20, the predicted value is $(10 + 15 + 20)/3 = 15$. This method is useful for numerical predictions.

The choice of $k$, distance metric, and feature preprocessing are important to k-NN's performance. Features with larger ranges can dominate distance calculations, so scaling or normalization is recomended. Furthermore, k-NN can be computationally intensive for large datasets, as it requires calculating distances for all training

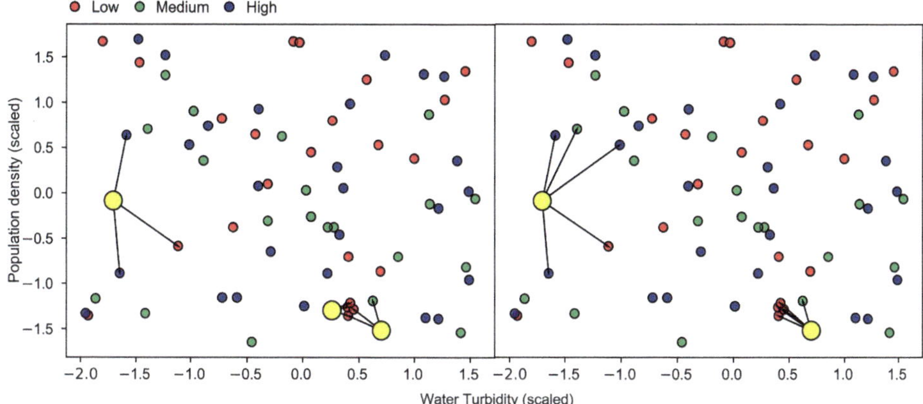


**Fig. 8.10** k-NN classification of example hypothetical contamination. This figure displays the classification of contamination levels using the k-NN algorithm. Training data points are color-coded, with red indicating "Low" contamination, green representing "Medium" contamination, and blue for "High" contamination. Test data points, marked in yellow, are classified based on the majority class among their nearest neighbors. Black arrows connect each test point to its three closest neighbors in the training set, visually illustrating the influence these neighbors have on the classification outcome. The axes of the plot represent "Water Turbidity" and "Population Density," which are the features used to determine the contamination levels. The spatial arrangement of the points and their connecting arrows conveys how the k-NN algorithm uses proximity in feature space to make predictions

data points during prediction. Techniques like KD-trees or approximate nearest neighbor algorithms can address this challenge [66, 67].

Overall, k-NN is a flexible and easy-to-understand algorithm, particularly useful in scenarios where simplicity and interpretability are valued (Fig. 8.10). However, careful hyperparameter tuning and preprocessing are necessary to maximize its effectiveness.

### 8.2.11 Deep Learning

Deep learning is inspired by the network of neurons found in the human brain and uses artificial neural networks to process and analyze data. These networks consist of layers of interconnected nodes, or artificial neurons, that are organized into three key types of layers: an input layer, which receives raw data; multiple hidden layers, which perform complex transformations; and an output layer, which delivers the final predictions [8, 68]. The interconnected nodes are linked by weighted edges, with weights fine-tuned during training to minimize prediction errors (Fig. 8.11).

Deep learning distinguishes itself by using deep neural networks, which feature numerous hidden layers capable of capturing highly complex and abstract patterns in data. Each node integrates inputs, multiplying each by a corresponding weight and adding a bias term, and processes the aggregate through an activation function [69]. Common activation functions include sigmoid, hyperbolic tangent (tanh), and Rectified Linear Unit (ReLU), which introduce nonlinearity into the network and enable it to model relationships within the data [70].

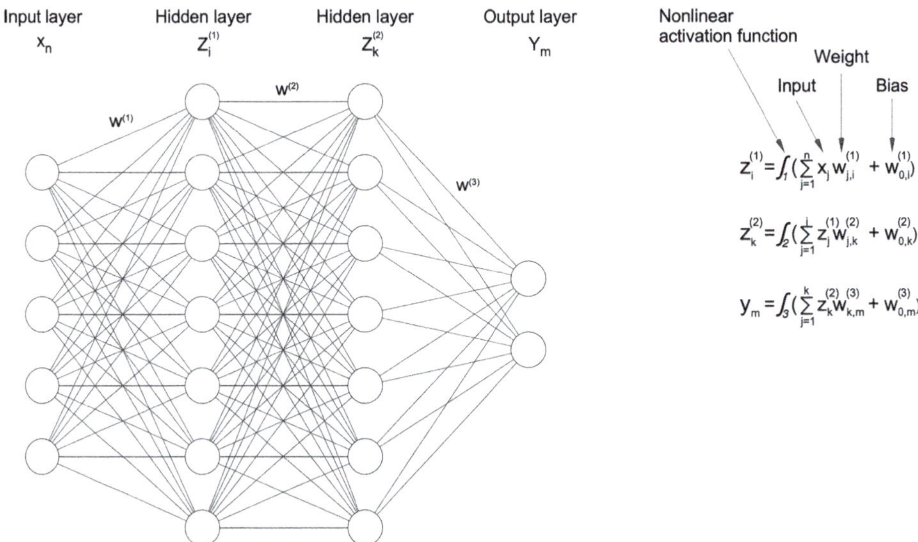




**Fig. 8.11** Schematic representation of a feedforward neural network for predictive modeling. The figure illustrates the architecture of a feedforward artificial neural network. The network consists of an input layer, multiple hidden layers, and an output layer. Each layer contains neurons (nodes) represented by circles. The input layer receives the feature data ($x$), which is then processed through one or more hidden layers using weighted connections ($w$) and nonlinear activation functions to capture complex patterns and relationships within the data. The final output layer ($Y$) produces the prediction outcome based on the processed information. The equations below each layer describe the mathematical operations performed, including the summation of weighted inputs and the application of a bias term ($w_0$), followed by a nonlinear activation function to generate the output for each layer. This structure exemplifies the key components and operations of feedforward artificial neural network, showing their capability for modeling nonlinear relationships in high-dimensional data

Specialized architectures within deep learning target specific data types and tasks. Convolutional Neural Networks (CNNs) are designed for analyzing grid-like data structures such as images, transforming fields like facial recognition, medical imaging, and autonomous driving by identifying spatial hierarchies in data [71]. Recurrent Neural Networks (RNNs), along with their advanced variant, Long Short-Term Memory (LSTM) networks, are designed to process sequential data [72, 73]. These networks power applications such as speech recognition, real-time transcription, and voice-controlled technologies by retaining information about prior inputs in a sequence. The Transformer architecture processes sequential data in parallel and efficiently captures long-range dependencies [74]. It has widely been used in Natural Language Processing (NLP) tasks, including language translation, summarization, and question-answering, surpassing the performance of traditional sequential models [75]. Deep reinforcement learning combines deep learning with reinforcement learning principles, enabling neural networks to learn through sequential decision-making [76]. Deep reinforcement learning has achieved superhuman performance in strategic games and is increasingly applied in robotics and autonomous systems.

Training deep learning models involves backpropagation, a technique where errors are propagated backward through the network to iteratively adjust the weights of connections [77]. While this process demands extensive computational resources and large datasets, the resulting improvements in model accuracy and generalization justify these requirements in many applications.

Deep learning extends the capabilities of traditional machine learning. For example, while supervised learning maps inputs to outputs using labeled data and unsupervised learning identifies hidden patterns in unlabeled data, deep learning is used in both paradigms and beyond. In unsupervised contexts, techniques like generative adversarial networks (GANs), dimensionality reduction, and anomaly detection are applications of deep learning [78–80]. Moreover, self-supervised learning leverage large unlabeled datasets by generating pseudo-labels from data itself, reducing reliance on labeled data and further expanding the range of applications [81].

Deep learning is well suited for recognizing patterns, modeling high-dimensional data, and adapting to diverse domains. It is applied across tasks involving images, text, audio, or sequential data, and has expanded what is possible in many areas of artificial intelligence.

## 8.3 Unsupervised Learning

Unsupervised learning focuses on identifying patterns or structures in unlabeled data [5]. It is useful in genomics for clustering genes with similar expression patterns or in identifying novel biomarkers. Unsupervised learning supports the discovery of hidden patterns, analysis of genetic variations, and help in understanding disease mechanisms [82]. These applications are relevant in areas like personalized medicine and drug development, and they contribute to the study of complex biological systems. However, the interpretability of results and the complexity of genomic data can limit how easily these techniques are applied. Unsupervised learning covers several methods such as clustering, dimensionality reduction, association rules, anomaly detection, generative models, topic modeling, and self-supervised learning (◘ Fig. 8.12). We will briefly discuss several of them

### 8.3.1 Clustering

Clustering is an unsupervised machine learning technique used to organize datasets into subsets or clusters. In clustering, objects within the same cluster show higher similarity to each other compared to objects in other clusters [83]. The primary goal is to discover inherent groupings in the data, guided by similarity or distance measures between data points. Clustering has widespread applications, including gene expression analysis in biology, customer segmentation in marketing, and document organization in information retrieval [84]. The choice of similarity measures (e.g., Euclidean distance, Manhattan distance, cosine similarity) and clustering algorithms used affect the outcome. While common clustering algorithms including K-means, hierarchical clustering, and Density-Based Spatial Clustering of Applications with Noise (DBSCAN) are widely used in biomedical research, clustering methods can be briefly subdivided into four groups (◘ Fig. 8.13).

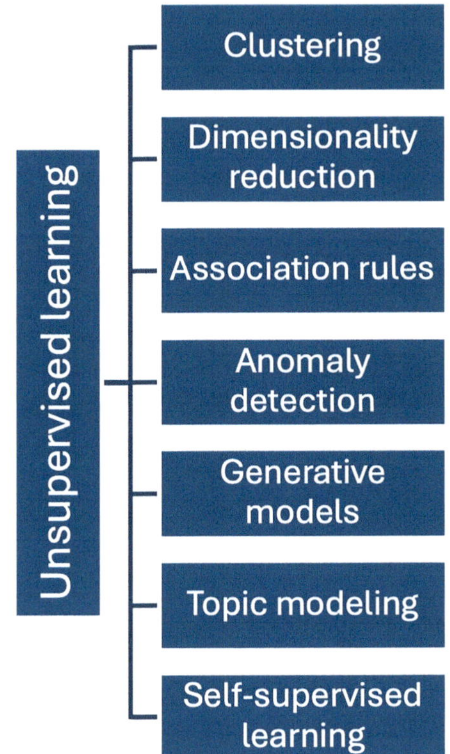


**Fig. 8.12** The main categories of unsupervised learning. Unsupervised learning mainly contains clustering, dimensionality reduction, association rules, anomaly detection, generative models, topic modeling, and self-supervised learning. These categories represent diverse techniques used to uncover patterns, structures, and relationships in data without relying on labeled outputs



The K-means clustering algorithm partitions data into $k$ groups by iteratively assigning points to the nearest cluster mean and updating the cluster means accordingly (Fig. 8.14a). This process continues until the clusters stabilize, or a maximum number of iterations is reached. While K-means is simple and computationally efficient, it is best suited for spherical clusters, requires the number of clusters to be specified beforehand, and is sensitive to the initial placement of centroids [85].

Conversely, hierarchical clustering builds a tree-like hierarchy of clusters using either an agglomerative (bottom-up) approach, where each data point starts in its own cluster, and clusters are merged iteratively, or a divisive (top-down) approach, where all points begin in one cluster and are recursively split [86, 87]. This method does not require predefining the number of clusters and provides a dendrogram to visually represent the clustering process (Fig. 8.14b). However, it is computationally expensive for large datasets.

DBSCAN groups points based on density by initiating a cluster around any point with a sufficient number of neighbors within a specified radius (epsilon) [88]. It incrementally adds points to the cluster if they fall within the density threshold. DBSCAN can identify clusters of arbitrary shapes, is resilient to outliers, and does not require predefining the number of clusters. However, it depends heavily on the choice of epsilon and the minimum number of points (minPts) required to form a dense region (Fig. 8.14c).

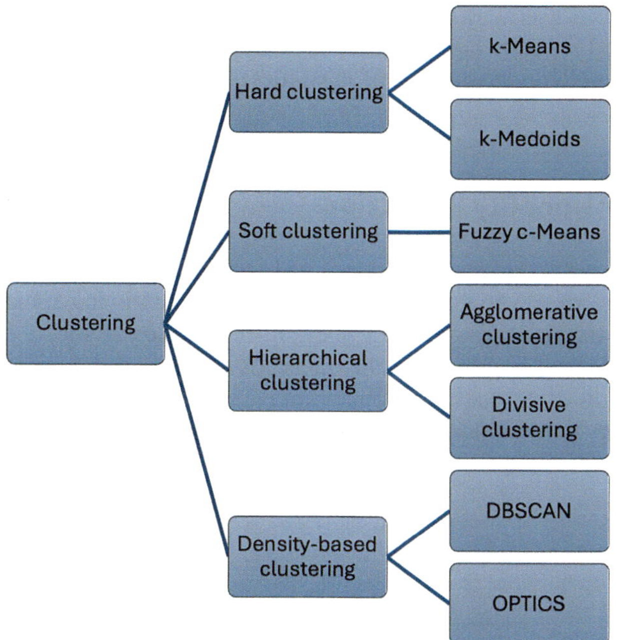


**Fig. 8.13** Different types of clustering. Clustering methods can be broadly divided into hard clustering, soft clustering, hierarchical clustering, and density-based clustering. Hard clustering assigns each data point to a single cluster, exemplified by methods like K-means and k-Medoids. Soft clustering assigns probabilities of cluster membership, as in Fuzzy c-Means. Hierarchical clustering creates a tree-like structure of clusters, further divided into agglomerative (bottom-up) and divisive (top-down) approaches. Density-based clustering methods, such as DBSCAN and OPTICS, group data points based on density, enabling the identification of clusters with arbitrary shapes and resilience to outliers

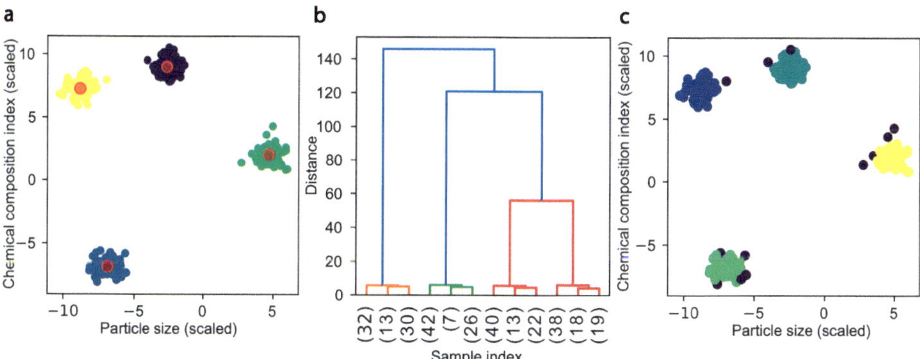


**Fig. 8.14** Comparative analysis of hypothetical particle characteristics across different samples. (**a**) The scaled distribution of particle sizes relative to sample indices displayed the variability in particle size distribution. (**b**) A detailed breakdown of particle sizes measured against a specific metric, emphasizing the observed range within the sample set. (**c**) The relationship between particle size and another unspecified characteristic

Clustering is a useful analytical tool for finding patterns and structures in data without the need for prior labels. Selecting the appropriate algorithm and parameters requires a thorough understanding of the data's characteristics and the specific application needs.

### 8.3.2 Dimensionality Reduction

Dimensionality reduction is a technique aimed at simplifying the complexity of data by reducing its number of variables. This process involves identifying a smaller set of principal variables that capture the most relevant information from the original dataset (◘ Fig. 8.15). Several of dimensionality methods are discussed in different chapters.

Techniques like principal component analysis (PCA) linearly transform the data to a lower-dimensional space, showing patterns and structures. t-distributed stochastic neighbor embedding (t-SNE) focuses on preserving the local structure of data,



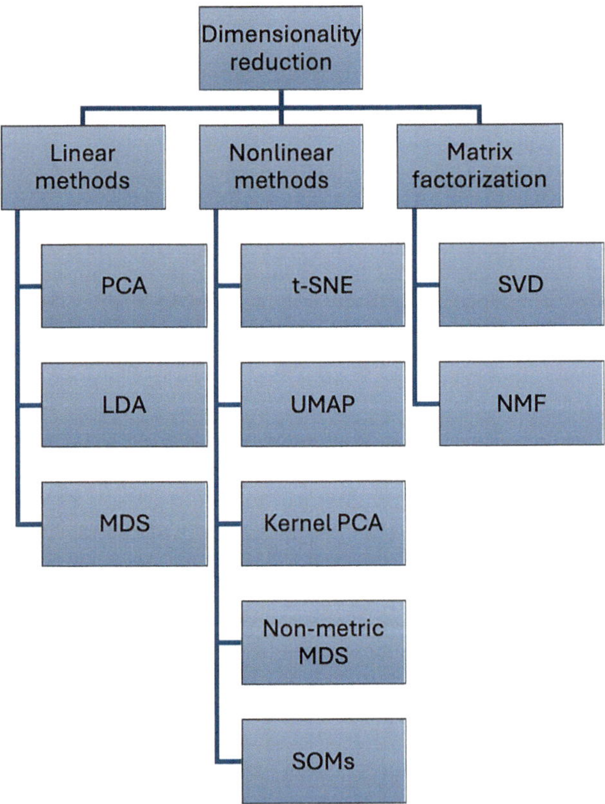


◘ **Fig. 8.15** Widely used dimensionality reduction algorithms. Widely used dimensionality reduction algorithms can be subdivided into linear methods, nonlinear methods, and matrix factorizations. Linear methods include principal component analysis (PCA), linear discriminant analysis (LDA), and multidimensional scaling (MDS). Various nonlinear methods include t-distributed stochastic neighbor embedding (t-SNE), uniform manifold approximation and projection (UMAP), kernel PCA, nonmetric MDS, and self-organizing maps (SOMs). Matrix factorization methods such as nonnegative matrix factorization (NMF) and singular value decomposition (SVD) are also used making it ideal for visualizing high-dimensional data in two or three dimensions. These methods improve data processing efficiency and interpretability, required for various machine learning tasks.

### 8.3.3 Association Rule Learning

Association rule learning is designed to uncover relationships, dependencies, or patterns among variables in datasets [89]. It identifies rules of the form "if-then" to determine co-occurrences or associations between items. Commonly applied in market basket analysis, it helps determine which products are frequently purchased together, enabling strategic decisions such as product placement, bundling, and recommendation systems. Key algorithms include the Apriori algorithm [90], Eclat [91], and FP-Growth [92], which efficiently mine frequent item sets and generate meaningful rules based on metrics like support, confidence, and lift (◘ Fig. 8.16). This method is widely applicable in retail, healthcare, and other domains for analyzing transactional or categorical data.

### 8.3.4 Anomaly Detection

Anomaly detection is an unsupervised learning technique aimed at identifying data points or patterns that deviate from the expected behavior or norm [51, 80]. These anomalies, or outliers, may indicate rare events, such as fraudulent transactions, network intrusions, equipment malfunctions, or data quality issues. Common approaches include statistical methods (e.g., *Z*-scores, Gaussian mixture models), distance-based methods (e.g., k-nearest neighbors), density-based techniques (e.g., local outlier factor), and machine learning algorithms like Isolation Forests (◘ Fig. 8.17). Anomaly

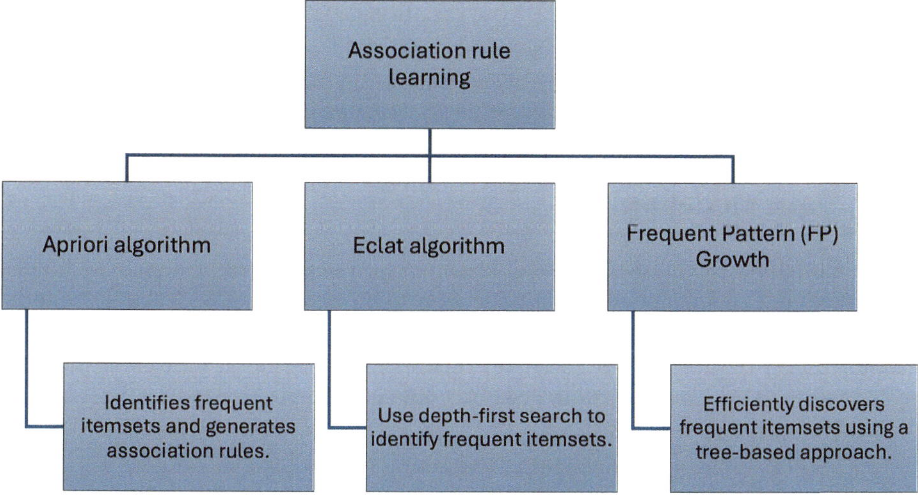


◘ **Fig. 8.16** Different association rule learning types include Apriori algorithm, Eclat algorithm, frequent pattern growth, etc

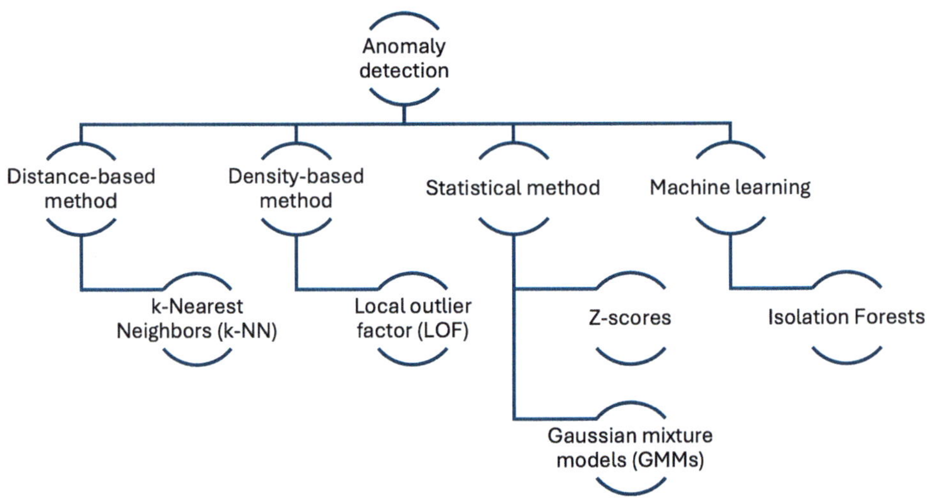


**Fig. 8.17** Anomaly detection algorithms. Widely used anomaly detection algorithms can be divided into four groups depending on the method used to detect anomalies



detection is widely applied in industrial quality control, healthcare monitoring, and other fields where identifying deviations is important for decision-making or risk mitigation.

### 8.3.5 Generative Models

Generative models are unsupervised learning techniques designed to learn the underlying distribution of a dataset and generate new data points that closely resemble the original data. Modeling the joint probability of features, these models can simulate realistic examples or fill in missing data. Common types of generative models include restricted Boltzmann machines (RBMs), variational autoencoders (VAEs), and generative adversarial networks (GANs) (Fig. 8.18) [78, 79, 93, 94]. They are widely used in applications such as image synthesis, data augmentation, text generation, and drug discovery, where generating high-quality, realistic data is required.

### 8.3.6 Topic Modeling

Topic modeling is an unsupervised learning technique used to uncover hidden abstract topics or themes within a collection of documents [95]. By analyzing the co-occurrence of words across the dataset, it groups words into topics, enabling the categorization of documents based on their thematic content. Common methods include latent Dirichlet allocation (LDA), nonnegative matrix factorization (NMF), and latent semantic analysis (LSA) (Fig. 8.19). Topic modeling is widely applied in text mining, natural language processing, and information retrieval for tasks such as document classification, summarization, and content recommendation [96].

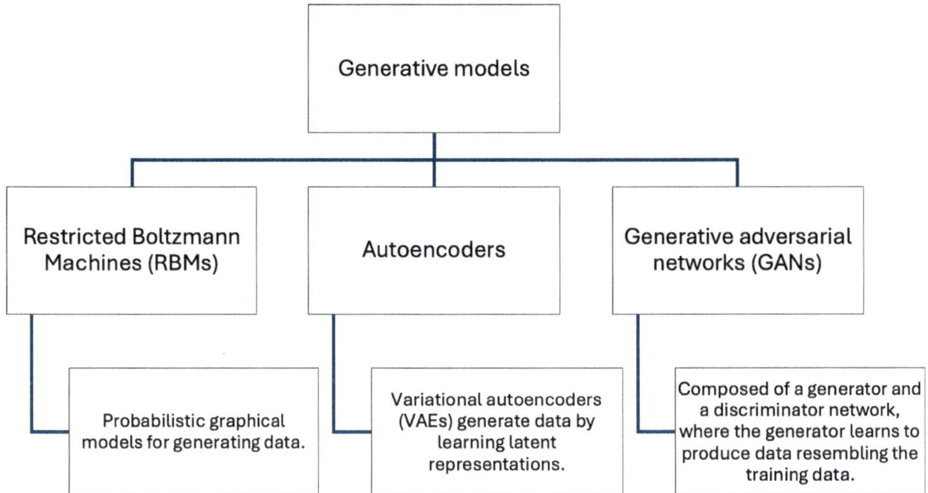


**Fig. 8.18** Classifications of generative models. Generative models aim to learn the underlying data distribution and generate new data points that resemble the original dataset. Within several types of generative models, RBMs, autoencoders, and GANs are widely used

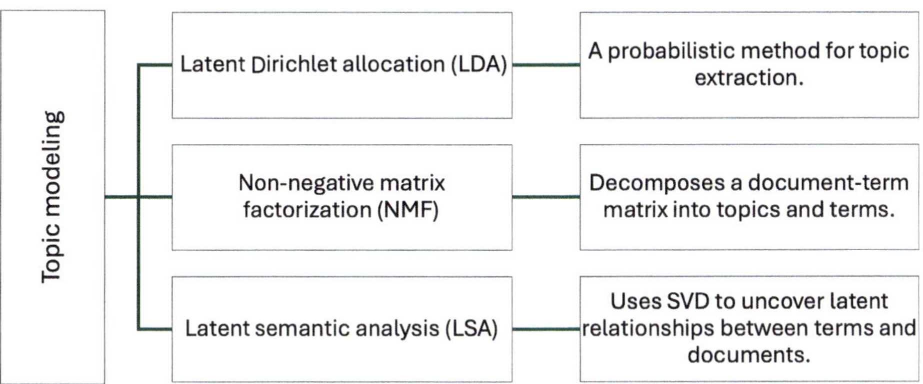


**Fig. 8.19** Topic modeling algorithms. Topic modeling algorithms such as LDA, NMF, and LSA identify abstract topics or themes within a collection of documents

### 8.3.7 Self-Supervised Learning

Self-supervised learning is a type of machine learning where the system creates its own labels from the data, so models can learn meaningful representations without requiring manually labeled data [81, 97]. In this approach, the algorithm sets up prediction tasks using part of the data as input and part as the target for learning. For example, in image analysis, the model might be trained to predict missing sections of an image using the visible parts, or to identify the order of shuffled segments. In language models, the system might be tasked with predicting the next word or filling in missing words in a sentence. These automatic tasks, called pretext tasks, allow the model to extract features and build internal representations from large amounts of

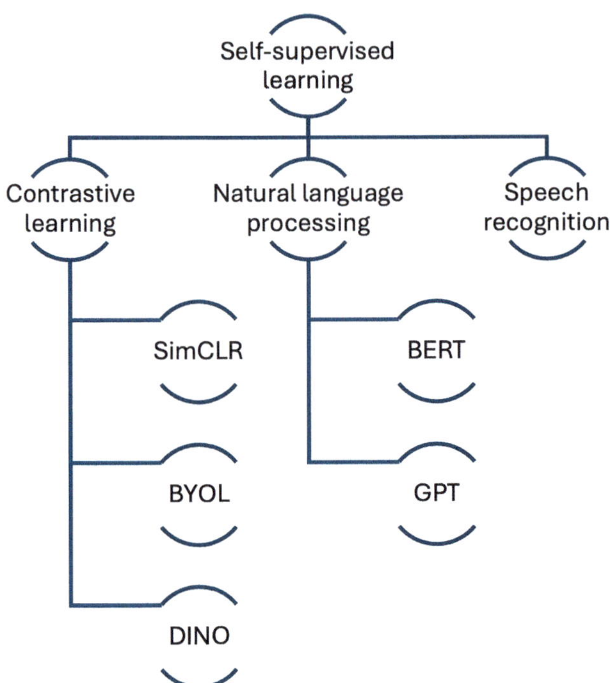


**Fig. 8.20** Different types of self-supervised learning. Self-supervised learning can be considered an emerging category where the data itself generates pseudo-labels. It is often associated with unsupervised approaches and used for representation learning

raw data. Although self-supervised learning does not use external labels, it uses the structure within the data to form prediction goals. This places it between unsupervised and supervised learning, as it takes advantage of unlabeled data but creates a supervised learning setup by generating its own targets [98]. Self-supervised learning is particularly valuable for extracting meaningful features from complex data, and is used in areas such as contrastive learning, natural language processing, and speech recognition, where labeled data may be limited but large datasets are available (Fig. 8.20).

## 8.4 Semi-supervised Learning

Semi-supervised learning is an approach that combines elements of supervised and unsupervised learning by using a small amount of labeled data alongside a larger pool of unlabeled data to train machine learning models [6]. In practice, semi-supervised learning works by first using the labeled examples to provide some initial guidance for the model. The model then tries to learn from the patterns present in the unlabeled data, often by assigning provisional labels or by identifying data structure that supports the initial learning from the labeled set. This approach is particularly useful in biomedicine, where labeled data is often scarce and expensive to obtain, but vast amounts of unlabeled data, such as genomic sequences, medical images, or electronic health records, are readily available. By combining both types of data, semi-supervised learning can improve model performance [99–102].

One of the key applications of semi-supervised learning in biomedicine is in disease diagnosis and classification, especially when labeled datasets are limited. For instance, in medical imaging, where expert annotation of images can be time-intensive and costly, semi-supervised learning can use a small set of labeled scans to guide the learning process while incorporating unlabeled scans to improve model generalization [101]. Similarly, in genomics, semi-supervised learning can be applied to classify genetic variants or predict gene functions by using available sequence data, only a fraction of which may have been labeled with known biological functions [99]. Semi-supervised learning techniques can also play a role in drug discovery, where labeled data on drug efficacy or toxicity is limited [11, 103]. Combining these sparse labeled datasets with abundant unlabeled molecular and biological data, semi-supervised learning can accelerate the identification of drug candidates and reduce the cost and time associated with drug development.

In addition to classification tasks, semi-supervised learning is also applied in tasks like clustering and anomaly detection in biomedicine. For example, in patient stratification, semi-supervised learning can use limited labeled patient data alongside unlabeled clinical records to identify subgroups of patients with similar characteristics, which can inform personalized treatment strategies [104]. Similarly, in anomaly detection, such as identifying rare diseases or adverse drug reactions, semi-supervised learning can leverage a small number of known cases along with large-scale unlabeled data to improve detection sensitivity and accuracy.

Despite its advantages, semi-supervised learning comes with challenges. The quality and representativeness of the labeled data are important, as biased or limited labeled datasets can adversely impact model performance. Furthermore, the choice of algorithms and assumptions about the relationship between labeled and unlabeled data (e.g., the assumption that data points with similar features belong to the same class) must align with the nature of the problem at hand. Furthermore, ensuring the interpretability and robustness of semi-supervised learning models is required, especially in healthcare applications where trust and transparency are needed.

## 8.5 Reinforcement Learning

Reinforcement learning is a branch of machine learning where algorithms learn to make sequences of decisions through trial and error to maximize long-term rewards [7]. While its use in biomedicine is currently less common compared to other machine learning techniques, reinforcement learning can be applied in areas requiring dynamic, adaptive, and personalized decision-making [105]. Its ability to optimize strategies over time makes it particularly suited for complex medical scenarios such as personalized treatment regimens, adaptive therapies, and clinical decision support. However, the implementation of reinforcement learning in healthcare poses unique challenges, including the need for high-quality data, ensuring patient safety, addressing ethical concerns, and improving the interpretability of the models to build trust among clinicians.

One of the most promising applications of reinforcement learning in biomedicine is the development of personalized treatment regimens. For instance, in chronic diseases like diabetes or cancer, where treatment responses vary across patients and over time, reinforcement learning can dynamically optimize treatment strategies based on continuous patient data [106, 107]. Algorithms can adjust insulin doses in real time for diabetes patients, considering factors like glucose levels and lifestyle data, or recommend adaptive chemotherapy plans for cancer patients to maximize efficacy while minimizing side effects. Reinforcement learning also holds promise in clinical decision support systems, where it can assist clinicians in sequential decision-making, such as determining the optimal sequence of diagnostic tests or treatments [108]. In intensive care units, reinforcement learning-based systems could recommend the timing and dosage of medications for septic patients, continuously adapting recommendations based on real-time patient data and outcomes [109, 110].

Beyond treatment and diagnostics, reinforcement learning is transforming other areas of biomedicine, such as drug discovery, adaptive radiotherapy, and robotics [111, 112]. In drug development, reinforcement learning can optimize the iterative design and testing of drug formulations, guiding molecular modifications to improve efficacy and reduce toxicity [113, 114]. Similarly, in oncology, reinforcement learning can dynamically adjust radiotherapy plans based on changes in tumor size and patient anatomy, improving treatment precision and reducing harm to healthy tissue [115, 116]. Reinforcement learning can also improve robotic systems, from improving the precision and safety of surgical robots through continuous learning to adapting rehabilitation robotics to the evolving needs of patients during physical therapy [117, 118]. For example, reinforcement learning-enabled robotic arms can learn patient-specific exercises and adjust therapy as the patient progresses [118].



## 8.6 Data Preprocessing for Machine Learning Models

Data preprocessing involves preparing and transforming raw data into a suitable format that can be effectively used by machine learning algorithms to identify patterns, make predictions, and derive relevant information.

### 8.6.1 Data Cleaning

The data cleaning step addresses issues such as missing values, outliers, and errors within the dataset. For example, missing data in certain samples can be addressed using statistical methods (e.g., Iterative Imputation, mean/median imputation, K-nearest neighbors) or may be omitted based on the study's specific requirements [119, 120]. Outliers, which might arise from measurement errors or anomalies in sample collection, must be identified and addressed (e.g., removed or corrected) to prevent skewed analyses. Statistical thresholds (e.g., $Z$-score, Interquartile Range (IQR) methods) are utilized to identify and either remove or adjust outlier values that do not reflect typical conditions [121]. The $Z$-score method identifies outliers by measuring how many standard deviations an observation is from the mean [122]. Observations with a $Z$-score exceeding a defined threshold (typically 3 or −3) are considered outliers. The IQR, defined as the difference between the 75th and 25th percentiles of the data, helps further in outlier identification. Data points that fall below the 25th percentile minus 1.5 times the IQR or above the 75th percentile plus 1.5 times the IQR are considered outliers [123, 124]. Box plot analysis provides a visual representation of the data distribution and aids in outlier identification. Points located outside the whiskers of the box plot are typically considered outliers. This approach ensures that the dataset is refined and optimized for subsequent analyses, facilitating more accurate and reliable outcomes.

### 8.6.2 Data Normalization

Data normalization and standardization are important preprocessing steps in machine learning and data analysis. These processes adjust the scale of input variables to ensure consistency across datasets, which is particularly vital when data points are recorded using varying scales and units [125]. Achieving this consistency is required for meaningful comparison, analysis, and effective training of machine learning models. Normalization refers to the process of rescaling the values of an attribute to a standard range, typically [0, 1] or [−1, 1].

For example, min-max normalization:

$$x_{scaled} = \frac{x - x_{\min}}{x_{\max} - x_{\min}}.$$

This method is beneficial when dealing with parameters that exhibit diverse ranges. For instance, in an arbitrary study, concentrations of an element may be reported in parts per million (ppm) in one study and micrograms per liter (μg/L) in another. Without normalization, these differences in scale could bias the analysis, giving undue weight to certain features. By applying normalization, each concentration value is proportionately scaled down to a comparable range, ensuring that no single measurement unduly influences the model's performance [38].

Standardization, on the other hand, modifies the data to have a mean of zero and a standard deviation of one. This process, also known as *Z*-score normalization, is useful for algorithms that assume the data to be normally distributed, such as many types of neural networks and support vector machines. When data from different studies are standardized, it not only facilitates a more straightforward comparative analysis but also aids in the efficient convergence of algorithms. This is because standardized data helps in smoothing the optimization landscape, making it easier for algorithms to find minima. Scaling concentration levels across different studies to a uniform range, we can aggregate data more effectively, making it possible to perform broader analyses that might reveal patterns and correlations not apparent before standardization.

Several tools and libraries support data normalization and standardization, including Pandas and Scikit-learn in Python. These libraries offer built-in functions, such as MinMaxScaler for normalization and StandardScaler for standardization, which can be easily integrated into data processing pipelines [126].

### 8.6.3 Feature Engineering

Feature engineering plays a key role in preparing data for machine learning models, involving the transformation or creation of new features from existing data to improve the model's predictive power [127]. This process is relevant in studies where data comes with numerous variables. Python libraries (e.g., pandas, NumPy, scikit-learn) can be used for feature engineering. These tools provide functions for calculating ratios, statistical summaries, temporal transformations, and interaction terms, among other capabilities. Feature engineering is an iterative process that requires domain knowledge, data exploration, and an understanding of the modeling objectives.

### 8.6.4 Data Augmentation



Data augmentation addresses the challenges posed by imbalanced datasets or those lacking in diversity, which can skew model predictions and hinder the generalizability of the models. Through data augmentation, synthetic data points are generated, improving the dataset's robustness and enabling models to learn from a more general representation of the problem space. The Synthetic Minority Over-sampling Technique (SMOTE) is a popular method for generating synthetic samples in the feature space [31, 128]. It operates by selecting examples that are close in the feature space, drawing a line between the examples in the feature space, and generating new samples along that line [129]. Similar to SMOTE, ADASYN (Adaptive Synthetic Sampling) focuses on generating synthetic data for minority classes but with an added algorithmic feature that automatically decides the number of synthetic samples to produce for each minority sample by considering its level of difficulty in learning. This technique allows a more refined approach to balancing the dataset, which can improve model performance.

## 8.7 Technique to Optimize Model Parameters

Optimizing model parameters ensures they perform optimally on the task at hand. This process involves fine-tuning the model's internal settings to achieve the best balance between underfitting and overfitting, thereby maximizing its predictive accuracy on unseen data. Several techniques are employed to optimize these parameters, with some of the most prominent including Grid Search, Random Search, Bayesian Optimization, Gradient-Based Optimization, Evolutionary Algorithms, and Optuna [130].

### 8.7.1 Grid Search

Grid Search is a brute-force method that systematically explores a manually specified subset of the hyperparameter space of a learning algorithm. It trains the model with every combination of parameters in the grid to find the set that produces the best model performance, typically measured by a cross-validated score. While Grid Search is thorough and easy to implement, its main drawback is computational inefficiency, especially when dealing with large parameter spaces or when the parameters influence the model in a nonlinear manner [131].

### 8.7.2 Random Search

Random Search addresses some of the inefficiencies of Grid Search by randomly sampling the parameter space and evaluating models with these parameter sets. It offers a more efficient exploration of the parameter space and can sometimes find a good set of parameters more quickly than Grid Search. Random Search is particularly effective when only a few parameters influence the model's performance, as it does not waste time evaluating unimportant parameters [131].

### 8.7.3 Bayesian Optimization

Bayesian Optimization is a probabilistic model-based optimization technique. It builds a probabilistic model of the function mapping from hyperparameter values to the objective evaluated on a validation set. Then, it uses this model to select the most promising hyperparameters to evaluate in the true objective function. This approach is more efficient than Grid and Random Searches, as it uses the results of past evaluations to inform future selections, reducing the number of iterations needed to find optimal parameters [132].

### 8.7.4 Gradient-Based Optimization

Gradient-Based Optimization techniques, such as Gradient Descent, are used for optimizing differentiable objective functions [133]. In the context of deep learning, where models are often differentiable with respect to their parameters, these techniques can adjust parameters in the direction that minimally decreases the loss function [134]. This method is highly effective for continuous parameter spaces and is the backbone of neural network training.

### 8.7.5 Evolutionary Algorithms

Evolutionary Algorithms (EAs) are inspired by the process of natural selection, where the fittest individuals are selected for reproduction in order to produce offspring of the next generation. In the context of hyperparameter optimization, EAs can be used to evolve the model parameters over time [135]. These algorithms, including Genetic Algorithms, start with a population of random parameter sets and iteratively apply operations such as selection, crossover, and mutation to evolve the parameters toward better performance on the objective function.

### 8.7.6 Optuna

Optuna is an open-source hyperparameter optimization framework that automates the process of finding the best hyperparameters for machine learning models [130]. It is designed to be efficient and flexible, supporting a wide range of optimization techniques, including those based on Bayesian optimization, evolutionary algorithms, and grid search. Optuna utilizes efficient search strategies such as Tree-Structured Parzen Estimator (TPE) and CMA-ES (Covariance Matrix Adaptation Evolution Strategy), which can speed up the search process compared to traditional methods like Grid Search and Random Search. It provides an automatic trial pruning feature that can stop unpromising trials early, saving computational resources and time. This feature is particularly useful when working with large datasets or complex models, where each trial can be time-consuming.



## 8.8 Techniques for Evaluating a Machine Learning Model

The evaluation of a machine learning model is important for understanding its performance, generalizability, and applicability to real-world scenarios. This process involves a series of techniques, each providing different perspectives on the model's effectiveness and areas for improvement. The core objective is to ascertain how well the model has learned the underlying patterns in the data and its ability to make accurate predictions on unseen data. Here, we discuss key evaluation techniques, their importance and applications [136].

### 8.8.1 Training and Testing Split

One of the most basic evaluation strategies is dividing the dataset into training and testing sets. The model is trained on the training set, which contains a large portion of the data, and evaluated on the testing set, a separate portion unseen by the model during training. This approach helps in assessing the model's generalization capabilities, that is, its ability to perform well on new, unseen data [137].

### 8.8.2 Cross-Validation

Cross-validation is used to ensure the robustness and reliability of the model evaluation. The most common form is $k$-fold cross-validation, where the data is divided into $k$ equal-sized parts. The model is trained on $k-1$ parts and tested on the remaining part. This process is repeated k times, with each part serving as the testing set once. Cross-validation provides a more accurate measure of a model's performance by utilizing the entire dataset for both training and testing, reducing the bias associated with a single train-test split [138].

### 8.8.3 Confusion Matrix

A confusion matrix is a specific table layout that allows visualization of the performance of an algorithm, typically a supervised learning one. It is especially useful for classification tasks. The matrix compares the actual target values with those predicted by the model, allowing identification of the true positives (TP), true negatives (TN), false positives (FP), and false negatives (FN).

- True Positives (TP): The cases in which the model correctly predicts the positive class.
- True Negatives (TN): The cases in which the model correctly predicts the negative class.
- False Positives (FP): The cases in which the model incorrectly predicts the positive class.
- False Negatives (FN): The cases in which the model incorrectly predicts the negative class.

This matrix forms the basis for calculating various performance metrics, such as accuracy, precision, recall, and $F1$ score.

### 8.8.4 Accuracy

Accuracy is the most straightforward performance measure, and it is simply a ratio of correctly predicted observations to the total observations. It is suitable for binary and multiclass classification problems. However, accuracy alone can be misleading if the data is imbalanced; that is, when the number of observations in different classes varies.

$$\text{Accuracy} = \frac{\text{Number of Correct Predictions}}{\text{Total Number of Predictions}}.$$

More specifically, when considering elements of a confusion matrix:

$$\text{Accuracy} = \frac{\text{TP} + \text{TN}}{\text{TP} + \text{TN} + \text{FP} + \text{FN}}.$$

### 8.8.5 Precision (Positive Predictive Value)

Precision measures the accuracy of the positive predictions made by the model. It quantifies the proportion of positive identifications that were actually correct. Precision is particularly important in scenarios where the cost of false positives is high [139].

$$\text{Precision} = \frac{\text{TP}}{\text{TP} + \text{FP}}.$$

### 8.8.6 Sensitivity or Recall

Sensitivity measures the ability of the model to capture the actual positive cases. It quantifies the proportion of actual positives that were correctly identified by the model. Sensitivity is useful in situations where the cost of false negatives is high [139].



$$\text{Sensitivity} = \frac{\text{TP}}{\text{TP} + \text{FN}}.$$

### 8.8.7 *F*1 Score

The $F1$ score is the harmonic mean of precision and recall, taking both metrics into account. It is particularly useful when the class distribution is imbalanced. The $F1$ score is a better measure than accuracy for imbalanced classes, because it considers both false positives and false negatives [139].

$$F1\text{score} = \frac{2 \times \text{TP}}{2 \times \text{TP} + \text{FP} + \text{FN}}.$$

### 8.8.8 Negative Predictive Value (NPV)

Negative Predictive Value (NPV) measures the proportion of negative instances correctly identified as negative among all instances predicted as negative. NPV is especially important in contexts where the ability to accurately identify negative cases is important. It helps in assessing the model's effectiveness in predicting the absence of a condition, attribute, or characteristic. The NPV complements the precision by focusing on the accuracy of the negative predictions. It is particularly relevant in medical testing, where a high NPV indicates that individuals who test negative are indeed likely not to have the disease.

$$\text{NPV} = \frac{\text{TN}}{\text{TN} + \text{FN}}.$$

### 8.8.9 Balanced Accuracy

Similar to F1 score, balanced accuracy is useful in contexts where the classes are imbalanced. It is defined as the average of sensitivity obtained on each class. This metric calculates the accuracy for each class individually, then averages these accuracies, giving equal weight to both classes in binary classification scenarios. Balanced accuracy amends the issue with traditional accuracy where imbalanced class distributions can lead to misleadingly high scores simply by favoring the majority class.

$$\text{Balanced accuracy} = \frac{1}{2} \times \left( \frac{\text{TP}}{\text{TP} + \text{FN}} + \frac{\text{TN}}{\text{TN} + \text{FP}} \right).$$

### 8.8.10 Matthews Correlation Coefficient (MCC)

The MCC provides a measure of the quality of predictions, incorporating true and false positives and negatives into a single score. It ranges from −1 to +1, where +1 indicates perfect prediction, 0 indicates no better than random prediction, and −1 indicates total disagreement between prediction and observation [140]. The MCC is particularly useful because it is balanced: it can be used even if the classes are of very different sizes, making it useful in situations where precision, recall, *F*1 score, or accuracy might provide a biased view due to class imbalance.

$$\text{MCC} = \frac{(\text{TP} \times \text{TN}) - (\text{FP} \times \text{FN})}{\sqrt{(\text{TP} + \text{FP}) \times (\text{TP} + \text{FN}) \times (\text{TN} + \text{FP}) \times (\text{TN} + \text{FN})}}.$$

### 8.8.11 ROC Curve and AUC

The Receiver Operating Characteristic (ROC) curve is a graphical plot that illustrates the diagnostic ability of a binary classifier system as its discrimination threshold is varied. The curve plots two parameters: True Positive Rate (TPR) and False Positive Rate (FPR). The Area Under the Curve (AUC) represents a measure of separability. A higher AUC indicates a better model at distinguishing between the positive and negative classes [141].

### 8.8.12 Machine Learning Model Building and Evaluation

The evaluation of machine learning models involves a set of techniques, each providing different perspectives on the model's performance. By employing a combination of these methods, we can gain a holistic view of the model's strengths and weaknesses, enabling informed decisions on model selection, tuning, and deployment. The choice of evaluation metrics should align with the specific objectives of the machine learning task and the nature of the data. Understanding and properly applying these techniques is paramount for achieving models that are not only accurate but also relevant and practical for real-world applications.

## 8.9 Implementing Simple Machine Learning Models in Python

In biomedicine, machine learning models play important roles. These models facilitate the analysis of complex biological data, enabling advancements in drug discovery, target identification, and personalized medicine. Python serves as an ideal platform for developing these models. This section aims to provide an understanding of how basic machine learning models can be built and applied within the biomedical sphere using Python.

### 8.9.1 Key Python Libraries for Machine Learning in Biomedicine

The Python libraries simplify data handling, model building, training, and evaluation, enabling researchers to tackle complex biomedical problems effectively. Below, we summarize widely used Python libraries for machine learning in biomedicine and their relevance to this domain.

**scikit-learn**: scikit-learn is a free and open-source machine learning library renowned for its simplicity and accessibility, making it an excellent choice for both beginners and experts in biomedicine [142]. It provides a wide range of supervised and unsupervised learning algorithms, including classification, regression, clustering, and dimensionality reduction, which are useful for biomedical tasks such as disease classification and patient stratification. Furthermore, scikit-learn includes tools for cross-validation and performance evaluation, ensuring the reliability and accuracy of predictive models. Its pipeline creation feature allows integration of data preprocessing, model training, and evaluation, enabling simplified workflows for complex biomedical datasets.

**TensorFlow**: TensorFlow, developed by Google Brain, is a widely used open-source library for numerical computation and large-scale machine learning [143]. It is particularly useful for deep learning, providing advanced neural network capabilities that are increasingly applied in biomedical fields like genomics, drug discovery, and medical image analysis. TensorFlow is designed for scalability and flexibility, making it well-suited for handling large datasets, such as those used in whole-genome sequencing. Its advanced tools, such as TensorFlow Probability, enable probabilistic reasoning and uncertainty estimation, improving model interpretability and trust in clinical applications.

**PyTorch**: PyTorch, developed by Facebook AI Research, is a flexible and dynamic machine learning library widely used in research for its adaptability [144]. It features dynamic computational graphs, allowing for the design of complex models that are ideal for experimental biomedical applications. With an active research community, PyTorch benefits from a robust ecosystem of tools and extensions, making it a popular choice for cutting-edge biomedical research.

**SciPy**: SciPy is an open-source Python library that builds on NumPy and provides advanced mathematical algorithms and convenience functions for scientific computing in biomedicine [145]. It provides a wide array of mathematical tools for optimization, signal processing, and statistical analysis, making it useful for processing complex biomedical data. SciPy integrates with other libraries like NumPy,

Pandas, and Matplotlib, supporting workflows for biomedical data analysis and visualization.

## 8.10 Implementing Machine Learning Models in Python

Implementing machine learning models in Python, particularly in biomedicine, involves several systematic steps: data preprocessing, model training, prediction, and evaluation. These steps form the foundation for developing robust and interpretable models. In this section, we demonstrate this process through predicting drug sensitivity using a logistic regression model. The steps are designed to handle typical datasets and include performance visualization to assess model effectiveness.

### 8.10.1 Predicting Drug Sensitivity Using LogisticRegression

**Notebook: Section 8.10.1. Predicting Drug Sensitivity Using LogisticRegression**
This notebook provides code examples demonstrating drug sensitivity prediction using LogisticRegression.
Link to the GitHub repository:
► https://github.com/sn-code-inside/BioPy
Go to: Chap. 8—example data—► Sect. 8.10.1

Imagine a dataset where each row represents a patient's genetic profile, and the target variable indicates their sensitivity to a specific drug (encoded as 1 for sensitive and 0 for resistant). The dataset is stored in a Pandas DataFrame named `data`, with features labeled `feature_1, feature_2, ..., feature_n` and the target variable labeled as `target`.

Step 1: Import required libraries

To begin, we import Python libraries for data manipulation, model training, evaluation, and visualization:

```
import pandas as pd
from sklearn.model_selection import train_test_split
from sklearn.linear_model import LogisticRegression
from sklearn.metrics import classification_report
from sklearn.metrics import confusion_matrix
import matplotlib.pyplot as plt
import seaborn as sns
```

Step 2: Data preprocessing

Data preprocessing ensures the dataset is clean and ready for modeling. Here, we split the data into features (X) and the target variable (y), followed by dividing it into training and test sets:

```
# Load dataset (modify the path as per your dataset)
data = pd.read_csv('example_data/data.csv', index_col=0)

# Split the dataset into features (X) and target variable (y)
X = data.drop('target', axis=1)

# data standardization
scalar = StandardScaler()
X = scalar.fit_transform(X)

y = data['target']

# Split data into training and test sets
X_train, X_test, y_train, y_test = train_test_split(X, y,
test_size=0.3, random_state=42)
```



Step 3: Train the logistic regression model

A logistic regression model is initialized and trained on the preprocessed training data:

```
# Initialize the Logistic Regression model
log_reg = LogisticRegression()

# Fit the model to the training data
log_reg.fit(X_train, y_train)
```

Step 4: Make predictions and evaluate the model

The trained model is used to predict outcomes for the test set, and its performance is evaluated using metrics like classification reports and confusion matrices:

```
# Making predictions on the test set
y_pred = log_reg.predict(X_test)

# Evaluating the model
print(classification_report(y_test, y_pred))

# Confusion Matrix
conf_matrix = confusion_matrix(y_test, y_pred)
```

Step 5: Visualization of model performance

Visualizing the confusion matrix helps interpret the model's performance and identify misclassifications:

```
# Visualizing the confusion matrix
sns.heatmap(conf_matrix, annot=True, fmt='g')
plt.title('Confusion Matrix')
plt.ylabel('Actual Label')
plt.xlabel('Predicted Label')
plt.show()
```

### 8.10.2 Predicting Drug Sensitivity Using RandomForest

**Notebook: Section 8.10.2. Predicting Drug Sensitivity Using RandomForest**

This notebook provides code examples demonstrating drug sensitivity prediction using RandomForest.

Link to the GitHub repository:

▶ https://github.com/sn-code-inside/BioPy

Go to: Chap. 8—example data—▶ Sect. 8.10.2

Random Forest, an ensemble learning method, can be used to predict biomedical outcomes. In this example, we implement a Random Forest classifier to analyze the clinical and genetic data of patients and predict a health outcome, such as the presence or absence of a disease. The dataset consists of patient records stored in a Pandas DataFrame named `data`, with clinical and genetic features labeled `feature_1, feature_2, ..., feature_n` and the target variable labeled as the `target`, encoded as 0 (absence of the disease) and 1 (presence of the disease).

Step 1: Import required libraries

First, import the relevant Python libraries for data manipulation, model training, evaluation, and visualization:

```
import pandas as pd
from sklearn.model_selection import train_test_split
from sklearn.ensemble import RandomForestClassifier
from sklearn.metrics import classification_report, confusion_matrix
import matplotlib.pyplot as plt
import seaborn as sns
```

Step 2: Data preprocessing

Preprocessing ensures that the dataset is clean and ready for modeling. The steps include loading the data, splitting it into features (X) and the target variable (y), and dividing it into training and test sets:

```
# Load dataset (modify the path as per your dataset)
data = pd.read_csv('path_to_dataset.csv')

# Split the dataset into features (X) and target variable (y)
X = data.drop('target', axis=1)
y = data['target']

# Split data into training and test sets
X_train, X_test, y_train, y_test = train_test_split(X, y, test_
size=0.3, random_state=42)
```



Step 3: Train the Random Forest model

Initialize the Random Forest classifier with 100 estimators and train it on the preprocessed training data:

```
# Initialize the RandomForest Classifier
random_forest = RandomForestClassifier(n_estimators=100,
random_state=42)

# Fit the model to the training data
random_forest.fit(X_train, y_train)
```

Step 4: Make predictions and evaluate the model

The trained model is used to make predictions on the test set, and its performance is evaluated using a classification report and confusion matrix:

```
# Making predictions on the test set
y_pred = random_forest.predict(X_test)

# Evaluating the model
print(classification_report(y_test, y_pred))

# Confusion Matrix
conf_matrix = confusion_matrix(y_test, y_pred)
```

Step 5: Visualization of model performance

The confusion matrix is visualized using a heatmap to examine model's performance and misclassifications:

```
# Visualizing the confusion matrix
sns.heatmap(conf_matrix, annot=True, fmt='g')
plt.title('Confusion Matrix for RandomForest Model')
plt.ylabel('Actual Label')
plt.xlabel('Predicted Label')
plt.show()
```

### 8.10.3 Predicting Drug Sensitivity Using AlphaML

AlphaML is a machine learning platform that integrates a wide range of tools and methodologies to help with the model building and evaluation process, making it useful for predicting drug sensitivity. It supports both unsupervised and supervised feature selection, hyperparameter optimization, and data preprocessing techniques like under-sampling, over-sampling, and normalization. The platform offers support for 15 different machine learning algorithms and includes advanced tools for model interpretation, such as ROC-AUC and accuracy plots, permutation feature importance, SHAP plots, and LIME plots. One of its features is a custom metric for hyperparameter search, which balances training and validation scores to minimize overfitting and underfitting. With its user-friendly graphical interface, AlphaML is accessible even to users with limited programming experience, making it a useful tool in biomedical applications [146].

To predict drug sensitivity using AlphaML, follow these steps:

Step 1. Set up the environment

It is recommended to use the Anaconda environment for managing dependencies. Refer to ▶ Chap. 1 for instructions on setting up Anaconda.

Step 2. Install `AlphaML`

Open the Anaconda Powershell Prompt and install AlphaML by running the following command:

```
pip install alphaml
```

Step 3. Run the AlphaML GUI

Open the AlphaML graphical user interface (GUI) for model building by executing:

```
python -c "from alphaml import guir"
```

This command will launch the main AlphaML window, where you can configure and build machine learning models tailored to your dataset.

Step 4. Perform predictions
Use the prediction GUI to make predictions on your data by running:

```
python -c "from alphaml import guip"
```

This will open the prediction interface, where you can upload your dataset and execute drug sensitivity prediction models.

## 8.11 Exercises and Questions

In this section, we provide exercises and questions designed to reinforce the material covered in this chapter.



### Exercises

The following exercises will improve your understanding of the concepts and techniques discussed in this chapter:

1. Classifying cancer types using logistic regression: Use a dataset of gene expression profiles from cancer patients to build a logistic regression model that classifies cancer subtypes (e.g., breast cancer vs. lung cancer). Evaluate the model's accuracy and precision.
2. Predicting drug sensitivity with random forest: Implement a Random Forest model to predict drug sensitivity in cancer cell lines based on genomic features. Use a public dataset like the Genomics of Drug Sensitivity in Cancer (GDSC) and compare the model's performance with other classifiers.
3. Clustering patients using K-means: Apply K-means clustering to group patients based on clinical data (e.g., blood pressure, BMI, cholesterol levels). Visualize the clusters and interpret the differences between the groups.
4. Predicting disease risk with support vector machines (SVM): Build an SVM model to predict the risk of cardiovascular disease based on patient data (e.g., age, cholesterol, smoking status). Use cross-validation to tune the model's hyperparameters.
5. Implementing principal component analysis (PCA): Perform PCA on a high-dimensional dataset of gene expression profiles. Reduce the data to two principal components and create a 2D scatter plot to visualize the data structure.
6. Reinforcement learning for personalized medicine: Simulate a reinforcement learning environment where an agent learns to optimize a treatment plan for a patient with diabetes. Use different reward functions to simulate treatment success based on blood glucose levels.
7. Comparing supervised and unsupervised learning: Use the same biomedical dataset (e.g., a diabetes dataset) to build both a supervised model (e.g., decision tree) and an unsupervised model (e.g., K-means clustering). Compare their outputs and discuss the differences.

8. Building a neural network for image classification: Use a dataset of medical images (e.g., X-rays or MRIs) and build a deep learning model using a neural network to classify the images (e.g., normal vs. diseased). Use a convolutional neural network (CNN) for image processing.
9. Model evaluation and validation: Using a dataset of patient records, implement a machine learning model (e.g., logistic regression or random forest) and evaluate its performance using cross-validation, confusion matrix, precision, recall, and *F*1 score.
10. Handling missing data in machine learning: Use a biomedical dataset with missing values (e.g., patient lab results) and apply different imputation techniques (mean, median, or k-NN imputation). Train a model on the dataset and evaluate the impact of imputation on the model's accuracy.

## Questions to Be Answered

1. What are the key differences between supervised, unsupervised, and reinforcement learning? Provide examples of when each would be used in biomedicine.
2. Why is model evaluation and validation required in machine learning? How do techniques like cross-validation help in assessing model performance?
3. How does Principal Component Analysis (PCA) help in reducing dimensionality in high-dimensional biomedical datasets, and what are its limitations?
4. Explain the difference between overfitting and underfitting in machine learning models. How can cross-validation and regularization help in avoiding these issues?
5. How can machine learning contribute to personalized medicine? Provide examples of how predictive models can be used to customize treatment strategies.
6. Compare and contrast Random Forest and Logistic Regression. In what biomedical contexts would one be more appropriate than the other?
7. What challenges do machine learning models face when applied to high-dimensional biomedical data, and how can these challenges be addressed?
8. How does reinforcement learning differ from other machine learning techniques, and how can it be applied in healthcare for optimizing treatment plans?
9. Explain the role of hyperparameter tuning in machine learning and discuss its importance in optimizing the performance of models like SVM and Random Forest.
10. How can ethical considerations, such as bias in data and fairness in model predictions, impact the deployment of machine learning in clinical settings?

**Acknowledgement** The language of the human-generated text was corrected with the assistance of artificial intelligence (AI) tools [GPT-3.5 and GTP-4 from OpenAI]. GitHub Co-Pilot was used to check the correctness of the codes. The text underwent subsequent human revision to ensure its accuracy.

## References


1. Sarker IH. Machine learning: algorithms, real-world applications and research directions. SN Comput Sci. 2021;2(3):160. https://doi.org/10.1007/s42979-021-00592-x.
2. Archana R, Jeevaraj PSE. Deep learning models for digital image processing: a review. Artif Intell Rev. 2024;57(1). https://doi.org/10.1007/s10462-023-10631-z.
3. Lauriola I, Lavelli A, Aiolli F. An introduction to deep learning in natural language processing: models, techniques, and tools. Neurocomputing. 2022;470: 443–56, https://doi.org/10.1016/j.neucom.2021.05.103.
4. Cunningham P, Cord M, Delany SJ. Supervised learning. In: Cord M, Cunningham P, editors. Machine learning techniques for multimedia. Cognitive technologies. Berlin, Heidelberg: Springer; 2008. https://doi.org/10.1007/978-3-540-75171-7_2.
5. Ghahramani Z. Unsupervised learning. In: Bousquet O, von Luxburg U, Rätsch G, editors. Advanced lectures on machine learning. ML 2003. Lecture Notes in Computer Science, vol 3176. Berlin, Heidelberg: Springer; 2004. https://doi.org/10.1007/978-3-540-28650-9_5.
6. van Engelen JE, Hoos HH. A survey on semi-supervised learning. Mach Learn. 2020; 109:373–440. https://doi.org/10.1007/s10994-019-05855-6.
7. Bhatnagar S, Prasad H, Prashanth L. Reinforcement learning. In: Stochastic recursive algorithms for optimization. Lecture Notes in Control and Information Sciences, vol 434. London: Springer; 2013. https://doi.org/10.1007/978-1-4471-4285-0_11.
8. Rafique R, Islam SMR, Kazi JU. Machine learning in the prediction of cancer therapy. Comput Struct Biotechnol J. 2021;19:4003–17. https://doi.org/10.1016/j.csbj.2021.07.003.
9. Shah K, Ahmed M, Kazi JU. The Aurora kinase/β-catenin axis contributes to dexamethasone resistance in leukemia. NPJ Precis Oncol. 2021;5(1):13. https://doi.org/10.1038/s41698-021-00148-5.
10. Nasimian A, Ahmed M, Hedenfalk I, Kazi JU. A deep tabular data learning model predicting cisplatin sensitivity identifies BCL2L1 dependency in cancer. Comput Struct Biotechnol J. 2023;21:956–64. https://doi.org/10.1016/j.csbj.2023.01.020.
11. Obaido G, Mienye ID, Egbelowo OF, Emmanuel ID, Ogunleye A, Ogbuokiri B, Mienye P, Aruleba K. Supervised machine learning in drug discovery and development: algorithms, applications, challenges, and prospects, Mach Learn Appl. 2024;17:100576. https://doi.org/10.1016/j.mlwa.2024.100576.
12. Shobha G, Rangaswamy S. Chapter 8 – Machine learning. In: Gudivada VN, Rao CR, editors. Handbook of statistics, vol. 38. Amsterdam: Elsevier; 2018. p. 197–228.
13. James G, Witten D, Hastie T, Tibshirani R, Taylor J. Linear regression. In: An introduction to statistical learning. Springer Texts in Statistics. Cham: Springer; 2023. https://doi.org/10.1007/978-3-031-38747-0_3.
14. Demir-Kavuk O, Kamada M, Akutsu T, Knapp EW. Prediction using step-wise L1, L2 regularization and feature selection for small data sets with large number of features. BMC Bioinformatics. 2011;12:412.
15. Hoerl AE, Kennard RW. Ridge regression: biased estimation for nonorthogonal problems. Technometrics. 1970;12:55–67.
16. Tibshirani R. Regression shrinkage and selection via the lasso. J R Stat Soc Ser B Stat Method. 1996;58:267–88.
17. Zou H, Hastie T. Regularization and variable selection via the elastic net. J R Stat Soc Ser B Stat Method. 2005;67:301–20.
18. Yoo W, Ference BA, Cote ML, Schwartz A. A comparison of logistic regression, logic regression, classification tree, and random forests to identify effective gene-gene and gene-environmental interactions. Int J Appl Sci Technol. 2012;2(7):268.
19. Sperandei S. Understanding logistic regression analysis. Biochem Med (Zagreb). 2014;24(1):12–8. https://doi.org/10.11613/BM.2014.003.
20. Lasko TA, Bhagwat JG, Zou KH, Ohno-Machado L. The use of receiver operating characteristic curves in biomedical informatics. J Biomed Inform. 2005;38(5): 404–15. https://doi.org/10.1016/j.jbi.2005.02.008.
21. Bradley AP. The use of the area under the ROC curve in the evaluation of machine learning algorithms. Pattern Recogn. 1997;30(7):1145–59. https://doi.org/10.1016/S0031-3203(96)00142-2.

22. Zhao S, Zhang B, Yang J, et al. Linear discriminant analysis. Nat Rev Methods Primers. 2024;4(70). https://doi.org/10.1038/s43586-024-00346-y.
23. James G, Witten D, Hastie T, Tibshirani R, Taylor J. Classification. In: An introduction to statistical learning. Springer Texts in Statistics. Cham: Springer; 2023. https://doi.org/10.1007/978-3-031-38747-0_4.
24. Kingsford C, Salzberg SL. What are decision trees? Nat Biotechnol. 2008;26(9):1011–3.
25. Aliferis C, Simon G. Overfitting, underfitting and general model overconfidence and underperformance pitfalls and best practices in machine learning and AI. In: Simon GJ, Aliferis C, editors. Artificial intelligence and machine learning in health care and medical sciences: best practices and pitfalls. Cham: Springer International Publishing; 2024. p. 477–524.
26. Breiman L, Friedman JH, Olshen RA, Stone CJ. Classification and regression trees. 1st ed. Abingdon: Routledge; 1984. p. 1–368.
27. Quinlan JR. Induction of decision trees. Mach Learn. 1986;1(1):81–106.
28. Quinlan JR. Improved use of continuous attributes in C4.5. J Artif Intell Res. 1996;4:77–90.
29. Breiman L. Random forests. Mach Learn. 2001;45(1):5–32.
30. Friedman JH. Ensembles on random patches. In: Machine learning and knowledge discovery in databases, ECML PKDD 2012. Lecture notes in computer science, vol. 7523. Berlin, Heidelberg: Springer Berlin Heidelberg; 2012.
31. Liu X-Y, Zhou Z-H. Ensemble methods for class imbalance learning. In: Imbalanced learning; 2013. p. 61–82. https://doi.org/10.1002/9781118646106.ch4.
32. Freund Y, Schapire RE. A decision-theoretic generalization of on-line learning and an application to boosting. J Comput Syst Sci. 1997;55(1):119–39.
33. Friedman JH. Greedy function approximation: a gradient boosting machine. Ann Stat. 2001;29(5):1189–232.
34. Chen T, Guestrin C. XGBoost: a scalable tree boosting system. In: Proceedings of the 22nd ACM SIGKDD international conference on knowledge discovery and data mining. San Francisco: Association for Computing Machinery; 2016. p. 785–94.
35. Ke G, Meng Q, Finley T, Wang T, Chen W, Ma W, et al. LightGBM: a highly efficient gradient boosting decision tree. In: Proceedings of the 31st international conference on neural information processing systems. Long Beach: Curran Associates Inc.; 2017. p. 3149–57.
36. Dorogush AV, Ershov V, Gulin A. CatBoost: gradient boosting with categorical features support. ArXiv 2018;abs/1810.11363.
37. Wolpert DH. Stacked generalization. Neural Netw. 1992;5(2):241–59. https://doi.org/10.1016/S0893-6080(05)80023-1.
38. Hsu CW, Lin CJ. A comparison of methods for multiclass support vector machines. IEEE Trans Neural Netw. 2002;13(2):415–25.
39. Scholkopf B, Smola AJ, Williamson RC, Bartlett PL. New support vector algorithms. Neural Comput. 2000;12(5):1207–45.
40. Coretes C, Vapnik VN. Support-vector networks. Mach Learn. 1995;20:273–97.
41. Muller KR, Mika S, Ratsch G, Tsuda K, Scholkopf B. An introduction to kernel-based learning algorithms. IEEE Trans Neural Netw. 2001;12(2):181–201.
42. scikit-learn. "1.4. Support Vector Machines." Scikit-Learn.org, 2018, scikit-learn.org/stable/modules/svm.html.
43. Sabzekar M, Hasheminejad SMH. Robust regression using support vector regressions. Chaos, Solitons Fractals. 2021;144:110738.
44. Yang FJ. An implementation of Naive Bayes classifier. In: 2018 International Conference on Computational Science and Computational Intelligence (CSCI). Las Vegas, NV, USA; 2018. p. 301–6. https://doi.org/10.1109/CSCI46756.2018.00065.
45. Soria D, Garibaldi JM, Ambrogi F, Biganzoli EM, Ellis IO. A 'non-parametric' version of the naive Bayes classifier. Knowl.-Based Syst. 2011;24(6):775–84. https://doi.org/10.1016/j.knosys.2011.02.014.
46. Wolfson J, Bandyopadhyay S, Elidrisi M, Vazquez-Benitez G, Vock DM, Musgrove D, Adomavicius G, Johnson PE, O'Connor PJ. A Naive Bayes machine learning approach to risk prediction using censored, time-to-event data. Stat Med. 2015;34(21):2941–57. https://doi.org/10.1002/sim.6526.
47. Pajila PJB, Sheena BG, Gayathri A, Aswini J, Nalini M, Subramanian RS. Comprehensive survey on Naive Bayes algorithm: advantages, limitations and applications. In: 2023 4th Interna-

tional Conference on Smart Electronics and Communication (ICOSEC). Trichy, India; 2023. p. 1228–34. https://doi.org/10.1109/ICOSEC58147.2023.10276274.
48. Ben-Gal I. Bayesian networks. In: Ruggeri F, Kenett RS, Faltin FW, editors. Encyclopedia of statistics in quality and reliability. 2008. https://doi.org/10.1002/9780470061572.eqr089.
49. Digitale JC, Martin JN, Glymour MM. Tutorial on directed acyclic graphs. J Clin Epidemiol. 2022;142:264–7. https://doi.org/10.1016/j.jclinepi.2021.08.001.
50. Mihaljević B, Bielza C, Larrañaga P. Bayesian networks for interpretable machine learning and optimization. Neurocomputing. 2021;456:648–65, https://doi.org/10.1016/j.neucom.2021.01.138.
51. Rashidi L, Hashemi S, Hamzeh A. Anomaly detection in categorical datasets using Bayesian networks. In: Deng H, Miao D, Lei J, Wang FL, editors. Artificial Intelligence and Computational Intelligence (AICI) 2011. Lecture Notes in Computer Science, vol 7003. Berlin, Heidelberg: Springer; 2011. https://doi.org/10.1007/978-3-642-23887-1_78.
52. Puga J, Krzywinski M, Altman N. Bayesian networks. Nat Methods . 2015; 12:799–800. https://doi.org/10.1038/nmeth.3550.
53. Fernández A, Salmerón A. Extension of Bayesian network classifiers to regression problems. In: Geffner H, Prada R, Machado Alexandre I, David N, editors. Advances in artificial intelligence – IBERAMIA 2008. Lecture Notes in Computer Science, vol 5290. Berlin, Heidelberg: Springer; 2008. https://doi.org/10.1007/978-3-540-88309-8_9.
54. Rasmussen CE. Gaussian processes in machine learning. In: Bousquet O, von Luxburg U, Rätsch, G, ediotrs. Advanced lectures on machine learning. ML 2003. Lecture Notes in Computer Science, vol 3176. Berlin, Heidelberg: Springer; 2004. https://doi.org/10.1007/978-3-540-28650-9_4.
55. García-Fernández AF, Tronarp F, Särkkä S. Gaussian process classification using posterior linearization. IEEE Signal Processing Lett. 2019;26(5): 735–9. https://doi.org/10.1109/LSP.2019.2906929.
56. Wang J. An intuitive tutorial to Gaussian process regression. Comput Sci Eng. 2023;25(4): 4–11. https://doi.org/10.1109/MCSE.2023.3342149.
57. Swain PS, Stevenson K, Leary A, Montano-Gutierrez LF, Clark IB, Vogel J, Pilizota T. Inferring time derivatives including cell growth rates using Gaussian processes. Nat Commun. 2016;7:13766. https://doi.org/10.1038/ncomms13766.
58. Berns F, Hüwel J, Beecks C. Automated model inference for Gaussian processes: an overview of state-of-the-art methods and algorithms. SN Comput Sci. 2022;3:300. https://doi.org/10.1007/s42979-022-01186-x.
59. Kramer O. K-Nearest neighbors. In: Dimensionality reduction with unsupervised nearest neighbors. Intelligent Systems Reference Library, vol 51. Berlin, Heidelberg: Springer; 2013. https://doi.org/10.1007/978-3-642-38652-7_2.
60. Zhang Z. Introduction to machine learning: k-nearest neighbors. Ann Transl Med. 2016;4(11):218. https://doi.org/10.21037/atm.2016.03.37.
61. Dokmanic I, Parhizkar R, Ranieri J, Vetterli M. Euclidean distance matrices: essential theory, algorithms, and applications. IEEE Signal Process Mag. 2015;32(6):12–30. https://doi.org/10.1109/MSP.2015.2398954.
62. Nixon MS, Aguado AS, 12 - Distance, classification and learning. In: Nixon MS, Aguado AS, editors. Feature extraction and image processing for computer vision (4th ed.), Academic Press; 2020. p. 571–604. https://doi.org/10.1016/B978-0-12-814976-8.00012-9.
63. Liu X, Chen W, Peng L, Luo D, Jia L, Xu G, Chen X, Liu X. Secure computation protocol of Chebyshev distance under the malicious model. Sci Rep. 2024;14(1):17115. https://doi.org/10.1038/s41598-024-67907-9.
64. Hennig C. Minkowski distances and standardisation for clustering and classification of high dimensional data. arXiv 2019. https://doi.org/10.48550/arXiv.1911.13272.
65. Dodge Y. Mahalanobis distance. In: The Concise Encyclopedia of Statistics. New York: Springer; 2008. p. 325–6. https://doi.org/10.1007/978-0-387-32833-1_240.
66. "KDTree — SciPy V1.16.0 Manual." Scipy.org, 2025, docs.scipy.org/doc/scipy-1.16.0/reference/generated/scipy.spatial.KDTree.html. Accessed 31 July 2025.
67. Aumüller M, Bernhardsson E, Faithfull A. ANN-Benchmarks: a benchmarking tool for approximate nearest neighbor algorithms. Inf Syst. 2020;87:101374. https://doi.org/10.1016/j.is.2019.02.006.
68. LeCun Y, Bengio Y, Hinton G. Deep learning. Nature . 2015;521:436–44. https://doi.org/10.1038/nature14539.

69. Vakalopoulou M, Christodoulidis S, Burgos N, Colliot O, Lepetit V. Deep learning: basics and convolutional neural networks (CNNs). In: Colliot O, editor. Machine learning for brain disorders. Neuromethods, vol 197. New York: Humana; 2023. https://doi.org/10.1007/978-1-0716-3195-9_3.
70. Dubey SR, Singh SK, Chaudhuri BB. Activation functions in deep learning: a comprehensive survey and benchmark. Neurocomputing. 2022;503:92–108. https://doi.org/10.1016/j.neucom.2022.06.111.
71. Yamashita R, Nishio M, Do RKG, et al. Convolutional neural networks: an overview and application in radiology. Insights Imaging. 2018;9:611–29. https://doi.org/10.1007/s13244-018-0639-9.
72. Mienye ID, Swart TG, Obaido G. Recurrent neural networks: a comprehensive review of architectures, variants, and applications. Information. 2024;15:517. https://doi.org/10.3390/info15090517.
73. Sherstinsky A. Fundamentals of recurrent neural network (RNN) and long short-term memory (LSTM) network. Phys D: Nonlinear Phenom. 2020;404:132306. https://doi.org/10.1016/j.physd.2019.132306.
74. Turner RE. An introduction to transformers. arXiv, 2023. https://doi.org/10.48550/arXiv.2304.10557.
75. Jain SM. Introduction to transformers. In: Introduction to transformers for NLP. Berkeley: Apress; 2022. https://doi.org/10.1007/978-1-4842-8844-3_2.
76. Terven J. Deep reinforcement learning: a chronological overview and methods. AI. 2025;6:46. https://doi.org/10.3390/ai6030046.
77. Nielsen MA. Chapter 2 – How the backpropagation algorithm works in "Neural Networks and Deep Learning." Determination Press; 2019. Neuralnetworksanddeeplearning.com; neuralnetworksanddeeplearning.com/chap2.html.
78. Creswell A, White T, Dumoulin V, Arulkumaran K, Sengupta B, Bharath AA. Generative adversarial networks: an overview. IEEE Signal Process Mag. 2018;35(1):53–65. https://doi.org/10.1109/MSP.2017.2765202.
79. Berahmand K, Daneshfar F, Salehi ES, et al. Autoencoders and their applications in machine learning: a survey. Artif Intell Rev. 2024;57:28. https://doi.org/10.1007/s10462-023-10662-6.
80. Pang G, Shen C, Cao L, Hengel AVD. Deep learning for anomaly detection: a review. ACM Comput. Surv. 2022;54(2):38. https://doi.org/10.1145/3439950.
81. Gui J, et al. A survey on self-supervised learning: algorithms, applications, and future trends. IEEE Trans Pattern Anal Mach Intell. 2024;46(12):9052–71. https://doi.org/10.1109/TPAMI.2024.3415112.
82. Oyelade J, Isewon I, Oladipupo F, Aromolaran O, Uwoghiren E, Ameh F, Achas M, Adebiyi E. Clustering algorithms: their application to gene expression data. Bioinform Biol Insights. 2016;10:237–53. https://doi.org/10.4137/BBI.S38316.
83. Greene D, Cunningham P, Mayer R. Unsupervised learning and clustering. In: Cord M, Cunningham P, editors. Machine learning techniques for multimedia. Cognitive Technologies. Berlin, Heidelberg: Springer; 2008. https://doi.org/10.1007/978-3-540-75171-7_3.
84. Ezugwu AE, et al. A comprehensive survey of clustering algorithms: State-of-the-art machine learning applications, taxonomy, challenges, and future research prospects. Eng Appl Artif Intell. 2022;110:104743. https://doi.org/10.1016/j.engappai.2022.104743.
85. Ikotun AM, et al. K-means clustering algorithms: A comprehensive review, variants analysis, and advances in the era of big data. Inf Sci. 2023;622:178–210, https://doi.org/10.1016/j.ins.2022.11.139.
86. Nielsen F. Hierarchical clustering. In: Introduction to HPC with MPI for data science. Undergraduate Topics in Computer Science. Cham: Springer; 2016. https://doi.org/10.1007/978-3-319-21903-5_8.
87. Mojena R. Hierarchical grouping methods and stopping rules: an evaluation. Comput J. 1977;20(4):359-63. https://doi.org/10.1093/comjnl/20.4.359.
88. Deng D. DBSCAN clustering algorithm based on density. In: 2020 7th International Forum on Electrical Engineering and Automation (IFEEA). Hefei, China; 2020. p. 949-53, https://doi.org/10.1109/IFEEA51475.2020.00199.
89. Liu Z. Association rule learning. In: Artificial intelligence for engineers. Cham: Springer; 2025. https://doi.org/10.1007/978-3-031-75953-6_13.

90. Toivonen H. Apriori algorithm. In: Sammut C, Webb GI, editors. Encyclopedia of machine learning. Boston: Springer; 2011. https://doi.org/10.1007/978-0-387-30164-8_27.
91. Zaki MJ, Parthasarathy S, Li W. A localized algorithm for parallel association mining. In: SPAA '97: Proceedings of the ninth annual ACM symposium on Parallel algorithms and architectures. 1997. p. 321–30. https://doi.org/10.1145/258492.258524.
92. Han J, Pei J, Yin Y. Mining frequent patterns without candidate generation. SIGMOD Rec. 29, 2 2000;29(2):1–12. https://doi.org/10.1145/335191.335372.
93. Fischer A, Igel C. An introduction to restricted Boltzmann machines. In: Alvarez L, Mejail M, Gomez L, Jacobo J, editors. Progress in pattern recognition, image analysis, computer vision, and applications. CIARP 2012. Lecture Notes in Computer Science, vol 7441. Berlin, Heidelberg: Springer; 2012. https://doi.org/10.1007/978-3-642-33275-3_2.
94. Kingma DP, Welling M. An Introduction to Variational Autoencoders. Found Trends Mach Learn. 2019;12(4):307–92. https://doi.org/10.1561/2200000056.
95. Vayansky I, Kumar SAP. A review of topic modeling methods. Inf Syst. 2020;94:101582. https://doi.org/10.1016/j.is.2020.101582.
96. Liu L, Tang L, Dong W, Yao S, Zhou W. An overview of topic modeling and its current applications in bioinformatics. Springerplus. 2016;5(1):1608. https://doi.org/10.1186/s40064-016-3252-8.
97. Rani V, Nabi ST, Kumar M, Mittal A, Kumar K. Self-supervised learning: a succinct review. Arch Comput Methods Eng. 2023;30(4):2761–75. https://doi.org/10.1007/s11831-023-09884-2.
98. Singh A. Self-supervised learning of molecular representations. Nat Methods. 2025;22:1395. https://doi.org/10.1038/s41592-025-02757-5.
99. Mourad R. Semi-supervised learning improves regulatory sequence prediction with unlabeled sequences. BMC Bioinformatics. 2023;24(1):186. https://doi.org/10.1186/s12859-023-05303-2.
100. Wang J, Xu Q, Lin H, Yang Z, Li Y. Semi-supervised method for biomedical event extraction. Proteome Sci. 2013;11(Suppl 1):S17. https://doi.org/10.1186/1477-5956-11-S1-S17.
101. Huynh T, Nibali A, He Z. Semi-supervised learning for medical image classification using imbalanced training data. Comput Methods Programs Biomed. 2022;216:106628. https://doi.org/10.1016/j.cmpb.2022.106628.
102. Beaulieu-Jones BK, Greene CS. Pooled resource open-access ALS clinical trials consortium. Semi-supervised learning of the electronic health record for phenotype stratification. J Biomed Inform. 2016;64:168–78. https://doi.org/10.1016/j.jbi.2016.10.007.
103. Chen J, Si YW, Un CW, Siu SWI. Chemical toxicity prediction based on semi-supervised learning and graph convolutional neural network. J Cheminform. 2021;13(1):93. https://doi.org/10.1186/s13321-021-00570-8.
104. Hsu TC, Lin C. Learning from small medical data—robust semi-supervised cancer prognosis classifier with Bayesian variational autoencoder. Bioinform Adv. 2023;3(1):vbac100. https://doi.org/10.1093/bioadv/vbac100.
105. Jayaraman P, Desman J, Sabounchi M, et al. A Primer on reinforcement learning in medicine for clinicians. NPJ Digit Med. 2024;7:337. https://doi.org/10.1038/s41746-024-01316-0.
106. Wang G, Liu X, Ying Z, Yang G, Chen Z, Liu Z, Zhang M, Yan H, Lu Y, Gao Y, Xue K, Li X, Chen Y. Optimized glycemic control of type 2 diabetes with reinforcement learning: a proof-of-concept trial. Nat Med. 2023;29(10):2633–42. https://doi.org/10.1038/s41591-023-02552-9.
107. Eastman B, Przedborski M, Kohandel M. Reinforcement learning derived chemotherapeutic schedules for robust patient-specific therapy. Sci Rep. 2021;11:17882. https://doi.org/10.1038/s41598-021-97028-6.
108. Liu S, See KC, Ngiam KY, Celi LA, Sun X, Feng M. Reinforcement learning for clinical decision support in critical care: comprehensive review. J Med Internet Res. 2020;22(7):e18477. https://doi.org/10.2196/18477.
109. Bologheanu R, Kapral L, Laxar D, Maleczek M, Dibiasi C, Zeiner S, Agibetov A, Ercole A, Thoral P, Elbers P, Heitzinger C, Kimberger O. Development of a reinforcement learning algorithm to optimize corticosteroid therapy in critically Ill patients with sepsis. J Clin Med. 2023;12(4):1513. https://doi.org/10.3390/jcm12041513.
110. Choi Y, Oh S, Huh JW, et al. Deep reinforcement learning extracts the optimal sepsis treatment policy from treatment records. Commun Med. 2024;4:245. https://doi.org/10.1038/s43856-024-00665-x.

111. Ma R, Vanstrum EB, Lee R, Chen J, Hung AJ. Machine learning in the optimization of robotics in the operative field. Curr Opin Urol. 2020;30(6):808–16. https://doi.org/10.1097/MOU.0000000000000816.
112. Tan RK, Liu Y, Xie L. Reinforcement learning for systems pharmacology-oriented and personalized drug design. Expert Opin Drug Discov. 2022;17(8):849–63. https://doi.org/10.1080/17460441.2022.2072288.
113. Popova M, Isayev O, Tropsha A. Deep reinforcement learning for de novo drug design. Sci Adv. 2018;4(7):eaap7885. https://doi.org/10.1126/sciadv.aap7885.
114. Zhou Z, Kearnes S, Li L, et al. Optimization of molecules via deep reinforcement learning. Sci Rep. 2019;9:10752. https://doi.org/10.1038/s41598-019-47148-x.
115. Ebrahimi S, Lim GJ. A reinforcement learning approach for finding optimal policy of adaptive radiation therapy considering uncertain tumor biological response. Artif Intell Med. 2021;121:102193. https://doi.org/10.1016/j.artmed.2021.102193.
116. Madondo M, et al. Patient-specific deep reinforcement learning for automatic replanning in head-and-neck cancer proton therapy. arXiv 2025. https://doi.org/10.48550/arXiv.2506.10073.
117. Qian C, Ren H. Deep Reinforcement Learning in Surgical Robotics: Enhancing the Automation Level, arXiv 2023. https://doi.org/10.48550/arXiv.2309.00773.
118. Luo S, Androwis G, Adamovich S, Su H, Nunez E, Zhou X. Reinforcement learning and control of a lower extremity exoskeleton for squat assistance. Front Robot AI. 2021;8:702845. https://doi.org/10.3389/frobt.2021.702845.
119. Mousafi Alasal L, Hammarlund EU, Pienta KJ, Rönnstrand L, Kazi JU. XeroGraph: enhancing data integrity in the presence of missing values with statistical and predictive analysis. Bioinform Adv. 2025;5(1):vbaf035. https://doi.org/10.1093/bioadv/vbaf035.
120. Younus S, Rönnstrand L, Kazi JU. Xputer: bridging data gaps with NMF, XGBoost, and a streamlined GUI experience. Front Artif Intell. 2024;7:1345179. https://doi.org/10.3389/frai.2024.1345179.
121. Thériault R, Ben-Shachar MS, Patil I, Lüdecke D, Wiernik BM, Makowski D. Check your outliers! An introduction to identifying statistical outliers in R with easystats. Behav Res Methods. 2024;56(4):4162–72.
122. Leys C, Klein O, Dominicy Y, Ley C. Detecting multivariate outliers: use a robust variant of the Mahalanobis distance. J Exp Soc Psychol. 2018;74:150–6.
123. Vinutha HP, Poornima B, Sagar BM. Detection of outliers using interquartile range technique from intrusion dataset. In: Information and decision sciences; 2018. p. 511–8. https://doi.org/10.1007/978-981-10-7563-6_53.
124. Simpson RJ, Johnson TA, Amara IA. The box-plot: an exploratory analysis graph for biomedical publications. Am Heart J. 1988;116(6, Part 1):1663–5.
125. Federico A, Serra A, Ha MK, Kohonen P, Choi JS, Liampa I, et al. Transcriptomics in toxicogenomics, part II: preprocessing and differential expression analysis for high quality data. Nanomaterials (Basel). 2020;10(5):903.
126. McKinney W. Data structures for statistical computing in python. In: Proceedings of the 9th python in science conference. Austin; 2010. p. 56–61. https://doi.org/10.25080/Majora-92bf1922-00a.
127. Butcher B, Smith BJ. Feature engineering and selection: a practical approach for predictive models. Am Stat. 2020;74(3):308–9.
128. Hoens TR, Chawla NV. Imbalanced datasets: from sampling to classifiers. In: Imbalanced learning; 2013. p. 43–59. https://doi.org/10.1002/9781118646106.ch3.
129. Fernández A, Garcia S, Herrera F, Chawla N. SMOTE for learning from imbalanced data: progress and challenges, marking the 15-year anniversary. J Artif Intell Res. 2018;61:863–905.
130. Akiba T, Sano S, Yanase T, Ohta T, Koyama M. Optuna: a next-generation hyperparameter optimization framework. In: KDD '19: proceedings of the 25th ACM SIGKDD international conference on knowledge discovery & data mining. New York: Association for Computing Machinery; 2019. p. 2623–31.
131. Dimgba M, Andreas R. Optimal hyperparameter search strategies: benchmarking grid search, random search, and genetic algorithms across regression, classification, and clustering tasks. 2024. https://doi.org/10.13140/RG.2.2.24957.37603.
132. Victoria AH, Maragatham G. Automatic tuning of hyperparameters using Bayesian optimization. Evol Syst. 2021;12(1):217–23.

133. Maurya M, Yadav N, editors. A comparative analysis of gradient-based optimization methods for machine learning problems. In: Proceedings on international conference on data analytics and computing. Singapore: Springer Nature Singapore; 2023.
134. Hassan E, Shams MY, Hikal NA, Elmougy S. The effect of choosing optimizer algorithms to improve computer vision tasks: a comparative study. Multimed Tools Appl. 2023;82(11): 16591–633.
135. Tani, L., Rand, D., Veelken, C. et al. Evolutionary algorithms for hyperparameter optimization in machine learning for application in high energy physics. Eur Phys J C . 2021;81:170. https://doi.org/10.1140/epjc/s10052-021-08950-y.
136. Somogyi Z. Performance evaluation of machine learning models. In: Somogyi Z, editor. The application of artificial intelligence: step-by-step guide from beginner to expert. Cham: Springer International Publishing; 2021. p. 87–112.
137. Muraina I. Ideal dataset splitting ratios in machine learning algorithms: general concerns for data scientists and data analysts. In: 7th International mardin artuklu scientific researches conference. Mardin; 2022.
138. Zhong E, Fan W, Yang Q, Verscheure O, Ren J. Cross validation framework to choose amongst models and datasets for transfer learning. In: Machine learning and knowledge discovery in databases. Berlin, Heidelberg: Springer Berlin Heidelberg; 2010.
139. Powers D. Evaluation: from precision, recall and F-factor to ROC, informedness, markedness & correlation. J Mach Learn Technol. 2011;2(1):37–63.
140. Brodersen KH, Ong CS, Stephan K, Buhmann J. The balanced accuracy and its posterior distribution. In: 2010 20th international conference on pattern recognition. Istanbul. p. 3121–4. https://doi.org/10.1109/ICPR.2010.764.
141. Grau J, Grosse I, Keilwagen J. PRROC: computing and visualizing precision-recall and receiver operating characteristic curves in R. Bioinformatics. 2015;31(15):2595–7.
142. Pedregosa F, et al. Scikit-learn: machine learning in Python, J Mach Learn Res. 2011;12:2825-30
143. Abadi M, et al. TensorFlow: large-scale machine learning on heterogeneous systems. 2015. Software available from tensorflow.org. https://doi.org/10.5281/zenodo.4724125.
144. Paszke A, et al. PyTorch: an imperative style, high-performance deep learning library. arXiv, 2019. https://doi.org/10.48550/arXiv.1912.0170.
145. Virtanen P, Gommers R, Oliphant TE, Haberland M, Reddy T, Cournapeau D, Burovski E, Peterson P, Weckesser W, Bright J, van der Walt SJ, Brett M, Wilson J, Millman KJ, Mayorov N, Nelson ARJ, Jones E, Kern R, Larson E, Carey CJ, Polat İ, Feng Y, Moore EW, VanderPlas J, Laxalde D, Perktold J, Cimrman R, Henriksen I, Quintero EA, Harris CR, Archibald AM, Ribeiro AH, Pedregosa F, van Mulbregt P, SciPy 1.0 Contributors. SciPy 1.0: fundamental algorithms for scientific computing in Python. Nat Methods. 2020;17(3):261–72. https://doi.org/10.1038/s41592-019-0686-2.
146. Nasimian A, Younus S, Tatli O, Hammarlund EU, Pienta KJ, Ronnstrand L, et al. AlphaML: a clear, legible, explainable, transparent, and elucidative binary classification platform for tabular data. Patterns (N Y). 2024;5(1):100897.


8

# Image Processing in Biomedical Research

## Contents




J. U. Kazi, *Python Essentials for Biomedical Data Analysis: An Introductory Textbook*,
https://doi.org/10.1007/978-3-031-85600-6_9

Images serve as a medium for understanding complex biological structures and processes in biomedical research. From the microscopic details of cellular mechanisms to the macroscopic views provided by MRI and CT scans, image processing has become a key component in advancing medical knowledge and improving patient care. The ability to analyze and interpret these images supports the work of researchers, clinicians, and anyone involved in the biomedical field [1, 2]. In this chapter we discuss the basic concepts of image processing and demonstrate how they are applied within biomedical research contexts. Using Python tools, we will explore widely used methods for extracting meaningful information from biomedical images.

**Learning Goals**

This chapter will help you to understand why image processing matters in biomedical research. You will become familiar with basic concepts, such as pixels, resolution, and color models, which are required for interpreting and manipulating biomedical images. The chapter will also discuss widely used Python libraries for image analysis, including OpenCV, scikit-image, SimpleITK, Mahotas and Napari.



## 9.1 Overview of Image Processing

Image processing refers to a suite of techniques that apply various operations to images to enhance their quality or extract relevant information. It is comparable to refining a photograph or isolating key features from it for further analysis. The process starts with an image as the input and generates either an improved version of the image or specific data extracted from the image as the output [3]. While traditional image processing methods relied on analog techniques, modern applications predominantly utilize digital methods, leveraging computer algorithms to process digital images. This shift has transformed research and practical applications across various fields.

A digital image represents a physical image in numerical form, usually modeled in multiple dimensions. Unlike the familiar two-dimensional images, many images, especially in biomedical fields, extend into three or more dimensions (e.g., time-lapse imaging or multispectral data). This multidimensional approach enables more advanced analyses such as pattern recognition, resolution enhancement, and object detection, which are important in biomedical research, remote sensing, and computer vision [4].

The image processing workflow comprises several key steps, as illustrated in ◘ Fig. 9.1.

*Image acquisition*: This is the first stage, where images are captured using biomedical imaging devices such as MRI, CT scans, or microscopy. The acquired visual data must faithfully represent the biological structures or functions under investigation.

*Preprocessing*: Preprocessing enhances image quality by reducing noise, adjusting contrast, and applying normalization techniques. Methods such as filtering (e.g.,

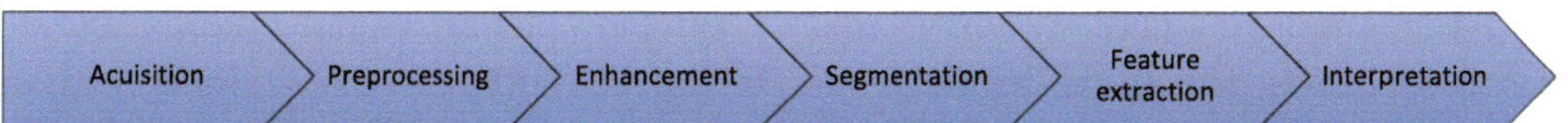


**Fig. 9.1** Image processing typically involves a multistep approach. The key stages include, but are not limited to, image acquisition, preprocessing, enhancement, segmentation, feature extraction, and interpretation

Gaussian or median filters) and normalization help clarify the images, facilitating easier identification of patterns and anomalies.

*Image enhancement*: In this stage, techniques like histogram equalization and sharpening filters are applied to further improve the visual representation of images. Enhancement often focuses on displaying important features within the image that may not be immediately apparent in raw data, improving visibility for subsequent steps.

*Segmentation*: Segmentation is the process of isolating specific regions or structures within an image. This step is particularly vital in biomedical applications, where segmentation techniques like thresholding, region-based segmentation, or edge detection can help identify structures such as tumors, organs, or tissues by delineating their boundaries.

*Feature extraction*: Feature extraction involves identifying and quantifying specific attributes within the image, such as shape, texture, intensity, or size. These extracted features provide quantitative data required for further analysis. For instance, shape analysis might help in classifying benign versus malignant tumors based on morphological differences.

*Interpretation*: The final step involves analyzing the processed images to draw meaningful conclusions. Interpretation may involve various techniques, from basic statistical analysis to advanced machine learning algorithms. These methods help decipher biological phenomena, such as disease progression or cellular behavior, through pattern recognition or predictive modeling.

Advances in image processing technology have enabled more accurate and detailed imaging, allowing for early detection of diseases, a better understanding of cellular processes, and enhanced precision in surgical planning and execution. With the growing use of machine learning and artificial intelligence, automated image analysis has become more powerful, supporting high-throughput data analysis and providing deeper interpretations of complex biological phenomena.

### 9.1.1 Importance and Impact of Image Analysis

The image analysis in biomedicine serves as a basis for diagnostic and therapeutic procedures, providing support on patient care and treatment strategies [5]. Image analysis enables the early detection and diagnosis of various diseases, including cancer, neurological disorders, and cardiovascular conditions. Assessing changes in tissues and organs over time, clinicians can make more precise diagnoses, predict disease progression, and monitor the efficacy of treatments.

For example, in oncology, image analysis can help identify tumor boundaries and assessing tumor growth or shrinkage in response to therapy [6]. Techniques such as magnetic resonance imaging (MRI), computed tomography (CT), and positron emission tomography (PET) scans enable oncologists to monitor tumors' metabolic and structural changes over time [7–9]. In neurology, it helps detect structural abnormalities in the brain associated with disorders like Alzheimer's disease, multiple sclerosis, or epilepsy. Functional imaging techniques like functional MRI (fMRI) and diffusion tensor imaging (DTI) provide information about brain connectivity and neural activity [10]. In cardiology, image analysis is employed to evaluate arterial blockages, visualize blood flow, or assess myocardial tissue damage after cardiac events, such as heart attacks, using technologies like echocardiography, angiography, and cardiac MRI.

In surgical contexts, image processing can be used for preoperative planning and intraoperative guidance. Surgeons use three-dimensional reconstructions of anatomical structures, often generated from CT or MRI scans, to plan the safest and most effective surgical approach [11, 12]. Real-time imaging systems, such as intraoperative MRI or CT and augmented reality-based guidance systems, assist surgeons in maintaining precision during procedures, thereby reducing risks to patients and improving outcomes [13].



In biomedical research and drug development, image analysis can be used for visualizing cellular and molecular responses to diseases at a microscopic level. Techniques such as fluorescence microscopy, confocal microscopy, and electron microscopy allow researchers to monitor how therapeutics impact biological structures over time [14, 15]. This supports the evaluation of drug efficacy and toxicity in preclinical studies. High-throughput image analysis platforms further accelerate drug discovery by automating the processing and interpretation of large datasets, helping identify promising drug candidates more efficiently [16].

Image processing technologies also play an important role in medical education by providing high-resolution visualizations and simulation tools. Three-dimensional models of anatomical structures improve the comprehension of complex human anatomy, while virtual reality (VR) and augmented reality (AR) environments allow students and professionals to practice procedures in risk-free, simulated settings [17, 18].

Moreover, large-scale public health screening programs often rely on image analysis for early disease detection. For instance, in breast cancer screening, analyzing mammograms enables the detection of early-stage tumors [19, 20]. In lung cancer screening, radiologists use low-dose CT scans to identify lung nodules that may be indicative of malignancy [21]. In ophthalmology, retinal imaging techniques are used to screen for diabetic retinopathy, a common complication of diabetes that can lead to blindness [22]. Furthermore, PET scans are widely used in detecting and staging various cancers across the body by pinpointing areas of increased metabolic activity, which are often indicative of malignant cells [23].

Overall, image analysis is integral not only to advancing medical science and research but also to support clinical practices, improving patient outcomes, and contributing to public health initiatives. As technology continues to evolve, with advancements in artificial intelligence and machine learning, image analysis is expected to become even more useful in healthcare, driving innovations in diagnostic precision, personalized treatment, and population health management.

## 9.2 Fundamentals of Image Processing

Image processing involves a detailed understanding of how images are formed and manipulated to extract meaningful information. This knowledge is needed for interpreting complex visual data and for developing techniques that can aid in diagnosis, research, and treatment strategies. Despite the challenges inherent in biomedical imaging, such as variability in data quality and the complexity of biological structures, advancements in this area are revolutionizing medical science [24].

### 9.2.1 Image Representation

**Notebook: Section 9.2.1. Image Representation**
This notebook provides code examples to generate images presented in Fig. 9.2.
Link to the GitHub repository:
► https://github.com/sn-code-inside/BioPy
Go to: Chap. 9—example data—► Sect. 9.2.1

At the core of image processing is the digital representation of an image. A digital image is a two-dimensional array of values, where each value, often referred to as a pixel, corresponds to the intensity or color of a specific point within the image. In grayscale images, pixel values represent varying degrees of brightness, whereas in color images, each pixel contains multiple components (e.g., red, green, and blue) that correspond to different wavelengths of light. A common approach is to allocate 8 bits per pixel, which allows for $2^8$ or 256 discrete intensity values (0 to 255). Moreover, binary images consist of pixels represented by just two possible values, typically 0 or 1, where 0 often represents the background, and 1 denotes the presence of a feature (◘ Fig. 9.2).

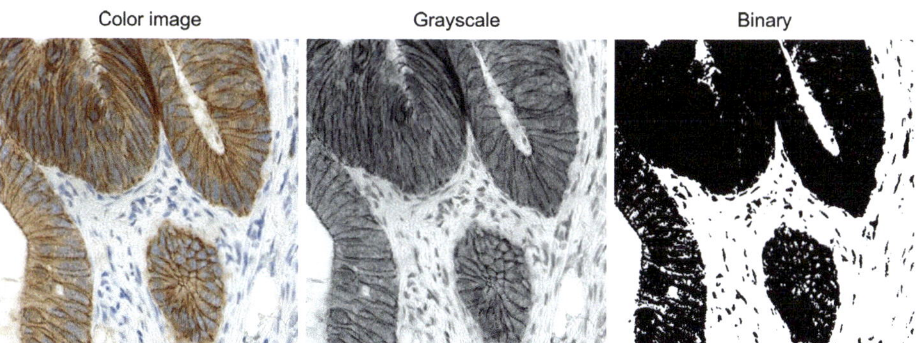


◘ **Fig. 9.2** Image representation. The immunohistochemistry (IHC) image from scikit-image was converted into different representations

*Pixels*: Pixels are the basic building blocks of a digital image. These tiny dots collectively form the visual content we perceive. Each pixel holds data about the color and intensity of a specific point in the image. In biomedical imaging, pixel values correspond to various data points, such as tissue density in MRI scans, the brightness of specific biomarkers in fluorescence microscopy, or the density of structures in X-rays. The value of each pixel depends on the imaging modality and the physical properties being measured, such as absorption, reflection, or emission of energy from the biological sample.

*Spatial resolution*: Resolution refers to the amount of detail an image can display and is usually expressed as pixels per inch (PPI) or dots per inch (DPI). Higher resolution signifies that more pixels are used to represent the image, resulting in finer details and improved clarity. In medical imaging, high resolution is particularly important for identifying small and complex structures. For instance, detecting microcalcifications in mammograms, tiny calcium deposits that may indicate early signs of breast cancer, requires images of sufficiently high resolution to visualize these minute features clearly [25].



*Color models*: Color models define how color information is represented in an image. Two common models include RGB (Red, Green, Blue) and grayscale. RGB is used for color images, where varying intensities of red, green, and blue light are combined to generate a wide range of colors. This model is widely used in visual representations where color differentiation is required, such as in pathology or anatomical imaging. Grayscale, on the other hand, is used for black-and-white images, where pixel values represent intensity variations without any color information. Grayscale images are common in X-rays, MRI, CT, and ultrasound imaging, where color is not necessary to depict the structures or features of interest. In biomedical imaging, pseudocoloring is often applied to grayscale images (such as X-rays or ultrasound scans) to enhance contrast. Pseudocoloring involves assigning different colors to intensity values, which can highlight specific anatomical or pathological structures that might be difficult to discern in the original grayscale image. This technique helps clinicians identify abnormalities, such as tumors or lesions, by making subtle differences in intensity more visually pronounced [26].

*Bit depth*: Bit depth refers to the number of bits used to represent the intensity or color of each pixel in an image. The more bits you use per pixel, the greater the range of possible values, and therefore, the more precise the representation of intensity or color. For example, 1-bit depth allows only 2 possible values (since $2^1 = 2$), typically black or white, which is common in binary images. An 8-bit depth allows 256 different values ($2^8 = 256$), meaning each pixel can represent 256 levels of gray in a grayscale image, ranging from black (0) to white (255). In color images that use the RGB model, each color channel (red, green, blue) can also have 256 levels. When combined, this results in over 16 million possible colors ($256 \times 256 \times 256$). A 16-bit depth allows for 65,536 possible values ($2^{16} = 65{,}536$), which is useful in applications that require very fine gradations in intensity, such as medical imaging or high dynamic range (HDR) photography.

### 9.2.2 Image Preprocessing

**Notebook: Section 9.2.2. Image Preprocessing**
This notebook provides code examples demonstrating image preprocessing.
Link to the GitHub repository:
▸ https://github.com/sn-code-inside/BioPy
Go to: Chap. 9—example data—▸ Sect. 9.2.2

Image preprocessing helps prepare images for further analysis by improving image quality and making important features more discernible. Preprocessing techniques aim to reduce noise, correct distortions, and improve contrast. The choice of preprocessing methods depends on the specific problems in the image that need to be addressed, such as noise, poor contrast, or variations in intensity. Preprocessing can improve the results of image segmentation, feature extraction, or machine learning tasks that follow (Fig. 9.3).

*Noise reduction*: Noise is an unwanted variation in pixel intensity, often introduced by the imaging sensor or environmental conditions. Noise reduction is required for improving image quality, especially in biomedical images where fine details are important for diagnosis. Common noise reduction techniques include the following:

- Gaussian filtering: Applies a Gaussian kernel to the image, which smooths the image by averaging pixel values with neighboring pixels, reducing noise but slightly blurring the image.
- Median filtering: Replaces each pixel value with the median value of the surrounding pixels, preserving edges better than Gaussian filtering.
- Wiener filtering: A more advanced filter that reduces noise by taking both the local mean and variance into account, which can better handle images with varying noise levels.

Example code to process IHC image:

```
import matplotlib.pyplot as plt
from skimage import img_as_float, color, io
from skimage.filters import median, gaussian
from skimage.morphology import disk
from scipy.signal import wiener
from skimage import io
# Load color image
color_image = io.imread('example_data/ihc.png')
# Convert the color image to grayscale
image = color.rgb2gray(color_image)
# Apply Gaussian filter for noise reduction
gaussian_filtered = gaussian(image, sigma=1)
# Apply Median filter for noise reduction
median_filtered = median(image, disk(3))
# Apply Wiener filter for noise reduction
wiener_filtered = wiener(image)
```

```
# Display the original and filtered images
fig, ax = plt.subplots(1, 4, figsize=(12, 3))
ax[0].imshow(image, cmap='gray')
ax[0].set_title('Original Image')
ax[0].axis('off')
ax[1].imshow(gaussian_filtered, cmap='gray')
ax[1].set_title('Gaussian Filtered')
ax[1].axis('off')
ax[2].imshow(median_filtered, cmap='gray')
ax[2].set_title('Median Filtered')
ax[2].axis('off')
ax[3].imshow(wiener_filtered, cmap='gray')
ax[3].set_title('Wiener Filtered')
ax[3].axis('off')
plt.tight_layout()
plt.show()
```

Output:

```
Figure 9.3
```



*Normalization*: Normalization adjusts the intensity values of an image so that they fit within a specific range or match a predefined standard. This step is important when images from different sources or modalities have different brightness levels or intensity scales, which could affect the performance of downstream algorithms (Fig. 9.4).

- Min-Max normalization: Scales the pixel values to a specific range (e.g., 0 to 1) by linearly transforming the data.
- Z-score normalization: Adjusts the image so that the mean is 0 and the standard deviation is 1. This is particularly useful for ensuring consistent contrast across images in a dataset.

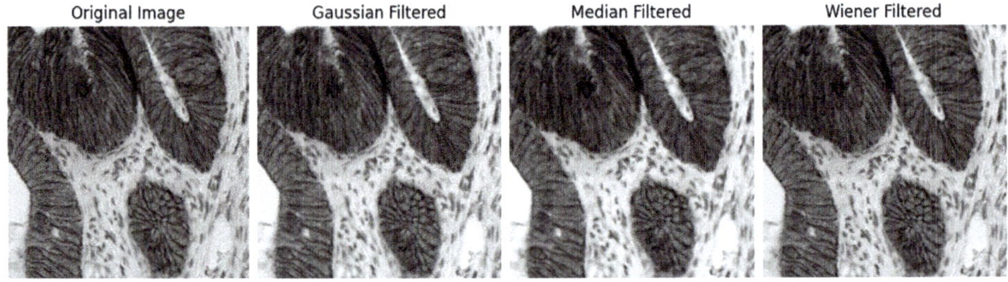


**Fig. 9.3** Noise reduction examples. The noise was filtered using three different filtering methods

Example code for normalization:

```
import matplotlib.pyplot as plt
import numpy as np
from skimage import exposure, color
from skimage import io
# Load color image
color_image = io.imread('example_data/ihc.png')
# Convert the color image to grayscale
image = color.rgb2gray(color_image)
# Min-Max Normalization to scale pixel values between 0 and 1
min_max_normalized = (image - np.min(image)) / (np.max(image) - np.min(image))
# Z-score normalization (mean = 0, std = 1)
z_score_normalized = (image - np.mean(image)) / np.std(image)
# Display original and normalized images
fig, ax = plt.subplots(1, 3, figsize=(9, 3))
ax[0].imshow(image, cmap='gray')
ax[0].set_title('Original Image')
ax[0].axis('off')
ax[1].imshow(min_max_normalized, cmap='gray')
ax[1].set_title('Min-Max normalization')
ax[1].axis('off')
ax[2].imshow(z_score_normalized, cmap='gray')
ax[2].set_title('Z-score normalization')
ax[2].axis('off')
plt.tight_layout()
plt.show()
```

Output:

```
Figure 9.4
```

*Contrast enhancement*: Contrast enhancement techniques improve the visibility of important features in an image by increasing the difference in intensity between light

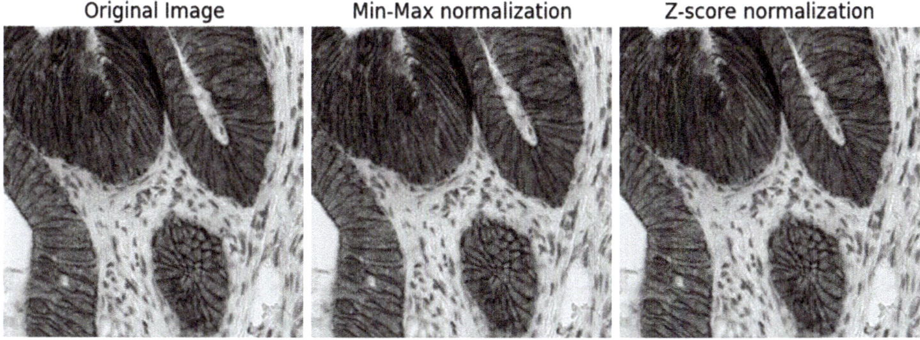


**Fig. 9.4** Image normalization. Min-Max and Z-score normalization were applied on grayscale IHC images

and dark regions. This is especially useful in images where key features are difficult to distinguish due to poor contrast (Fig. 9.5).

- Global histogram equalization: Redistributes the intensity values of an image to increase the contrast. It is particularly useful for improving the contrast in low-light or low-contrast images.
- Contrast-limited adaptive histogram equalization (CLAHE): A variant of histogram equalization, adaptive histogram equalization (AHE) works by applying histogram equalization to small, local regions (tiles) within the image, rather than to the entire image at once. This allows for local contrast enhancement in different parts of the image, which is especially useful in cases where global histogram equalization might fail to bring out important features in localized regions. However, a problem with basic AHE is that it can overly amplify noise in some regions, especially in homogeneous areas. CLAHE solves the problem of noise amplification by introducing a contrast-limiting step. It limits the amplification of noise by "clipping" the histogram at a predefined value, preventing any pixel intensity from being overly exaggerated. This contrast clipping is controlled by a parameter called clip limit, which prevents noise from being overamplified in areas of the image where the pixel intensity variations are low (like smooth or homogeneous regions). This technique is more effective when different parts of the image require varying levels of contrast enhancement.



Example code:

```
from skimage import exposure, color
import matplotlib.pyplot as plt
from skimage import io
# Load color image
color_image = io.imread('example_data/ihc.png')
# Convert the color image to grayscale
image = color.rgb2gray(color_image)
# Apply global histogram equalization
hist_eq_image = exposure.equalize_hist(image)
# Apply adaptive histogram equalization (CLAHE)
adaptive_eq_image = exposure.equalize_adapthist(image, clip_limit=0.03)
# Display the original, global histogram equalized, and CLAHE results
fig, ax = plt.subplots(1, 3, figsize=(15, 5))
ax[0].imshow(image, cmap='gray')
ax[0].set_title('Original Image')
ax[0].axis('off')
ax[1].imshow(hist_eq_image, cmap='gray')
ax[1].set_title('Global Histogram Equalization')
ax[1].axis('off')
ax[2].imshow(adaptive_eq_image, cmap='gray')
ax[2].set_title('Adaptive Histogram Equalization (CLAHE)')
ax[2].axis('off')
plt.tight_layout()
plt.show()
```

Output:

```
Figure 9.5
```

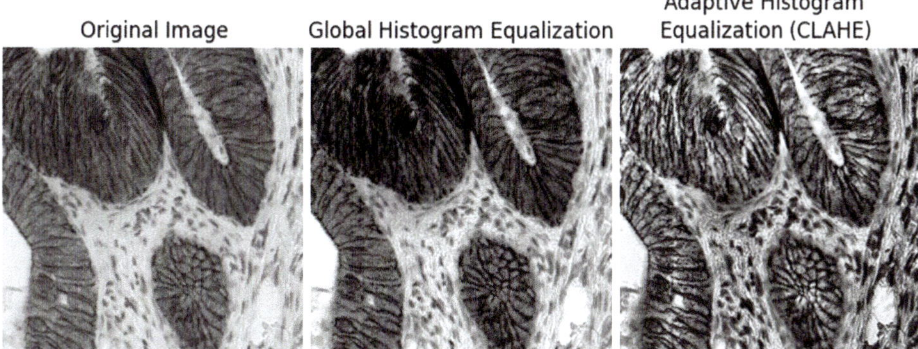


**Fig. 9.5** Image contrast enhancement. Global Histogram Equalization and Adaptive Histogram Equalization (CLAHE) methods were applied on grayscale IHC images

### 9.2.3 Image Enhancement Techniques

**Notebook: Section 9.2.3. Image Enhancement Techniques**
This notebook provides code examples demonstrating image enhancement techniques.
Link to the GitHub repository:
► https://github.com/sn-code-inside/BioPy
Go to: Chap. 9—example data—► Sect. 9.2.3

Image enhancement techniques improve the visual quality of images by enhancing features, improving contrast, or reducing noise. In biomedical imaging, such enhancements help in better visualization of structures such as tissues, cells, or organs. The two common techniques discussed here are filtering and histogram equalization.

*Filtering*: Filtering is used to enhance an image by either smoothing (to reduce noise) or sharpening (to enhance edges and features). In biomedical imaging, filters are especially useful for improving the clarity of images (Fig. 9.6). Some common types of filters include the following:

- Gaussian blur: This filter is used to smooth an image by reducing noise and detail. It works by averaging pixel values with their neighbors, weighted by a Gaussian function. This is useful in applications where noise needs to be suppressed, such as in low-light microscopy images.
- Sharpening: Sharpening filters emphasize edges, making boundaries between different regions of an image more pronounced. These filters display important features in medical images, such as the boundaries of tumors or other abnormal structures.

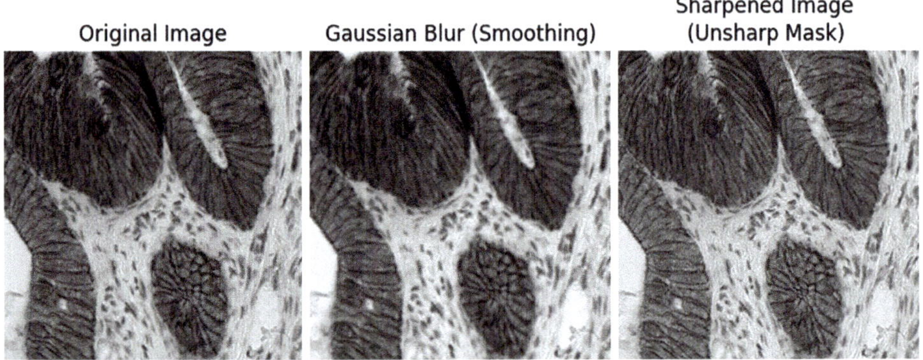


Fig. 9.6 Image enhancement by noise reduction. IHC image was processed using Gaussian blur and an unsharp mask

Example codes for image enhancement:



```
import matplotlib.pyplot as plt
from skimage import data, filters, color, io
from skimage.filters import gaussian, unsharp_mask
# Load a sample image (grayscale)
image = io.imread('example_data/ihc.png')
# Convert to grayscale if needed
image_gray = color.rgb2gray(image)
# Apply Gaussian blur for smoothing
gaussian_blur = gaussian(image_gray, sigma=1)
# Apply Unsharp Mask for sharpening
sharpened_image = unsharp_mask(image_gray, radius=1, amount=1)
# Display the original, blurred, and sharpened images
fig, ax = plt.subplots(1, 3, figsize=(10, 3))
ax[0].imshow(image_gray, cmap='gray')
ax[0].set_title('Original Image')
ax[0].axis('off')
ax[1].imshow(gaussian_blur, cmap='gray')
ax[1].set_title('Gaussian Blur (Smoothing)')
ax[1].axis('off')
ax[2].imshow(sharpened_image, cmap='gray')
ax[2].set_title('Sharpened Image (Unsharp Mask)')
ax[2].axis('off')
plt.tight_layout()
plt.show()
```

Output:

```
Figure 9.6
```

*Histogram equalization*: Histogram equalization is a technique that enhances the contrast of an image by redistributing pixel intensity values. In biomedical imaging, this is especially useful in medical images such as mammograms or MRIs, where it can be used to differentiate between structures like tissues, tumors, and other anomalies. We have discussed global histogram equalization and adaptive histogram equalization in "contrast enhancement."

### 9.2.4 Image Segmentation

**Notebook: Section 9.2.4. Image Segmentation**
This notebook provides code examples demonstrating image segmentation.
Link to the GitHub repository:
▸ https://github.com/sn-code-inside/BioPy
Go to: Chap. 9—example data—▸ Sect. 9.2.4

Image segmentation involves partitioning an image into different regions or objects, allowing for more detailed analysis of specific structures or areas. Segmentation techniques are particularly important in biomedical imaging for identifying and isolating key features such as organs, cells, or pathological regions like tumors. Two common image segmentation techniques are thresholding and edge detection.

*Thresholding*: Thresholding is one of the simplest and most widely used segmentation methods. It works by converting a grayscale image into a binary image, where pixels are assigned a value of 0 (black) or 1 (white) based on whether their intensity is below or above a certain threshold (Fig. 9.7). In biomedical imaging, thresholding is particularly useful for isolating regions of interest, such as separating tumors from healthy tissue or segmenting blood vessels.

There are several types of thresholding:

- Global thresholding: A single intensity value is chosen as the threshold for the entire image.
- Adaptive thresholding: The threshold value is computed for smaller regions of the image, which is useful when there are varying lighting conditions across the image.
- Otsu's thresholding: A method that automatically selects the optimal threshold by minimizing the variance within the foreground and background pixels.

Example code:

```
import matplotlib.pyplot as plt
from skimage import data, filters, color, io
from skimage.color import rgb2gray
from skimage.filters import threshold_otsu, threshold_local
# Load a sample image
image = io.imread('example_data/ihc.png')
# Convert to grayscale if necessary
gray_image = rgb2gray(image)
# Apply Global Thresholding (using a fixed threshold value)
global_thresh_value = 0.5  # Example fixed threshold value
global_binary_image = gray_image > global_thresh_value
# Apply Adaptive Thresholding (using a local threshold)
block_size = 255
# Size of the local region used for calculating threshold
adaptive_thresh_image = threshold_local(gray_image, block_size,
offset=0)
adaptive_binary_image = gray_image > adaptive_thresh_image
# Apply Otsu's Thresholding
otsu_thresh_value = threshold_otsu(gray_image)
otsu_binary_image = gray_image > otsu_thresh_value
# Display original grayscale image and three thresholded binary images
fig, ax = plt.subplots(1, 4, figsize=(20, 5))
ax[0].imshow(gray_image, cmap='gray')
ax[0].set_title('Original Grayscale Image')
ax[0].axis('off')
ax[1].imshow(global_binary_image, cmap='gray')
ax[1].set_title(f'Global Thresholding (T = {global_thresh_value:.2f})')
ax[1].axis('off')
ax[2].imshow(adaptive_binary_image, cmap='gray')
ax[2].set_title('Adaptive Thresholding')
ax[2].axis('off')
ax[3].imshow(otsu_binary_image, cmap='gray')
ax[3].set_title(f'Otsu\'s Thresholding (T = {otsu_thresh_value:.2f})')
ax[3].axis('off')
plt.tight_layout()
plt.show()
```



Output:

```
Figure 9.7
```

*Edge detection*: Edge detection is another technique for image segmentation that helps identify the boundaries of objects or structures in an image. In biomedical imaging, this is particularly useful for delineating the outlines of organs, blood vessels, or tumors. Edge detection algorithms work by detecting sharp changes in intensity, which often correspond to object boundaries (Fig. 9.8).

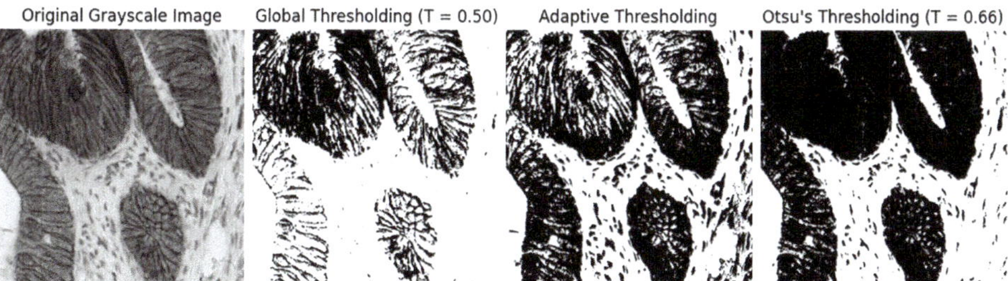


**Fig. 9.7** Different thresholding methods. In the global thresholding method, a fixed threshold value of 0.5 was used, turning all pixels with intensities greater than 0.5 white, and the others black. In adaptive thresholding, the block_size defines the size of the local neighborhood used to calculate the threshold, while the offset adjusts the threshold value by a constant factor. Otsu's method automatically calculates the optimal threshold by minimizing the intra-class variance (the variance within the foreground and background), making it ideal when a fixed threshold is not available and optimal separation between regions is desired

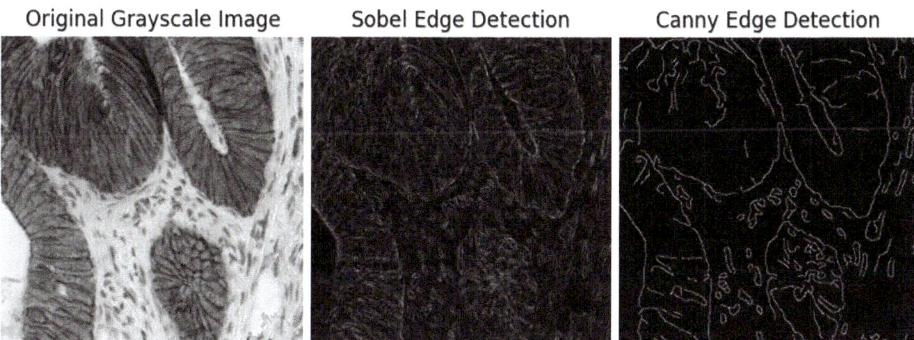


**Fig. 9.8** Different edge detection methods. Sobel Edge Detection computes gradients in the image to highlight regions of rapid intensity change, which correspond to edges in the image. Canny Edge Detection is a more sophisticated method that includes several steps to detect edges while minimizing noise, making it ideal for detecting fine structures like blood vessels or small tumors

Common edge detection techniques include the following:

- Sobel edge detector: This technique computes the gradient of the image intensity at each pixel, showing regions with high spatial frequency (i.e., edges).
- Canny edge detector: A multistep process that includes noise reduction, gradient calculation, nonmaximum suppression, and edge tracking by hysteresis. It is one of the most popular edge detection methods because of its ability to detect edges with high accuracy while minimizing noise.

Example code:

```
import matplotlib.pyplot as plt
from skimage import data, filters, color, io
from skimage.color import rgb2gray
from skimage.filters import sobel
from skimage.feature import canny

# Load a sample image
image = io.imread('example_data/ihc.png')
gray_image = rgb2gray(image)

# Apply Sobel edge detection
sobel_edges = sobel(gray_image)

# Apply Canny edge detection
canny_edges = canny(gray_image, sigma=3)

# Display original and edge-detected images
fig, ax = plt.subplots(1, 3, figsize=(10, 3))

ax[0].imshow(gray_image, cmap='gray')
ax[0].set_title('Original Grayscale Image')
ax[0].axis('off')


ax[1].imshow(sobel_edges, cmap='gray')
ax[1].set_title('Sobel Edge Detection')
ax[1].axis('off')

ax[2].imshow(canny_edges, cmap='gray')
ax[2].set_title('Canny Edge Detection')
ax[2].axis('off')

plt.tight_layout()
plt.show()
```

Output:

```
Figure 9.8
```

### 9.2.5 Feature Extraction and Pattern Recognition

**Notebook: Section 9.2.5. Feature Extraction and Pattern Recognition**
This notebook provides code examples demonstrating feature extraction and pattern recognition.

Link to the GitHub repository:
▶ https://github.com/sn-code-inside/BioPy
Go to: Chap. 9—example data—▶ Sect. 9.2.5

Feature extraction and pattern recognition are two important steps in image analysis. These techniques enable the identification of relevant structures and attributes within images that can be used for diagnostic purposes or for research. In biomedical imaging, feature extraction and pattern recognition help automate complex tasks such as disease detection, cell classification, and identification of pathological patterns.

*Feature extraction*: Feature extraction is the process of identifying and quantifying important attributes or characteristics of objects within an image. In biomedical imaging, features might include the shape, size, texture, and intensity of cells, tissues, or organs. These features can be used for understanding biological phenomena, diagnosing diseases, or quantifying changes in cellular structures over time.

Common types of features include the following:

- Shape features: Includes attributes like the area, perimeter, and eccentricity of cells or organs. These are useful in identifying abnormal growths (e.g., tumors) or changes in organ morphology.
- Texture features: Describes the surface properties of the tissue or organ, which may reveal underlying pathological conditions. Techniques such as the Gray-Level Co-occurrence Matrix (GLCM) can be used to quantify texture.
- Intensity features: Related to the pixel intensity values in the image. These are useful in medical images such as MRIs or X-rays, where the intensity values correspond to tissue density or composition.

Example code for feature extraction (shape and texture):

```
import matplotlib.pyplot as plt
from skimage import data, measure, color, io, feature
from skimage.color import rgb2gray
from skimage.filters import threshold_otsu
from skimage.measure import regionprops
import numpy as np
# Load a sample image and convert to grayscale
image = io.imread('example_data/ihc.png')
gray_image = rgb2gray(image)
# Apply Otsu's thresholding to create a binary image for segmentation
```

```
thresh_value = threshold_otsu(gray_image)
binary_image = gray_image > thresh_value
# Label connected regions
label_image = measure.label(binary_image)
# Extract region properties (shape features) from labeled regions
regions = regionprops(label_image)
# Display results for first region (as an example)
for region in regions:
    print(f"Area: {region.area}")
    print(f"Perimeter: {region.perimeter}")
    print(f"Eccentricity: {region.eccentricity}")
    print(f"Centroid: {region.centroid}")
# Feature extraction: Texture using GLCM (Gray Level Co-occurrence
Matrix)
from skimage.feature import graycomatrix, graycoprops
# Define GLCM texture properties
glcm = graycomatrix((gray_image * 255).astype('uint8'), distances=[5],
           angles=[0], symmetric=True, normed=True)
contrast = graycoprops(glcm, 'contrast')
dissimilarity = graycoprops(glcm, 'dissimilarity')
print(f"Texture Contrast: {contrast[0, 0]}")
print(f"Texture Dissimilarity: {dissimilarity[0, 0]}")
# Display original and binary image
fig, ax = plt.subplots(1, 2, figsize=(10, 5))
ax[0].imshow(gray_image, cmap='gray')
ax[0].set_title('Original Grayscale Image')
ax[0].axis('off')
ax[1].imshow(binary_image, cmap='gray')
ax[1].set_title('Binary Image (Otsu\'s Thresholding)')
ax[1].axis('off')
plt.tight_layout()
plt.show()
```



*Pattern recognition*: Pattern recognition involves classifying data based on patterns found in the features extracted from an image. In biomedical imaging, pattern recognition can be used to classify cells and tissues, or identify disease markers. For instance, machine learning models can be trained to recognize specific types of cancer cells or to differentiate between healthy and diseased tissue based on extracted features.

Pattern recognition generally involves the following steps:

- Feature selection: Choosing which features (e.g., shape, texture, intensity) to use for classification.
- Training a model: Using a labeled dataset to train a machine learning algorithm to recognize patterns in the extracted features.
- Classification: Applying the trained model to new images to classify objects (e.g., cells or tissues) based on the patterns it has learned.

Example code for pattern recognition using a simple classifier:

Below is an example where we use a simple machine learning classifier (Support Vector Machine) to recognize patterns based on shape and texture features extracted from the image.

Example code using MNIST dataset:

```
from sklearn import datasets
from sklearn.model_selection import train_test_split
from sklearn.preprocessing import StandardScaler
from sklearn.svm import SVC
from sklearn.metrics import classification_report, accuracy_score
import matplotlib.pyplot as plt
# Load the MNIST dataset from sklearn
digits = datasets.load_digits()
# The images are stored as 8x8 pixel grids
print("Image data shape:", digits.images.shape)
# Plot some example digits from the dataset
fig, axes = plt.subplots(2, 5, figsize=(10, 5))
for ax, image, label in zip(axes.ravel(), digits.images, digits.target):
    ax.imshow(image, cmap='gray')
    ax.set_title(f'Label: {label}')
    ax.axis('off')
plt.tight_layout()
plt.show()
# Flatten the image data for pattern recognition (feature extraction)
# Each image is 8x8 pixels, so we flatten them into 64-pixel vectors
X = digits.images.reshape((len(digits.images), -1))
y = digits.target  # Target labels (the digits 0-9)
# Split data into training and test sets
X_train, X_test, y_train, y_test = train_test_split(X, y, test_size=0.3,
random_state=42)
# Standardize the feature values (important for SVM)
scaler = StandardScaler()
X_train = scaler.fit_transform(X_train)
X_test = scaler.transform(X_test)
# Train a Support Vector Classifier (SVC)
svm_classifier = SVC(kernel='linear', random_state=42)
svm_classifier.fit(X_train, y_train)
# Make predictions on the test set
y_pred = svm_classifier.predict(X_test)
# Evaluate the classifier
print(f"Accuracy: {accuracy_score(y_test, y_pred):.2f}")
print(classification_report(y_test, y_pred))
```

### 9.2.6 Challenges in Biomedical Image Processing

Biomedical image processing plays a role in diagnosis, treatment planning, and biomedical research. However, there are several challenges that need to be addressed to ensure the effectiveness, accuracy, and reliability of image analysis. Below are some of the key challenges encountered in biomedical image processing:

*Variability in data*: Biomedical images are produced by a range of imaging modalities such as MRI, CT, ultrasound, and microscopy, each with different quality, resolution, and contrast characteristics. For example, MRI scans may have high contrast but might suffer from noise or artifacts, while ultrasound images are often noisy and have lower resolution [27]. This variability in image quality, coupled with differences in acquisition parameters (e.g., different hospitals or scanners), poses a challenge in developing standardized image processing techniques that can be universally applied. As a result, algorithms that work well for one type of data might fail when applied to another, so image processing methods often need to be adapted or fine-tuned to handle different imaging modalities and conditions [27, 28].

*Complexity of biological structures*: Biological structures are complex and highly variable. The shape and size of organs, cells, or tissues can differ significantly between individuals, and pathological changes such as tumors often alter these structures in unpredictable ways [29, 30]. Furthermore, biological structures can overlap or appear in various orientations, which makes it difficult to accurately segment or identify key features within an image [31]. Developing algorithms that can capture and interpret the full range of biological variations is extremely challenging. This is particularly true in cases where abnormalities, such as tumors or lesions, do not follow regular patterns, making detection more difficult [32, 33].



*High dimensionality of data*: The development of modern imaging technologies like 3D imaging, confocal microscopy, and whole-slide imaging (WSI) has led to a dramatic increase in the amount of data generated from a single scan. For instance, a single whole-slide histopathology image can contain billions of pixels, and 3D or 4D MRI scans produce time-series volumetric data [34]. This high dimensionality presents both computational and analytical challenges, as algorithms need to be optimized to process these large datasets in a timely manner without sacrificing accuracy.

*Need for accuracy and reliability*: Biomedical image processing is often used in settings where the results directly impact diagnosis, treatment planning, and patient outcomes. Errors or inaccuracies in image processing algorithms can lead to misdiagnoses, inappropriate treatments, or even patient harm [35, 36]. Therefore, there is an inherent need for algorithms to be highly accurate and reliable. This requirement for precision presents a challenge, particularly in edge cases where biological structures are ambiguous, image quality is poor, or the disease exhibits uncommon characteristics [33, 37].

*Ethical and privacy concerns*: The increasing use of artificial intelligence (AI) and machine learning in biomedical image processing introduces several ethical and privacy concerns. Medical images often contain sensitive patient information, and improper use or sharing of this data could violate patient confidentiality [38]. Moreover, bias in AI models trained on imbalanced or nonrepresentative datasets can lead to unequal treatment outcomes [36]. There is also concern about the transparency of AI algorithms, particularly in healthcare, where the reasoning behind decisions (e.g., why a specific diagnosis was made) must be explainable and trustworthy [39, 40].

## 9.3 Python Libraries for Image Analysis

Python offers a large collection of tools to simplify image manipulation, analysis, and interpretation. Below is an overview of widely used Python libraries for image analysis. Each library offers unique features that make it suitable for different types of image analysis tasks.

### 9.3.1 OpenCV

**Notebook: Section 9.3.1. OpenCV**
This notebook provides code examples demonstrating the use of OpenCV.
Link to the GitHub repository:
▸ https://github.com/sn-code-inside/BioPy
Go to: Chap. 9—example data—▸ Sect. 9.3.1

OpenCV (Open-Source Computer Vision Library) is designed for real-time computer vision and image processing [41]. OpenCV includes a wide array of functions that cover image manipulation, object detection, feature extraction, and video analysis. The Python interface of OpenCV is particularly popular in research and real-world applications.

Some of the key features of OpenCV include the following:

- Image filtering (blurring, sharpening, edge detection)
- Morphological operations (dilation, erosion)
- Object detection and tracking
- Image segmentation (e.g., watershed, contour detection)
- Real-time video processing
- Machine learning support for tasks such as object recognition and pattern classification.

In biomedical imaging, OpenCV is applied in tasks like the following:

- Segmentation: Identifying regions of interest such as cells or tumors in an MRI or CT scan.
- Edge Detection: Detecting the boundaries of structures in X-ray or ultrasound images.
- Morphological Operations: Increasing the visibility of small structures like blood vessels or cell membranes.
- Real-Time Processing: Used in tasks like image-guided surgery or live cell imaging, where images need to be processed in real time.

Example code using OpenCV for edge detection (Fig. 9.9):

```
import cv2
import matplotlib.pyplot as plt
# Load a sample biomedical image
image = cv2.imread('example_data/ihc.png', cv2.IMREAD_GRAYSCALE)
# Replace with your image
# Apply GaussianBlur to reduce noise
blurred_image = cv2.GaussianBlur(image, (5, 5), 0)
# Use Canny edge detection
edges = cv2.Canny(blurred_image, threshold1=100, threshold2=200)
# Display the original image and the edges detected
plt.figure(figsize=(10, 5))
plt.subplot(1, 2, 1)
plt.imshow(image, cmap='gray')
plt.title('Original Image')
plt.axis('off')
plt.subplot(1, 2, 2)
plt.imshow(edges, cmap='gray')
plt.title('Edges Detected with Canny')
plt.axis('off')
plt.show()
```



Output:

```
Figure 9.9
```

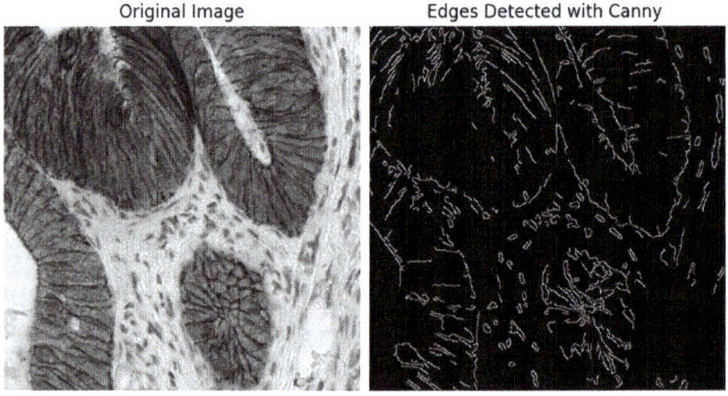


**Fig. 9.9** Edge detection using OpenCV. OpenCV is used to apply Gaussian blurring to reduce noise, followed by the Canny edge detection technique, which helps identify the boundaries of structures in the image

### 9.3.2 scikit-image

scikit-image is part of the SciPy ecosystem and provides a collection of image processing algorithms. Unlike OpenCV, which is more general-purpose, scikit-image is specifically designed for scientific image analysis. It uses NumPy and SciPy as base that allows easy integration to scientific computing stack.

Some of the key features of scikit-image include the following:

- Image filtering (e.g., median, Gaussian, and bilateral filters)
- Segmentation algorithms (e.g., thresholding, watershed)
- Geometric transformations (e.g., scaling, rotation)
- Object measurement (e.g., region properties such as area, perimeter)
- Histogram equalization and contrast adjustment
- Color space manipulation

In biomedical research, scikit-image is widely used in tasks such as:

- Cell morphology quantification: Analyzing the size, shape, and texture of cells in microscopy images.
- Neuron tracking: Monitoring the growth of neurons over time, which can be used in developmental neuroscience.
- Histopathology analysis: Automating the detection of abnormalities in tissue samples, such as cancerous cells in histology slides.

Examples provided in ► Sect. 9.2 mainly used scikit-image (skimage).

### 9.3.3 SimpleITK in Biomedical Image Processing

**Notebook: Section 9.3.3. SimpleITK in Biomedical Image Processing**
This notebook provides code examples demonstrating the use of SimpleITK in biomedical image processing.

Link to the GitHub repository:
► https://github.com/sn-code-inside/BioPy
Go to: Chap. 9—example data—► Sect. 9.3.3

SimpleITK is a simplified layer built on top of the Insight Toolkit (ITK), a widely used open-source software system for medical image processing [42]. SimpleITK is designed to make ITK more accessible to a broader range of users, particularly those using Python and other high-level languages. SimpleITK is ideal for tasks such as image segmentation, registration, and advanced medical image analysis, offering tools for handling medical images in 2D, 3D, and even 4D (time series) formats [43]. While SimpleITK simplifies ITK's interface, it retains much of ITK's power and flexibility, making it particularly useful for processing DICOM (Digital Imaging and Communications in Medicine) images and volumetric data in medical applications.

Key features of SimpleITK

- Support for multidimensional medical images: SimpleITK can handle 2D, 3D, and 4D images, including common medical imaging formats such as DICOM, NIfTI, and NRRD, which are used in MRI, CT, PET, and other types of scans.
- Image registration: One of SimpleITK's strongest features is its support for image registration, a process of aligning multiple images from different sources or modalities (e.g., aligning an MRI and a CT scan of the same patient).
- Segmentation tools: SimpleITK provides numerous segmentation algorithms (e.g., region growing, threshold-based, level sets) to delineate regions of interest like tumors, organs, or blood vessels.
- Image filtering: SimpleITK includes common filters for noise reduction, edge detection, and smoothing. This is useful in preprocessing steps before segmentation or analysis.
- Integration with ITK: Although SimpleITK simplifies the ITK interface, it retains access to ITK's core functionalities, enabling high-precision image processing.

SimpleITK is designed for handling complex biomedical imaging tasks that require working with volumetric data, such as:

- Medical image segmentation: SimpleITK provides algorithms for delineating regions of interest in 3D medical scans, such as segmenting organs in CT or MRI scans.
- Image registration: SimpleITK's registration capabilities allow aligning images from different modalities or time points, a common requirement in clinical workflows.
- 3D image processing: SimpleITK is highly suited for 3D image processing, such as volumetric rendering, slicing, and 3D transformations.
- Handling medical data formats: SimpleITK's support for DICOM and other medical formats useful for clinical and research applications, where data is typically stored in specialized formats.



Example code for image segmentation using SimpleITK (Fig. 9.10):

```
import SimpleITK as sitk
import matplotlib.pyplot as plt
from skimage import io
import numpy as np
# Load a grayscale PNG image using skimage
png_image = io.imread('example_data/Tr-no_0015.jpg', as_gray=True)
# Convert the 2D PNG image to a SimpleITK image
image = sitk.GetImageFromArray(png_image)
# Inspect the intensity range of the image
image_array = sitk.GetArrayViewFromImage(image)
min_intensity = np.min(image_array)
max_intensity = np.max(image_array)
print(f"Intensity range: {min_intensity} to {max_intensity}")
# Set threshold values based on the intensity range
# You should adjust these based on the intensity range
lower_threshold = min_intensity + (max_intensity - min_intensity) * 0.25
upper_threshold = max_intensity - (max_intensity - min_intensity) * 0.25
print(f"Using threshold values: {lower_threshold} to {upper_threshold}")
# Apply a threshold to segment the image
```

```
segmented_image = sitk.BinaryThreshold(image,
lowerThreshold=lower_threshold,
upperThreshold=upper_threshold,
insideValue=1, outsideValue=0)
# Convert both images back to NumPy arrays for visualization
original_slice = sitk.GetArrayViewFromImage(image)
segmented_slice = sitk.GetArrayViewFromImage(segmented_image)
# Create a figure to display both the original and segmented images
plt.figure(figsize=(8, 4))
# Display the original image
plt.subplot(1, 2, 1)
plt.imshow(original_slice, cmap='gray')
plt.title('Original Image')
plt.axis('off')
# Display the segmented image
plt.subplot(1, 2, 2)
plt.imshow(segmented_slice, cmap='gray')
plt.title('Segmented Image')
plt.axis('off')
# Show the figure
plt.tight_layout()
plt.show()
```

Output:

```
Figure 9.10
```

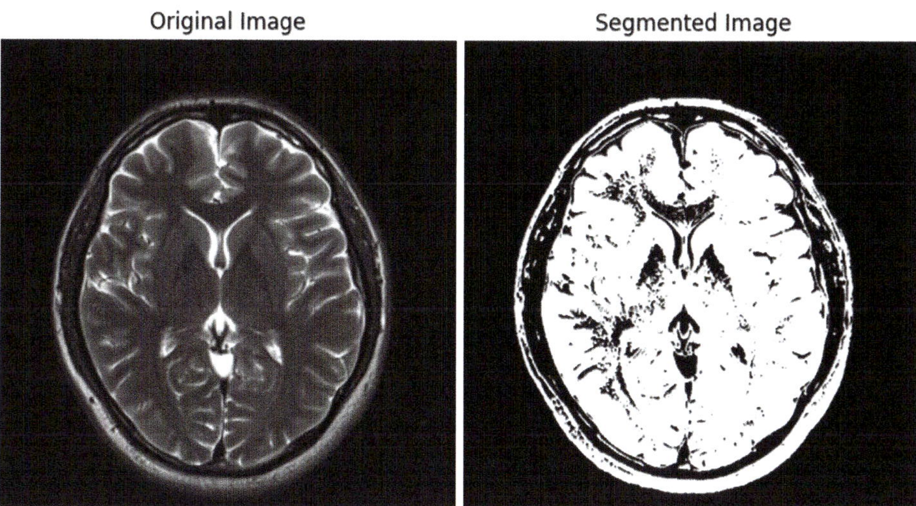


**Fig. 9.10** Brain MRI segmentation using SimpleITK. The original MRI image was downloaded from ► https://www.kaggle.com/datasets/masoudnickparvar/brain-tumor-mri-dataset/data (Tr-no_0015.jpg)

### 9.3.4 Mahotas in Biomedical Image Processing

**Notebook: Section 9.3.4. Mahotas in Biomedical Image Processing**
This notebook provides code examples demonstrating the use of Mahotas in biomedical image processing.

Link to the GitHub repository:
▶ https://github.com/sn-code-inside/BioPy
Go to: Chap. 9—example data—▶ Sect. 9.3.4

Mahotas is a Python library for computer vision and image processing that provides a variety of functions for basic and advanced image analysis. While not as extensive as OpenCV or SimpleITK, Mahotas stands out for its efficient implementation of specific image processing algorithms, especially for morphological operations, segmentation, and texture analysis. Built on top of NumPy, Mahotas is fast and suitable for large image datasets, making it useful in biomedical research where performance and simplicity matter.



Mahotas is commonly used in areas such as microscopy, radiology, and digital pathology, where efficient and accurate image processing is required. Its simplicity, coupled with specialized functions, makes it an excellent choice for researchers handling biomedical images such as cell structures, tissues, or tumors.

Key features of Mahotas:

- Morphological operations: Mahotas provides an efficient implementation of morphological operations such as erosion, dilation, and opening, which are required in many biomedical applications for enhancing structures in images.
- Segmentation algorithms: Mahotas includes several segmentation techniques, such as watershed segmentation, thresholding, and connected components, making it easier to delineate regions of interest like cells or tissues in biomedical images.
- Texture analysis: The library offers advanced tools for texture analysis, including Haralick features (from the Gray-Level Co-occurrence Matrix), which are useful for distinguishing between different tissue types or detecting abnormalities.
- Speed and performance: Mahotas is designed for efficiency and uses C++ under the hood, making it particularly fast when handling large images or datasets. This is an advantage for high-throughput analysis, such as in microscopy image processing.

Mahotas is widely applied in biomedical imaging tasks where morphological operations, segmentation, and feature extraction are needed. Examples include the following:

- Cell segmentation: Identifying and isolating individual cells in microscopy images.
- Tissue analysis: Segmenting and quantifying different regions in histopathology images, such as identifying cancerous regions in tissue samples.
- Texture analysis: Extracting texture features from medical images, which can be used for tissue classification or disease diagnosis.

Example code for image segmentation and texture analysis using Mahotas (Fig. 9.11):

```
import numpy as np
import mahotas as mh
import matplotlib.pyplot as plt
# Load a sample grayscale image
image = mh.imread('example_data/ihc.png', as_grey=True)
# Replace with your biomedical image
# Apply Gaussian filter to smooth the image
smoothed_image = mh.gaussian_filter(image, 0.25)
# Compute the gradient (edge detection)
gradient = mh.sobel(smoothed_image)
# Perform watershed segmentation
seeds, num_seeds = mh.label(gradient > gradient.mean())
segmentation = mh.cwatershed(gradient, seeds)
# Display the original image and the segmented image
fig, ax = plt.subplots(1, 2, figsize=(8, 4))
ax[0].imshow(image, cmap='gray')
ax[0].set_title('Original Image')
ax[0].axis('off')
ax[1].imshow(segmentation, cmap='nipy_spectral')
ax[1].set_title('Watershed Segmentation')
ax[1].axis('off')
plt.tight_layout()
plt.show()
```

Output:

```
Figure 9.11
```

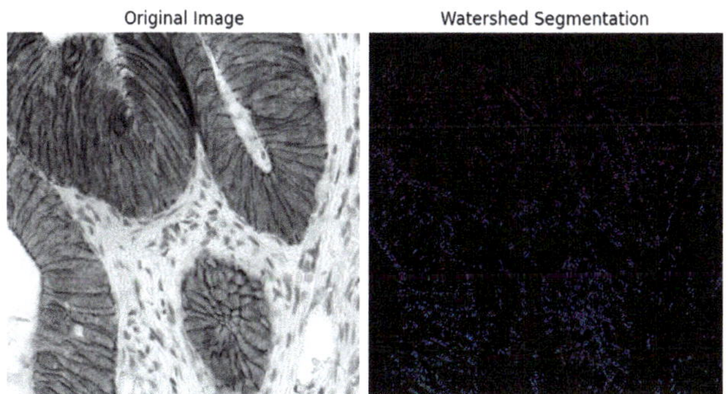


**Fig. 9.11** Watershed segmentation using Mahotas. IHC was smoothed using a Gaussian filter

### 9.3.5 Napari in Biomedical Image Processing

**Notebook: Section 9.3.5. Napari in Biomedical Image Processing**

This notebook provides code examples demonstrating the use of Napari in biomedical image processing.

Link to the GitHub repository:

► https://github.com/sn-code-inside/BioPy

Go to: Chap. 9—example data—► Sect. 9.3.5

Napari is a multidimensional image viewer for Python, specifically designed for fast and interactive visualization and analysis of large biomedical image datasets. It is particularly useful for handling images in various formats, such as 2D, 3D, and even 4D images (which may include time or multichannel data). Napari's primary focus is on being user-friendly and interactive for scientific and biomedical imaging workflows, especially when dealing with large volumes of data from microscopy, radiology, or whole-slide imaging.



Key features of Napari:

- Multidimensional image support: Napari supports viewing and processing 2D, 3D, and 4D images, which is useful for many biomedical applications, such as time-lapse imaging, volumetric scans (e.g., MRI, CT), or multichannel fluorescence microscopy data.
- Interactive visualization: Users can explore image layers interactively, adjust contrast and brightness, and visualize overlays, segmentations, and annotations in real time. This is useful for visual inspection, exploratory analysis, and validating results.
- Layered architecture: Napari allows users to work with multiple layers of data simultaneously, including image layers, segmentation masks, and annotations, all in the same view. This is ideal for biomedical researchers working with complex data, such as combining different imaging modalities (e.g., MRI with histopathology).
- Plugin support: Napari is designed to be highly extensible via plugins. Researchers can develop custom plugins to extend the capabilities of Napari, allowing them to incorporate domain-specific functionality into their workflows.
- Integration with Python libraries: Napari integrates well with other Python libraries such as NumPy, SciPy, scikit-image, Pandas, and Dask, making it an excellent tool for both interactive image visualization and advanced processing workflows in Python.

Napari is particularly suited for biomedical imaging, where researchers need to visualize and process large, multidimensional datasets. Some key applications include the following:

- Microscopy imaging: Napari is widely used for analyzing large multichannel fluorescence microscopy datasets. Researchers can visualize live or fixed cell images in 3D and 4D, apply segmentation algorithms, and track cell movement over time.

- Radiological imaging: Napari's multidimensional support makes it ideal for visualizing volumetric medical images such as MRI, CT, and PET scans. It also supports overlaying segmentation masks to mark areas of interest, such as tumors or abnormal tissues.
- Histopathology: Napari can handle large whole-slide imaging (WSI) datasets, making it useful in digital pathology for visualizing high-resolution tissue samples and identifying pathological features.
- Time-lapse imaging: For biological processes observed over time (e.g., cell division or neuron growth), Napari can visualize these datasets interactively, allowing users to view how biological structures change across time points.

Integration with scikit-image and OpenCV

Napari can be integrated with libraries like scikit-image and OpenCV to extend its image processing capabilities. For instance:

- scikit-image can be used for advanced segmentation, feature extraction, and image enhancement, while the results can be visualized in real time using Napari.
- OpenCV can be utilized for real-time image manipulation (e.g., blurring, edge detection), with Napari providing an interactive viewer for visualizing the results of those transformations on large datasets.

Here is a simple example showing how to load and interactively view a biomedical image using Napari:

```
import napari
from skimage import io, filters
import numpy as np
# Load a multi-dimensional biomedical image (replace with actual image
# path)
image = io.imread('example_data/ihc.png')
# Apply some basic image processing using scikit-image (e.g., Gaussian
# filtering)
smoothed_image = filters.gaussian(image, sigma=1)
# Create a Napari viewer
viewer = napari.Viewer()
# Add the original image and the processed (smoothed) image as layers
viewer.add_image(image, name='Original Image')
viewer.add_image(smoothed_image, name='Smoothed Image')
# Start Napari interactive viewer
napari.run()
```

In this example, Napari is used as an interactive image viewer to visualize both the original and processed versions of a biomedical image. The `napari.Viewer()` function opens the graphical interface where users can interactively explore the images, adjust contrast, and overlay additional data layers (◘ Fig. 9.12). scikit-image is used to apply Gaussian filtering to the image as an example of image processing.

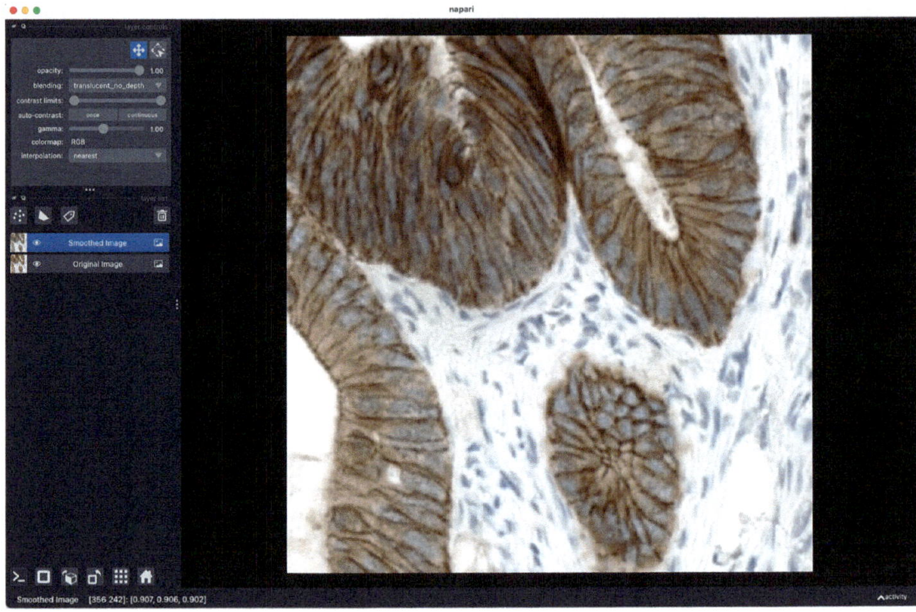


**Fig. 9.12** Napari graphical user interface (GUI). Images can be processed by the GUI

Advantages of Napari in Biomedical Imaging

- Handling large datasets: Napari can handle large multidimensional datasets efficiently, which is particularly useful in fields like histopathology and radiology where images can be gigabytes in size.
- Real-time feedback: Napari allows real-time visualization of image manipulations, which is needed for ensuring accurate image processing workflows in biomedical research.
- Customizability: The ability to create custom plugins makes Napari highly adaptable for specific research needs in biomedical image analysis.
- Ease of use: Napari is designed to be intuitive and user-friendly, allowing researchers with minimal programming experience to quickly visualize and explore complex datasets.

### 9.3.6 Other Python Applications in Biomedical Image Processing

In addition to the well-known libraries like SimpleITK, OpenCV, Mahotas, Napari, and scikit-image, there are several other Python tools and libraries that play specific roles in biomedical image processing. These tools include deep learning frameworks, libraries for handling specific medical imaging formats, and tools for advanced image analysis and feature extraction. Each of these tools offers specialized capabilities for various tasks in biomedical image processing:

*Deep learning-based image processing (PyTorch and TensorFlow)*: Deep learning frameworks like PyTorch and TensorFlow have transformed image processing, particularly in biomedical research. These libraries enable the development of Convolutional Neural Networks (CNNs), U-Net architectures, and other deep learning models, which are highly effective for image classification, segmentation, and object detection tasks. In biomedical image processing, these methods are used for automated diagnosis (e.g., detecting tumors in MRI scans), segmentation of cells in microscopy images, and increasing image resolution.

PyTorch is known for its flexibility, making it popular in research, while TensorFlow is widely used in production environments. Both frameworks support multidimensional data processing, allowing for 2D, 3D, and even 4D biomedical images to be used in deep learning workflows.

*NiBabel for neuroimaging data*: NiBabel is a specialized Python library for working with neuroimaging data in formats like NIfTI, Analyze, and MINC. It allows researchers to load, manipulate, and save complex 3D or 4D neuroimaging datasets, such as MRI or fMRI scans. NiBabel integrates with other scientific Python tools and supports advanced data manipulations, including affine transformations and voxel-based operations. This makes it useful in neuroscience research, where brain imaging data must be processed for tasks such as functional connectivity analysis, brain mapping, or volumetric assessments.

*Nilearn (machine learning for neuroimaging)*: Building on NiBabel, Nilearn provides machine learning tools specifically for neuroimaging data. It offers functionality to easily implement and visualize machine learning models for tasks such as brain decoding, classification of cognitive states, and resting-state functional connectivity analysis. Nilearn is particularly useful for combining traditional machine learning algorithms with neuroimaging data, enabling researchers to uncover patterns related to cognitive function or neurological disorders. It provides high-level functions to process and visualize brain atlases, statistical maps, and brain activation patterns in an accessible way.

*Python imaging library (PIL or Pillow)*: Pillow, the modern version of the Python Imaging Library (PIL), is a library for basic image processing tasks, such as image resizing, cropping, and format conversion. While it is not tailored for medical images, it plays a vital role in biomedical workflows by providing the foundation for initial image manipulation, format conversions (e.g., from PNG to TIFF), and basic color space adjustments. Pillow is frequently used in preprocessing pipelines before more complex medical image analysis tasks.

*DICOMpyler for radiation therapy*: DICOMpyler is a Python library specifically designed for handling DICOM-RT files, which are used in radiation therapy. DICOMpyler enables radiation oncologists and medical physicists to analyze and visualize radiation therapy plans, doses, and structures from DICOM files. It offers specialized tools to interact with radiation therapy data, including dose-volume histograms (DVH), contour analysis, and 3D visualization of treatment plans.

*PyRadiomics for feature extraction in radiomics*: PyRadiomics is a tool for extracting quantitative features from medical images, particularly for the field of radiomics. Radiomics is the process of converting medical images (e.g., CT, MRI) into a large number of features that can be used to build predictive models for diagnosis, prognosis, or treatment response. PyRadiomics extracts features related to shape, intensity, texture, and wavelet transformations from segmented regions of interest (ROIs) within an image. These features can be used to develop machine learning models for personalized medicine, such as predicting how a tumor will respond to treatment based on its radiomic profile.

## 9.4 Exercises and Questions

In this section, we include exercises and questions to complement the material covered in this chapter.

### Exercises

The following exercises will enhance your understanding of the materials we covered in this chapter.

1. Thresholding on medical images: Write a Python script to load a grayscale brain MRI image and apply different thresholding techniques (global, adaptive, and Otsu's thresholding) to segment brain regions.
2. Image smoothing using filters: Use scikit-image to apply Gaussian, median, and bilateral filters to a lung CT image. Compare the results and explain how each filter reduces noise while preserving edges.
3. Morphological operations: Load a binary image of cells, and apply dilation, erosion, and opening operations using Mahotas or scikit-image. Visualize the impact of each operation on the cell boundaries.
4. Feature extraction with PyRadiomics: Use PyRadiomics to extract shape, texture, and intensity features from a segmented tumor region in a CT scan. Analyze how these features can be used for predictive modeling in radiomics.
5. Image registration with SimpleITK: Implement a registration algorithm to align two MRI brain scans (a fixed and moving image) using SimpleITK. Perform mutual information-based registration and visualize the aligned results.
6. Deep learning for image classification: Build a simple convolutional neural network (CNN) using PyTorch or TensorFlow to classify MRI images of brain tumors into benign or malignant categories.
7. Handling DICOM data: Write a script to load a series of DICOM images representing slices of a lung CT scan using pydicom. Convert the DICOM series to a 3D NIfTI volume and visualize the volume using Matplotlib or Napari.
8. Texture analysis using Haralick features: Extract Haralick texture features from a histopathology image using Mahotas. Use the extracted features to differentiate between healthy and cancerous tissue.
9. Segmentation of cells in a fluorescence microscopy image: Use scikit-image to segment nuclei in a fluorescence microscopy image of cells. Apply watershed segmentation to improve the results after an initial thresholding step.
10. Image enhancement using histogram equalization: Load a mammogram image and apply histogram equalization using scikit-image to enhance the contrast of the image. Visualize the before and after results.

## Questions to Be Answered

1. Explain the role of thresholding in image segmentation. How does global thresholding differ from adaptive thresholding?
2. What are the advantages of using deep learning frameworks like PyTorch or TensorFlow for biomedical image processing?
3. Describe the difference between dilation and erosion in morphological operations. How do these operations affect an image?
4. What are the key challenges associated with biomedical image registration, and how can these be addressed using mutual information-based registration?
5. In feature extraction, what is the importance of texture features like Haralick features? Provide an example where texture analysis is required.
6. Discuss the significance of NIfTI format in medical imaging. How does it differ from other image formats like PNG or JPEG?
7. Explain how PyRadiomics can be used in radiomics to predict tumor behavior. What are the common features extracted using PyRadiomics?
8. What are the ethical considerations when using AI and machine learning algorithms for biomedical image analysis?
9. Describe how DICOM files store medical imaging data. Why is it important to handle DICOM data in clinical settings?
10. How do histogram equalization techniques enhance medical images? Provide an example where this technique improves the diagnosis.

**Acknowledgement** The language of the human-generated text was corrected with the assistance of artificial intelligence (AI) tools [GPT-3.5 and GTP-4 from OpenAI]. GitHub Co-Pilot was used to check the correctness of the codes. The text underwent subsequent human revision to ensure its accuracy.

## References

1. Windhager J, Zanotelli VRT, Schulz D, Meyer L, Daniel M, Bodenmiller B, et al. An end-to-end workflow for multiplexed image processing and analysis. Nat Protoc. 2023;18(11):3565–613.
2. Qi Y, Yang Z, Sun W, Lou M, Lian J, Zhao W, et al. A comprehensive overview of image enhancement techniques. Arch Comput Methods Eng. 2022;29(1):583–607.
3. Lu Z-M, Guo S-Z. Chapter 1 – Introduction. In: Lu Z-M, Guo S-Z, editors. Lossless information hiding in images: syngress; 2017. p. 1–68.
4. Pertusa JF, Morante-Redolat JM. Main steps in image processing and quantification: the analysis workflow. Methods Mol Biol. 2019;2040:3–21.
5. Nattkemper TW. Multivariate image analysis in biomedicine. J Biomed Informatics. 2004;37(5):380–91.
6. Cai WL, Hong GB. Quantitative image analysis for evaluation of tumor response in clinical oncology. Chronic Dis Transl Med. 2018;4(1):18–28.
7. Zhu A, Lee D, Shim H. Metabolic positron emission tomography imaging in cancer detection and therapy response. Semin Oncol. 2011;38(1):55–69.
8. Guimaraes MD, Schuch A, Hochhegger B, Gross JL, Chojniak R, Marchiori E. Functional magnetic resonance imaging in oncology: state of the art. Radiol Bras. 2014;47(2):101–11.
9. Fahim Ul H, Cook GJ. PET/CT in oncology. Clin Med (Lond). 2012;12(4):368–72.
10. Filippi M, Agosta F. Diffusion tensor imaging and functional MRI. Handb Clin Neurol. 2016;136:1065–87.

11. Le Moal J, Peillon C, Dacher JN, Baste JM. Three-dimensional computed tomography reconstruction for operative planning in robotic segmentectomy: a pilot study. J Thorac Dis. 2018;10(1):196–201. https://doi.org/10.21037/jtd.2017.11.144.
12. Mankovich NJ, Samson D, Pratt W, Lew D, Beumer J 3rd. Surgical planning using three-dimensional imaging and computer modeling. Otolaryngol Clin North Am. 1994;27(5):875–89. PMID: 7816436.
13. Dell'Oglio P, Mazzone E, Buckle T, Maurer T, Navab N, van Oosterom MN, Schilling C, Witjes MJ, Vahrmeijer AL, Klode J, Vojnovic B, Mottrie A, van der Poel HG, Hamdy F, van Leeuwen FW. Precision surgery: the role of intra-operative real-time image guidance – outcomes from a multidisciplinary European consensus conference. Am J Nucl Med Mol Imaging. 2022;12(2):74–80. PMID: 35535122.
14. Costanzo M, Carton F, Marengo A, Berlier G, Stella B, Arpicco S, Malatesta M. Fluorescence and electron microscopy to visualize the intracellular fate of nanoparticles for drug delivery. Eur J Histochem. 2016;60(2):2640. https://doi.org/10.4081/ejh.2016.2640.
15. Cardellini J, Balestri A, Montis C, Berti D. Advanced static and dynamic fluorescence microscopy techniques to investigate drug delivery systems. Pharmaceutics. 2021;13(6):861. https://doi.org/10.3390/pharmaceutics13060861.
16. Simm J, Klambauer G, Arany A, Steijaert M, Wegner JK, Gustin E, Chupakhin V, Chong YT, Vialard J, Buijnsters P, Velter I, Vapirev A, Singh S, Carpenter AE, Wuyts R, Hochreiter S, Moreau Y, Ceulemans H. Repurposing high-throughput image assays enables biological activity prediction for drug discovery. Cell Chem Biol. 2018;25(5):611–18.e3. https://doi.org/10.1016/j.chembiol.2018.01.015.
17. Kumar, A., Saudagar, A.K.J., Kumar, A. et al. Innovating medical education using a cost effective and scalable VR platform with AI-Driven haptics. Sci Rep. 15, 26360 (2025). https://doi.org/10.1038/s41598-025-10543-8.
18. Tene T, Vique López DF, Valverde Aguirre PE, Orna Puente LM, Vacacela Gomez C. Virtual reality and augmented reality in medical education: an umbrella review. Front Digit Health. 2024;6:1365345. https://doi.org/10.3389/fdgth.2024.1365345.
19. Zhang D, Dihge L, Bendahl PO, Arvidsson I, Dustler M, Ellbrant J, Gulis K, Hjärtström M, Ohlsson M, Rejmer C, Schmidt D, Zackrisson S, Edén P, Rydén L. Deep learning on routine full-breast mammograms enhances lymph node metastasis prediction in early breast cancer. NPJ Digit Med. 2025;8(1):425. https://doi.org/10.1038/s41746-025-01831-8.
20. Champaign JL, Cederbom GJ. Advances in breast cancer detection with screening mammography. Ochsner J. 2000;2(1):33–5. PMID: 21765659.
21. Lancaster HL, Heuvelmans MA, Oudkerk M. Low-dose computed tomography lung cancer screening: Clinical evidence and implementation research. J Intern Med. 2022;292(1):68–80. https://doi.org/10.1111/joim.13480.
22. Goh JK, Cheung CY, Sim SS, Tan PC, Tan GS, Wong TY. Retinal imaging techniques for diabetic retinopathy screening. J Diabetes Sci Technol. 2016;10(2):282–94. https://doi.org/10.1177/1932296816629491
23. Schrevens L, Lorent N, Dooms C, Vansteenkiste J. The role of PET scan in diagnosis, staging, and management of non-small cell lung cancer. Oncologist. 2004;9(6):633–43. https://doi.org/10.1634/theoncologist.9-6-633.
24. Huang Z, Li Q, Lu J, Feng J, Hu J, Chen P. Recent advances in medical image processing. Acta Cytol. 2021;65(4):310–23.
25. Willekens I, Van de Casteele E, Buls N, Temmermans F, Jansen B, Deklerck R, de Mey J. High-resolution 3D micro-CT imaging of breast microcalcifications: a preliminary analysis. BMC Cancer. 2014;14:9. https://doi.org/10.1186/1471-2407-14-9.
26. Gunasundari C, Selva Bhuvaneswari K. A novel approach for the detection of brain tumor and its classification via independent component analysis. Sci Rep. 2025;15(1):8252. https://doi.org/10.1038/s41598-025-87934-4.
27. Abbasi S, Lan H, Choupan J, Sheikh-Bahaei N, Pandey G, Varghese B. Deep learning for the harmonization of structural MRI scans: a survey. Biomed Eng Online. 2024;23(1):90. https://doi.org/10.1186/s12938-024-01280-6.
28. Keenan KE, Delfino JG, Jordanova KV, Poorman ME, Chirra P, Chaudhari AS, Baessler B, Winfield J, Viswanath SE, deSouza NM. Challenges in ensuring the generalizability of image quantitation methods for MRI. Med Phys. 2022;49(4):2820–35. https://doi.org/10.1002/mp.15195.

29. Hatton IA, Galbraith ED, Merleau NSC, Miettinen TP, Smith BM, Shander JA. The human cell count and size distribution. Proc Natl Acad Sci U S A. 2023;120(39):e2303077120. https://doi.org/10.1073/pnas.2303077120.
30. Lipinski KA, Barber LJ, Davies MN, Ashenden M, Sottoriva A, Gerlinger M. Cancer Evolution and the Limits of Predictability in Precision Cancer Medicine. Trends Cancer. 2016;2(1):49–63. https://doi.org/10.1016/j.trecan.2015.11.003.
31. Xu Y, Quan R, Xu W, Huang Y, Chen X, Liu F. Advances in medical image segmentation: a comprehensive review of traditional, deep learning and hybrid approaches. Bioengineering (Basel). 2024;11(10):1034. https://doi.org/10.3390/bioengineering11101034.
32. Bi WL, Hosny A, Schabath MB, Giger ML, Birkbak NJ, Mehrtash A, Allison T, Arnaout O, Abbosh C, Dunn IF, Mak RH, Tamimi RM, Tempany CM, Swanton C, Hoffmann U, Schwartz LH, Gillies RJ, Huang RY, Aerts HJWL. Artificial intelligence in cancer imaging: clinical challenges and applications. CA Cancer J Clin. 2019;69(2):127–57. https://doi.org/10.3322/caac.21552.
33. Panayides AS, Amini A, Filipovic ND, Sharma A, Tsaftaris SA, Young A, Foran D, Do N, Golemati S, Kurc T, Huang K, Nikita KS, Veasey BP, Zervakis M, Saltz JH, Pattichis CS. AI in medical imaging informatics: current challenges and future directions. IEEE J Biomed Health Inform. 2020;24(7):1837–57. https://doi.org/10.1109/JBHI.2020.2991043.
34. Dimitriou N, Arandjelović O, Harrison DJ. Magnifying Networks for histopathological images with billions of pixels. Diagnostics (Basel). 2024;14(5):524. https://doi.org/10.3390/diagnostics14050524.
35. Evans H, Snead D. Understanding the errors made by artificial intelligence algorithms in histopathology in terms of patient impact. NPJ Digit Med. 2024;7(1):89. https://doi.org/10.1038/s41746-024-01093-w.
36. Koçak B, Ponsiglione A, Stanzione A, Bluethgen C, Santinha J, Ugga L, Huisman M, Klontzas ME, Cannella R, Cuocolo R. Bias in artificial intelligence for medical imaging: fundamentals, detection, avoidance, mitigation, challenges, ethics, and prospects. Diagn Interv Radiol. 2025;31(2):75–88. https://doi.org/10.4274/dir.2024.242854.
37. Herath HMSS, Herath HMKKMB, Madusanka N, Lee BI. A Systematic Review of Medical Image Quality Assessment. J Imaging. 2025;11(4):100. https://doi.org/10.3390/jimaging11040100.
38. Ziller A, Mueller TT, Stieger S. et al. Reconciling privacy and accuracy in AI for medical imaging. Nat Mach Intell. 2024;6:764–74. https://doi.org/10.1038/s42256-024-00858-y.
39. Kiseleva A, Kotzinos D, De Hert P. Transparency of AI in Healthcare as a Multilayered System of Accountabilities: Between Legal Requirements and Technical Limitations. Front Artif Intell. 2022;5:879603. https://doi.org/10.3389/frai.2022.879603.
40. Amann J, Blasimme A, Vayena E. et al. Explainability for artificial intelligence in healthcare: a multidisciplinary perspective. BMC Med Inform Decis Mak. 2020;20:310. https://doi.org/10.1186/s12911-020-01332-6.
41. Bradski G. The OpenCV library. Dr Dobb's J Softw Tools. 2000;4:2236121.
42. McCormick M, Liu X, Jomier J, Marion C, Ibanez L. ITK: enabling reproducible research and open science. Front Neuroinform. 2014;8:13. https://doi.org/10.3389/fninf.2014.00013.
43. Lowekamp BC, DChen DT, Ibáñez L, Blezek D. The Design of SimpleITK, Front Neuroinform. 2013;7:45. https://doi.org/10.3389/fninf.2013.00045.

# Genomic Data Analysis

Contents




J. U. Kazi, *Python Essentials for Biomedical Data Analysis: An Introductory Textbook*,
https://doi.org/10.1007/978-3-031-85600-6_10

This chapter provides an overview of genomics data analysis. It briefly describes common data formats and tools used in genomics. The chapter includes a short guide on managing and working with large-scale genomic datasets, emphasizing data processing and analysis techniques. It also includes some basic methods used in genomics data analysis, covering approaches for genomic sequencing, annotation, and comparative genomics.

**Learning Goals**

The goal of this chapter is to provide students with an overview of the basics of genomic data analysis using Python and bioinformatics tools. Students will develop an understanding in processing, analyzing, and visualizing genomic datasets, including variant calling, RNA-seq analysis, population genomics, and epigenomic data interpretation. They will also learn how bioinformatics pipelines work and how to use external tools like BWA, GATK, and SAMtools.

## 10.1 Introduction to Genomic Data Analysis

Here, we provide a brief overview of key concepts, basic development, and broad applications of genomic data analysis. This section also describes commonly used genomic data types and highlights their relevance in biomedical research.

### 10.1.1 Genomic Data and Its Importance

Genomic data refers to the information encoded within the DNA of an organism [1]. It includes the full set of genes, their sequences, and the regulatory regions that govern gene expression [2]. The goal of genomic data analysis is to extract meaningful biological information from this data by studying the structure, function, and evolution of genomes. This type of data can come from a wide range of organisms, from bacteria to humans, and can be used to explore questions about biology, health, and disease.

With the development of high-throughput sequencing technologies, genomic data analysis has become central to fields such as medicine, evolutionary biology, agriculture, and personalized healthcare [3]. For instance, genomic data analysis allows researchers to:

- Identify genetic variants associated with diseases, such as cancer or heart disease.
- Investigate how certain genes are regulated and expressed under different conditions.
- Understand the evolutionary relationships between species through comparative genomics.
- Develop personalized medicine approaches by identifying patient-specific genomic variations.

Given the massive size of modern genomic datasets, computational tools and techniques are widely used to efficiently analyze, interpret, and visualize this data.

### 10.1.2 Key Types of Genomic Data

Genomic data includes a wide range of information obtained from the DNA of different organisms. This type of data helps us to understand genetic variations, evolutionary relationships, and the molecular mechanisms underlying various biological processes and diseases. Common types of genomic data are whole genome sequencing (WGS), whole exome sequencing (WES), targeted gene sequencing, genotyping and single nucleotide polymorphism (SNP) analysis, copy number variation (CNV) analysis, structural variation data, methylation sequencing, chromatin immunoprecipitation sequencing (ChIP-Seq), Hi-C and chromosome conformation capture techniques, metagenomic sequencing, ancient DNA sequencing, and different kinds of functional genomic data.

Whole genome sequencing determines the complete DNA sequence of an organism's genome at a single time, including all coding (exons) and noncoding regions (introns, regulatory elements, intergenic regions) [4]. It is used for analyses, such as identifying genetic variants associated with diseases, studying population genetics, and exploring evolutionary biology. Whole exome sequencing focuses on sequencing all the exonic regions of the genome, which are the parts that are transcribed and translated into proteins [5]. This method is a cost-effective alternative to WGS for identifying mutations that affect protein-coding genes and is often used in clinical diagnostics for genetic disorders.



Targeted gene sequencing sequences specific areas of the genome known to be of interest, such as genes associated with a particular disease [6]. It is used in diagnostics and research to detect mutations in specific genes, allowing for focused and efficient analysis. Genotyping and SNP analysis identify differences in the genetic makeup by examining individual genetic variants, often focusing on SNPs. This approach is employed in genome-wide association studies to find genetic variations linked to specific traits or diseases [7, 8].

Copy number variation analysis detects changes in the number of copies of a particular gene or genomic region [9]. This is important for understanding genetic diversity and has implications in diseases like cancer and developmental disorders. Structural variation data examines larger changes in the genome structure, such as insertions, deletions, inversions, translocations, and duplications [10]. Studying these variations helps in identifying genetic causes of diseases that result from structural alterations in chromosomes.

Methylation sequencing studies the methylation patterns of DNA, an epigenetic modification that can regulate gene expression without changing the DNA sequence [11]. This type of data helps us to understand how epigenetic regulation in development, aging, and diseases like cancer. Chromatin immunoprecipitation sequencing combines chromatin immunoprecipitation with sequencing to identify binding sites of DNA-associated proteins [12]. It is used to map global binding sites for transcription factors and other proteins, providing information about the gene regulation mechanisms.

Hi-C and chromosome conformation capture techniques analyze the three-dimensional organization of genomes by identifying physical interactions between chromosomal regions [13]. These methods are important for understanding how genome architecture influences gene expression and genome stability. Metagenomic sequencing involves sequencing genetic material recovered directly from environmental samples without isolating individual organisms [14]. This approach is used to study microbial communities in various environments, such as soil, oceans, and the human gut microbiome. The diversity of genomic data types reflects the complexity of genetic information and its regulation. Advances in sequencing technologies have made it possible to explore genomes in detail. This progress has provided a deeper understanding of biological system, contributed to medical research, and clarified patterns in evolutionary studies. Knowing these different types of genomic data is important for researchers and clinicians want to understand the genetic contributions to health and disease.

## 10.2 Overview of Python Tools for Genomic Data Analysis

Python is widely used by bioinformaticians and researchers in genomics because of its flexibility, clear syntax, and the large selection of libraries available for genomic data analysis. These libraries allow users to handle large-scale data, perform sequence analysis, statistical modeling, and visualization, among many other tasks. This section provides an overview of Python tools widely used for genomic data analysis.

### 10.2.1 Why Python for Genomic Data Analysis?

Python has emerged as a widely-used programming language in bioinformatics and genomics due to several key advantages. Its straightforward and readable syntax makes it accessible to researchers and biologists who may not have extensive programming backgrounds, accelerating learning and allowing for rapid development and prototyping of analytical tools. Python offers a rich collection of libraries and frameworks specifically designed for biological and genomic data analysis. Notable examples include Biopython, which provides tools for computational biology and bioinformatics such as parsers for common bioinformatics file formats, interfaces to online databases, and functions for sequence analysis; PyVCF and cyvcf2, which facilitate the manipulation and analysis of Variant Call Format (VCF) files used to store gene sequence variations; and HTSeq, which offers infrastructure to process data from high-throughput sequencing assays like RNA-Seq, enabling tasks such as read counting and differential expression analysis.

Python works well with other programming languages and tools commonly used in genomics, such as R (via libraries like rpy2), C/C++, and command-line tools like Bowtie, BWA, and GATK. This interoperability allows Python to act as a "glue" language, orchestrating complex bioinformatics pipelines that combine multiple tools and languages. It has a vibrant and active community of developers and researchers, with continuous contributions of new tools, libraries, and updates that ensure Python stays at the forefront of genomic research needs. This collaborative environment also means extensive documentation and community assistance are readily available.

Python is capable of handling tasks ranging from basic sequence manipulation to advanced data analysis and machine learning applications. Libraries like NumPy, pandas, and scikit-learn enhance Python's ability to process and analyze large genomic datasets efficiently. With these advantages, Python enables researchers to perform a wide array of genomic analyses within a single, cohesive platform. Whether it is for developing custom scripts for data processing, integrating various bioinformatics tools, or conducting complex statistical analyses, Python provides the flexibility and power needed in modern genomic research.

### 10.2.2 Python Libraries for Genomic Data

Python offers a multitude of libraries that are useful for genomic data analysis. One of the most widely used libraries is Biopython, which provides tools for computational biology and bioinformatics [15]. Biopython includes modules for reading and writing sequence file formats (like FASTA and GenBank), performing sequence alignments, accessing online biological databases, and working with phylogenetic trees. Its collection of functions simplifies tasks such as parsing biological files, calculating sequence statistics, and manipulating sequences. Libraries such as PyVCF and cyvcf2 are designed specifically for working with VCF files that store gene sequence variations [16, 17]. These tools allow users to read, query, and write VCF files easily, facilitating the analysis of genetic variants across different samples. They are also useful for tasks like filtering variants based on quality metrics, annotating variants with additional information, and integrating variant data into larger analysis pipelines. HTSeq is an important library for processing high-throughput sequencing data [18]. It provides infrastructure to handle data in common formats such as SAM/BAM files and offers tools for counting how many reads map to each gene, which is needed for quantifying gene expression levels. HTSeq simplifies the process of read alignment counting, differential expression analysis, and quality control, making it a valuable tool for transcriptomic studies.



For specialized genomic data formats, pysam offers an interface for reading, writing, and manipulating SAM/BAM files, which are commonly used for storing aligned sequence data. Pysam enables efficient access to alignment information, allowing for tasks like extracting read counts, identifying alignment positions, and performing custom filtering of sequencing data.

Pybedtools wraps the BEDTools suite, providing genomic arithmetic operations on intervals stored in BED files. It allows for efficient querying and manipulation of genomic intervals, which is useful for tasks like finding overlaps between genomic features, intersecting datasets, and annotating genomic regions with functional information. Another useful library is gffutils, which provides an interface for working with GFF and GTF files that store gene annotation data. Gffutils allows users to create local databases of genomic features, enabling fast queries and manipulations of gene models, exons, introns, and other genomic elements. In addition, DendroPy is a library for phylogenetic computing, providing tools for manipulating tree and character data, which can be used for evolutionary biology studies. It supports a variety of tree operations, including reading and writing tree file formats, tree traversal, and calculating tree statistics.

## 10.3 Introduction to DNA Sequence Data

DNA, or deoxyribonucleic acid, carries the genetic instructions used in the growth, development, functioning, and reproduction of all known living organisms. DNA sequence data is widely used in genomics to understand genes, their functions, and their regulatory mechanisms. The analysis of DNA sequences involves reading, interpreting, and extracting meaningful information from raw nucleotide sequences, typically represented as strings of the nucleotides A (adenine), T (thymine), C (cytosine), and G (guanine). Here we will briefly discuss DNA sequence data, their formats, and how Python can be used to manipulate and analyze DNA sequences.

### 10.3.1 FASTA Format: Structure and Parsing DNA Sequences

**Notebook: Section 10.3.1. FASTA Format: Structure and Parsing DNA Sequences**
This notebook provides code examples demonstrating the FASTA format for structure and parsing DNA sequences.
Link to the GitHub repository:
► https://github.com/sn-code-inside/BioPy
Go to: Chap. 10—example data—► Sect. 10.3.1

One of the most common file formats for storing DNA sequences is FASTA. The FASTA format is simple and consists of a header line that begins with >, followed by one or more lines of the nucleotide sequence.

Example of a FASTA sequence:

```
>seq1
AGCTGATCGATCGTAGCTAGCTAGCTA
```

The header line (starting with >) contains a description of the sequence (e.g., its identifier or name), and the subsequent lines contain the actual sequence. FASTA files can contain multiple sequences, and they are commonly used in bioinformatics for tasks like sequence alignment, searching for homologs, or annotating genes.

*Parsing FASTA files with Biopython:* Biopython can be used to read, write, and manipulate FASTA files. The SeqIO module in Biopython is designed to handle sequence input/output and can be used to parse and extract sequences from FASTA files.

Example code to parse a FASTA file (previously described in ▶ Sect. 3.4.5):

```
from Bio import SeqIO

# Parse the FASTA file and read sequences
with open("example_data/example.fasta") as fasta_file:
    for record in SeqIO.parse(fasta_file, "fasta"):
        print(f"ID: {record.id}")
        print(f"Sequence: {record.seq}")
```

This script reads a FASTA file (`example.fasta`), parses it, and prints each sequence's identifier (record.id) and the corresponding nucleotide sequence (`record.seq`).

*Writing sequences to FASTA files:* You can also write sequences back into a FASTA file after performing modifications.

```
from Bio.Seq import Seq
from Bio.SeqRecord import SeqRecord
from Bio import SeqIO

# Example sequence
seq = Seq("AGCTGATCGATCGTAGCTAGCTAGCTA")
seq_record = SeqRecord(seq, id="new_sequence",
description="Example sequence")

# Write to a new FASTA file
with open("output.fasta", "w") as output_file:
 SeqIO.write(seq_record, output_file, "fasta")
```



This code generates a new FASTA file with the provided sequence and metadata (ID and description).

### 10.3.2 Working with DNA Sequences

**Notebook: Section 10.3.2. Working with DNA Sequences**
This notebook provides code examples demonstrating working with DNA sequences.
Link to the GitHub repository:
▶ https://github.com/sn-code-inside/BioPy
Go to: Chap. 10—▶ Sect. 10.3.2

DNA sequences can be represented as strings in Python. Using Biopython, you can perform various operations on these sequences, such as finding reverse complements, translating DNA to protein, and searching for motifs.

### Basic sequence manipulation

*Reverse complement:* DNA sequences are double-stranded, and for many analyses, it is important to compute the reverse complement of a sequence (where A ↔ T and C ↔ G).

Example:

```
from Bio.Seq import Seq

dna_sequence = Seq("AGCTGATCGATCGTAGCTAGCTAGCTA")
reverse_complement = dna_sequence.reverse_complement()
print(f"Reverse Complement: {reverse_complement}")
```

Output:

```
Reverse Complement: TAGCTAGCTAGCTACGATCGATCAGCT
```

*Transcription and translation:* DNA can be transcribed to RNA (replace thymine with uracil) and translated into protein sequences using the genetic code.

Example:

```
dna_sequence = Seq("AGCAGATCGATCGTAGCTAGCAAGCTA")
rna_sequence = dna_sequence.transcribe()
print(f"RNA Sequence: {rna_sequence}")

protein_sequence = dna_sequence.translate()
print(f"Protein Sequence: {protein_sequence}")
```

Output:

```
RNA Sequence: AGCAGAUCGAUCGUAGCUAGCAAGCUA
Protein Sequence: SRSIVASKL
```

*Searching for motifs:* DNA motifs, such as specific regulatory sequences or conserved regions, can be searched within DNA sequences using string methods or pattern-matching tools from libraries like Biopython.

Example:

```
dna_sequence = Seq("AGCAGATCGATGGTAGCTAGCAAGCTA")
motif = "ATG"
positions = [i for i in range(len(dna_sequence))
             if dna_sequence[i:i+len(motif)] == motif]
print(f"Motif found at positions: {positions}")
```

Output:

```
Motif found at positions: [9]
```

### 10.3.3 Sequence Alignment

**Notebook: Section 10.3.3. Sequence Alignment**
This notebook provides code examples demonstrating sequence alignment.
Link to the GitHub repository:
▶ https://github.com/sn-code-inside/BioPy
Go to: Chap. 10—▶ Sect. 10.3.3

Sequence alignment is used to identify regions of similarity between sequences, which can indicate functional, structural, or evolutionary relationships. There are two main types of sequence alignment:

- Pairwise alignment: Aligns two sequences.
- Multiple sequence alignment: Aligns three or more sequences simultaneously.



*Pairwise sequence alignment:* Biopython provides the pairwise2 module to perform pairwise sequence alignment. You can use it to align two DNA sequences and visualize their similarities and differences.

Example code for global pairwise alignment:

```
from Bio.Align import PairwiseAligner

# Example sequences
seq1 = "AGCTGATCGATCGT"
seq2 = "AGCTCATCGATCGA"

# Create an aligner instance
aligner = PairwiseAligner()

# Perform global alignment
alignments = aligner.align(seq1, seq2)

# Print the best alignment
for alignment in alignments:
    print(alignment)
    break # Only print the first (best) alignment
```

Output:

```
target  0 AGCTG-ATCGATCGT- 14
        0 ||||  --  ||||||||-- 16
query   0 AGCT-CATCGATCG-A 14
```

Multiple sequence alignment: For multiple sequence alignment, tools like ClustalW or MAFFT can be called from Python using Biopython's interface. This allows you to align multiple sequences efficiently and understand their evolutionary relationships.

### 10.3.4 DNA Motif and Pattern Searching

**Notebook: Section 10.3.4. DNA Motif and Pattern Searching**

This notebook provides code examples demonstrating DNA motif and pattern searching.

Link to the GitHub repository:

▶ https://github.com/sn-code-inside/BioPy

Go to: Chap. 10—example data—▶ Sect. 10.3.4

Motifs are short, recurring sequences in DNA that have a biological significance, such as transcription factor binding sites. Identifying these motifs is required in regulatory genomics. Python can be used to search for these motifs across multiple sequences or in a specific genome region.

Motif searching in genomic data: Biopython can help identify known motifs or search for conserved patterns in a set of sequences. Below is an example of motif searching using a simple regular expression pattern:

```
import re
from Bio import SeqIO

# Define the motif (e.g., promoter region pattern)
motif_pattern = re.compile(r"ATG")

# Parse FASTA file and search for the motif in each sequence
with open("example_data/example_gene.fasta") as fasta_file:
    for record in SeqIO.parse(fasta_file, "fasta"):
        matches = motif_pattern.finditer(str(record.seq))
        for match in matches:
          print(f"Motif found in {record.id} at position {match.start()}")
```

Output:

```
Motif found in NM_005030.6 at position 44
Motif found in NM_005030.6 at position 201
Motif found in NM_005030.6 at position 335
...
Motif found in NM_001278717.2 at position 322
Motif found in NM_001278717.2 at position 395
```

This example searches for the motif `ATG` (the start codon) in all sequences within a FASTA file.

### 10.3.5 Identifying Genes in Genomic Sequences

**Notebook: Section 10.3.5. Identifying Genes in Genomic Sequences**
This notebook provides code examples demonstrating identifying genes in genomic sequences.
Link to the GitHub repository:
▶ https://github.com/sn-code-inside/BioPy
Go to: Chap. 10—example data—▶ Sect. 10.3.5



One common task in DNA sequence analysis is identifying gene locations within genomic sequences. This can involve searching for start codons (`ATG`), stop codons (`TAA`, `TAG`, `TGA`), and other signals that define gene boundaries.

Gene identification example.

```
from Bio import SeqIO

def find_genes(sequence):
    start_codon = "ATG"
    stop_codons = ["TAA", "TAG", "TGA"]
    genes = []
    # Ensure the sequence is a string in uppercase
    sequence = str(sequence).upper()

    # Search for start codon
    for start_pos in range(len(Sequence) - 2):
        if sequence[start_pos:start_pos + 3] == start_codon:
            # Search for the next stop codon
            for stop_pos in range(start_pos + 3, len(sequence) - 2, 3):
                codon = sequence[stop_pos:stop_pos + 3]
                if codon in stop_codons:
                    gene = sequence[start_pos:stop_pos + 3]
                    genes.append(gene)
                    break  # Stop after finding the first stop codon
    return genes
```

```
# Specify the path to your FASTA file
fasta_file = "example_data/example_gene.fasta"

# Read sequences from the FASTA file and find genes
for record in SeqIO.parse(fasta_file, "fasta"):
    dna_sequence = record.seq
    genes = find_genes(dna_sequence)
    print(f"Sequence ID: {record.id}")
    print(f"Identified genes: {genes}")
```

This code scans a DNA sequence to identify genes by locating start and stop codons and extracting the gene sequences between them.

## 10.4 Variant Calling and SNP Analysis

Variant calling and SNPs allow the identification of genetic variations between individuals or populations. Genetic variants, such as SNPs, insertions and deletions (indels), and structural variants, can have useful implications for health, disease susceptibility, evolution, and population diversity. Variant calling refers to the process of identifying these differences by comparing a sample's DNA sequence to a reference genome. We will briefly discuss the concepts of variant calling, Python tools to handle variant data, and techniques for analyzing and visualizing SNPs and other genomic variants.

### 10.4.1 Variants and Single Nucleotide Polymorphisms (SNPs)

A genetic variant refers to a difference in the DNA sequence when compared to a reference genome [19]. These variations are the basis of genetic diversity among individuals and populations. Variants can influence traits, susceptibility to diseases, and responses to medications. They occur at various scales and can be classified into several types based on the nature and size of the change in the DNA sequence.

SNPs are the most common type of genetic variation among people. They involve a change in a single DNA base, adenine [A], thymine [T], cytosine [C], or guanine [G], at a specific position in the genome [20]. For example, an SNP may change a DNA sequence from AAGGCT to AACGCT, where the third nucleotide has changed from G to C. SNPs occur approximately once every 300 nucleotides on average, which means there are roughly ten million SNPs in the human genome [21]. They can be found in coding regions (exons), noncoding regions (introns), or intergenic regions between genes. While many SNPs have no effect on health or development, some can influence how genes function. SNPs in coding regions may lead to synonymous mutations (no change in amino acid) or nonsynonymous mutations (change in amino acid), altering protein structure and function. SNPs in regulatory regions can affect gene expression levels. SNPs serve as biological markers, helping scientists locate genes associated with disease. For example, genome-wide association studies (GWAS) identify SNPs linked to complex diseases like diabetes, cancer, and heart disease. SNPs also play a role in pharmacogenomics, influencing individual responses to drugs.

Indels are genetic variants where one or more nucleotides are inserted into (insertion) or removed from (deletion) the DNA sequence [22]. They can range in size from a single nucleotide to large segments of chromosomes. Small indels involving one to a few nucleotides are known microindels. For example, inserting an extra adenine (A) in a sequence might change ACTG to ACATG. Furthermore, indels in coding regions that are not in multiples of three nucleotides can shift the reading frame of the gene, leading to entirely different amino acid sequences downstream and often resulting in nonfunctional proteins; such mutations are referred to as frameshift mutations. Indels can disrupt gene function by truncating proteins, altering active sites, or affecting regulatory elements. They are associated with various genetic disorders, such as cystic fibrosis (a three-nucleotide deletion) and certain forms of muscular dystrophy.

Structural variants involve larger alterations in the genome's architecture and can impact genome function and stability [23]. Types of structural variants include copy number variations (CNVs), which are duplications or deletions of large DNA segments leading to variations in the number of copies of particular genes; inversions, where a chromosome segment breaks off, flips around, and reattaches, reversing the gene order; translocations, where segments from one chromosome break off and attach to another chromosome, which can be reciprocal or nonreciprocal; duplications, in which copies of a genomic region are inserted adjacent to the original or elsewhere in the genome; and large insertions and deletions, where substantial segments of DNA are inserted or deleted, disrupting multiple genes [24]. Structural variants can alter gene dosage, disrupt gene integrity, or create fusion genes with new functions, and they are implicated in developmental disorders, intellectual disabilities, and cancers—for example, the Philadelphia chromosome is a translocation between chromosomes 9 and 22, leading to chronic myeloid leukemia [25].



### 10.4.2 Variant Calling Workflow

Variant calling is the process of identifying genetic variants by comparing sequencing reads from a sample (such as an individual's genome) to a reference genome [26]. This process helps in understanding genetic diversity, identifying disease-associated alleles, and conducting evolutionary studies. The typical steps involved in variant calling include:

*Read alignment:* The first step involves aligning the sequencing reads to a reference genome using alignment tools like BWA (Burrows-Wheeler Aligner) or Bowtie. These tools map short DNA sequences obtained from high-throughput sequencing to their corresponding positions in the reference genome. Accurate alignment is important because errors at this stage can lead to false variant calls. The alignment accounts for possible sequencing errors and variations, ensuring that reads are correctly positioned even in regions with mutations or structural variations.

*Sorting and indexing:* After alignment, the aligned reads are sorted based on their genomic coordinates using tools like SAMtools. Sorting organizes the data, which is necessary for efficient access and processing in downstream analyses. Creating an index for the sorted alignment file (usually in BAM format) allows rapid retrieval of reads from specific genomic regions. Indexing is especially important when dealing with large datasets, as it speeds up data querying and visualization in genome browsers.

Variant calling: In this step, variants are identified by comparing the aligned reads to the reference genome using variant calling tools such as GATK (Genome Analysis Toolkit), FreeBayes, or bcftools. These programs analyze the pileup of reads at each genomic position to detect differences like SNPs, indels, and other structural variants. The tools use statistical models to distinguish true variants from sequencing errors, considering factors like read depth, base quality scores, and allele frequency.

Annotation: Once variants are called, they are annotated with functional information to understand their biological impact. Annotation tools like ANNOVAR or Ensembl's Variant Effect Predictor (VEP) add context by providing details such as the variant's effect on genes (e.g., whether it is in a coding region, causes a synonymous or nonsynonymous change), known associations with diseases, population frequency data from databases like 1000 Genomes or gnomAD, and predictions about the variant's impact on protein function. This step helps prioritize variants that may be clinically relevant or of particular interest for further research.

### 10.4.3 Working with VCF Files in Python

The results of variant calling are typically stored in VCF files. A VCF file records detailed information about each variant, including its position in the genome, reference and alternate alleles, quality metrics, and additional annotations. VCF is a standardized format widely used in genomics because it can represent both small-scale variations (like SNPs and indels) and large structural variants.

Python offers libraries to parse and manipulate VCF files. For example, PyVCF and cyvcf2 provides a simple interface for reading and writing VCF files, filtering variants, extracting specific fields, and performing variant annotations. VCFPy is an alternative to PyVCF.

### 10.4.4 Annotation of Genomic Variants

Variant annotation provides information about the possible function of genetic variants, helping to interpret their possible effects on genes, proteins, and overall biological processes. This process helps in understanding the biological significance of variants identified in sequencing data, especially in the context of disease research, personalized medicine, and evolutionary studies.

For example, some variants may be located in coding regions and cause amino acid changes (missense mutations), altering protein function, while others may be synonymous and not affect the amino acid sequence. Variants can also occur in noncoding regions, affecting regulatory elements like promoters, enhancers, or splice sites, which can influence gene expression levels or RNA processing.

Several bioinformatics tools have been developed to annotate genetic variants by providing detailed information about their possible impact. Notable tools include:

ANNOVAR: A widely used tool that annotates genetic variants detected from diverse genomes. ANNOVAR can identify variants that cause protein coding changes, affect splicing, or are located in regulatory regions. It integrates multiple databases to provide annotations, including population allele frequencies, known disease associations, and predicted functional impacts.

Variant Effect Predictor (VEP): Developed by Ensembl, VEP predicts the functional effects of genomic variants. It provides information such as the affected gene, the consequence of the variant (e.g., synonymous, missense, nonsense), SIFT and PolyPhen scores (predicting the impact on protein function), and whether the variant is known in population databases like dbSNP or ClinVar.

### 10.4.5 Visualization of Genomic Variants

Visualizing genetic variants is important for understanding their distribution across the genome, their effects on biological functions, and their associations with phenotypic traits or diseases. Effective visualization techniques help researchers identify patterns, detect loci, and generate hypotheses for further investigation. Common visualization methods for genomic variants include Manhattan plots, volcano plots, and other genome-wide association study (GWAS) plots.



Manhattan plot: A Manhattan plot is a specialized scatter plot used to display data with a large number of data points, such as the millions of single nucleotide polymorphisms (SNPs) tested in a genome-wide association study (GWAS). In this plot, the $x$-axis represents the genomic coordinates, typically organized by chromosomes laid out sequentially from chromosome 1 to the X and Y chromosomes; within each chromosome, variants are ordered based on their position in base pairs (bp). The $y$-axis displays the negative logarithm of the $p$-value ($-\log_{10} p$-value) obtained from statistical tests assessing the association between each SNP and the trait or disease under study, which transforms small $p$-values (indicating strong associations) into more visible values. Each point on the plot represents an SNP, and the color of the points often alternates between chromosomes to improve visual distinction and clarity. A horizontal line indicates the genome-wide significance threshold (e.g., a $p$-value of $5 \times 10^{-8}$), adjusted for multiple testing, and SNPs above this line are considered statistically significant. Interpreting Manhattan plots involves looking at tall peaks, which indicate genomic regions where SNPs are associated with the trait or disease and are candidates for containing genes or regulatory elements influencing the phenotype. Clusters of significant SNPs suggest linkage disequilibrium (LD), where nearby SNPs are inherited together due to physical proximity, and variations in the density of significant SNPs across chromosomes may reflect differences in gene density or recombination rates.

*Genome-wide association study (GWAS) analysis:* GWAS aim to identify genetic variants associated with traits or diseases by scanning the genomes of many individuals. The process involves gathering a large cohort of individuals, including both cases (with the trait or disease) and controls (without), and using high-throughput genotyping technologies to detect SNPs across the genome. Quality control steps are required to filter out unreliable data, such as SNPs with low call rates or individuals with high missing data. Statistical analysis is performed by conducting association tests (e.g., logistic regression) for each SNP to evaluate its association with the trait, adjusting for confounders by including covariates like age, sex, and population stratification to reduce false positives, and applying multiple testing correction methods like Bonferroni correction or False Discovery Rate (FDR) to control for the large number of tests performed. Visualization tools like Manhattan and QQ plots help researchers visualize results, assess the distribution of $p$-values, identify true associations by distinguishing significant SNPs from background noise, detect systematic biases such as population stratification or batch effects, and effectively communicate findings with the scientific community. Significant findings are then validated in independent cohorts to confirm associations and functional annotation is performed to investigate the biological relevance by annotating significant SNPs with gene information, expression data, and known pathways.

## 10.5 Epigenomics Data Analysis

Epigenomics is the study of changes in gene expression and cellular function that are not caused by alterations in the DNA sequence itself but rather by chemical modifications to the DNA or histones. These modifications can include DNA methylation, histone modifications, and chromatin accessibility changes, which regulate gene expression [27]. Understanding epigenetic modifications helps explain processes such as cellular differentiation, disease progression (such as cancer), and environmental influences on gene regulation.

### 10.5.1 Key Epigenetic Mechanisims

Epigenomic modifications do not alter the underlying DNA sequence but play a role in gene regulation. These modifications can be influenced by environmental factors and are often reversible, making epigenomics a key area of study in understanding diseases like cancer, metabolic disorders, and neurological conditions. Key epigenetic mechanisms include DNA methylation, the addition of a methyl group to the DNA molecule, usually at cytosine bases, which often results in gene silencing; histone modifications, post-translational modifications (e.g., acetylation, methylation) to histone proteins that influence chromatin structure and gene expression; and chromatin accessibility, the degree to which DNA is accessible to transcription factors, often studied using ATAC-seq (Assay for Transposase-Accessible Chromatin using sequencing).

### 10.5.2 Common File Formats and Handling Epigenomic Data

Epigenomic data is typically large-scale and requires specialized formats and tools for processing and analysis [28]. Common file formats used in epigenomic studies include:

BED: Used for storing genomic intervals such as DNA methylation sites, histone modification peaks, or transcription factor binding sites. BED files are simple, tab-delimited text files that contain information about genomic regions, including chromosome, start and end positions, and optionally, additional data like name, score, and strand.

BigWig: An indexed binary format for efficient storage and retrieval of dense, continuous data like read coverage or signal intensity from experiments such as ChIP-seq or RNA-seq. BigWig files allow rapid access to subsets of the data without loading the entire file into memory, which is particularly useful for large datasets.

GFF/GTF: General Feature Format (GFF) and Gene Transfer Format (GTF) are used for storing gene and genomic feature annotations. These formats include information about genes, exons, introns, regulatory elements, and other genomic features, along with their positions and attributes.

*ChIP-seq data:* ChIP-seq is a technique used to identify DNA-binding sites of proteins, such as transcription factors or histone modifications. The process involves immunoprecipitating DNA-protein complexes using specific antibodies against the protein of interest and then sequencing the associated DNA fragments. ChIP-seq allows researchers to map protein-DNA interactions across the genome, providing an overview of gene regulation, chromatin states, and epigenetic modifications. ChIP-seq data is often used to map transcription factor binding sites, identify regions with specific histone modifications (e.g., H3K27ac, H3K4me3) and study chromatin architecture and accessibility indirectly. To analyze ChIP-seq data in Python, you can use pybedtools to manipulate and perform operations on genomic intervals derived from peak-calling results.

*Intersection of genomic features:* One common task in epigenomics is intersecting genomic features to find overlaps between different datasets. For example, you might want to identify which ChIP-seq peaks (representing transcription factor binding sites) overlap with known gene promoters or enhancers. This can help determine the regulatory impact of transcription factors on gene expression. Similarly, pybedtools can be used to identify intersections between ChIP-seq peaks and gene promoters.



### 10.5.3 Peak Calling in ChIP-seq Data

Peak calling in ChIP-seq analysis involves identifying regions of the genome that are enriched with sequencing reads, which correspond to DNA-binding sites of proteins such as transcription factors or histone modifications. These enriched regions, or "peaks," indicate where proteins interact with DNA.

While Python itself does not have dedicated peak-calling tools built into its standard library, it can be used to automate the peak-calling process by interfacing with external tools like MACS2 (Model-based Analysis for ChIP-Seq). MACS2 is widely used due to its statistical models and ability to handle various types of ChIP-seq data, including narrow peaks (e.g., transcription factors) and broad peaks (e.g., histone modifications).

By using Python's subprocess module, you can call MACS2 from within a Python script, allowing for integration into larger analysis pipelines and automation of repetitive tasks. After running MACS2, you can further analyze and visualize the results using Python libraries such as `pybedtools`, `pandas`, and `matplotlib`.

### 10.5.4 Analysis of Differential Methylation Patterns

DNA methylation typically occurs at CpG sites, where a cytosine nucleotide is followed by a guanine nucleotide in the DNA sequence. Abnormal DNA methylation patterns are often linked to diseases such as cancer, neurological disorders, and autoimmune conditions. Analyzing DNA methylation data involves identifying differentially methylated regions (DMRs), genomic regions where methylation levels differ between conditions (e.g., healthy vs. diseased tissues). Detecting DMRs helps researchers understand epigenetic alterations associated with disease states or environmental factors.

DNA methylation data can be processed using Python libraries like pandas for data manipulation and numpy and scipy for statistical analysis. For specialized bisulfite sequencing data analysis, Python package such as methylpy, or other such as Bismark (although Bismark is typically used via the command line) can be utilized. To identify DMRs, you can perform statistical tests at each CpG site or region to compare methylation levels between conditions. Since methylation data often involves proportions (e.g., percentage of methylated reads at a site), appropriate statistical methods should be used.

### 10.5.5 Identifying Differentially Methylated Regions in Cancer

DNA methylation changes are a hallmark of cancer, and identifying DMRs can be useful in the diagnosis, prognosis, and treatment of various cancers. These epigenetic alterations can affect gene expression and contribute to tumor development and progression. Identifying DMRs in cancer involves analyzing DNA methylation data to detect regions showing different methylation levels between cancerous and normal tissues. The process typically begins with obtaining methylation data, such as from bisulfite sequencing or methylation arrays, formatted as methylation percentages or beta values. The next step is to load this data into Python using libraries such as pandas. Statistical methods, commonly t-tests or non-parametric tests like the Wilcoxon test, are then applied using scipy.stats to compare methylation levels between cancer and normal samples. After testing, results are adjusted for multiple comparisons using methods such as Benjamini-Hochberg to control false discovery rates. DMRs identified by these statistical analyses are visualized through heatmaps or clustering plots using Python visualization libraries like seaborn or matplotlib, providing clear graphical comparisons between cancer and normal tissues.

## 10.6 Population Genomics Analysis

Population genomics is the study of genetic variation within and between populations of organisms. It is used to study evolutionary processes, genetic diversity, population structure, natural selection, and adaptation [29]. Analyzing large-scale genomic data from populations, researchers can identify the genetic basis of phenotypic traits, understand evolutionary history, and assess the impact of demographic events such as migration and bottlenecks.

### 10.6.1 Population Genomics Data

Population genomics builds on traditional population genetics but focuses on large-scale genomic data to study genetic diversity, population structure, evolutionary processes, and genotype-phenotype associations [30, 31]. It involves understanding how diverse populations are at the genetic level, identifying subpopulations or genetic clusters within larger populations, detecting signals of selection, genetic drift, and migration, and investigating how genetic variants correlate with phenotypic traits such as disease susceptibility or adaptation to environmental factors. Common types of data used in population genomics include VCF files containing information about genetic variants like SNPs and indels, haplotype data representing phased variants across individuals, and genotype matrices that are binary or categorical representations of individual genotypes.

### 10.6.2 Working with Population Genomic Data

Working with population genomic data involves handling large datasets that often span thousands of individuals and millions of genetic variants, necessitating efficient data handling and analysis tools specifically designed for genomic data. One of the most widely used Python libraries for this purpose is scikit-allel, which is designed for analyzing large-scale VCF files and computing population genetic statistics.



### 10.6.3 Population Structure and Phylogenetic Analysis

Understanding population structure helps in the identification of genetic subpopulations or clusters within a population, which can be inferred using techniques such as principal component analysis (PCA) and admixture analysis. PCA is a dimensionality reduction technique that identifies genetic similarities and differences between individuals by reducing the genotype matrix to a set of principal components that capture the majority of genetic variation. Admixture analysis estimates the proportion of an individual's ancestry from different ancestral populations and is commonly used to study gene flow between populations or the genetic mixing of distinct populations. Although Python doesn't have direct tools for admixture analysis, external tools like ADMIXTURE or fastStructure can be integrated with Python to automate this process.

### 10.6.4 Computing Population Genetics Statistics

Computing population genetics statistics helps to study genetic diversity, selection, and population differentiation. Tools like scikit-allel support this analysis by providing measures such as allele frequencies, fixation index (F_ST), and Tajima's D. Allele frequency, the proportion of a particular allele in a population, provides information about genetic diversity and selection. The F_ST measures genetic differentiation between populations, where a high F_ST value indicates population differentiation due to factors like geographic isolation or natural selection. Tajima's D is a statistic used to detect deviations from neutral evolution; positive values indicate balancing selection or population bottlenecks, while negative values suggest population expansion or purifying selection.

### 10.6.5 Analyzing Genetic Diversity in Human Populations

Analyzing genetic diversity in human populations involves several steps that transform raw genomic data into meaningful biological overviews. The process starts with acquiring and preparing genetic variation data, often in VCF format, which is loaded and managed using scikit-allel. Once the data are loaded, quality control and filtering are performed to remove low-frequency variants or those with missing data, ensuring the dataset is suitable for further analysis. Next, genetic diversity metrics such as nucleotide diversity (π), heterozygosity, and fixation indices (F_ST) are calculated to measure variation within and between populations. These statistics help reveal the degree of genetic differentiation, which is important for understanding population structure and evolutionary patterns. To examine genetic relationships and population substructure, dimensionality reduction techniques like PCA are used, making it easier to visualize genetic clusters among individuals. Python's plotting libraries are then used to visualize and interpret the results.

## 10.7 Visualization Techniques for Genomic Data

Visualization of genomic data analysis helps researchers interpret complex data and identify meaningful patterns. Effective visualization techniques enable the representation of large datasets, such as gene expression profiles, genomic variants, and population structure, in a way that is intuitive and easy to understand. In genomics, visualizations can range from simple plots, such as bar graphs, to more advanced techniques, such as heatmaps and genome browser-like visualizations.

### 10.7.1 Heatmaps for Gene Expression Analysis

> **Notebook: Section 10.7.1. Heatmaps for Gene Expression Analysis**
> This notebook provides code examples demonstrating the FASTA format for structure and parsing DNA sequences.
> Link to the GitHub repository:
> ▶ https://github.com/sn-code-inside/BioPy
> Go to: Chap. 10—example data—▶ Sect. 10.7.1

Heatmaps are widely used in genomics to visualize gene expression data. They provide a color-coded representation of the expression levels of genes across different samples or conditions. Heatmaps can be particularly useful for identifying clusters of genes that exhibit similar expression patterns.

*Creating a heatmap of gene expression data:* Python's Seaborn library provides an easy-to-use interface for creating heatmaps. Gene expression data, typically in the form of a matrix with genes as rows and samples as columns, can be visualized using a heatmap to highlight differentially expressed genes.

Example code to create a heatmap using gene expression data:

```
import pandas as pd
import seaborn as sns
import matplotlib.pyplot as plt

# Load gene expression data
# Rows are samples, and columns are genes
# (e.g., gene expression counts or normalized values)
expression_data = pd.read_csv('example_data/gene_data.csv',
index_col=0)

# Create a heatmap
plt.figure(figsize=(10, 8))
sns.heatmap(expression_data.T, cmap='viridis')
plt.title('Heatmap of Gene Expression Data')
plt.xlabel('Samples')
plt.ylabel('Genes')
plt.show()
```



This heatmap visualizes the expression of genes across samples, with color intensity representing the level of expression (◘ Fig. 10.1).

Customizing heatmaps: Heatmaps can be further customized by adding hierarchical clustering to group similar genes or samples together:

Clustering to the heatmap

```
import pandas as pd
import seaborn as sns
import matplotlib.pyplot as plt

# Load gene expression data
# Rows are samples, and columns are genes
# (e.g., gene expression counts or normalized values)
expression_data = pd.read_csv('example_data/gene_data.csv',
index_col=0)

sns.clustermap(expression_data.T, cmap='viridis',
metric="correlation", method="average")
plt.title('Clustered Heatmap of Gene Expression Data')
plt.show()
```

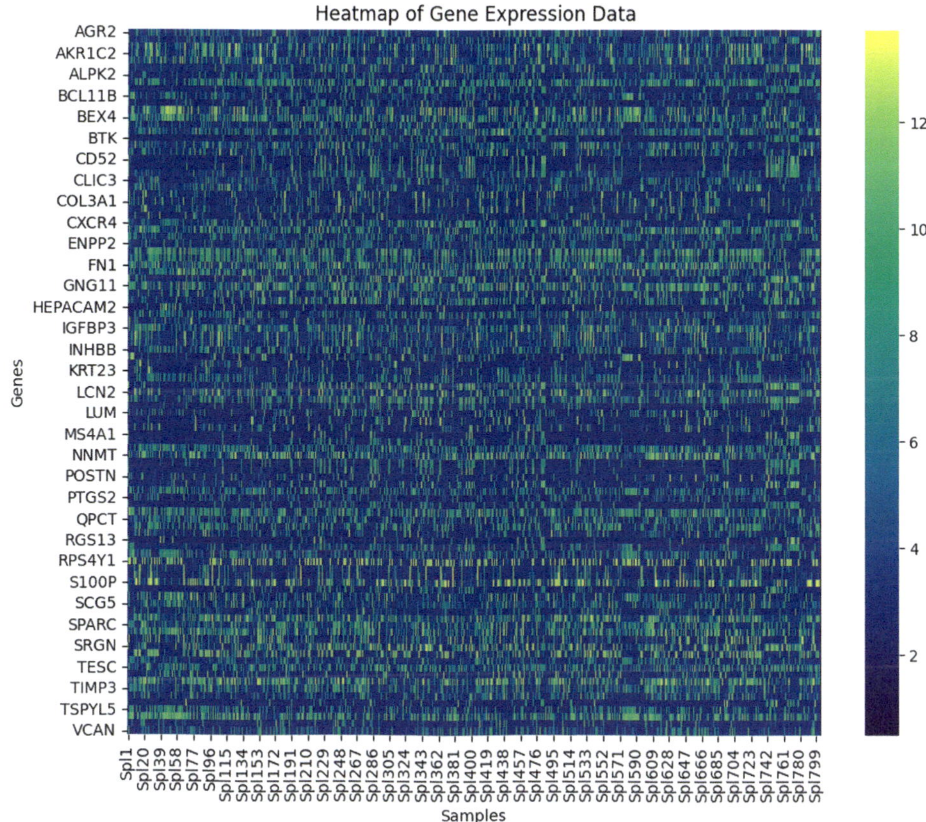


**Fig. 10.1** Heatmap of gene expression data. Highly variable 100 genes were plotted for around 800 samples

This code adds hierarchical clustering to the heatmap, which groups genes and samples based on their similarity, making it easier to identify patterns in gene expression (Fig. 10.2).

## 10.7.2 Genome Browser-Like Visualization

Genome browser-like visualizations allow researchers to visualize genomic data alongside annotations, such as genes, regulatory elements, and variants, across genomic coordinates. Tools like PyGenomeTracks enable the visualization of multiple tracks (e.g., gene annotations, ChIP-seq peaks, RNA-seq coverage) aligned to a reference genome, similar to genome browsers like UCSC Genome Browser or Ensembl. PyGenomeTracks is a Python library that generates genome browser-like visualizations. It can display a variety of data types, including gene annotations, coverage plots, and peaks from ChIP-seq or RNA-seq experiments.

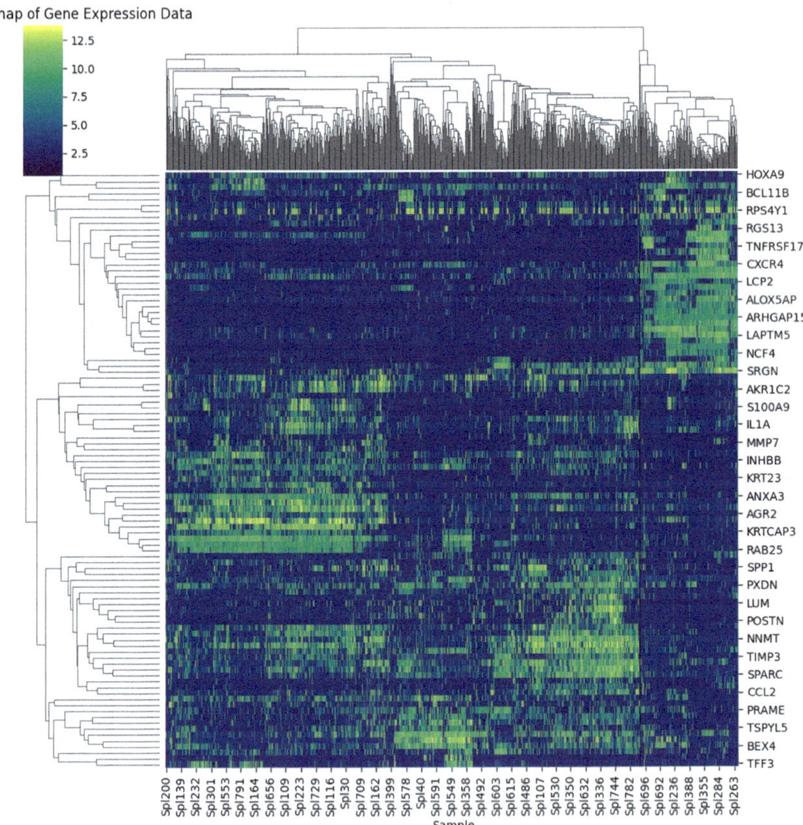


**Fig. 10.2** Clustered heatmap of gene expression data. Clustering provides additional information

## 10.7.3 Plotting Genomic Variants and Association Studies

Manhattan plots and volcano plots are commonly used to visualize the results of GWAS and variant analysis. These plots help researchers identify associations between genomic variants and phenotypic traits or diseases.

A Manhattan plot is used to display the *p*-values of SNPs across the genome. It helps identify genomic regions with SNPs that are associated with a trait or disease.

Example of creating a Manhattan plot using GWAS data (example file "gwas_results.csv" is a place holder for your data):

```
import pandas as pd
import matplotlib.pyplot as plt

# Load GWAS results (chromosome, position, and p-value)
gwas_data = pd.read_csv('gwas_results.csv')
```

```
# Create a Manhattan plot
plt.scatter(gwas_data['position'], -np.log10(gwas_
data['pvalue']), c=gwas_data['chromosome'])
plt.xlabel('Genomic Position')
plt.ylabel('-log10(p-value)')
plt.title('Manhattan Plot')
plt.show()
```

In this code generated plot, each point represents an SNP, and the $y$-axis shows the negative logarithm of the $p$-value, highlighting SNPs with strong associations to the trait of interest.

### 10.7.4 3D Visualization of Genomic Data

With the development of Hi-C and other chromosome conformation capture technologies, researchers can visualize the three-dimensional structure of the genome. Hi-C data shows how chromatin is folded within the nucleus, revealing interactions between different genomic regions.

Visualizing Hi-C data in 3D: Hi-C data can be visualized using 3D plots that show interactions between genomic regions in three-dimensional space. Tools like matplotlib's 3D plotting capabilities can be used for this. Other specialized tools include HiGlass, PyGenomeTracks with Hi-C support, and cooler with HiCExplorer.

Example of visualizing Hi-C data using 3D plotting:

```
from mpl_toolkits.mplot3d import Axes3D
import matplotlib.pyplot as plt
import numpy as np

# Example Hi-C interaction data (x, y, z coordinates of interactions)
hic_data = np.random.rand(100, 3) # Simulated data for demonstration

# Create a 3D scatter plot
fig = plt.figure()
ax = fig.add_subplot(111, projection='3d')
ax.scatter(hic_data[:, 0], hic_data[:, 1], hic_data[:, 2])
# Customize the plot
ax.set_xlabel('X Position')
ax.set_ylabel('Y Position')
ax.set_zlabel('Z Position')
ax.set_title('3D Visualization of Hi-C Data')
plt.show()
```

This example simulates Hi-C interaction data and visualizes it in 3D, demonstrating how genomic interactions between distant regions can be mapped in three-dimensional space.

## 10.8 Integration with Bioinformatics Tools and Pipelines

In genomic research, it is often required to combine different tools and pipelines to efficiently handle, process, and analyze large-scale data. Many bioinformatics tools are command-line-based and developed in languages like C, C++, or R. However, Python can serve as a glue language to combine these tools into reproducible, automated pipelines. These pipelines make it easier to integrate the various steps in genomic data analysis, from raw data preprocessing to advanced statistical analysis and visualization.

### 10.8.1 Bioinformatics Pipelines

A bioinformatics pipeline is a series of computational steps that take raw biological data and process it to generate meaningful outputs, including tasks such as quality control of sequencing data, alignment of reads to a reference genome, variant calling and annotation, and statistical analysis and visualization [32]. Python's ability to call external tools and manage files makes it an ideal language for building bioinformatics pipelines that automate the analysis of large genomic datasets. Key benefits of using pipelines include reproducibility, as pipelines ensure that analyses are reproducible by documenting every step of the process; scalability, as they can handle large datasets efficiently by parallelizing tasks and managing compute resources; and automation, as pipelines automate repetitive tasks, reducing the likelihood of human error and speeding up analyses.



### 10.8.2 Workflow Management Tools

Python is often used in combination with workflow management tools like Snakemake and Nextflow to build complex, reproducible pipelines [33, 34]. These tools enable the automation of multistep workflows, allowing users to define dependencies between tasks, specify input and output files, and parallelize computations across multiple cores or nodes in a high-performance computing (HPC) environment.

Snakemake is a Python-based workflow management system that allows users to define workflows in simple, readable Snakefiles. Each rule in a Snakefile defines a task, its input and output, and the command to execute.

Example of a simple Snakemake pipeline for variant calling (all files are placeholders):

```
# Snakefile
rule all:
    input:
        "variants.vcf"

rule fastqc:
    input:
        "reads.fastq"
    output:
        "fastqc_report.html"
    shell:
        "fastqc {input} -o ."

rule align:
    input:
        "reads.fastq"
    output:
        "aligned_reads.bam"
    shell:
       "bwa mem reference.fasta {input} | samtools sort -o {output}"

rule variant_calling:
    input:
        "aligned_reads.bam"
    output:
        "variants.vcf"
    shell:
       "gatk HaplotypeCaller -I {input} -O {output} -R reference.fasta"
```

In this example, the pipeline includes three steps: quality control using FastQC, alignment of sequencing reads using BWA, and variant calling using GATK. Each rule specifies the input, output, and command for a particular step. Snakemake manages the dependencies and ensures that each step is run in the correct order.

To run the pipeline, execute the following command in the terminal:

```
snakemake --cores 4
```

This command runs the pipeline using four cores in parallel, making the workflow more efficient for large datasets.

Nextflow supports distributed execution on cloud platforms and HPC environments. Although Nextflow is not Python-based, it can integrate Python scripts and tools within its workflows.

## 10.9 Exercises and Questions

In this section, we include practical exercises and questions to complement the material covered in the chapter.

### Exercises

The following exercises will enhance your understanding of the materials we covered in this chapter.

1. Parsing and analyzing a FASTA file: Write a Python script using Biopython to read a FASTA file containing genomic sequences and count the number of A, T, C, and G nucleotides in each sequence.
2. Performing RNA-seq read alignment using BWA: Use BWA to align paired-end RNA-seq reads to a reference genome. Write a Python script to automate the alignment process and sort the aligned reads using SAMtools.
3. Generating a heatmap for gene expression data: Load a gene expression matrix (genes x samples) using Pandas, then use Seaborn to generate a clustered heatmap to visualize differential gene expression across samples.
4. Calculating allele frequencies from VCF data: Use scikit-allel to load a VCF file, extract genotype calls, and calculate allele frequencies across the population. Visualize the distribution of allele frequencies using a histogram.
5. Performing PCA on genomic data: Load genotype data using scikit-allel, perform PCA, and visualize the population structure in a 2D scatter plot. Label the clusters based on population groups.
6. Annotating genomic variants using VEP: Write a Python script that calls VEP (Variant Effect Predictor) to annotate variants in a VCF file. Filter the variants based on their predicted impact (e.g., missense or nonsense mutations).
7. Building a simple Snakemake pipeline: Write a Snakefile that automates a bioinformatics workflow, including read alignment, sorting, and variant calling using BWA, SAMtools, and GATK.
8. Creating a Manhattan plot for GWAS data: Load genome-wide association study (GWAS) results from a CSV file using Pandas, then create a Manhattan plot to highlight associations between SNPs and a phenotypic trait.
9. Visualizing genome regions with PyGenomeTracks: Use PyGenomeTracks to create a genome browser-like visualization for a specific genomic region. Include gene annotations, RNA-seq coverage, and ChIP-seq peaks in the visualization.



### ? Questions to Be Answered

1. What are the main differences between DNA sequencing and RNA sequencing? How are these technologies applied in biomedical research?
2. Describe the role of VCF (Variant Call Format) in genomic data analysis.
3. What types of information are stored in a VCF file?
4. Explain the concept of Principal Component Analysis (PCA) in the context of population genomics. What insights can PCA provide about genetic diversity and population structure?
5. How does Tajima's D help in detecting deviations from neutral evolution?

6. What might positive and negative Tajima's D values indicate?
7. What is variant annotation, and why is it important in clinical genomics? How do tools like ANNOVAR or VEP facilitate this process?
8. How does ChIP-seq help in understanding gene regulation?
9. Describe how peak calling is performed in ChIP-seq data analysis.
10. In what situations might you use Snakemake or Nextflow for genomic data analysis pipelines? What are the advantages of using these tools?

**Acknowledgements** The language of the human-generated text was corrected with the assistance of artificial intelligence (AI) tools [GPT-3.5 and GTP-4 from OpenAI]. GitHub Co-Pilot was used to check the correctness of the codes. The text underwent subsequent human revision to ensure its accuracy.

## References

1. Goldman AD, Landweber LF. What is a genome? PLoS Genet. 2016;12(7):e1006181.
2. Stamatoyannopoulos JA. What does our genome encode? Genome Res. 2012;22(9):1602–11.
3. Hassan M, Awan FM, Naz A, deAndres-Galiana EJ, Alvarez O, Cernea A, et al. Innovations in genomics and big data analytics for personalized medicine and health care: a review. Int J Mol Sci. 2022;23(9):4645.
4. Bagger FO, Borgwardt L, Jespersen AS, Hansen AR, Bertelsen B, Kodama M, et al. Whole genome sequencing in clinical practice. BMC Med Genet. 2024;17(1):39.
5. Warr A, Robert C, Hume D, Archibald A, Deeb N, Watson M. Exome sequencing: current and future perspectives. G3 (Bethesda). 2015;5(8):1543–50. https://doi.org/10.1534/g3.115.018564.
6. Pei XM, Yeung MHY, Wong ANN, Tsang HF, Yu ACS, Yim AKY, Wong SCC. Targeted sequencing approach and its clinical applications for the molecular diagnosis of human diseases. Cells. 2023;12(3):493. https://doi.org/10.3390/cells12030493.
7. Uffelmann E, Huang QQ, Munung NS, et al. Genome-wide association studies. Nat Rev Methods Primers. 2021;1:59. https://doi.org/10.1038/s43586-021-00056-9.
8. Jiang Y, Zhang H. Empowering genome-wide association studies via a visualizable test based on the regional association score. Proc Natl Acad Sci U S A. 2025;122(9):e2419721122. https://doi.org/10.1073/pnas.2419721122.
9. Pos O, Radvanszky J, Buglyo G, Pos Z, Rusnakova D, Nagy B, et al. DNA copy number variation: main characteristics, evolutionary significance, and pathological aspects. Biom J. 2021;44(5):548–59.
10. Yang L. A practical guide for structural variation detection in the human genome. Curr Protoc Hum Genet. 2020;107(1):e103. https://doi.org/10.1002/cphg.103.
11. Abakir A, Reik W. Direct methylation sequencing. Nat Chem Biol. 2023;19(8):932–3.
12. Park PJ. ChIP-seq: advantages and challenges of a maturing technology. Nat Rev Genet. 2009;10(10):669–80. https://doi.org/10.1038/nrg2641.
13. Belton JM, McCord RP, Gibcus JH, Naumova N, Zhan Y, Dekker J. Hi-C: a comprehensive technique to capture the conformation of genomes. Methods. 2012;58(3):268–76.
14. Zhang L, Chen F, Zeng Z, Xu M, Sun F, Yang L, Bi X, Lin Y, Gao Y, Hao H, Yi W, Li M, Xie Y. Advances in metagenomics and its application in environmental microorganisms. Front Microbiol. 2021;12:766364. https://doi.org/10.3389/fmicb.2021.766364.
15. Cock PJ, Antao T, Chang JT, Chapman BA, Cox CJ, Dalke A, et al. Biopython: freely available Python tools for computational molecular biology and bioinformatics. Bioinformatics. 2009;25(11):1422–3.
16. Casbon J. and the PyVCF development team. PyVCF: A Variant Call Format parser for Python, 2013–2016. https://github.com/jamescasbon/PyVCF.
17. Pedersen BS, Quinlan AR. cyvcf2: fast, flexible variant analysis with Python. Bioinformatics. 2017;33(12):1867–69. https://doi.org/10.1093/bioinformatics/btx057.
18. Putri GH, Anders S, Pyl PT, Pimanda JE, Zanini F. Analysing high-throughput sequencing data in Python with HTSeq 2.0. Bioinformatics. 2022;38(10):2943–5.

19. Eichler EE. Genetic variation, comparative genomics, and the diagnosis of disease. N Engl J Med. 2019;381(1):64–74.
20. Nickerson DA, Kolker N, Taylor SL, Rieder MJ. Sequence-based detection of single nucleotide polymorphisms. Methods Mol Biol. 2001;175:29–35.
21. Nelson MR, Marnellos G, Kammerer S, Hoyal CR, Shi MM, Cantor CR, Braun A. Large-scale validation of single nucleotide polymorphisms in gene regions. Genome Res. 2004;14(8):1664–8. https://doi.org/10.1101/gr.2421604.
22. Mullaney JM, Mills RE, Pittard WS, Devine SE. Small insertions and deletions (INDELs) in human genomes. Hum Mol Genet. 2010;19(R2):R131–6.
23. Collins RL, Brand H, Karczewski KJ, et al. A structural variation reference for medical and population genetics. Nature. 2020;581:444–51. https://doi.org/10.1038/s41586-020-2287-8.
24. De Coster W, Van Broeckhoven C. Newest methods for detecting structural variations. Trends Biotechnol. 2019;37(9):973–82. https://doi.org/10.1016/j.tibtech.2019.02.003.
25. Sampaio MM, Santos MLC, Marques HS, Goncalves VLS, Araujo GRL, Lopes LW, et al. Chronic myeloid leukemia-from the Philadelphia chromosome to specific target drugs: a literature review. World J Clin Oncol. 2021;12(2):69–94.
26. Koboldt DC. Best practices for variant calling in clinical sequencing. Genome Med. 2020;12(1):91.
27. Fazzari MJ, Greally JM. Introduction to epigenomics and epigenome-wide analysis. Methods Mol Biol. 2010;620:243–65.
28. Fingerman IM, McDaniel L, Zhang X, Ratzat W, Hassan T, Jiang Z, et al. NCBI Epigenomics: a new public resource for exploring epigenomic data sets. Nucleic Acids Res. 2011;39(Database issue):D908–12.
29. Leroy T, Rougemont Q. Introduction to population genomics methods. Methods Mol Biol. 2021;2222:287–324.
30. Luikart G, Kardos M, Hand BK, Rajora OP, Aitken SN, Hohenlohe PA. Population genomics: advancing understanding of nature. In: Rajora O, editor. Population Genomics. Population genomics. Concepts, approaches and applications. Cham: Springer; 2018. p. 3–79. https://doi.org/10.1007/13836_2018_60.
31. Pool JE, Hellmann I, Jensen JD, Nielsen R. Population genetic inference from genomic sequence variation. Genome Res. 2010;20(3):291–300.
32. SoRelle JA, Wachsmann M, Cantarel BL. Assembling and validating bioinformatic pipelines for next-generation sequencing clinical assays. Arch Pathol Lab Med. 2020;144(9):1118–30.
33. Koster J, Rahmann S. Snakemake-a scalable bioinformatics workflow engine. Bioinformatics. 2018;34(20):3600.
34. Di Tommaso P, Chatzou M, Floden EW, Barja PP, Palumbo E, Notredame C. Nextflow enables reproducible computational workflows. Nat Biotechnol. 2017;35(4):316–9.

# Pharmacokinetics and Pharmacodynamics Analysis

## Contents




J. U. Kazi, *Python Essentials for Biomedical Data Analysis: An Introductory Textbook*,
https://doi.org/10.1007/978-3-031-85600-6_11

Pharmacokinetics (PK) and pharmacodynamics (PD) are key concepts in drug discovery and development that describe how a drug moves through the body and it effects on biological systems. PK focuses on the processes of absorption, distribution, metabolism, and excretion (ADME), while PD deals with the relationship between drug concentration and its therapeutic or toxic effects. Understanding these areas helps to optimize drug dosing, predict therapeutic outcomes, and minimize adverse effects. Several Python tools are widely used for modeling and analyzing PK/PD data. This chapter provides an overview of PK/PD analysis using Python tools.

**Learning Goals**

This chapter aims to provide an overview of pharmacokinetics and pharmacodynamics, including the key concepts of drug absorption, distribution, metabolism, excretion, and the dose-response relationship. Students will develop basic skills to use Python for building PK/PD models, simulating drug behavior, and analyzing pharmacological data. Moreover, this chapter will highlight the importance of PK/PD modeling in drug development and clinical practice.

## 11.1 Pharmacokinetics and Pharmacodynamics

Pharmacokinetics (PK) and pharmacodynamics (PD) are two interconnected fields that describe how drugs interact with the human body. Together, they form the basis for understanding the behavior of drugs from administration to the production of therapeutic or adverse effects [1]. Pharmacokinetics focuses on the movement of drugs within the body, covering the processes of absorption, distribution, metabolism, and excretion (ADME). These processes determine the concentration of the drug in the bloodstream and tissues over time, which is required for optimizing drug dosing and minimizing toxicity. Pharmacodynamics, on the other hand, examines the biological effects of the drug at its site of action, often expressed as the relationship between drug concentration and its pharmacological response. PD provides insight into the efficacy and potency of a drug, as well as its possible therapeutic and toxic effects.

### 11.1.1 Pharmacokinetics

Pharmacokinetics is the study of the time course of drug absorption, distribution, metabolism, and excretion, often summarized by the acronym ADME [2]. Understanding these processes helps to determine how a drug reaches its target tissue, how long it remains in the body, and how it is eliminated. This knowledge informs the design of appropriate dosing regimens, maximizes therapeutic efficacy, and minimizes the risk of adverse effects.

*Absorption* refers to the process by which a drug enters the bloodstream from the site of administration, the step that determines the onset of the drug's action [3]. Several factors influence drug absorption, including the route of administration. Drugs can be administered orally, intravenously, intramuscularly, subcutaneously, transdermally, or via inhalation, each with unique absorption characteristics. For example, oral administration is convenient but may have variable absorption due to factors like gastrointestinal pH and the presence of food; intravenous administration ensures immediate and complete absorption by delivering the drug directly into the bloodstream; and transdermal patches provide slow, sustained absorption through the skin. Furthermore, the physicochemical properties of the drug, such as solubility, molecular size, ionization state, and lipophilicity, affect its ability to cross biological membranes. Formulation factors, like the design of the drug product in forms such as tablets, capsules, suspensions, or controlled-release formulations, can also influence the rate and extent of absorption. Lastly, physiological factors like blood flow to the absorption site, gastric emptying time, intestinal motility, and the presence of other substances can impact the efficiency of absorption.

*Distribution* describes how a drug disperses throughout the body's fluids and tissues after entering the systemic circulation, influenced by several key factors [4]. Organs with high blood flow, such as the liver, kidneys, and brain, typically receive more of the drug shortly after absorption, demonstrating the impact of blood flow and tissue perfusion. Drugs may also bind reversibly to plasma proteins like albumin; this bound fraction is usually inactive, serving as a reservoir from which the drug can be gradually released, illustrating the role of plasma protein binding. Furthermore, some drugs accumulate in specific tissues like fat or bone due to their affinity for certain tissue components, which affects both the duration and intensity of their action. Physiological barriers, such as the blood-brain barrier, restrict drug access to certain tissues, with only sufficiently lipophilic drugs or those with specific transport mechanisms able to cross these barriers. Finally, the distribution characteristics of a drug are also dictated by its properties; lipid-soluble drugs tend to distribute more widely than water-soluble drugs and can reach intracellular compartments.



*Metabolism* involves the chemical alteration of a drug by the body, primarily through enzymatic reactions in the liver, with additional contributions from organs like the kidneys and intestines [5]. The main purposes of drug metabolism are to enhance excretion by converting lipophilic drugs into more hydrophilic metabolites that are easily excreted by the kidneys, and to deactivate the drug by reducing or eliminating its pharmacological activity, although some metabolites may retain activity or even become toxic. Metabolic reactions are categorized into two phases: Phase I reactions, which include oxidation, reduction, and hydrolysis that introduce or expose functional groups on the drug molecule with cytochrome P450 enzymes playing a role, and Phase II reactions, which involve the conjugation of the drug or its Phase I metabolites with endogenous substances such as glucuronic acid, sulfate, or glycine to form more water-soluble conjugates [6, 7]. Several factors influence metabolism, including genetic variability where polymorphisms in metabolic enzymes like CYP450 isoforms can cause interindividual differences in drug metabolism rates; age and gender, as metabolic capacity can vary with age, with neonates and the elderly having reduced enzyme activity, and hormonal differences affecting metabolism; disease states such as hepatitis or cirrhosis which can impair metabolic functions; and drug interactions, where the concurrent administration of certain drugs can induce or inhibit metabolic enzymes, altering the metabolism of other drugs, such as enzyme inducers like rifampin or inhibitors like ketoconazole.

*Excretion*: Excretion is the process of removing the drug and its metabolites from the body, with several primary routes involved [8]. Renal excretion through the kidneys is the most common pathway, which includes mechanisms such as glomerular filtration, active tubular secretion, and passive reabsorption [9]. Factors such as urine pH and overall renal function influence excretion rates. Biliary excretion via the liver involves the secretion of drugs and metabolites into bile, followed by elimination in feces. This process can include enterohepatic recirculation, where the conjugated drug is deconjugated by intestinal flora and reabsorbed, extending the drug's presence in the body [10]. Other excretion routes include exhalation of volatile compounds, as well as elimination through sweat, saliva, tears, and breast milk, each of which can carry clinical implications for drug safety, particularly in breastfeeding. Impaired excretion, whether due to renal or hepatic dysfunction, can lead to drug accumulation and toxicity, often necessitating dose adjustments in affected patients to mitigate adverse effects.

### 11.1.2 Pharmacodynamics

Pharmacodynamics is the branch of pharmacology that studies the biochemical and physiological effects of drugs on the body, as well as the mechanisms of their action at the molecular and cellular levels [11]. It examines how a drug interacts with its target sites, typically receptors, enzymes, ion channels, or transport proteins, to produce therapeutic effects or adverse reactions. Understanding pharmacodynamics helps to optimize drug therapy, predict therapeutic outcomes, and minimize side effects.

*Drug-receptor interactions*: Most drugs exert their effects by binding to specific receptors, which are proteins located on the surface or inside cells that recognize and respond to growth factors, hormones or neurotransmitters [12, 13]. When a drug binds to a receptor, it can either mimic the action of the natural ligand as an agonist, block the action as an antagonist, or modulate the receptor activity in other ways, such as a partial agonist or inverse agonist. Efficacy refers to the maximum effect a drug can produce, independent of the dose; a drug with high efficacy can elicit a strong response upon binding to its receptor [14]. Potency, on the other hand, indicates the amount of drug needed to produce a specific effect, with more potent drugs achieving the desired effect at lower concentrations. Drug-receptor interactions include agonists, which activate receptors to produce a biological response; antagonists, which bind to receptors but do not activate them, thereby blocking the action of agonists; partial agonists, which activate receptors but produce a weaker response than full agonists; and inverse agonists, which bind to receptors and induce the opposite effect of an agonist. Theories explaining these interactions include the Lock and Key model, which suggests a high specificity where the drug (key) fits precisely into the receptor (lock), and the Induced Fit model, which proposes that the drug binding induces a conformational change in the receptor, enhancing binding affinity [15].

*Dose-response relationships*: Dose-response relationships describe how the magnitude of a drug's effect varies with its concentration or dose that help in understanding the therapeutic and toxic effects of drugs and determining the optimal dosing regimen [16]. Graded dose-response curves illustrate the response of a single biological unit, such as an organ or receptor, to varying drug doses, typically displaying a sigmoidal shape that reflects three phases: subtherapeutic (no effect), therapeutic (increasing effect), and a plateau (maximum effect) [17]. Quantal dose-response curves, on the other hand, depict the distribution of responses to different doses across a population and are useful for determining the dose at which a certain percentage of the population shows a specified response [18]. Key parameters in dose-response relationships include the threshold dose, which is the minimum dose required to produce a measurable effect; the ceiling effect, which is the dose beyond which no further increase in effect is observed; and the slope of the curve, which indicates the responsiveness of the effect to changes in dose.

*Key pharmacodynamic parameters*: Key pharmacodynamic parameters include $EC_{50}$ (Effective Concentration 50), $E_{max}$, the therapeutic window, and therapeutic index.

$EC_{50}$ is the concentration of a drug that produces 50% of its maximum effect and is used to compare the potencies of drugs acting on the same receptor. $E_{max}$ represents the maximum effect achievable with a drug, reflecting its efficacy. Drugs with the same $E_{max}$ but different $EC_{50}$ values have the same efficacy but different potencies.

The therapeutic window is the range of drug concentrations between the minimum effective concentration (MEC) and the minimum toxic concentration (MTC). Maintaining drug levels within this window maximizes efficacy while minimizing toxicity.

The therapeutic index (TI) is a ratio that compares the toxic dose of a drug to its effective dose, providing an estimate of its safety margin.



$$TI = \frac{TD_{50}}{ED_{50}}$$

where,

$TD_{50}$ = The dose that produces a toxic effect in 50% of the population, Toxic Dose 50.

$ED_{50}$ = The dose that produces the desired therapeutic effect in 50% of the population, Effective Dose 50.

A higher TI indicates a wider safety margin, meaning there is a larger gap between effective and toxic doses. Drugs with a low TI (e.g., digoxin, warfarin) require careful dosing and monitoring [19].

*Pharmacodynamic models*: Pharmacodynamic models describe the relationship between drug concentration at the site of action and the resulting effect. These models help predict the time course and intensity of drug effects, aiding in dose optimization. The linear models assume a direct proportionality between drug concentration and effect. Such models are suitable for drugs with effects that change linearly with concentration. On the other hand, the $E_{max}$ model describes a hyperbolic relationship where the effect approaches a maximum ($E_{max}$) as concentration increases [20].

*Factors affecting pharmacodynamics*: Various factors can influence a drug's pharmacodynamic properties, altering its efficacy and safety. Age is a factor, with pediatric patients often requiring dose adjustments due to different receptor sensitivity and expression, while elderly patients may experience age-related changes in receptor function that affect drug response [21]. Disease states can also alter pharmacodynamics; for example, conditions like renal or hepatic impairment can change receptor sensitivity, and diseases may upregulate or downregulate receptor expression. Genetic variability is another factor; pharmacogenomics explores how genetic differences influence drug response, with variations in genes encoding receptors or downstream signaling proteins affecting drug efficacy [22]. Drug interactions can lead to synergistic effects, where two drugs enhance each other's effects, or antagonistic effects, where one drug reduces the effect of another, and altered receptor sensitivity, where chronic drug use can result in tolerance (desensitization) or sensitization. Furthermore, tolerance, defined as a gradual decrease in response to a drug over time requiring higher doses, and tachyphylaxis, a rapid decrease in response after repeated doses in a short period, further complicate pharmacodynamics.

## 11.2 Data Collection and Types of Data

Accurate data collection is required for pharmacokinetics (PK) and pharmacodynamics (PD) analysis because it provides the information needed to model drug behavior and its effects on the body. High-quality PK/PD data helps to understand how a drug is absorbed, distributed, metabolized, and excreted (the ADME processes), as well as how it interacts with biological targets to produce therapeutic effects or adverse reactions. These datasets are often gathered from clinical trials, preclinical animal studies, and in vitro experiments. In this section, we will explore the sources of PK/PD data, the types of data typically collected, and the importance of time-course measurements in PK/PD modeling.

### 11.2.1 Sources of PK/PD Data

PK/PD data can come from several sources depending on the stage of drug development and specific research objectives. Common sources include human clinical trials, preclinical animal studies, and in vitro experiments. Clinical trials are the primary source of PK/PD data, providing information on how a drug behaves in human subjects [23, 24]. Phase I trials are conducted with a small group of healthy volunteers and focus on assessing the drug's safety profile, tolerability, and basic pharmacokinetics. Researchers collect data on absorption rates, distribution patterns, metabolism, excretion, and preliminary PD effects, helping to determine safe dosage ranges and identify side effects. Phase II trials involve a larger group of participants with the condition the drug aims to treat and evaluate the drug's efficacy and further assess its safety. PK/PD data collected in Phase II trials include dose-response relationships, optimal dosing regimens, and therapeutic effects, refining dosing strategies and confirming the drug's effectiveness. Phase III trials are large-scale studies that provide data on the drug's efficacy and monitor adverse reactions across diverse populations. PK/PD data from Phase III trials help confirm therapeutic benefits, detect less common side effects, and gather information on long-term use. Phase IV trials, or post-marketing surveillance, are conducted after a drug is approved and marketed, continuing to collect data on its effectiveness and safety in the general population. This phase monitors rare or long-term adverse effects, drug interactions, and gathers additional PK/PD data that may inform future research or regulatory actions.

Preclinical animal studies are conducted before initiating human trials and involve testing drugs in animal models to assess their PK/PD properties and safety profiles. Commonly used animals include rodents (mice and rats), rabbits, dogs, and nonhuman primates. These studies help predict human responses, identify toxicities, and outline mechanisms of action. Data collected can include organ-specific distribution, metabolic pathways, and preliminary efficacy in disease models.

In vitro studies involve studying the drug in controlled laboratory settings using isolated cells, tissues, or biochemical assays. Examples include cell culture assays, which are used to evaluate cytotoxicity, cellular uptake, and metabolic stability; receptor binding studies, which assess the affinity and specificity of the drug for its target receptors, required for understanding its pharmacodynamics; enzyme inhibition or induction tests, which determine how the drug interacts with metabolic enzymes, predicting possible drug-drug interactions; and transporter assays, which examine how the drug is transported across cell membranes, influencing absorption and distribution. While in vitro studies cannot replicate the complexity of a whole organism, they are useful for elucidating molecular mechanisms, screening drug candidates, and predicting pharmacological responses.

### 11.2.2 Types of Data



Different types of data are collected depending on the specific aspects of PK or PD being studied. Plasma concentration data is required for PK studies and involves measuring the concentration of the drug in blood plasma at various time points following administration. Plasma concentration-time profiles help determine key PK parameters such as the absorption rate constant ($K_a$), which indicates how quickly the drug enters systemic circulation; maximum concentration ($C_{max}$), the peak plasma concentration achieved, important for efficacy and toxicity assessments; time to reach $C_{max}$ ($T_{max}$), which helps in understanding the onset of action; elimination half-life ($t_{1/2}$), the time it takes for the plasma concentration to reduce by half, influencing dosing intervals; area under the curve (AUC), which represents the total drug exposure over time, correlating with efficacy and toxicity; clearance ($Cl$), the rate at which the drug is eliminated from the body, helping to determine maintenance doses; and volume of distribution ($V_d$), which reflects how extensively the drug disperses into body tissues.

Urinary excretion data involves measuring the amount of drug and its metabolites excreted in urine, helping to describe renal elimination processes. This data helps assess renal clearance, the efficiency of the kidneys in eliminating the drug; fraction excreted unchanged ($f_e$), the proportion of the administered dose eliminated unchanged, indicating the extent of metabolism; and metabolite identification, which involves detecting and quantifying metabolites to understand metabolic pathways and pharmacologically active or toxic metabolites.

Receptor occupancy data in PD studies assesses how much of the drug binds to its target receptors at various concentrations. Techniques such as positron emission tomography (PET) imaging can quantify receptor occupancy in vivo [25]. Receptor occupancy data is vital for determining effective dose ranges, establishing the minimum concentration needed for therapeutic effects, understanding saturation kinetics, identifying doses beyond which no additional therapeutic benefit occurs due to receptor saturation, and predicting therapeutic side effects by correlating receptor occupancy with clinical outcomes.

Biomarker data involves measuring biological markers to evaluate pharmacodynamic responses. Examples include biochemical markers, such as levels of enzymes, hormones, or other proteins that indicate biological activity; genomic and proteomic data, which involves changes in gene or protein expression profiles in response to the drug; and physiological parameters, such as blood pressure, heart rate, glucose levels, or tumor size in oncology studies. Adverse event data involves recording and analyzing side effects and adverse reactions to understand the safety profile, which is often correlated with PK parameters to assess dose-related toxicities.

Collecting these different types of data allows better understanding of the drug's pharmacokinetic and pharmacodynamic behavior. It facilitates the development of accurate models to predict clinical outcomes, optimize dosing regimens, and improve therapeutic efficacy while minimizing risks.

### 11.2.3 Time-Course Data in PK/PD

Time-course data, which involves measuring drug concentrations and biological responses at multiple time points, are necessary for PK/PD modeling [26]. Tracking how drug levels and effects change over time helps define key parameters and supports clinical decision-making. Time-course data provide information on absorption rates, showing how quickly a drug enters systemic circulation. Absorption can be affected by factors such as the drug formulation, route of administration (oral, intravenous, transdermal, etc.), and physiological conditions such as gastric pH and intestinal motility. Accurate absorption profiles help predict the onset of action and bioavailability. Identifying the peak concentration ($C_{max}$) is needed for efficacy, as therapeutic effects often correlate with $C_{max}$, and for safety assessments, as higher concentrations may increase the risk of adverse effects. The elimination half-life ($t_{1/2}$) determines how long the drug stays in the system and influences dosing frequency, with short half-life drugs often requiring multiple daily doses, while long half-life drugs may be dosed less frequently.

Time-course pharmacodynamic data help to define the onset of effect, or how quickly the drug begins to exert its therapeutic action, and the duration of effect, which determines how long the drug remains effective. This information is used to compare drugs can be prescribed for acute conditions versus those for chronic management. With repeated dosing, drugs may accumulate in the body, and time-course data can predict when steady-state concentrations are achieved, which is particularly important for drugs with narrow therapeutic windows. Some drugs may show a lag time or delayed effects between plasma concentration peaks and therapeutic responses due to slow distribution to the site of action or complex mechanisms, and time-course modeling helps to understand these delays [26, 27].

Time-course data provide the dynamic context needed for PK/PD analysis, enabling researchers to construct pharmacokinetic models using compartmental or noncompartmental approaches to describe drug movement through the body over time. It also helps establish pharmacodynamic relationships by correlating concentration data with observed effects to develop concentration-effect models. These data supports the optimization of therapeutic regimens, determining appropriate dosing intervals and amounts to maintain drug concentrations within the therapeutic window. It also allows researchers to assess variability in drug response due to factors such as genetics, age, gender, disease states, or concomitant medications and predict outcomes in special populations like children, the elderly, or patients with renal or hepatic impairment.

## 11.3 Python Libraries for PK/PD Analysis

Python tools are often used in PK and PD analysis to simplify data manipulation, mathematical modeling, and visualization. Python's versatility allows efficient handling of large datasets, performing complex calculations, and simulating drug behavior in biological systems. In this section, we will discuss several Python libraries that are commonly used for PK/PD analysis, including tools for data manipulation, modeling, and visualization.



### 11.3.1 NumPy and SciPy for Mathematical Modeling

NumPy and SciPy are particularly useful in PK/PD modeling for handling large datasets, performing mathematical operations, and solving differential equations involved in pharmacokinetic models.

*NumPy*: Provides support for multidimensional arrays, matrix operations, and a wide range of mathematical functions. In PK/PD modeling, NumPy is used to store and manipulate drug concentration data, calculate pharmacokinetic parameters, and efficiently perform vectorized operations on time-course data.

*SciPy*: Extends the capabilities of NumPy by providing additional tools for optimization, integration, and solving ordinary differential equations (ODEs). Many PK models involve differential equations (e.g., compartmental models), and SciPy's ODE solvers allow these models to be implemented easily.

## 11.4 Pharmacokinetic Modeling Using Python

PK modeling helps to describe and predict how drugs are absorbed, distributed, metabolized, and eliminated from the body. These models provide a quantitative framework for understanding the time course of drug concentrations in plasma and tissues, allowing researchers to optimize dosing regimens and predict therapeutic and toxic effects [28]. Python offers a range of tools to implement and simulate PK models, facilitating visualization and understanding drug behavior effectively.

In this section, we will discuss compartmental PK models, as well as modeling of absorption, distribution, metabolism, and excretion (ADME). We will demonstrate how to simulate first-order and zero-order kinetics using Python. We will provide example code to illustrate implementation of PK models.

### 11.4.1 Compartmental Models

Compartmental models represent the body as one or more interconnected compartments where the drug concentration is assumed to be uniformly distributed within each compartment [29–31]. These models simplify complex biological processes into manageable mathematical representations. The complexity of a compartmental model depends on the number of compartments used:

*One-compartment model*: The simplest model treats the entire body as a single homogeneous compartment [32]. After administration, the drug is absorbed into this compartment, distributed uniformly, and then eliminated. This model is suitable for drugs that rapidly equilibrate throughout the body.

Two-compartment model: This model divides the body into a central compartment (e.g., blood plasma and highly perfused organs) and a peripheral compartment (e.g., muscle and fat tissues) [33]. The drug moves between these compartments at different rates, allowing for a more accurate representation of drugs with slower distribution phases.

Multicompartment models: For drugs that distribute unevenly or have complex kinetics, models with more than two compartments are used [34]. Each additional compartment represents a group of tissues with similar pharmacokinetic properties.

### 11.4.2 Absorption, Distribution, Metabolism, and Excretion (ADME) Modeling

ADME modeling focuses on the processes that determine the fate of a drug in the body. These include:

*Absorption*: Absorption is the process by which a drug enters systemic circulation. This can be modeled using first-order kinetics where the absorption rate is proportional to the concentration of the drug or zero-order kinetics where absorption occurs at a constant rate.

*Distribution*: The dispersion of the drug throughout body fluids and tissues. Factors influencing distribution include blood flow, tissue permeability, and binding to plasma proteins.

*Metabolism*: The biochemical modification of the drug, primarily in the liver, transforming it into metabolites that are usually more water-soluble for easier excretion.

*Excretion*: The removal of the drug and its metabolites from the body, mainly through the kidneys (in urine) or biliary system (in feces).

Modeling these processes allows for the simulation of drug concentration-time profiles under different conditions.

### 11.4.3 First-Order and Zero-Order Kinetics

**Notebook: Section 11.4.3. First-Order and Zero-Order Kinetics**

This notebook provides code examples demonstrating the first-order and zero-order kinetics.

Link to the GitHub repository:

► https://github.com/sn-code-inside/BioPy

Go to: Chap. 11—► Sect. 11.4.3

The kinetics of drug absorption and elimination can be modeled using different types of equations:

*First-order kinetics*: The rate of drug absorption or elimination is proportional to the drug concentration. This type of kinetics is common for most drugs. Mathematically, it is expressed as:



$$\frac{dC}{dt} = -k \bullet C$$

where $C$ is the drug concentration, $t$ is the time and $k$ is the rate constant.

By solving the equation, we get:

$$C(t) = C_0 \bullet e^{-kt}$$

Where $C_0$ is the initial concentration.

Simulating first-order elimination demonstrates an exponential decay over time (◘ Fig. 11.1).

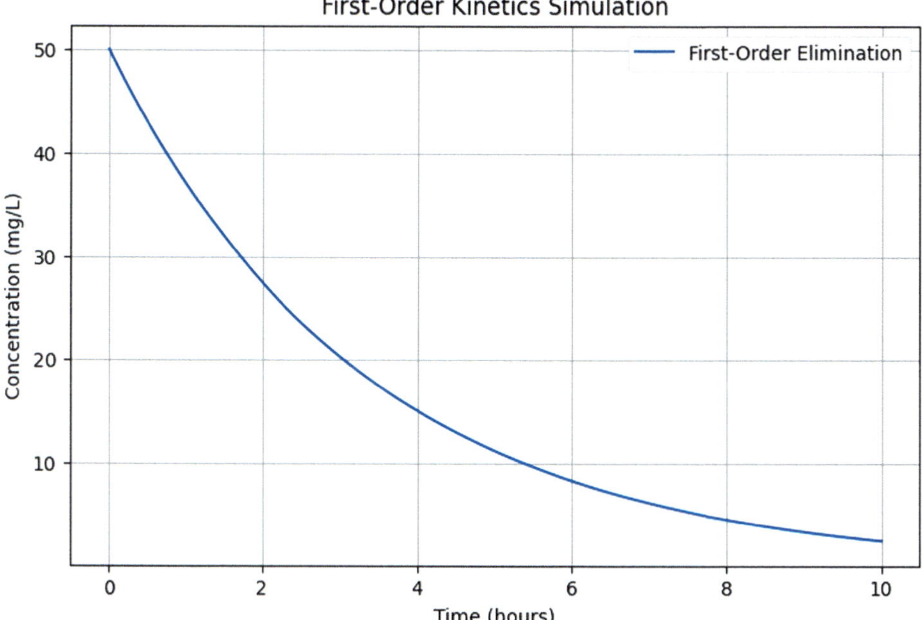


**Fig. 11.1** Simulation of first-order elimination kinetics showing an exponential decrease in drug concentration over time. The curve represents the decay of drug concentration (mg/L) in the central compartment as a function of time (hours) under first-order elimination. The rate of elimination is proportional to the remaining drug concentration, resulting in a smooth exponential decline

Codes:

```
import numpy as np
import matplotlib.pyplot as plt

# Parameters
kel = 0.3    # Elimination rate constant (per hour)
C0 = 50      # Initial concentration (mg/L)
time = np.linspace(0, 10, 100)

# Calculate concentration over time
concentration = C0 * np.exp(-kel * time)

# Plotting
plt.figure(figsize=(8, 5))
plt.plot(time, concentration, label='First-Order Elimination')
plt.xlabel('Time (hours)')
plt.ylabel('Concentration (mg/L)')
plt.title('First-Order Kinetics Simulation')
plt.legend()
plt.grid(True)
plt.show()
```

*Zero-order kinetics*: The rate of drug absorption or elimination is constant and independent of the drug concentration [35]. This type of kinetics occurs when the rate-limiting process is saturated (e.g., when drug absorption is limited by the formulation). It is expressed as:

$$\frac{dC}{dt} = -k$$

The solution of zero-order kinetics is

$$C(t) = C_0 - kt$$

Zero-order kinetics are less common but occur in processes like drug infusion at a constant rate or when metabolic pathways are saturated.

Simulating zero-order elimination demonstrated linear decay (◘ Fig. 11.2).

```
import numpy as np
import matplotlib.pyplot as plt
# Parameters
k0 = 5       # Zero-order rate constant (mg/L per hour)
C0 = 50      # Initial concentration (mg/L)
time = np.linspace(0, 10, 100)

# Calculate concentration over time
concentration = C0 - k0 * time
concentration[concentration < 0] = 0
# Concentration cannot be negative

# Plotting
plt.figure(figsize=(8, 5))
plt.plot(time, concentration, label='Zero-Order Elimination')
plt.xlabel('Time (hours)')
plt.ylabel('Concentration (mg/L)')
plt.title('Zero-Order Kinetics Simulation')
plt.legend()
plt.grid(True)
plt.show()
```



### 11.4.4 Implementation of PK Models

**Notebook: Section 11.4.4. Implementation of PK Models**

This notebook provides code examples demonstrating the implementation of PK models.

Link to the GitHub repository:

► https://github.com/sn-code-inside/BioPy

Go to: Chap. 11—► Sect. 11.4.4

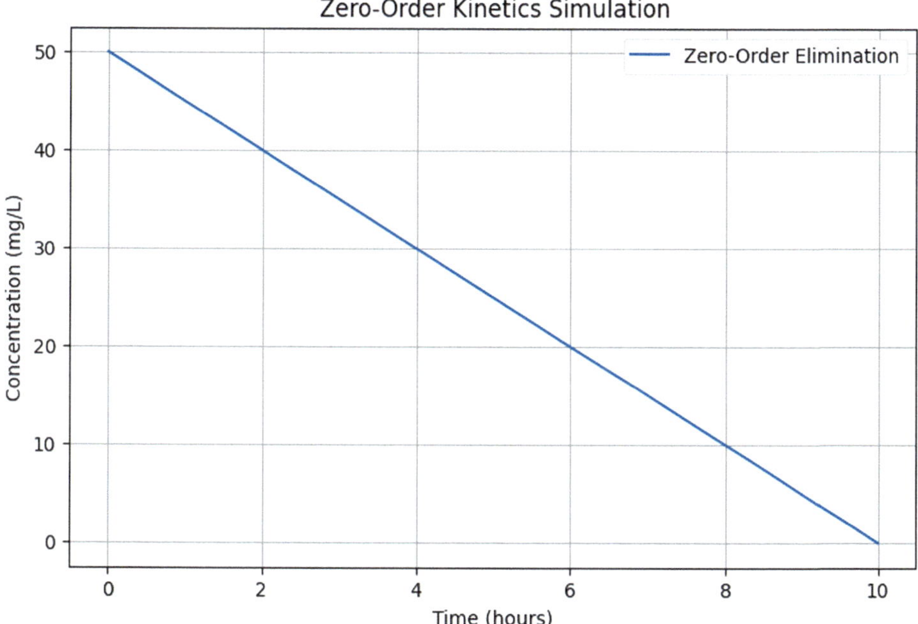


**Fig. 11.2** Simulation of zero-order elimination kinetics showing a linear decrease in drug concentration over time. The curve represents the drug concentration (mg/L) as a function of time (hours) under zero-order elimination, where the rate of drug elimination remains constant and does not depend on the remaining drug concentration. This linear decay is characteristic of zero-order processes, such as saturation kinetics or elimination via a fixed-capacity mechanism

Here, we will implement a simplified one-compartment pharmacokinetic (PK) model with first-order absorption and elimination using Python. This model simulates the concentration-time profile of a drug administered orally, considering the processes of absorption into the bloodstream and elimination from the body, both following first-order kinetics.

As the one-compartment model assumes that the body acts as a single, homogeneous compartment where the drug concentration is uniformly distributed, for oral administration, the drug is first absorbed from the gastrointestinal (GI) tract into the systemic circulation and then eliminated.

The rate of change of drug concentration in the central compartment (plasma) is described by the following differential equations:

Absorption phase:

$$\frac{dA_{\text{gut}}}{dt} = -k_a \bullet A_{\text{gut}}$$

where,

$A_{\text{gut}}$ is the amount of drug remaining in the GI tract.

$k_a$ is the first-order absorption rate constant.

Elimination phase:

$$\frac{dC_{\text{plasma}}}{dt} = \frac{k_a \bullet A_{\text{gut}}}{V_d} - k_e \bullet C_{\text{plasma}}$$

where,

$C_{\text{plasma}}$ is the drug concentration in plasma.

$V_d$ is the volume of distribution.

$k_e$ is the first-order elimination rate constant.

We assume that both absorption and elimination processes follow first-order kinetics, a single dose of drug is administrated, and the entire dose is available for absorption (bioavailability $F = 1$).

Example Code: One-compartment model with oral administration

```
import numpy as np
from scipy.integrate import odeint
import matplotlib.pyplot as plt


# Define the one-compartment model with first-order absorption and
# elimination
def one_compartment_model(y, t, ka, ke):
    A_gut, C_plasma = y
    dA_gut_dt = -ka * A_gut
    # Rate of change in the gut compartment
    dC_plasma_dt = (ka * A_gut) / Vd - ke * C_plasma
    # Rate of change in the plasma compartment
    return [dA_gut_dt, dC_plasma_dt]

# Parameters for the model
Dose = 500          # Dose in mg
ka = 1.0            # Absorption rate constant (1/hour)
ke = 0.5            # Elimination rate constant (1/hour)
Vd = 50             # Volume of distribution (L)

# Initial conditions
A_gut0 = Dose       # Initial amount in the gut (mg)
C_plasma0 = 0       # Initial plasma concentration (mg/L)
y0 = [A_gut0, C_plasma0]

# Time points (hours)
time = np.linspace(0, 24, 1000)

# Solve the differential equations using odeint
solution = odeint(one_compartment_model, y0, time, args=(ka, ke))
A_gut = solution[:, 0]
C_plasma = solution[:, 1]

# Plot the concentration vs. time curve
plt.figure(figsize=(10, 6))
plt.plot(time, C_plasma, label='Plasma Concentration')
plt.xlabel('Time (hours)')
```

```
plt.ylabel('Concentration (mg/L)')
plt.title('One-Compartment PK Model with Oral Administration')
plt.legend()
plt.grid(True)
plt.show()
```

In this example, the plot generated illustrates the plasma concentration of the drug over a 24-hour period following a single oral dose (◘ Fig. 11.3). In the absorption phase the initial rise in plasma concentration as the drug is absorbed from the GI tract. The highest plasma concentration is achieved, occurring when the rate of absorption equals the rate of elimination. Then in the elimination phase, the decline in plasma concentration as elimination surpasses absorption (which diminishes over time).

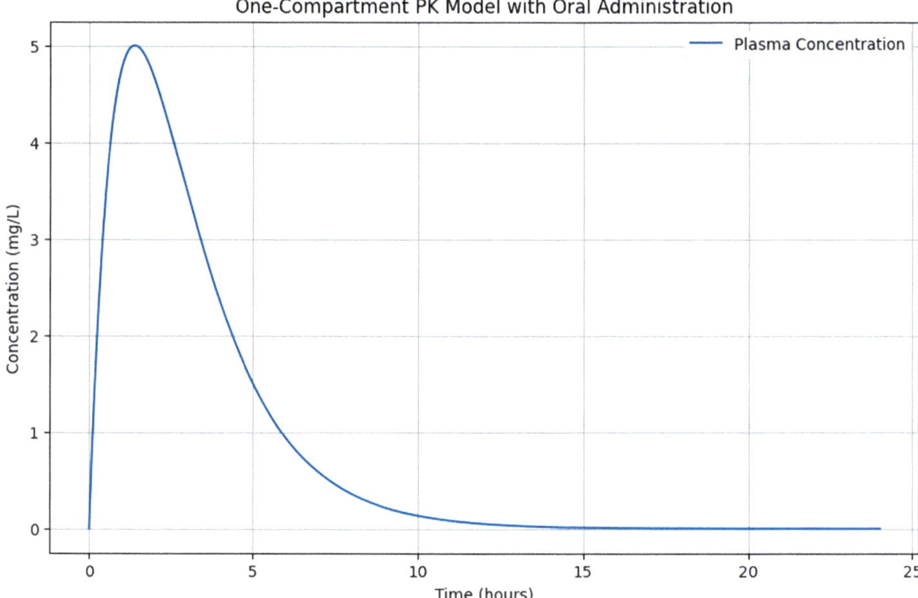


◘ **Fig. 11.3** Simulated plasma concentration-time profile for a one-compartment PK model with oral administration and first-order absorption and elimination. The curve shows an initial rise in plasma concentration as the drug is absorbed, reaching a peak concentration ($C_{max}$), followed by a decline due to elimination processes. This characteristic profile reflects the dynamics of drug absorption, distribution, and elimination in a one-compartment system. The x-axis represents time (hours), while the y-axis shows plasma drug concentration (mg/L)

## 11.5 Pharmacodynamic Modeling Using Python

PD examines the relationship between drug concentration and its biological effects. While PK describes how the drug concentration changes over time in the body, PD focuses on the drug's mechanism of action and how it produces therapeutic or toxic effects at various concentrations. PD modeling helps in understanding dose-response relationships and optimize drug efficacy while minimizing side effects. In this section, we will discuss concepts such as dose-response curves, $EC_{50}$, $E_{max}$, and Hill coefficients. We will also describe how to implement pharmacodynamic models using Python to simulate and analyze these relationships.

### 11.5.1 Dose-Response Curves

The dose-response curve is widely used in pharmacodynamics, illustrating how the biological effect of a drug varies with changes in drug concentration or dose. This curve typically shows the relationship between the drug dose (or concentration) and the observed effect, which helps to determine the optimal dose that produces the desired therapeutic effect without causing toxicity.

Key parameters in dose-response relationships include the following:

- $EC_{50}$ (Effective Concentration 50): The concentration of the drug that produces 50% of the maximum effect. It is often used as a measure of drug potency.
- Emax: The maximum effect that can be achieved with the drug, regardless of the dose. It represents the upper limit of the dose-response curve.
- Hill coefficient: Describes the steepness of the dose-response curve and reflects the cooperativity of drug binding to its target.



The standard dose-response curve may follow a sigmoidal shape, with a low response at low concentrations, a rapid increase in response over a certain range of concentrations, and a plateau at higher concentrations ($E_{max}$).

### 11.5.2 Linear Versus Nonlinear Models in PD

Pharmacodynamic responses can be modeled using either linear or nonlinear models. In practice, most dose-response relationships are nonlinear, especially at higher doses where the effect plateaus (due to saturation of receptors or pathways). Nonlinear models, such as the Hill equation, are commonly used to describe the sigmoidal shape of the dose-response curve.

The Hill equation is a classic model for dose-response relationships,

$$\text{Observed effect}, E = \frac{E_{max} \times C}{EC_{50} + C}$$

where $C$ is the concentration of the drug. The equation is referred to as the 2-parameter Hill equation on a linear scale where Hill coefficient is ignored or assumed as 1. When the drug effect is plotted against the concentration on a linear scale, with an $EC_{50}$ of 10, it produces a classic hyperbolic curve (Fig. 11.4).

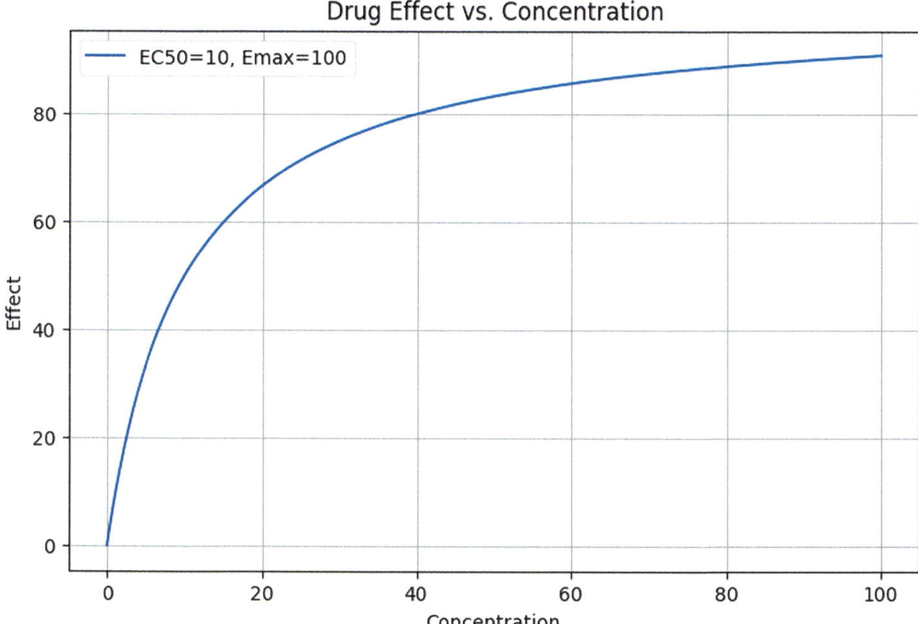


**Fig. 11.4** The dose-response curve shows the relationship between drug concentration and its effect. The curve represents a hyperbolic response where the effect increases rapidly at low concentrations and approaches a maximum ($E_{max}$) of 100 as the system saturates. The $EC_{50}$, the drug concentration required to achieve 50% of the maximum effect, is 10. This hyperbolic behavior is characteristic of noncooperative ligand-receptor interactions and follows the 2-parameter Hill equation. The x-axis represents the drug concentration, while the y-axis indicates the corresponding effect

The 2-parameter Hill equation can be transformed into its $\log_{10}$ form. In this case, the curve becomes sigmoidal, which is widely used for data visualization and analysis (Fig. 11.5).

$$E = \frac{E_{max}}{1+10^{\left(\log_{10}\mathrm{EC}_{50}-\log_{10}C\right)}}$$

The 2-parameter Hill equation can be extended to the 3-parameter Hill equation by introducing the Hill coefficient ($s$), which defines the slope of the curve and reflects the degree of cooperativity among ligand binding sites. When $s > 1$, the curve becomes steeper, indicating positive cooperativity, while $s < 1$ results in a shallower curve, indicating negative cooperativity (Fig. 11.6).

$$E = \frac{E_{max}}{1+10^{s\left(\log_{10}\mathrm{EC}_{50}-\log_{10}C\right)}}$$

Furthermore, an additional parameter, $E_{min}$, is introduced to the 3-parameter Hill equation to account for the baseline response. In this case, the equation is referred to as the 4-parameter Hill equation (Fig. 11.7).

$$E = E_{min} + \frac{E_{max} - E_{min}}{1+10^{s\left(\log_{10}\mathrm{EC}_{50}-\log_{10}C\right)}}$$

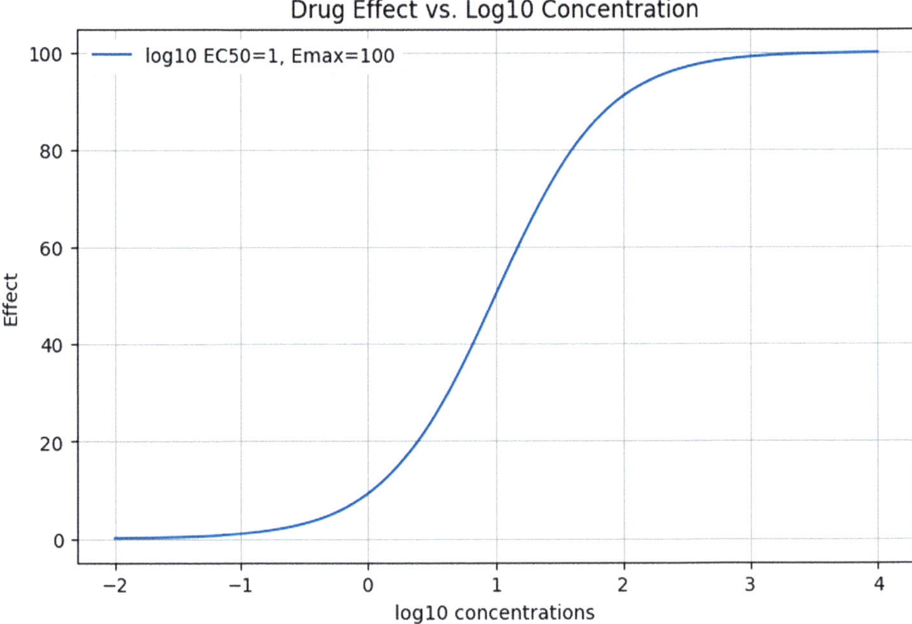


**Fig. 11.5** The dose-response curve shows the relationship between the drug effect and the logarithm ($\log_{10}$) of the drug concentration. The curve demonstrates a sigmoidal shape, which is characteristic of log-transformed dose-response data. The maximum effect ($E_{max}$) is 100, and the $\log_{10}$-transformed $EC_{50}$ is 1, corresponding to a concentration of 10 in a linear scale. The x-axis represents the $\log_{10}$-transformed drug concentrations, while the y-axis indicates the effect. This transformation allows better visualization of the response across a wide range of concentrations



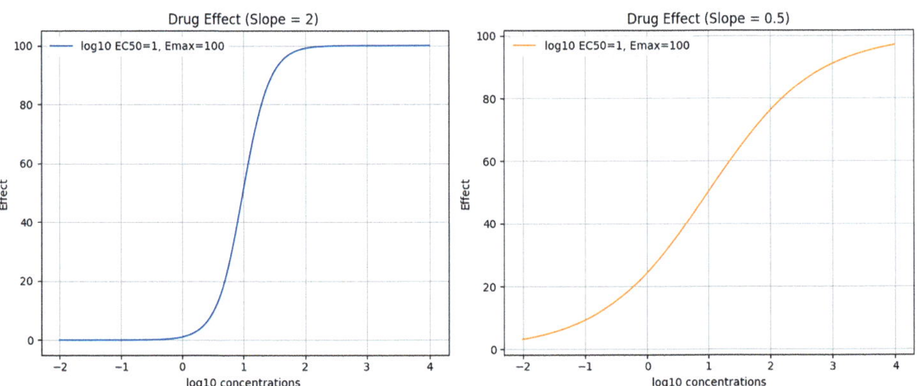


**Fig. 11.6** Dose-response curves showing the relationship between drug effect and log10-transformed drug concentrations at two different Hill slopes. The left panel depicts a curve with a Hill slope of 2, resulting in a steeper transition between low and high effects, indicative of high cooperativity. The right panel shows a curve with a Hill slope of 0.5, resulting in a shallower transition, indicative of low cooperativity. In both cases, the maximum effect ($E_{max}$) is 100, and the $\log_{10}$-transformed $EC_{50}$ is 1, corresponding to an $EC_{50}$ of 10 in linear scale. These sigmoidal curves demonstrate the impact of varying Hill slopes on the steepness of the dose-response relationship

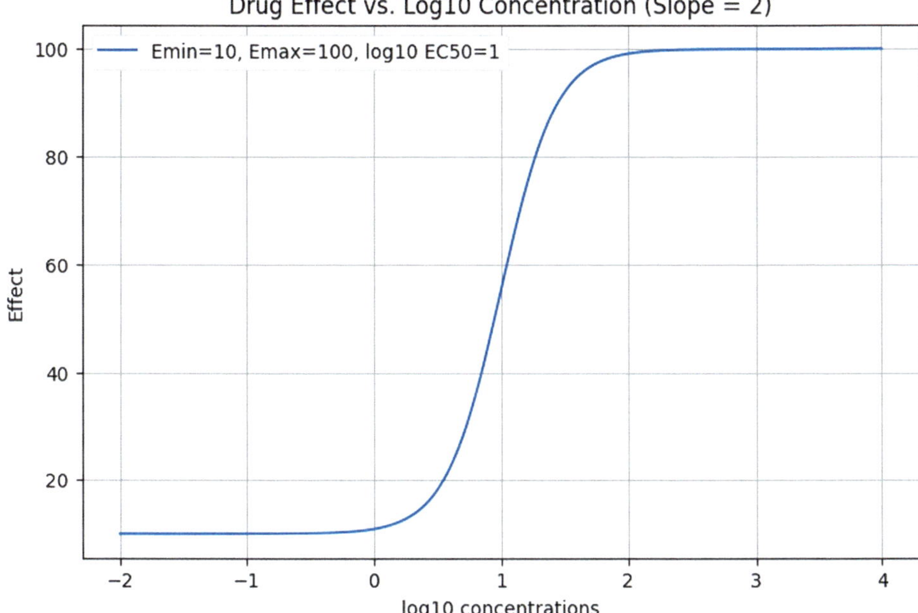


**Fig. 11.7** The dose-response curve was generated using the 4-parameter Hill equation, showing the relationship between drug effect and log10-transformed drug concentrations. The curve starts at a baseline effect ($E_{min}$) of 10 and asymptotically approaches the maximum effect ($E_{max}$) of 100. The $\log_{10}$ $EC_{50}$ is 1, corresponding to an $EC_{50}$ of 10 in linear scale, and the Hill slope is 2, resulting in a steep sigmoidal transition. The x-axis represents the $\log_{10}$-transformed drug concentrations, while the y-axis indicates the corresponding effect

## 11.6 Integrated PK/PD Modeling Using Python

PK and PD are often studied together to provide a better understanding of a drug's behavior and its effects on the body [36]. While PK models describe how the drug concentration changes over time through absorption, distribution, metabolism, and excretion, PD models explain how these concentrations translate into therapeutic or toxic effects. Integrated PK/PD models combine these two processes to capture the full relationship between drug concentration and its pharmacological response, helping optimize drug dosing and predict treatment outcomes. We will briefly discuss the principles of integrated PK/PD modeling, including linking PK and PD models, using effect compartment models to account for delayed drug effects, and implementing PK/PD simulations using Python. An example case study of a hypothetical drug will demonstrate how to apply these techniques.

### 11.6.1 Linking PK and PD Models

To develop an integrated PK/PD model, the pharmacokinetic profile of the drug is first simulated to predict drug concentration over time. This concentration-time profile is then used as input for the pharmacodynamic model, which translates drug concentration into a biological effect.

A basic approach to linking PK and PD models involves assuming a direct relationship between the plasma concentration of the drug and its effect, known as the direct link model [37]. In this model, the effect is directly related to the plasma concentration at any given time, often described by an $E_{max}$ model or Hill equation.

However, in many cases, the effect may lag behind the concentration changes due to factors such as drug distribution to the effect site, receptor binding kinetics, or downstream signaling pathways. In such cases, a more complex model that accounts for this delay is required, such as the effect compartment model or models incorporating indirect responses.

### 11.6.2 Effect Compartment Models

For drugs where the observed pharmacological effect is delayed relative to the plasma concentration, the effect compartment model can be used. This model introduces a hypothetical compartment (the effect compartment) that connects the central (plasma) compartment to the site of action, accounting for the time delay between plasma concentration and effect.

The differential equations governing the PK and effect compartments can be written as follows:

For the PK compartment we can reuse equations from 11.4.4.

Absorption phase:

$$\frac{dA_{\text{gut}}}{dt} = -k_a \bullet A_{\text{gut}}$$



Elimination phase:

$$\frac{dC_{\text{plasma}}}{dt} = \frac{k_a \bullet A_{\text{gut}}}{V_d} - k_e \bullet C_{\text{plasma}}$$

For the effect compartment, the rate of change of the drug concentration in the effect compartment ($C_{\text{effect}}$) is:

$$\frac{dC_{\text{effect}}}{dt} = k_{e0} \bullet \left(C_{\text{plasma}} - C_{\text{effect}}\right)$$

where $C_{\text{effect}}$ is the concentration in the effect compartment and $k_{e0}$ is the first-order rate constant for drug transfer between the plasma and effect compartments.

The PD effect is then related to the concentration in the effect compartment, often using a Hill equation, we can reuse the 3-parameter Hill equation from 11.5.2:

$$E = \frac{E_{\max} \times C_{\text{effect}}^S}{\text{EC}_{50}^S + C_{\text{effect}}^S}$$

This model allows for a delayed response, which is particularly useful for modeling drugs that require time to reach the site of action or show delayed pharmacodynamic effects due to slow receptor kinetics or signal transduction processes.

### 11.6.3 PK/PD Modeling of a Hypothetical Drug

**Notebook: Section 11.6.3. PK/PD Modeling of a Hypothetical Drug**
This notebook provides code examples demonstrating PK/PD modeling of a hypothetical drug.

Link to the GitHub repository:
▸ https://github.com/sn-code-inside/BioPy
Go to: Chap. 11—▸ Sect. 11.6.3

Let's consider an example of a hypothetical drug that displays delayed effects after oral administration. We will develop an integrated PK/PD model to simulate the drug concentration over time, and link this to the pharmacodynamic response using an effect compartment model.

We first model the amount of drug in the gastrointestinal tract ($A_{gut}$) with the first-order absorption. Then we model the plasma concentration ($C_{plasma}$) with first-order absorption from the gut and first-order elimination. This allows us to define the PK model. We then define the effect compartment model. We model the concentration in the effect compartment ($C_{effect}$) to capture the delay between plasma concentration and effect. We then define the PD model using the Hill equation to relate $C_{effect}$ to the pharmacodynamic effect.

Example Code: Integrated PK/PD model with effect compartment

```
import numpy as np
from scipy.integrate import odeint
import matplotlib.pyplot as plt

# Define the integrated PK/PD model (effect compartment)
def pk_pd_model(y, t, ka, ke, ke0, Vd):
    A_gut, C_plasma, C_effect = y  # Amount in gut, plasma
    # concentration, effect compartment concentration

    # PK: Amount in gut (A_gut)
    dA_gut_dt = -ka * A_gut

    # PK: Plasma concentration (C_plasma)
    dC_plasma_dt = (ka * A_gut) / Vd - ke * C_plasma

    # Effect compartment concentration (C_effect)
    dC_effect_dt = ke0 * (C_plasma - C_effect)

    return [dA_gut_dt, dC_plasma_dt, dC_effect_dt]
```

```
# Parameters for the model (example)
Dose = 500          # Dose in mg
ka = 1.0            # Absorption rate constant (1/hour)
ke = 0.5            # Elimination rate constant (1/hour)
ke0 = 0.2           # Transfer rate constant between plasma
# and effect compartment (1/hour)
Vd = 50             # Volume of distribution (L)
Emax = 100          # Maximum effect (%)
EC50 = 10           # Concentration producing 50% of maximum
effect (mg/L)
s = 1.5             # Hill coefficient

# Time points (hours)
time = np.linspace(0, 24, 200)

# Initial conditions: A_gut(0) = Dose, C_plasma(0) = 0,
# C_effect(0) = 0
initial_conditions = [Dose, 0, 0]

# Solve the system of ODEs using odeint
solution = odeint(pk_pd_model, initial_conditions, time,
args=(ka, ke, ke0, Vd))

A_gut = solution[:, 0]
C_plasma = solution[:, 1]
C_effect = solution[:, 2]

# Calculate the pharmacodynamic effect using the Hill equation
Effect = (Emax * C_effect**s) / (EC50**s + C_effect**s)

# Plot the concentration-time and effect-time profiles
fig, ax = plt.subplots(1, 3, figsize=(12, 3))

# Plot amount in gut vs. time
ax[0].plot(time, A_gut, label='Amount in Gut',
color='orange')
ax[0].set_xlabel('Time (hours)')
ax[0].set_ylabel('Amount (mg)')
ax[0].set_title('A. Amount in Gut Over Time')
ax[0].legend()
ax[0].grid(True)

# Plot plasma concentration vs. time
ax[1].plot(time, C_plasma, label='Plasma Concentration',
color='blue')
ax[1].set_xlabel('Time (hours)')
ax[1].set_ylabel('Concentration (mg/L)')
ax[1].set_title('B. Plasma Concentration-Time Profile')
ax[1].legend()
ax[1].grid(True)
```

```
        # Plot effect vs. time
        ax[2].plot(time, Effect, label='Effect (PD Response)',
        color='red')
        ax[2].set_xlabel('Time (hours)')
        ax[2].set_ylabel('Effect (%)')
        ax[2].set_title('C. Effect-Time Profile')
        ax[2].legend()
        ax[2].grid(True)

        plt.tight_layout()
        plt.show()
```

The plot of the effect-time profile shows how the pharmacological effect lags behind the plasma concentration due to the introduction of the effect compartment (Fig. 11.8). This is typical of drugs that take time to interact with their target or exhibit delayed

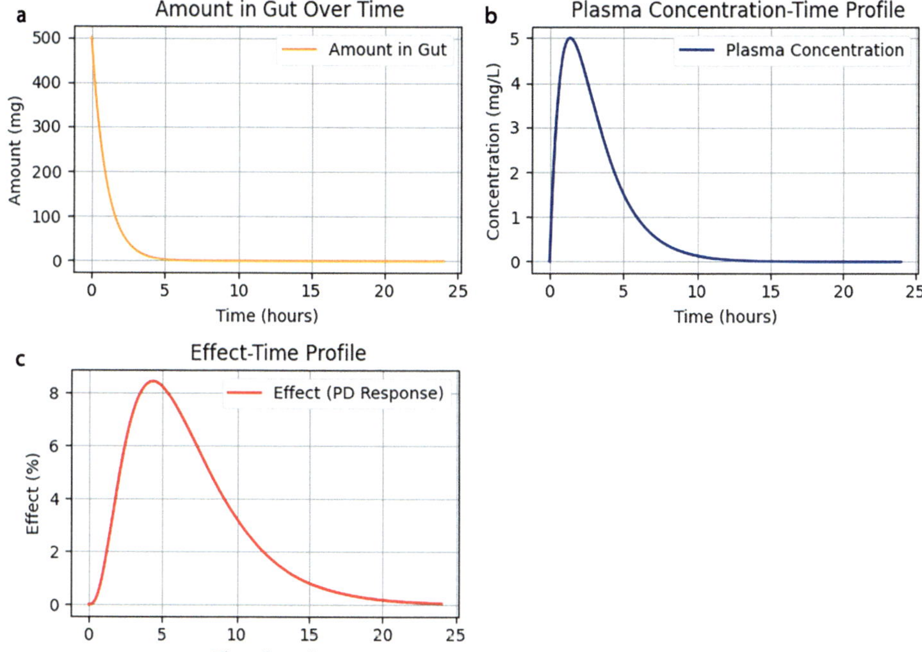


**Fig. 11.8** PK/PD modeling of a hypothetical drug using a one-compartment model with first-order absorption and elimination. (**a**) The amount of drug in the gut over time decreases exponentially as the drug is absorbed into systemic circulation. (**b**) The plasma concentration-time profile demonstrates a typical pharmacokinetic profile, with an initial rise due to absorption, reaching a peak plasma concentration ($C_{max}$), followed by an exponential decline as the drug is eliminated. (**c**) The effect-time profile (pharmacodynamic response) correlates with the plasma concentration, showing a delayed onset of action, a peak effect, and a gradual decline as drug levels fall below the therapeutic threshold. Each panel illustrates a distinct phase of the drug's pharmacokinetics and pharmacodynamics

signaling effects. Adjusting parameters like `ka`, `ke`, and `ke0` changes the time course of drug concentration and effect. Exploring these relationships helps optimize drug dosing strategies to achieve therapeutic efficacy with minimal side effects.

## 11.7 Parameter Estimation in PK/PD Models

Parameter estimation in PK/PD modeling allows fitting models to experimental data and extract important physiological parameters. These parameters include absorption and elimination rates in PK models and effective concentration ($EC_{50}$) or maximum effect ($E_{max}$) in PD models. The most common approach for estimating PK/PD parameters is through nonlinear regression, particularly nonlinear least squares (NLS). This section will discuss some techniques used for parameter estimation in PK/PD models, including nonlinear least squares regression, maximum likelihood estimation, and bootstrapping for confidence intervals.

### 11.7.1 Nonlinear Least Squares Regression for Parameter Estimation



**Notebook: Section 11.7.1. Nonlinear Least Squares Regression for Parameter Estimation**

This notebook provides code examples demonstrating nonlinear least squares regression for parameter estimation.

Link to the GitHub repository:

▸ https://github.com/sn-code-inside/BioPy

Go to: Chap. 11—▸ Sect. 11.7.1

Nonlinear least squares (NLS) regression is often used as a method for estimating PK/PD parameters. It involves minimizing the sum of the squared differences between observed data points and the predicted values from the model. The goal is to find the parameter values that best fit the experimental data.

Example: Estimating PK parameters using nonlinear least squares

Let's use Python to estimate the parameters of a simple one-compartment PK model with first-order elimination. Given experimental concentration-time data, we will fit the model to the data and estimate the elimination rate constant `ke` and absorption rate constant `ka` [38].

The plasma concentration C(t) after a single oral dose in a one-compartment model with first-order absorption and elimination is given by:

$$C(t) = \frac{F \cdot k_a \cdot D}{V_d (k_a - k_e)} \left( e^{-k_e t} - e^{-k_a t} \right)$$

where,

$F$ is the bioavailability (fraction of the dose that reaches systemic circulation), assumed to be 1 for simplicity.

$D$ is the administered dose (mg).

$V_d$ is the volume of distribution (L).

$k_a$ is the absorption rate constant ($h^{-1}$).

$k_e$ is the elimination rate constant ($h^{-1}$).

$t$ is time after dosing (h).

Example code to estimate $k_a$ and $k_e$.

```
import numpy as np
from scipy.optimize import curve_fit
import matplotlib.pyplot as plt

# Define the PK model (one-compartment, first-order absorption and
# elimination)
def pk_model(t, ka, ke):
    F = 1.0         # Bioavailability (fraction), assumed to be 1
    D = 500         # Dose in mg
    Vd = 50         # Volume of distribution in L
    C = (F * D * ka) / (Vd * (ka - ke)) * (np.exp(-ke * t) -
    np.exp(-ka * t))
    return C

# Experimental data (time in hours, concentration in mg/L)
time_data = np.array([0.5, 1, 2, 4, 6, 8, 12, 24])
concentration_data = np.array([5.1, 8.0, 9.3, 7.2, 4.8, 3.1, 1.5, 0.6])

# Perform nonlinear least squares fitting
initial_guesses = [1.0, 0.1]  # Initial guesses for ka and ke
params, covariance = curve_fit(pk_model, time_data, concentration_
data, p0=initial_guesses)

# Extract fitted parameters
ka_fitted, ke_fitted = params
print(f"Fitted ka: {ka_fitted:.4f} 1/hour")
print(f"Fitted ke: {ke_fitted:.4f} 1/hour")

# Generate fitted concentration-time profile
fitted_concentration = pk_model(time_data, ka_fitted, ke_fitted)

# Plot experimental data and fitted curve
plt.scatter(time_data, concentration_data, label='Experimental
Data', color='red')
plt.plot(time_data, fitted_concentration, label='Fitted Model',
color='blue')
plt.xlabel('Time (hours)')
plt.ylabel('Concentration (mg/L)')
plt.title('PK Model Fitting')
plt.legend()
plt.grid(True)
plt.show()
```

Output:

```
Fitted ka: 1.7425 1/hour
Fitted ke: 0.1331 1/hour
Figure 11.9
```

This example demonstrates how to fit a PK model to experimental data, and estimate the parameters using nonlinear least squares regression. The fitted model is shown in Fig. 11.9, which represents the output generated by the code above.

### 11.7.2 Maximum Likelihood Estimation (MLE) in PK/PD

**Notebook: Section 11.7.2. Maximum Likelihood Estimation (MLE) in PK/PD**
This notebook provides code examples demonstrating Maximum likelihood estimation (MLE) in PK/PD.

Link to the GitHub repository:
▶ https://github.com/sn-code-inside/BioPy
Go to: Chap. 11—▶ Sect. 11.7.2



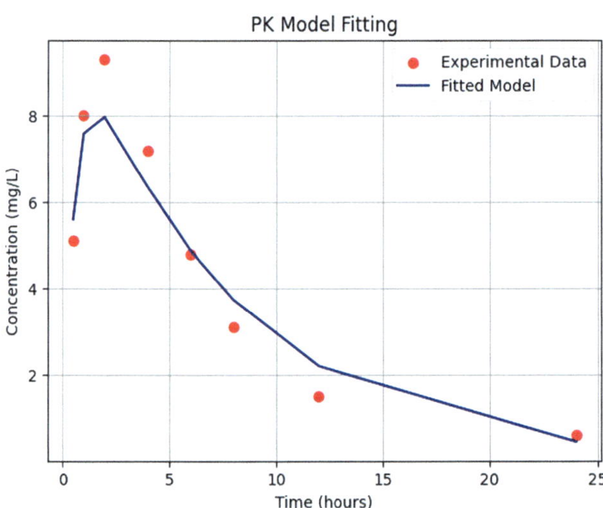


**Fig. 11.9** Nonlinear least squares fitting of a PK model. The red dots represent the experimental (hypothetical) data points, showing drug concentrations (mg/L) at various time points (hours). The blue curve represents the fitted one-compartment PK model with first-order elimination, using estimated parameters obtained via nonlinear least squares regression. The close alignment between the experimental data and the fitted model indicates a good fit, allowing for accurate estimation of key PK parameters

Maximum likelihood estimation (MLE) is another method used to estimate PK/PD parameters, especially when the data is noisy or follows a specific probability distribution. MLE involves maximizing the likelihood function, which represents the probability of observing the given data given a set of parameter values. While MLE can be more computationally intensive than NLS, it provides a robust method for parameter estimation, particularly when dealing with complex models or nonnormal error distributions.

Example: Estimating PK Parameters Using MLE

We will implement MLE in Python using optimization functions from SciPy. The key idea is to define the negative log-likelihood function for the PK model and then use optimization techniques to find the parameter values that minimize this function.

```
import numpy as np
from scipy.optimize import minimize
import matplotlib.pyplot as plt


# Define the PK model (as before)
def pk_model(t, ka, ke):
    F = 1.0
    D = 500
    Vd = 50
    C = (F * D * ka) / (Vd * (ka - ke)) * (np.exp(-ke *
    t) - np.exp(-ka * t))
    return C


# Negative log-likelihood function assuming normally
distributed errors
def negative_log_likelihood(params, t, concentration_obs):
    ka, ke, sigma = params
    concentration_pred = pk_model(t, ka, ke)
    residuals = concentration_obs - concentration_pred
    n = len(t)
    nll = ((n / 2) * np.log(2 * np.pi * sigma**2) +
    np.sum(residuals**2) / (2 * sigma**2))
    return nll


# Initial guesses for ka, ke, and sigma (standard
# deviation of errors)
initial_guesses = [1.0, 0.1, 1.0]


# Bounds to ensure sigma is positive
bounds = [(0, None), (0, None), (1e-6, None)]


# Experimental data (time in hours, concentration in mg/L)
time_data = np.array([0.5, 1, 2, 4, 6, 8, 12, 24])
concentration_data = np.array([5.1, 8.0, 9.3, 7.2, 4.8,
3.1, 1.5, 0.6])
```

```
# Perform MLE using minimize (SciPy)
result = minimize(negative_log_likelihood, initial_guesses,
args=(time_data, concentration_data), bounds=bounds)

# Extract fitted parameters
ka_mle, ke_mle, sigma_mle = result.x
print(f"MLE Estimated ka: {ka_mle:.4f} 1/hour")
print(f"MLE Estimated ke: {ke_mle:.4f} 1/hour")
print(f"Estimated sigma (error SD): {sigma_mle:.4f}")

# Generate fitted concentration-time profile
fitted_concentration_mle = pk_model(time_data, ka_mle, ke_mle)

# Plot experimental data and MLE fitted curve
plt.scatter(time_data, concentration_data,
label='Experimental Data', color='red')
plt.plot(time_data, fitted_concentration_mle, label='MLE
Fitted Model', color='green')
plt.xlabel('Time (hours)')
plt.ylabel('Concentration (mg/L)')
plt.title('PK Model Fitting using MLE')
plt.legend()
plt.grid(True)
plt.show()
```



Output:

```
MLE Estimated ka: 1.7425 1/hour
MLE Estimated ke: 0.1331 1/hour
Estimated sigma (error SD): 0.6931
Figure 11.10
```

We assume that the residuals (differences between observed and predicted concentrations) are normally distributed with mean zero and standard deviation $\sigma$. The negative log-likelihood is minimized to find the best-fitting parameters and $\sigma$ is included as a parameter to estimate the variability in the data. MLE allows the estimation of additional parameters like $\sigma$, can handle different error distributions by modifying the likelihood function, and provides a statistical framework for hypothesis testing and model comparison. The fitted model is shown in Fig. 11.10, which represents the output generated by the code above.

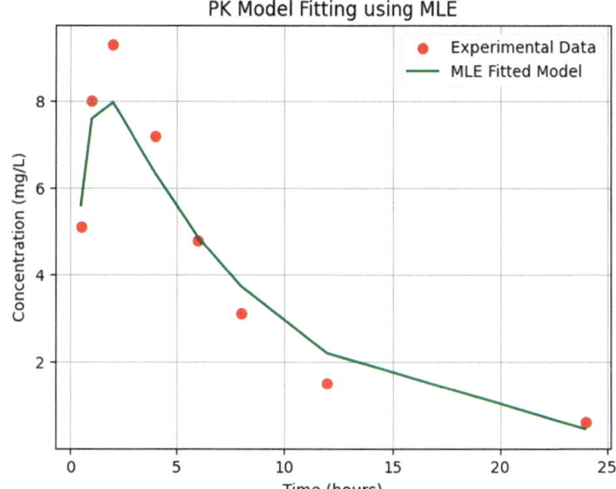


**Fig. 11.10** PK model fitting using maximum likelihood estimation (MLE). The red dots represent the experimental concentration-time data for a drug, while the green line represents the fitted pharmacokinetic (PK) model obtained using MLE. The model provides a close fit to the data, estimating parameters such as the absorption rate constant ($k_a$) and elimination rate constant ($k_e$). The x-axis represents time (hours), and the y-axis represents drug concentration (mg/L). This plot demonstrates the utility of MLE in accurately capturing the dynamics of drug concentration over time

### 11.7.3 Bootstrapping and Confidence Intervals in PK/PD Models

**Notebook: Section 11.7.3. Bootstrapping and Confidence Intervals in PK/PD Mod els**

This notebook provides code examples demonstrating bootstrapping and confidence intervals in PK/PD models.

Link to the GitHub repository:
▶ https://github.com/sn-code-inside/BioPy
Go to: Chap. 11—▶ Sect. 11.7.3

Estimating the uncertainty of parameters is as important as estimating the parameters themselves. Bootstrapping allows to calculate confidence intervals for estimated parameters by resampling the data multiple times and performing parameter estimation on each sample.

Example: Bootstrapping for confidence interval estimation. The bootstrap distribution plots of $k_a$ and $k_e$ are shown in Fig. 11.11, which represents the output generated by the code below.

```
import numpy as np
from scipy.optimize import curve_fit
import matplotlib.pyplot as plt

# Define the PK model
def pk_model(t, ka, ke):
    F = 1.0
    D = 500
    Vd = 50
    if ka == ke:
        return np.zeros_like(t)
    C = ((F * D * ka) / (Vd * (ka - ke)) * (np.exp(-ke * t) -
    np.exp(-ka * t)))
    return C

# Experimental data
time_data = np.array([0.5, 1, 2, 4, 6, 8, 12, 24])
concentration_data = np.array([5.1, 8.0, 9.3, 7.2, 4.8, 3.1, 1.5, 0.6])

# Initial parameter estimates
initial_guesses = [1.0, 0.5]

# Fit the original model
params, covariance = curve_fit(pk_model, time_data,
concentration_data, p0=initial_guesses,
bounds=([0.01, 0.01], [10, 10]))
ka_original, ke_original = params
print(f"Original ka: {ka_original:.4f} 1/hour")
print(f"Original ke: {ke_original:.4f} 1/hour")

# Calculate residuals
fitted_concentration = pk_model(time_data, ka_original, ke_original)
residuals = concentration_data - fitted_concentration

# Number of bootstrap samples
n_bootstraps = 1000

# Storage for bootstrap estimates
ka_bootstrap = []
ke_bootstrap = []

# Perform residual bootstrapping
for _ in range(n_bootstraps):
    # Resample residuals with replacement
    resampled_residuals = np.random.choice(residuals,
size=len(residuals), replace=True)
```

```
        # Generate new concentration data
        concentration_resample = fitted_concentration + resampled_residuals
        # Ensure no negative concentrations
        concentration_resample = np.maximum(concentration_resample, 0.01)

        # Fit the model to the resampled data with bounds
        try:
            params_boot, _ = curve_fit(pk_model, time_data,
            concentration_resample, p0=initial_guesses,
            bounds=([0.01, 0.01], [10, 10]))
            ka_bootstrap.append(params_boot[0])
            ke_bootstrap.append(params_boot[1])
        except RuntimeError:
            continue

    # Calculate 95% confidence intervals
    ka_confidence_interval = np.percentile(ka_bootstrap, [2.5, 97.5])
    ke_confidence_interval = np.percentile(ke_bootstrap, [2.5, 97.5])

    print(f"95% Confidence Interval for ka: {ka_confidence_interval}")
    print(f"95% Confidence Interval for ke: {ke_confidence_interval}")

    # Plot histograms of bootstrap estimates
    fig, ax = plt.subplots(1, 2, figsize=(12, 5))

    # Histogram for ka
    ax[0].hist(ka_bootstrap, bins=30, color='skyblue',
    edgecolor='black')
    ax[0].set_title('Bootstrap Distribution of ka')
    ax[0].set_xlabel('ka (1/hour)')
    ax[0].set_ylabel('Frequency')

    # Histogram for ke
    ax[1].hist(ke_bootstrap, bins=30, color='salmon',
    edgecolor='black')
    ax[1].set_title('Bootstrap Distribution of ke')
    ax[1].set_xlabel('ke (1/hour)')
    ax[1].set_ylabel('Frequency')

    plt.tight_layout()
    plt.show()
```

Output:

```
Original ka: 1.7425 1/hour
Original ke: 0.1331 1/hour
95% Confidence Interval for ka: [1.39149917 2.48465844]
95% Confidence Interval for ke: [0.10647499 0.15342977]

Figure 11.11.
```

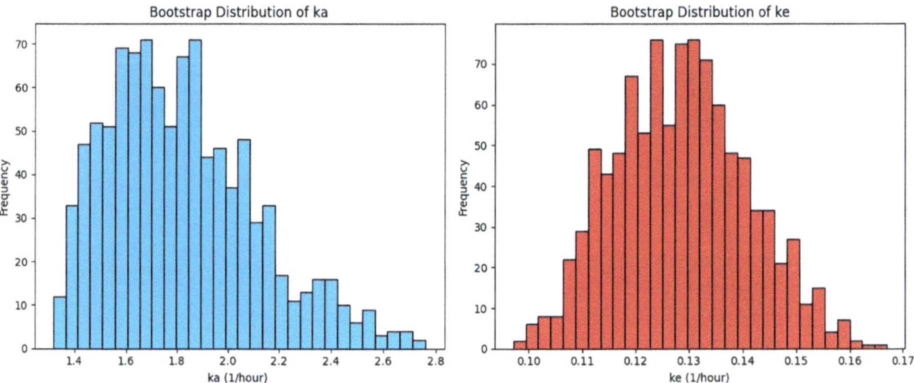


**Fig. 11.11** Bootstrap distributions of the absorption rate constant ($k_a$) and elimination rate constant ($k_e$) in a pharmacokinetic (PK) model. The left panel shows the bootstrap distribution of $k_a$ (1/h), and the right panel shows the bootstrap distribution of $k_e$ (1/h). These histograms illustrate the variability and uncertainty in the parameter estimates derived from the PK model through bootstrapping. The shape and spread of the distributions help describe the confidence intervals for the parameters, with the peak indicating the most frequent estimate. Bootstrapping is a robust statistical method to quantify the uncertainty and reliability of parameter estimates in PK/PD modeling

## 11.8 Sensitivity Analysis and Model Validation in PK/PD

Sensitivity analysis in PK/PD modeling helps determine how much the model's output changes in the input parameters, making it possible to identify which variables influence on model's predictions. Model validation, on the other hand, assesses how well the model fits the observed data and predicts future outcomes, ensuring that the model can be used with confidence in decision-making processes. In this section, we will discuss sensitivity analysis and model validation techniques in PK/PD modeling.

### 11.8.1 Sensitivity Analysis of PK/PD Parameters

**Notebook: Section 11.8.1. Sensitivity Analysis of PK/PD Parameters**

This notebook provides code examples demonstrating sensitivity analysis of PK/PD parameters.

Link to the GitHub repository:

► https://github.com/sn-code-inside/BioPy

Go to: Chap. 11—► Sect. 11.8.1

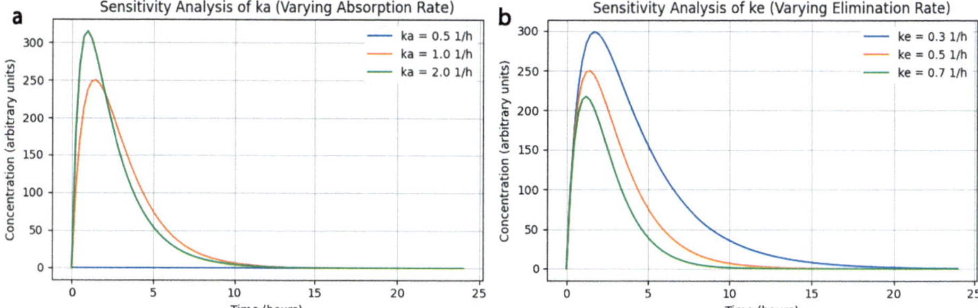


**Fig. 11.12** Sensitivity analysis of PK parameters on drug concentration-time profiles. (**a**) Sensitivity analysis of the absorption rate constant ($k_a$) shows how varying $k_a$ values (0.5, 1.0, and 2.0 1/h) influence the plasma drug concentration. Faster absorption rates result in earlier and higher peak concentrations ($C_{max}$) with a shorter time to peak. (**b**) Sensitivity analysis of the elimination rate constant ($k_e$) demonstrates how varying *ke* values (0.3, 0.5, and 0.7 1/h) affect the plasma drug concentration. Higher elimination rates lead to a more rapid decline in concentration and lower overall drug exposure

Sensitivity analysis examines how variations in model parameters affect the model's output. In the context of PK/PD, this involves systematically changing parameters like the absorption rate constant ($k_a$), elimination rate constant ($k_e$), or the $EC_{50}$ (effective concentration) to observe how these changes influence the concentration-time or dose-response profiles.

There are different types of sensitivity analysis such as the local sensitivity analysis that involves changing one parameter at a time, while keeping others constant, to assess its effect on the model. The global sensitivity analysis simultaneously varies multiple parameters to evaluate the combined effects on the model output.

Example: Local sensitivity analysis in PK modeling (we will ignore volume of distribution, $V_d$ and bioavailability, F for simplicity)

Let's perform a local sensitivity analysis by varying the absorption rate constant ($k_a$) and the elimination rate constant ($k_e$) in a one-compartment PK model. This analysis helps assess which parameter has a greater impact on drug concentration over time (Fig. 11.12).

```
import numpy as np
import matplotlib.pyplot as plt


# Define the one-compartment PK model
def pk_model(t, ka, ke, dose):
    if ka == ke:
        # Avoid division by zero
        return np.zeros_like(t)
    concentration = ((dose * ka / (ka - ke)) * (np.exp(-ke * t) -
    np.exp(-ka * t)))
    return concentration


# Time points
time = np.linspace(0, 24, 100)
```

```
# Dose in mg
dose = 500

# Vary ka and ke for sensitivity analysis
ka_values = [0.5, 1.0, 2.0]  # Absorption rate constants (1/hour)
ke_values = [0.3, 0.5, 0.7]  # Elimination rate constants (1/hour)

# Plot the concentration-time curves for different ka values
fig, ax = plt.subplots(1, 2, figsize=(12, 4))
for ka in ka_values:
    concentration = pk_model(time, ka, 0.5, dose)
    # Keep ke constant at 0.5 1/hour
    ax[0].plot(time, concentration, label=f'ka = {ka} 1/h')

ax[0].set_xlabel('Time (hours)')
ax[0].set_ylabel('Concentration (arbitrary units)')
ax[0].set_title(
'A. Sensitivity Analysis of ka (Varying Absorption Rate)')
ax[0].legend()
ax[0].grid(True)

# Plot the concentration-time curves for different ke values
for ke in ke_values:
    concentration = pk_model(time, 1.0, ke, dose)
    # Keep ka constant at 1.0 1/hour
    ax[1].plot(time, concentration, label=f'ke = {ke} 1/h')

ax[1].set_xlabel('Time (hours)')
ax[1].set_ylabel('Concentration (arbitrary units)')
ax[1].set_title(
'B. Sensitivity Analysis of ke (Varying Elimination Rate)')
ax[1].legend()
ax[1].grid(True)

plt.tight_layout()
plt.show()
```



### 11.8.2 Goodness-of-Fit Tests for Model Validation

**Notebook: Section 11.8.2. Goodness-of-Fit Tests for Model Validation**
This notebook provides code examples demonstrating goodness-of-fit tests for model validation.

Link to the GitHub repository:
► https://github.com/sn-code-inside/BioPy
Go to: Chap. 11—► Sect. 11.8.2

Model validation is the process of assessing how well the PK/PD model fits the experimental data. It ensures that the model accurately represents the underlying biological processes and can be used for prediction. Several goodness-of-fit tests and diagnostic plots are commonly used to validate PK/PD models:

Residual plots: Show the difference between observed and predicted values. Ideally, residuals should be randomly distributed around zero, indicating a good model fit.

Coefficient of determination ($R^2$): Measures the proportion of the variance in the observed data that is explained by the model. An $R^2$ value closer to 1 indicates a better fit.

Akaike information criterion (AIC): Used to compare models; a lower AIC value indicates a better model fit while penalizing model complexity [39].

The AIC is defined as:

$$\text{AIC} = 2k - 2\ln(L)$$

where,

$k$ is the number of parameters in the model.

$L$ is the maximum likelihood of the model.

For models with normally distributed residuals which is a common assumption in least squares fitting, the likelihood $L$ can be expressed in terms of the residual sum of squares (RSS) [40]. The AIC can then be calculated using:

$$\text{AIC} = n \bullet \ln\left(\frac{\text{RSS}}{n}\right) + 2k$$

where,

$n$ is the number of observations (data points).

RSS is the residual sum of squares.

$k$ is the number of estimated parameters in the model.

Example: Model validation using goodness-of-fit test

Let's validate the fit of a one-compartment PK model by calculating the residuals, computing

$R^2$, AIC and creating a residual plot (Fig. 11.13).

```
import numpy as np
from scipy.optimize import curve_fit
from sklearn.metrics import r2_score
import matplotlib.pyplot as plt

# Experimental data (time in hours, concentration in mg/L)
time_data = np.array([0.5, 1, 2, 4, 6, 8, 12, 24])
concentration_data = np.array([5.1, 8.0, 9.3, 7.2, 4.8, 3.1, 1.5, 0.6])

# Define the PK model for fitting
def pk_model(t, ka, ke):
    F = 1.0
    D = 500
    Vd = 50
```

```
    if ka == ke:
        return np.zeros_like(t)
    C = ((F * D * ka) / (Vd * (ka - ke)) * (np.exp(-ke * t) -
    np.exp(-ka * t)))
    return C

# Perform nonlinear regression to fit the model
initial_guesses = [1.0, 0.1]  # Initial guesses for ka and ke
params, covariance = curve_fit(pk_model, time_data, concentration_data,
p0=initial_guesses)
ka_fitted, ke_fitted = params

print(f"Fitted ka: {ka_fitted:.4f} 1/hour")
print(f"Fitted ke: {ke_fitted:.4f} 1/hour")

# Predict concentrations using the fitted model
predicted_concentration = pk_model(time_data, ka_fitted, ke_fitted)

# Calculate residuals (observed - predicted)
residuals = concentration_data - predicted_concentration

# Calculate Residual Sum of Squares (RSS)
RSS = np.sum(residuals**2)

# Number of observations
n = len(concentration_data)

# Number of parameters (ka and ke)
k = 2

# Calculate AIC
AIC = n * np.log(RSS / n) + 2 * k

# Alternatively, you can calculate AICc (corrected AIC for small
# sample sizes)
AICc = AIC + (2 * k * (k + 1)) / (n - k - 1)

# Plot residuals
plt.figure(figsize=(8, 5))
plt.scatter(time_data, residuals, color='purple')
plt.axhline(0, color='red', linestyle='--')
plt.xlabel('Time (hours)')
plt.ylabel('Residuals (mg/L)')
plt.title('Residual Plot for Model Validation')
plt.grid(True)
plt.show()
```

```
# Calculate R-squared
r2 = r2_score(concentration_data, predicted_concentration)
print(f"R-squared: {r2:.4f}")
```

```
# Print AIC and AICc
print(f"Akaike Information Criterion (AIC): {AIC:.4f}")
print(f"Corrected AIC (AICc): {AICc:.4f}")
```

Output:

```
Fitted ka: 1.7425 1/hour
Fitted ke: 0.1331 1/hour
R-squared: 0.9431
Akaike Information Criterion (AIC): -1.8645
Corrected AIC (AICc): 0.5355

Figure 11.13.
```

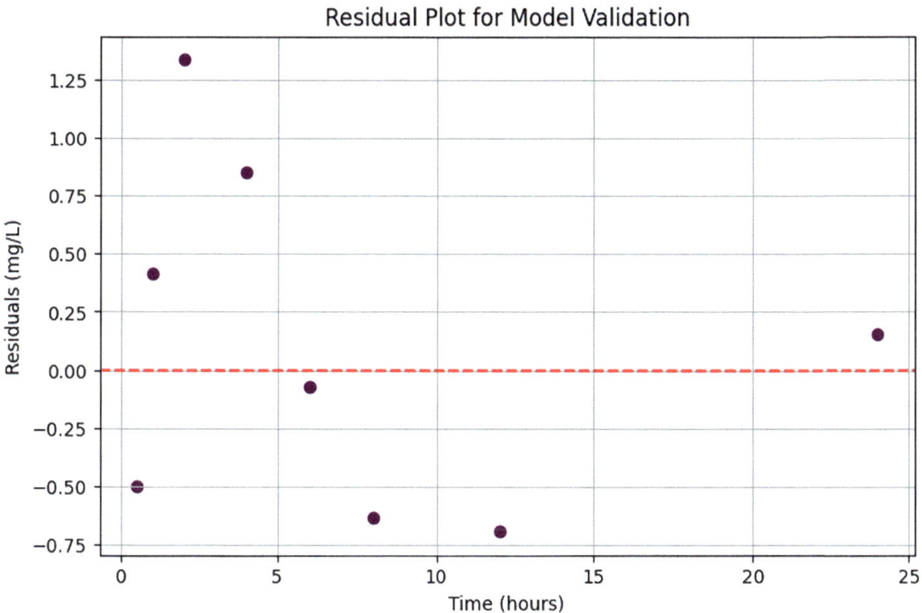


**Fig. 11.13** Residual plot for model validation in a goodness-of-fit test. The residuals, calculated as the difference between observed and model-predicted concentrations, are plotted against time (hours). The red dashed line at zero represents perfect agreement between the model and observed data. Points scattered randomly around the zero line indicate no systematic bias in the model predictions, supporting its validity. Any noticeable patterns or trends in the residuals would suggest a lack of fit or model misspecification

## 11.9 Applications of PK/PD Modeling in Drug Development

PK/PD modeling is a valuable tool in drug development, helping to predict how a drug behaves in the body and how it exerts its effects. PK/PD models play important roles in optimizing dosing strategies, predicting therapeutic outcomes, minimizing adverse effects, and improving the efficiency of clinical trials. Integrating PK and PD models, researchers and clinicians can better understand the relationship between drug concentration and its biological effects, allowing for more informed decision-making during the drug development process. In this section, we will look at some applications of PK/PD modeling in drug development, including dose optimization, predicting efficacy and toxicity, and personalized medicine.

### 11.9.1 PK/PD in Dose Optimization

> **Notebook: Section 11.9.1. PK/PD in Dose Optimization**
> This notebook provides code examples demonstrating PK/PD in dose optimization.
> Link to the GitHub repository:
> ► https://github.com/sn-code-inside/BioPy
> Go to: Chap. 11—► Sect. 11.9.1



One important application of PK/PD modeling is dose optimization. Simulating how different doses of a drug affect its concentration in the body and its therapeutic effect, PK/PD models help determine the optimal dose that maximizes efficacy while minimizing toxicity. This is especially important in drugs with narrow therapeutic windows, where small variations in dose can lead to either subtherapeutic effects or adverse reactions.

Dose optimization example: Let's consider a hypothetical drug with known PK/PD parameters and simulate different dosing regimens to identify the optimal dose. The goal is to maintain drug concentration within the therapeutic range (between $C_{min}$ and $C_{max}$) while avoiding toxicity (◘ Fig. 11.14).

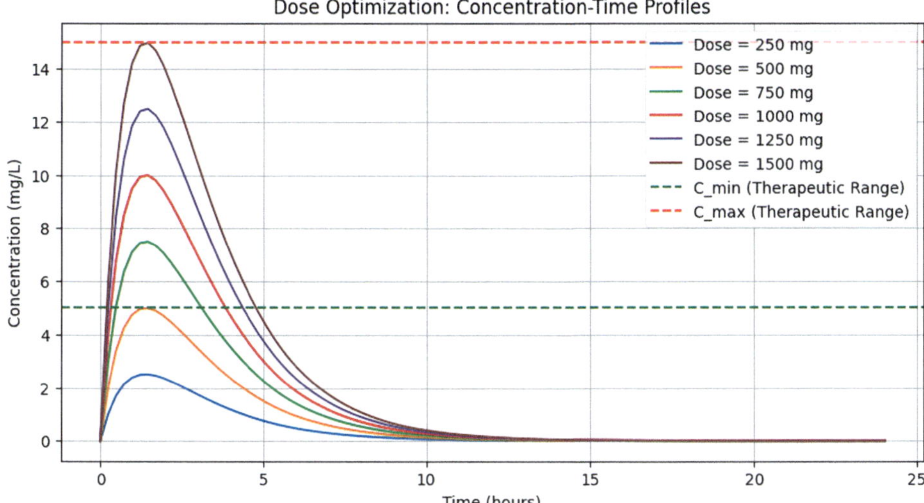


**Fig. 11.14** Dose optimization: Concentration-time profiles for varying drug doses. The curves represent plasma drug concentrations over time for doses ranging from 250 mg to 1500 mg. The green dashed line ($C_{min}$) represents the minimum effective concentration within the therapeutic range, while the red dashed line ($C_{max}$) indicates the maximum safe concentration. Doses below 750 mg fail to achieve $C_{min}$, suggesting subtherapeutic levels, while doses above 1250 mg approach or exceed $C_{max}$, increasing the risk of toxicity. The analysis shows the importance of dose selection to maintain concentrations within the therapeutic window for optimal efficacy and safety. The x-axis represents time (hours), and the y-axis represents drug concentration (mg/L)

Example Code: Simulating dose optimization (we will ignore bioavailability, F for simplicity).

```
import numpy as np
import matplotlib.pyplot as plt

# Define the PK model (one-compartment, first-order absorption and
# elimination)
def pk_model(t, ka, ke, dose, Vd):
    if ka == ke:
        return np.zeros_like(t)
    concentration = ((dose * ka) / (Vd * (ka - ke)) * (np.exp(-ke *
    t) - np.exp(-ka * t)))
    return concentration

# Time points (hours)
time = np.linspace(0, 24, 100)

# Dosing regimens (mg)
dose_values = [250, 500, 750, 1000, 1250, 1500]

# Parameters for the model
ka = 1.0     # Absorption rate constant (1/hour)
ke = 0.5     # Elimination rate constant (1/hour)
Vd = 50      # Volume of distribution (L)
```

```
# Therapeutic range (mg/L)
C_min = 5.0   # Minimum therapeutic concentration
C_max = 15.0  # Maximum therapeutic concentration

# Plot the concentration-time profiles for different doses
plt.figure(figsize=(10, 5))
for dose in dose_values:
    concentration = pk_model(time, ka, ke, dose, Vd)
    plt.plot(time, concentration, label=f'Dose = {dose} mg')

# Plot therapeutic range
plt.axhline(C_min, color='green', linestyle='--',
label='C_min (Therapeutic Range)')
plt.axhline(C_max, color='red', linestyle='--',
label='C_max (Therapeutic Range)')

plt.xlabel('Time (hours)')
plt.ylabel('Concentration (mg/L)')
plt.title('Dose Optimization: Concentration-Time Profiles')
plt.legend()
plt.grid(True)
plt.show()
```

### 11.9.2 Predicting Drug Efficacy and Toxicity



**Notebook: Section 11.9.2. Predicting Drug Efficacy and Toxicity**
This notebook provides code examples demonstrating how to predict drug efficacy and toxicity.

Link to the GitHub repository:
▶ https://github.com/sn-code-inside/BioPy
Go to: Chap. 11—▶ Sect. 11.9.2

PK/PD models can also be used to predict a drug's efficacy and toxicity based on the concentration-effect relationship. Modeling the drug's pharmacodynamic response, such as receptor occupancy or enzyme inhibition, PK/PD models can estimate the therapeutic effect at different concentrations. Moreover, these models can help predict the onset of toxic effects by simulating drug concentration changes in organs or tissues over time.

Example: Predicting efficacy and toxicity

In this example, we will model a drug's effect using the Hill equation and assess how different doses impact both efficacy and toxicity. We assume that the therapeutic effect occurs within a specific concentration range, while toxicity occurs at higher concentrations (◘ Fig. 11.15).

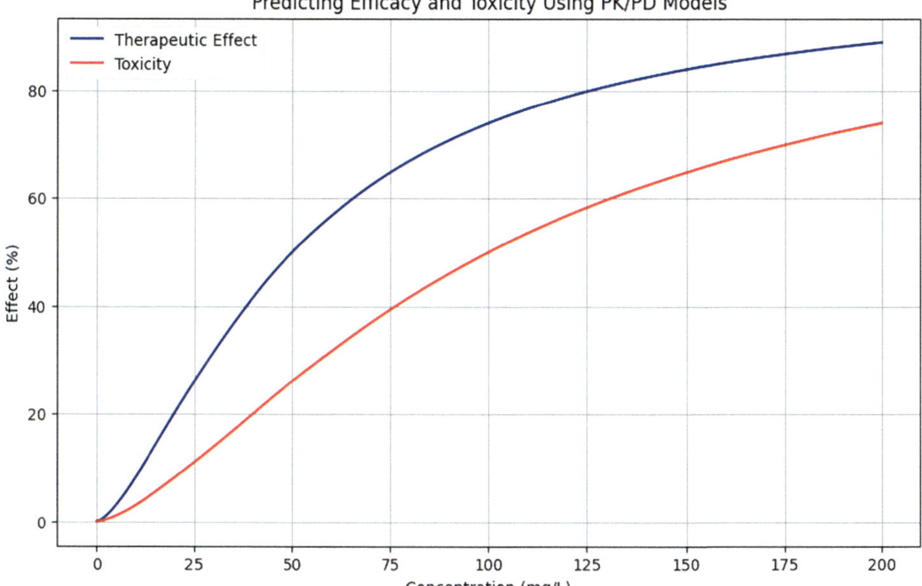


**Fig. 11.15** Predicting therapeutic efficacy and toxicity using PK/PD models. The blue curve represents the therapeutic effect (%) as a function of drug concentration (mg/L), showing an increase in effect with rising concentrations and plateauing at higher levels. The red curve represents toxicity (%), which also increases with concentration but at a lower rate than the therapeutic effect initially. At higher concentrations, the toxicity effect approaches the therapeutic effect, indicating a narrowing of the therapeutic window. This figure illustrates the importance of maintaining drug concentrations within a range that maximizes therapeutic efficacy while minimizing toxicity. The x-axis represents drug concentration (mg/L), and the y-axis represents the percentage of effect

```
import numpy as np
import matplotlib.pyplot as plt

# Define the Hill equation for therapeutic effect and toxicity
def hill_equation(C, Emax, EC50, n):
    return (Emax * C**n) / (EC50**n + C**n)

# Parameters for therapeutic effect
Emax_therapeutic = 100    # Maximum therapeutic effect (%)
EC50_therapeutic = 50     # EC50 for therapeutic effect (mg/L)

# Parameters for toxicity
Emax_toxicity = 100       # Maximum toxic effect (%)
EC50_toxicity = 100       # EC50 for toxicity (mg/L)

# Concentration range (0 to 200 mg/L)
concentration = np.linspace(0, 200, 1000)
```

```
# Calculate therapeutic effect and toxicity at each concentration
effect_therapeutic = hill_equation(concentration, Emax_therapeutic,
EC50_therapeutic, 1.5)
effect_toxicity = hill_equation(concentration, Emax_toxicity,
EC50_toxicity, 1.5)

# Plot therapeutic effect and toxicity
plt.figure(figsize=(10, 6))
plt.plot(concentration, effect_therapeutic, label='Therapeutic Effect',
color='blue')
plt.plot(concentration, effect_toxicity, label='Toxicity',
color='red')
plt.xlabel('Concentration (mg/L)')
plt.ylabel('Effect (%)')
plt.title('Predicting Efficacy and Toxicity Using PK/PD Models')
plt.legend()
plt.grid(True)
plt.show()
```

### 11.9.3 PK/PD in Personalized Medicine



**Notebook: Section 11.9.3. PK/PD in Personalized Medicine**
This notebook provides code examples demonstrating PK/PD in personalized medicine.

Link to the GitHub repository:
▶ https://github.com/sn-code-inside/BioPy
Go to: Chap. 11—▶ Sect. 11.9.3

One promising applications of PK/PD modeling is application in personalized medicine, where drug dosing is tailored to individual patients based on their unique characteristics (e.g., genetics, age, weight, disease state). Incorporating patient-specific data into PK/PD models, clinicians can predict how each patient will respond to a drug and adjust dosing accordingly. This approach helps to optimize treatment outcomes while minimizing the risk of adverse effects.

Example: Personalized PK/PD modeling (we will ignore bioavailability, F for simplicity)

Let's consider a scenario where two patients have different elimination rates due to genetic variations in drug-metabolizing enzymes. We will use a PK/PD model to simulate the concentration-time profiles for both patients and adjust the dosing regimen to achieve the desired therapeutic effect (◘ Fig. 11.16).

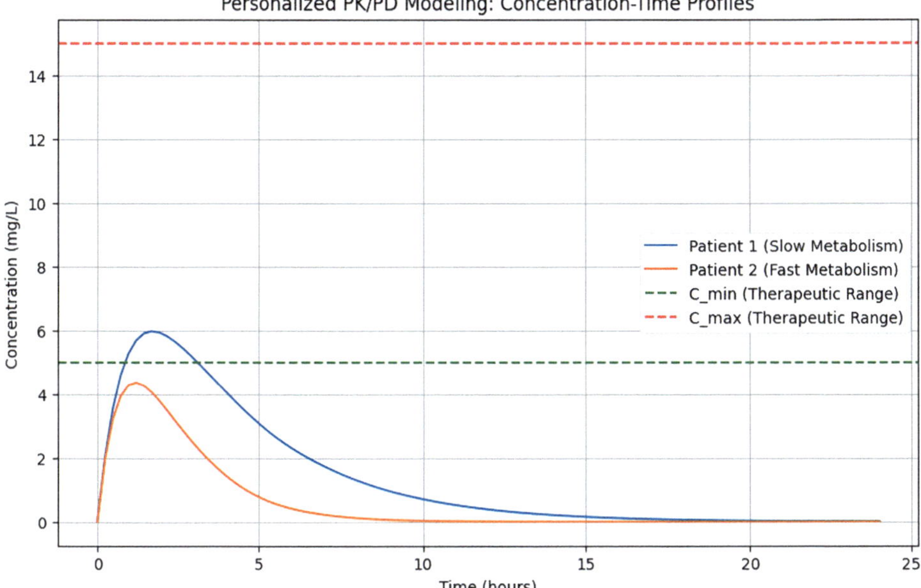


**Fig. 11.16** Personalized PK/PD modeling: Concentration-time profiles for two patients with different metabolic rates. The blue curve represents a patient with slow metabolism, resulting in higher peak plasma concentration ($C_{max}$) and prolonged drug exposure. The orange curve represents a patient with fast metabolism, showing a lower $C_{max}$ and faster drug elimination. The dashed green line ($C_{min}$) represents the minimum effective concentration within the therapeutic range, while the dashed red line ($C_{max}$) represents the maximum safe concentration. This figure shows the impact of interindividual variability on drug concentrations and the importance of personalized dosing to maintain drug levels within the therapeutic window for optimal efficacy and safety. The x-axis represents time (hours), and the y-axis represents drug concentration (mg/L)

```
import numpy as np
import matplotlib.pyplot as plt

# Define the PK model (including volume of distribution)
def pk_model(t, ka, ke, dose, Vd):
    if ka == ke:
        return np.zeros_like(t)
    concentration = ((dose * ka) / (Vd * (ka - ke)) * (np.exp(-ke *
    t) - np.exp(-ka * t)))
    return concentration

# Parameters for two patients with different elimination rates (ke)
ka = 1.0            # Absorption rate constant (1/hour)
ke_patient1 = 0.3 # Elimination rate for Patient 1 (slow metabolism)
ke_patient2 = 0.7 # Elimination rate for Patient 2 (fast metabolism)
Vd = 50             # Volume of distribution (L)
dose = 500          # Dose in mg

# Time points (hours)
time = np.linspace(0, 24, 100)
```

```
# Simulate concentration-time profiles for both patients
concentration_patient1 = pk_model(time, ka, ke_patient1, dose, Vd)
concentration_patient2 = pk_model(time, ka, ke_patient2, dose, Vd)

# Therapeutic range (mg/L)
C_min = 5.0   # Minimum therapeutic concentration
C_max = 15.0  # Maximum therapeutic concentration

# Plot concentration-time profiles
plt.figure(figsize=(10, 6))
plt.plot(time, concentration_patient1,
label='Patient 1 (Slow Metabolism)')
plt.plot(time, concentration_patient2,
label='Patient 2 (Fast Metabolism)')
plt.axhline(C_min, color='green', linestyle='--',
label='C_min (Therapeutic Range)')
plt.axhline(C_max, color='red', linestyle='--',
label='C_max (Therapeutic Range)')
plt.xlabel('Time (hours)')
plt.ylabel('Concentration (mg/L)')
plt.title('Personalized PK/PD Modeling: Concentration-Time Profiles')
plt.legend()
plt.grid(True)
plt.show()
```



### 11.9.4 Using PK/PD to Inform Clinical Trial Design

**Notebook: Section 11.9.4. Using PK/PD to Inform Clinical Trial Design**
This notebook provides code examples demonstrating the use of PK/PD to inform clinical trial design.

Link to the GitHub repository:

► https://github.com/sn-code-inside/BioPy

Go to: Chap. 11—► Sect. 11.9.4

PK/PD models can be used to design clinical trials by predicting how different dosing regimens will affect drug efficacy and safety. This helps in selecting the appropriate dose levels for early-phase clinical trials and reducing the likelihood of trial failure due to underdosing or overdosing.

Example: Simulating clinical trial doses (we will ignore bioavailability, F for simplicity)

Let's simulate the concentration-time profiles for three different doses (low, medium, and high) and assess which dose would be most appropriate for a Phase II clinical trial, aiming for a balance between efficacy and safety (◘ Fig. 11.17).

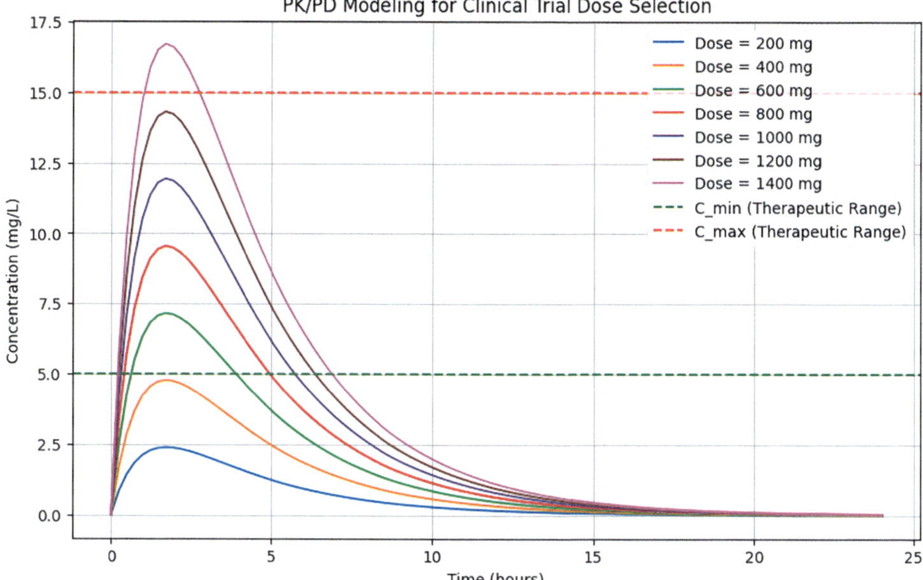


**Fig. 11.17** PK/PD modeling for clinical trial dose selection: Concentration-time profiles for varying doses ranging from 200 mg to 1400 mg. Each curve represents the plasma drug concentration over time for a specific dose. The green dashed line ($C_{min}$) indicates the minimum effective concentration required to achieve therapeutic efficacy, while the red dashed line ($C_{max}$) represents the maximum safe concentration to avoid toxicity. Lower doses (200–600 mg) fail to achieve $C_{min}$, suggesting subtherapeutic levels, while higher doses (1200–1400 mg) approach or exceed $C_{max}$, increasing the risk of toxicity. This figure shows the importance of selecting an optimal dose (e.g., 800–1000 mg) to maintain drug concentrations within the therapeutic range. The x-axis represents time (hours), and the y-axis represents plasma drug concentration (mg/L)

```
import numpy as np
import matplotlib.pyplot as plt

# Define the PK model (including volume of distribution)
def pk_model(t, ka, ke, dose, Vd):
    if ka == ke:
        return np.zeros_like(t)
    concentration = ((dose * ka) / (Vd * (ka - ke)) * (np.exp(-ke *
    t) - np.exp(-ka * t)))
    return concentration

# Dosing regimens for the clinical trial (mg)
dose_values = [200, 400, 600, 800, 1000, 1200, 1400]

# Parameters
ka = 1.0            # Absorption rate constant (1/hour)
ke_patient1 = 0.3 # Elimination rate constant for target patient group
Vd = 50             # Volume of distribution (L)
```

```
# Therapeutic range (mg/L)
C_min = 5.0   # Minimum therapeutic concentration
C_max = 15.0  # Maximum therapeutic concentration

# Time points (hours)
time = np.linspace(0, 24, 100)

# Plot concentration-time profiles for the different doses
plt.figure(figsize=(10, 6))
for dose in dose_values:
    concentration = pk_model(time, ka, ke_patient1, dose, Vd)
    plt.plot(time, concentration, label=f'Dose = {dose} mg')

# Plot therapeutic range
plt.axhline(C_min, color='green', linestyle='--',
label='C_min (Therapeutic Range)')
plt.axhline(C_max, color='red', linestyle='--',
label='C_max (Therapeutic Range)')

plt.xlabel('Time (hours)')
plt.ylabel('Concentration (mg/L)')
plt.title('PK/PD Modeling for Clinical Trial Dose Selection')
plt.legend()
plt.grid(True)
plt.show()
```



## 11.10 Exercises and Questions

The following examples and questions are designed to deepen students' understanding of pharmacokinetics and pharmacodynamics modeling using Python. They provide opportunities to practice coding, think critically about the challenges of drug development, and apply PK/PD concepts in various contexts.

### Exercises

The following exercises will enhance your understanding of the materials we covered in this chapter.

1. Simulating a one-compartment PK model: Write a Python script to simulate the concentration-time profile for a drug administered orally using a one-compartment PK model with first-order absorption and elimination. Vary the dose and absorption rate constant ($k_a$) to observe how they affect the concentration-time curve.
2. Two-compartment PK model: Develop a Python script to simulate a two-compartment PK model with first-order elimination. Compare the concentration-time profiles in both the central and peripheral compartments for different dosing regimens.
3. Dose-response curve using the Hill equation: Implement the Hill equation in Python to simulate a dose-response curve for a drug. Vary the $EC_{50}$ and Hill coefficient to see how they influence the shape of the curve.

4. Estimating PK parameters using nonlinear regression: Given a dataset of plasma drug concentrations over time, use Python's curve_fit function from SciPy to estimate the absorption rate constant ($k_a$) and elimination rate constant ($k_e$). Plot the experimental data and fitted curve.
5. Visualizing dose-response relationships: Create a Python script that uses Seaborn to visualize dose-response data for multiple patients, each with different $EC_{50}$ values. Explore how individual variability affects the observed therapeutic effects.
6. Sensitivity analysis in PK modeling: Perform a sensitivity analysis by varying both the absorption rate constant ($k_a$) and elimination rate constant ($k_e$) in a one-compartment PK model. Plot the concentration-time profiles to understand which parameter has a greater influence on drug concentration.
7. Simulating time-dependent PD models: Implement a time-dependent pharmacodynamic model in Python to simulate the delayed effect of a drug. Introduce a delay factor and plot how the pharmacological effect evolves over time.
8. Integrated PK/PD modeling: Write a Python script to integrate a PK model with a PD model using the Hill equation. Simulate both the concentration-time profile and the pharmacological effect over time.
9. Model validation and residual plots: Given a dataset of concentration-time data, fit a PK model and create residual plots to assess the goodness of fit. Calculate the $R^2$ value and discuss whether the model is a good fit for the data.

## Questions to Be Answered

1. What are the key differences between a one-compartment and a two-compartment PK model?
2. How can the Hill equation be used to model dose-response relationships? What are the key parameters involved?
3. Explain the concept of nonlinear regression in PK/PD modeling and its importance in parameter estimation.
4. What are the advantages of using integrated PK/PD models over separate PK and PD models?
5. In what ways can sensitivity analysis be useful in understanding the behavior of PK/PD models?
6. Why is model validation important in PK/PD modeling, and what are some of the key techniques used for validating models?
7. How does the absorption rate constant ($k_a$) influence the concentration-time profile of a drug?
8. How can PK/PD models contribute to personalized medicine and the optimization of drug therapy for individual patients?

**Acknowledgement** The language of the human-generated text was corrected with the assistance of artificial intelligence (AI) tools [GPT-3.5 and GTP-4 from OpenAI]. GitHub Co-Pilot was used to check the correctness of the codes. The text underwent subsequent human revision to ensure its accuracy.

## References


1. Gallo JM. Pharmacokinetic/pharmacodynamic-driven drug development. Mt Sinai J Med. 2010;77(4):381–8.
2. Fan J, de Lannoy IA. Pharmacokinetics. Biochem Pharmacol. 2014;87(1):93–120.
3. Jollow DJ, Brodie BB. Mechanisms of drug absorption and of drug solution pharmacology. 1972;8(1):21–32. https://doi.org/10.1159/000136324.
4. Onetto AJ, Sharif S. Drug distribution. 2023 Jun 26. In: StatPearls [Internet]. Treasure Island: StatPearls Publishing; 2025. PMID: 33620813.
5. Susa ST, Hussain A, Preuss CV. Drug metabolism. 2023 Aug 17. In: StatPearls [Internet]. Treasure Island: StatPearls Publishing; 2025. PMID: 28723052.
6. Zhao M, Ma J, Li M, Zhang Y, Jiang B, Zhao X, Huai C, Shen L, Zhang N, He L, Qin S. Cytochrome P450 enzymes and drug metabolism in humans. Int J Mol Sci. 2021;22(23):12808. https://doi.org/10.3390/ijms222312808.
7. Hossam Abdelmonem B, Abdelaal NM, Anwer EKE, Rashwan AA, Hussein MA, Ahmed YF, Khashana R, Hanna MM, Abdelnaser A. Decoding the role of CYP450 enzymes in metabolism and disease: a comprehensive review. Biomedicines. 2024;12(7):1467. https://doi.org/10.3390/biomedicines12071467.
8. Garza AZ, Park SB, Kocz R. Drug elimination. 2023 Jul 4. In: StatPearls [Internet]. Treasure Island: StatPearls Publishing; 2025. PMID: 31613442.
9. Bardal SK, Waechter JE, Martin DS. Chapter 2 - Pharmacokinetics. In: Applied pharmacology; 2011. p. 17–34., ISBN 9781437703108. https://doi.org/10.1016/B978-1-4377-0310-8.00002-6.
10. Dawson PA, Shneider BL, Hofmann AF. Bile formation and the enterohepatic circulation, physiology of the gastrointestinal tract. 4th ed. Academic Press; 2006. p. 1437–62., ISBN 9780120883943. https://doi.org/10.1016/B978-012088394-3/50059-3.
11. Marino M, Jamal Z, Zito PM. Pharmacodynamics. In: StatPearls. Treasure Island; 2024.
12. Watts SW, Townsend RR, Neubig RR. How new developments in Pharmacology Receptor theory Are changing (our understanding of) hypertension therapy. Am J Hypertens. 2024;37(4):248–260 https://doi.org/10.1093/ajh/hpad121.
13. Kazi JU, Rönnstrand L. FMS-like Tyrosine Kinase 3/FLT3: from basic science to clinical implications. Physiol Rev. 2019;99(3):1433–66. https://doi.org/10.1152/physrev.00029.2018.
14. Higham JP, Colquhoun D. The affinity-efficacy problem: an essential part of pharmacology education. R Soc Open Sci. 2024;11(7):240487. https://doi.org/10.1098/rsos.240487.
15. Mahapatra MK, Karuppasamy M. Fundamental considerations in drug design. Computer Aided Drug Design (CADD): from Ligand-based methods to structure-based approaches. 2022:17–55. https://doi.org/10.1016/B978-0-323-90608-1
16. Moini J, Logalbo A, Schnellmann JG. Chapter 4 - Pharmacodynamics, neuropsychopharmacology. Academic Press; 2023. p. 55–74., ISBN 9780323959742,. https://doi.org/10.1016/B978-0-323-95974-2.00012-8.
17. Waud DR. Analysis of dose-response curves. Trends Pharmacol Sci. 1981;2:52–5. https://doi.org/10.1016/0165-6147(81)90261-3.
18. Hewlett PS, Plackett RL. A Unified Theory for quantal responses to mixtures of drugs: non-interactive action. Biometrics. 1959;15(4):591–610. https://doi.org/10.2307/2527657.
19. Blix HS, Viktil KK, Moger TA, Reikvam A. Drugs with narrow therapeutic index as indicators in the risk management of hospitalised patients. Pharm Pract (Granada). 2010;8(1):50–5.
20. Holford N. Pharmacodynamic principles and the time course of immediate drug effects. Transl Clin Pharmacol. 2017;25(4):157–61. https://doi.org/10.12793/tcp.2017.25.4.157.
21. Ngcobo NN. Influence of ageing on the pharmacodynamics and pharmacokinetics of chronically administered medicines in geriatric patients: a review. Clin Pharmacokinet. 2025;64(3):335–367 https://doi.org/10.1007/s40262-024-01466-0.
22. Cacabelos R, Naidoo V, Corzo L, Cacabelos N, Carril JC. Genophenotypic factors and pharmacogenomics in adverse drug reactions. Int J Mol Sci. 2021;22(24):13302. https://doi.org/10.3390/ijms22241330.
23. Di Tonno D, Martena L, Taurisano M, Perlin C, Loiacono AC, Lagravinese S, et al. The requirements of managing Phase I Clinical Trials Risks: the British and Italian case studies. Epidemiologia (Basel). 2024;5(1):137–45.

24. Adashek JJ, LoRusso PM, Hong DS, Kurzrock R. Phase I trials as valid therapeutic options for patients with cancer. Nat Rev Clin Oncol. 2019;16(12):773–8.
25. Zhang Y, Fox GB. PET imaging for receptor occupancy: meditations on calculation and simplification. J Biomed Res. 2012;26(2):69–76. https://doi.org/10.1016/S1674-8301(12)60014-1.
26. Holford N. Pharmacodynamic principles and the time course of delayed and cumulative drug effects. Transl Clin Pharmacol. 2018;26(2):56–9.
27. Wright DF, Winter HR, Duffull SB. Understanding the time course of pharmacological effect: a PKPD approach. Br J Clin Pharmacol. 2011;71(6):815–23.
28. Zou H, Banerjee P, Leung SSY, Yan X. Application of pharmacokinetic-pharmacodynamic modeling in drug delivery: development and challenges. Front Pharmacol. 2020;11:997.
29. Fleishaker JC, Smith RB. Compartmental model analysis in pharmacokinetics. J Clin Pharmacol. 1987;27(12):922–6.
30. Bassingthwaighte JB, Butterworth E, Jardine B, Raymond GM. Compartmental modeling in the analysis of biological systems. Methods Mol Biol. 2012;929:391–438.
31. Wijnand HP. Pharmacokinetic model equations for the one- and two-compartment models with first-order processes in which the absorption and exponential elimination or distribution rate constants are equal. J Pharmacokinet Biopharm. 1988;16(1):109–28.
32. Norman J. One-compartment kinetics. Br J Anaesth. 1992;69(4):387–96.
33. Yan D, Wu X, Li J, Tang S. Statistical analysis of two-compartment pharmacokinetic models with drug non-adherence. Bull Math Biol. 2023;85(7):65.
34. Levy G, Gibaldi M, Jusko WJ. Multicompartment pharmacokinetic models and pharmacologic effects. J Pharm Sci. 1969;58(4):422–4.
35. Borowy CS, Ashurst JV. Physiology, zero and first order kinetics. 2022 Sep 19. In: StatPearls [Internet]. Treasure Island: StatPearls Publishing; 2025. PMID: 29763041.
36. Wang J, Chen T, Ruszaj DM, Mager DE, Straubinger RM. Integrated PK/PD modeling relates smoothened inhibitor biomarkers to the heterogeneous intratumor disposition of cetuximab in pancreatic cancer tumor models. J Pharm Sci. 2024;113(1):72–84.
37. Daryaee F, Tonge PJ. Pharmacokinetic-pharmacodynamic models that incorporate drug-target binding kinetics. Curr Opin Chem Biol. 2019;50:120–7.
38. Garrett ER. The Bateman function revisited: A critical reevaluation of the quantitative expressions to characterize concentrations in the one compartment body model as a function of time with first-order invasion and first-order elimination. J Pharmacokinet Biopharm. 1994;22:103–28. https://doi.org/10.1007/BF02353538.
39. Dziak JJ, Coffman DL, Lanza ST, Li R, Jermiin LS. Sensitivity and specificity of information criteria. Brief Bioinform. 2020;21(2):553–65.
40. Lescarbeau R, Kaplan DL. Correlating phosphoproteomic signaling with castration resistant prostate cancer survival through regression analysis. Mol Biosyst. 2014;10(3):605–12.

# Natural Language Processing (NLP) Basics

Contents




J. U. Kazi, *Python Essentials for Biomedical Data Analysis: An Introductory Textbook*,
https://doi.org/10.1007/978-3-031-85600-6_12

This chapter provides a brief introduction to natural language processing (NLP), an area within artificial intelligence and computational linguistics focused on how computers work with human language. The chapter discusses widely used Python tools for NLP, such as NLTK and spaCy. Furthermore, it explores the applications of NLP in biomedicine, particularly in drug discovery, patient care, and biomedical research. The chapter also briefly describes the role of NLP in interpreting clinical texts and extracting biomedical entities. The overall goal is to provide readers with a starting point of understanding NLP and its growing influence on the biomedical field.

### Learning Goals

In this chapter, students will gain a brief understanding of how NLP techniques are applied in biomedical research. They will explore how NLP can be used to process, analyze, and extract relevant information from various types of biomedical text data, such as clinical notes, research articles, and electronic health records (EHRs). This chapter also covers widely used NLP methods such as tokenization, named entity recognition (NER), relation extraction, sentiment analysis, and literature-based discovery (LBD). Moreover, students will learn how to handle domain-specific challenges in biomedical texts, such as dealing with specialized terminology and abbreviations in clinical data.

## 12.1 Introduction to NLP

NLP has become a valuable tool in biomedical research, allowing researchers and clinicians to extract useful information from large amounts of unstructured text data [1]. As biomedical literature, EHRs, clinical notes, and other text-based resources continue to grow rapidly, NLP offers practical methods to process, analyze, and interpret this complex information. From identifying disease-related genes in research papers to extracting patient information from clinical reports, NLP helps bridge the gap between textual data with practical outcomes in biomedicine [2].

In the biomedical applications, NLP tasks involve not only traditional language processing challenges but also domain-specific complexities such as handling specialized medical terminology, abbreviations, and ambiguous terms [3]. Advances in machine learning and deep learning techniques, along with the availability of domain-specific tools and libraries, have expanded the possibility for NLP in this field.

This chapter explores the basics of NLP, focusing on its application in biomedical research. We will introduce key NLP concepts, Python libraries often used for biomedical text processing, and examples to demonstrate how NLP can address biomedical problems, such as named entity recognition (NER), text classification, and information extraction.

### 12.1.1 NLP in Biomedical Research

NLP enables the efficient extraction and analysis of large amounts of unstructured text data. Biomedical research generates a wide range of textual data, including scientific articles, clinical trial reports, EHRs, drug labels, and patient narratives. These texts contain useful information about diseases, drugs, treatments, and outcomes, which can inform research, patient care, and clinical decision-making. However, manually extracting and analyzing this data is time-consuming, error-prone, and often infeasible given the sheer volume of information. NLP provides automated methods for processing and interpreting biomedical texts at scale.

NLP supports several key tasks that facilitate the analysis and interpretation of textual data. One such task is NER, which involves identifying important biomedical entities such as including genes, proteins, diseases, and drugs within the text. Due to the frequent use of specialized terminology and abbreviations, NER in biomedicine is often more complex than in general domains [4]. Another key task is text classification, which assigns predefined categories to biomedical documents [5]. For example, research papers can be classified based on disease categories, or clinical notes can be sorted by patient conditions, making it easier to organize and retrieve information from large databases. Information extraction focuses on pulling structured information from unstructured text, such as identifying relationships between genes and diseases or detecting drug-drug interactions [6]. This helps us to identify meaningful connections between biomedical concepts. Literature-based discovery (LBD) uses NLP to mine scientific literature and discover hidden connections between concepts, leading to new hypotheses and research directions [7]. For instance, LBD has been applied in drug repurposing efforts. Lastly, sentiment analysis involves analyzing opinions and sentiments expressed in patient reviews, clinical narratives, or social media posts related to health conditions or treatments [8]. This can help assess patient satisfaction, evaluate treatment effectiveness, and monitor public health trends.



### 12.1.2 Applications of NLP in Biomedical Research

NLP has been applied in a wide range of biomedical applications that support both research and clinical practices. One key application is clinical decision support, where NLP tools extract and analyze relevant information from clinical notes and EHRs to assist clinicians in making informed decisions about diagnoses and treatment options. For example, NLP can identify patterns in patient records that suggest early signs of diseases or adverse drug reactions [9].

NLP also plays a role in drug discovery and development. NLP can be used to mine biomedical literature for information on gene-drug interactions, drug efficacy, and adverse effects [10, 11]. This information aids in identifying novel therapeutic targets or repurposing existing drugs for new indications. In systematic reviews and meta-analyses, conducting systematic reviews often requires reviewing thousands of papers to extract relevant information. NLP tools can automate parts of this process by identifying and summarizing key findings from research papers, thereby speeding up evidence synthesis.

Furthermore, NLP is used in public health monitoring by analyzing social media posts, online forums, and patient reviews to identify emerging health trends, outbreaks, or public perceptions of vaccines and treatments. For instance, sentiment analysis of tweets during a flu season can help public health officials track public concerns and outbreaks [12].

NLP's ability to process unstructured data, identify relevant entities, extract relationships, and generate structured information from biomedical texts makes it a relevant tool for researchers and healthcare professionals. As new algorithms and techniques are developed to handle the unique challenges posed by biomedical text data, the applications of NLP in biomedicine continue to grow.

## 12.2 Understanding Biomedical Text Data

Biomedical text data is rich in content but often highly specialized, making it challenging to process and analyze. The nature of the language used in biomedical literature, clinical notes, EHRs, and other related sources differs from general text data. This section explores the types of biomedical text data, the challenges they pose for NLP, and the common data formats encountered in biomedical text mining.

### 12.2.1 Types of Biomedical Text Data

Biomedical texts come from a wide range of sources, each with its own structure, content, and focus. Understanding the types of text data available in biomedical research helps in selecting the appropriate NLP techniques. ◘ Table 12.1 provides an overview of the main types of biomedical text data, their structures, and how NLP can be applied to them.

### 12.2.2 Challenges in Biomedical Text Processing

Processing biomedical text data presents unique challenges that set it apart from general NLP tasks and often requires tailored approaches to develop effective models for biomedical applications (◘ Fig. 12.1). One major challenge is ambiguity, as biomedical texts frequently contain terms with multiple meanings depending on the context. For example, the abbreviation "BP" could signify "blood pressure" or "biological process," and NLP systems must accurately interpret such terms based on their contextual usage. Furthermore, biomedical language is rich in specialized terminology, including gene names, chemical compounds, and medical diagnoses, which are not typically encountered in general language corpora. This can lead to misinterpretation by standard NLP tools, such as mistaking the protein "p53" for a common word rather than recognizing it as a gene symbol.

The frequent use of acronyms and abbreviations further complicates the processing of biomedical text, as these are often domain-specific and may vary widely. For instance, "AML" could mean "acute myeloid leukemia" or "advanced machine learning," and NLP systems must determine the correct expansion based on context.

**Table 12.1** Overview of biomedical text data types and their NLP applications

| Type of text | Structure | Use case |
|---|---|---|
| Research papers and journal articles | Typically follow a structured format (introduction, methods, results, discussion) and contain highly technical language | NLP can extract key information from large volumes of biomedical literature, such as identifying gene-disease associations, summarizing findings, or classifying articles by topic |
| Clinical notes and EHRs | Unstructured or semi-structured, often written in free text by clinicians. Include patient history, diagnoses, medications, lab results, and treatment plans | NLP is used to extract patient-specific information (e.g., symptoms, diagnoses) from clinical notes, enabling decision support systems or personalized treatment recommendations |
| Drug labels and package inserts | Contain detailed information about drug composition, indications, contraindications, side effects, and dosage guidelines. Usually, semi-structured but involve dense medical terminology | NLP helps in extracting drug information for safety monitoring, clinical decision-making, and regulatory compliance |
| Clinical trial reports | Summarize the methodology, results, and conclusions of clinical studies. Highly structured with standardized sections like study design, patient population, and outcomes | NLP can extract data about trial outcomes, drug efficacy, and adverse effects to aid in systematic reviews and meta-analyses |
| Patents | Describe novel inventions related to drugs, medical devices, or therapies. Highly structured and written in formal, technical language | NLP is used to mine patents for information about new drug discoveries, formulations, or intellectual property claims |



Clinical notes add another layer of complexity due to their lack of standardized language. Healthcare providers may use informal expressions or shorthand that are not universally recognized, such as "pt" for "patient," "HTN" for "hypertension," and "fx" for "fracture," presenting additional hurdles for accurate interpretation.

Biomedical literature also tends to feature complex sentence structures, characterized by long sentences with multiple clauses and embedded phrases. These structures are particularly challenging for NLP algorithms tasked with extracting relationships, such as those between drugs and diseases. Furthermore, processing clinical texts from sources like EHRs brings important considerations regarding data privacy and ethics. These records often contain sensitive patient information, necessitating strict compliance with regulations like the health insurance portability and accountability act (HIPAA). Biomedical NLP systems need to address these challenges by ensuring the use of deidentified data and implementing strong privacy-preserving measures.

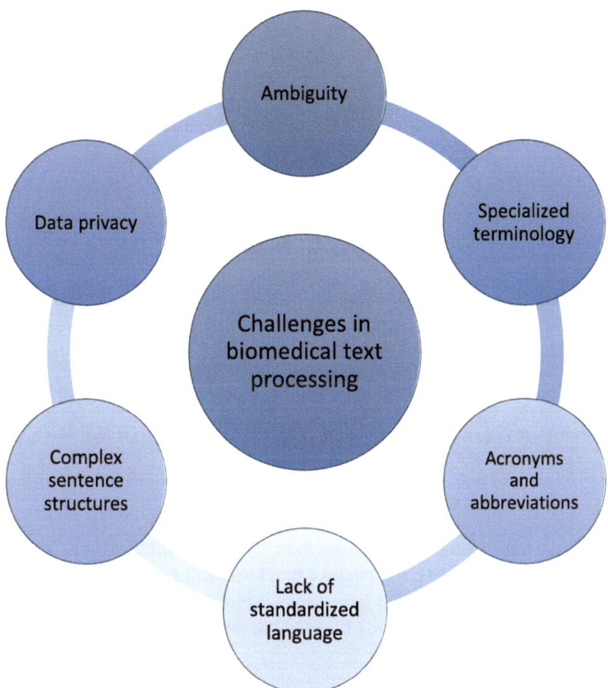


◘ **Fig. 12.1** Challenges in biomedical text processing. Various challenges are encountered when processing biomedical texts using NLP techniques including ambiguity, specialized terminology, acronyms and abbreviations, lack of standardized language, complex sentence structures, and data privacy. Each challenge is interconnected, highlighting the complexity and interrelated nature of these issues in the domain of biomedical NLP

### 12.2.3 Common Data Formats in Biomedical Text Mining

When working with biomedical text data, it is important to be familiar with the different formats in which these texts are stored and how they can be processed for NLP tasks (◘ Fig. 12.2).

Research papers, clinical notes, and drug labels are often stored as plain text files, which can be directly processed with standard NLP tools. However, the lack of structure in plain text means that custom methods are often needed to extract relevant sections and entities. In contrast, many biomedical databases and journals store data in XML or HTML formats, which provide a structured representation of text and metadata, such as article titles, authors, and section headers.

Biomedical data retrieved from APIs or electronic systems is frequently stored in JSON format, particularly for structured datasets like EHRs. JSON's hierarchical organization facilitates the extraction of specific fields of interest, making it useful for NLP tasks. A key example is PubMed, a widely used resource for biomedical literature, which offers data in XML format. This structured format includes metadata such as publication dates, abstracts, authors, and MeSH (medical subject headings) terms, which supports NLP applications like text classification and NER.

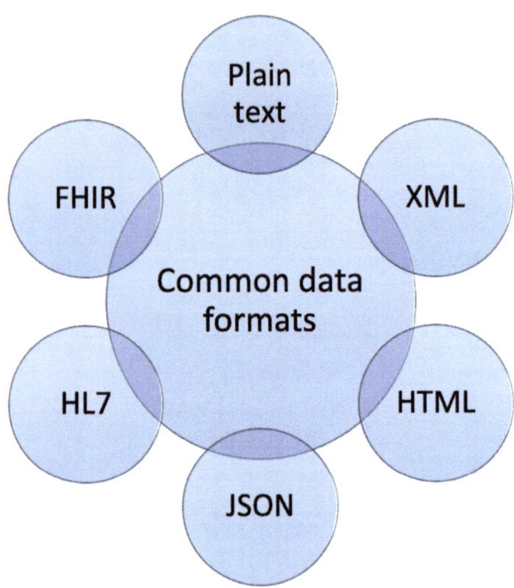


**Fig. 12.2** Representation of common data format. Common data formats include plain text, XML (extensible markup language), HTML (hypertext markup language), JSON (JavaScript object notation), HL7 (health level 7), and FHIR (fast healthcare interoperability resources). These formats facilitate data exchange and integration in various biomedical and healthcare applications

Healthcare settings often rely on clinical data exchange standards such as HL7 (health level 7) and FHIR (fast healthcare interoperability resources) for storing patient records and clinical data. These standards provide structured representations that can be processed with NLP tools to extract patient information or summarize clinical reports efficiently. Each format serves a distinct purpose, but collectively they support the diverse data needs of biomedical research and healthcare.



## 12.3 Basic Concepts in NLP

NLP involves several basic tasks that help to transform raw text into a format suitable for analysis. These tasks include tokenization, stemming, lemmatization, stopword removal, punctuation removal, lowercasing, special character removal, handling abbreviations and acronyms, normalization, sentence splitting, removing nonbiomedical text elements and domain-specific preprocessing. When working with biomedical text data, these preprocessing steps are especially important due to domain-specific challenges such as specialized terminology, frequent abbreviations, and complex sentence structures. We will discuss some of these fundamental NLP concepts, focusing on their application.

### 12.3.1 Tokenization and Text Preprocessing

Tokenization is the process of breaking text into smaller units called tokens, which can be individual words, phrases, or characters [13]. As the first step in most NLP workflows, tokenization enables subsequent tasks like entity recognition, text classification, and analysis. Proper tokenization helps to preserve structure and meaning of the text, which is particularly important in biomedicine where terminology and formatting can be highly specialized.

Tokenization of biomedical texts faces several challenges. For example, multi-word terms, such as "non-small cell lung cancer" or "tumor necrosis factor," need to be treated as single tokens to retain their semantic integrity. Splitting these into individual words would obscure their specific meaning. Similarly, hyphenated terms like "anti-inflammatory" or "T cell" must be processed carefully to avoid breaking important biomedical concepts into irrelevant fragments. Tailoring tokenization methods can better prepare data for downstream tasks and achieve more accurate and meaningful results.

Example: Tokenization in Python using NLTK.

```
import nltk
from nltk.tokenize import word_tokenize

nltk.download('punkt_tab') # For tokenization
# Example biomedical text
text = ("The patient was treated with anti-inflammatory"
"drugs for non-small cell lung cancer.")


# Tokenize the text
tokens = word_tokenize(text)
print(tokens)
```

Output:

```
['The', 'patient', 'was', 'treated', 'with', 'anti-
inflammatory', 'drugs', 'for', 'non-small', 'cell', 'lung',
'cancer', '.']
```

In this example, tokenization splits the sentence into individual words and handles punctuation. However, the hyphenated terms like "anti-inflammatory" and "non-small" remain intact, preserving their biomedical meaning.

### 12.3.2 Stemming and Lemmatization in Biomedical Texts

**Notebook: Section 12.3.2. Stemming and Lemmatization in Biomedical Texts**
This notebook provides code examples demonstrating stemming and lemmatization in biomedical texts.

Link to the GitHub repository:
► https://github.com/sn-code-inside/BioPy
Go to: Chap. 12—► Sect. 12.3.2.

Stemming and lemmatization are techniques used to reduce words to their base or root forms [14, 15]. This is useful in biomedical texts, where words may appear in various inflected forms (e.g., "inhibitors," "inhibit," "inhibition"). Reducing words to their base form allows for more effective comparison and analysis.

Stemming involves cutting off prefixes or suffixes to create the root form of a word. For example, "inhibiting," "inhibited," and "inhibition" may all be stemmed to "inhibit." Stemming is a rule-based process and can sometimes lead to nonstandard word forms. On the other hand, lemmatization is a more sophisticated approach that uses context and part-of-speech information to reduce words to their base form, or "lemma." For example, lemmatization would convert "cells" to "cell," or "was" to "be."

Example: Stemming using NLTK.



```
import nltk
from nltk.stem import PorterStemmer

nltk.download('punkt')  # For tokenization
nltk.download('stopwords')  # For stopword removal

# Create an instance of the PorterStemmer
stemmer = PorterStemmer()

# List of example words to stem
words = ["running", "runner", "runs", "easily", "happily",
"treatment", "treated", "treating"]

# Perform stemming on each word
stemmed_words = [stemmer.stem(word) for word in words]

# Print the original words and their stemmed versions
for original, stemmed in zip(words, stemmed_words):
    print(f"Original: {original} -> Stemmed: {stemmed}")
```

Output:

```
Original: running -> Stemmed: run
Original: runner -> Stemmed: runner
Original: runs -> Stemmed: run
Original: easily -> Stemmed: easili
Original: happily -> Stemmed: happili
Original: treatment -> Stemmed: treatment
Original: treated -> Stemmed: treat
Original: treating -> Stemmed: treat
```

Example: Lemmatization using NLTK.

```
import nltk
from nltk.stem import WordNetLemmatizer
# Download WordNet data (if not already done)
nltk.download('wordnet')

# Initialize the lemmatizer
lemmatizer = WordNetLemmatizer()

# Example biomedical words
words = ["cells", "inhibitors", "binding", "mice"]

# Lemmatize the words
lemmatized_words = [lemmatizer.lemmatize(word) for word in words]
print(lemmatized_words)
```

Output:

```
['cell', 'inhibitor', 'binding', 'mouse']
```

In these examples, Stemmer stemmed some words (like "easily" → "easili") may not resemble valid English words because stemming focuses on removing suffixes rather than producing linguistically correct roots. The lemmatizer correctly identifies the root form of each word, converting "cells" to "cell" and "mice" to "mouse," which is important in biomedical contexts where plural and singular forms need to be matched.

### 12.3.3 Removing Stopwords and Punctuation

**Notebook: Section 12.3.3. Removing Stopwords and Punctuation**
This notebook provides code examples demonstrating removing stopwords and punctuation.

Link to the GitHub repository:
► https://github.com/sn-code-inside/BioPy
Go to: Chap. 12—► Sect. 12.3.3.

Stopwords are common words like "the," "is," and "and" that do not typically carry meaning in a text and can be removed during preprocessing [16]. Removing stopwords reduces the dimensionality of the data and focuses attention on the more meaningful terms in a document. However, in biomedical texts, certain common words that might be treated as "stopwords" in general context, such as "inhibits" or "binds," may carry relevance depending on the analysis, so care must be taken in choosing which words to filter out.

Biomedical texts often contain punctuation and special characters (e.g., commas, parentheses, or percentages) that may need to be removed, depending on the analysis. Some punctuation marks, such as hyphens, may be part of meaningful biomedical terms, so these must be handled carefully.

Example: Removing stopwords and punctuation in Python using NLTK.



```
import string
import nltk
from nltk.corpus import stopwords
from nltk.tokenize import word_tokenize


# Download required NLTK data (stopwords and tokenizer)
nltk.download('stopwords')
nltk.download('punkt')


# Load stopwords
stop_words = set(stopwords.words('english'))


# Example biomedical text
text = ("The patient was treated with anti-inflammatory"
"drugs for non-small cell lung cancer.")


# Tokenize and filter stopwords
tokens = word_tokenize(text)
filtered_tokens = [word for word in tokens if word.lower()
not in stop_words and word not in string.punctuation]


print(filtered_tokens)
```

Output:

```
['patient', 'treated', 'anti-inflammatory', 'drugs', 'non-small',
'cell', 'lung', 'cancer']
```

In this example, common stopwords such as "The," "was," and "with" are removed from the text, leaving behind only the important biomedical terms. Punctuation such as periods is also removed, except for hyphens that are part of biomedical terms.

### 12.3.4 Handling Abbreviations and Acronyms

**Notebook: Section 12.3.4. Handling Abbreviations and Acronyms**
This notebook provides code examples demonstrating handling abbreviations and acronyms.
Link to the GitHub repository:
► https://github.com/sn-code-inside/BioPy
Go to: Chap. 12—► Sect. 12.3.4.

Biomedical texts are filled with abbreviations and acronyms that can have multiple meanings depending on the context. For example, as described previously as well, "BP" might represent "blood pressure" in a clinical setting or "biological process" in a molecular biology paper. Correctly resolving these terms is important for accurate NLP and downstream tasks like entity recognition or text classification.

Handling abbreviations and acronyms can involve several approaches, including the use of predefined abbreviation dictionaries, applying rule-based methods, or using machine-learning models for contextual disambiguation. Below, we explore some of these methods with examples.

*Using abbreviation dictionaries:* Predefined dictionaries map abbreviations to their expanded forms. While this method is straightforward, it relies on the completeness and accuracy of the dictionary.

```
import re
import string
# Define an abbreviation dictionary
abbreviation_dict = {
    "BP": "blood pressure",
    "HR": "heart rate",
    "ECG": "electrocardiogram",
    "AML": "acute myeloid leukemia"
}
```

```
# Function to expand abbreviations
def expand_abbreviation(text, abbr_dict):
    # Tokenize text using regex to separate words and punctuation
    tokens = re.findall(r'\b\w+\b|[^\s\w]', text)
    # Keeps punctuation as separate tokens
    # Expand abbreviations using the dictionary
    expanded_tokens = [abbr_dict.get(token, token) for token
    in tokens]
    # Reconstruct the text with spaces
    return ''.join(token if token in string.punctuation else
    ' ' + token for token in expanded_tokens).strip()

# Example biomedical text
text = "Patient has elevated BP and HR; ECG is abnormal."

# Expand abbreviations
expanded_text = expand_abbreviation(text, abbreviation_dict)
print(expanded_text)
```

Output:

```
Patient has elevated blood pressure and heart rate; electro-
cardiogram is abnormal.
```

*Contextual disambiguation:* Some abbreviations have multiple possible meanings. For example, "AML" could mean "acute myeloid leukemia" or "advanced machine learning." Disambiguating these requires understanding the context, often using machine learning models or context-aware embeddings like BERT.

Simplified example:

```
from transformers import pipeline

# Load a pre-trained model for masked language modeling
fill_mask = pipeline("fill-mask", model="bert-base-uncased",
device='mps')

# Example sentences
sentences = [
    "AML is a type of cancer affecting the blood and bone marrow.",
    "He is researching advanced machine learning (AML) techniques."
]

# Disambiguate "AML" using context
for sentence in sentences:
    print(f"Original: {sentence}")
    masked_sentence = sentence.replace("AML", "[MASK]")
    predictions = fill_mask(masked_sentence)
```

```
        print("Predictions for '[MASK]':")
        for pred in predictions[:3]:  # Show top 3 predictions
            print(f"  {pred['sequence']} ({pred['score']:.4f})")
        print()
```

Output:

```
    Original: AML is a type of cancer affecting the blood and
bone marrow.
    Predictions for '[MASK]':
      it is a type of cancer affecting the blood and bone marrow.
      (0.8779)
      this is a type of cancer affecting the blood and bone
      marrow. (0.0432)
      cancer is a type of cancer affecting the blood and bone
      marrow. (0.0287)

    Original: He is researching advanced machine learning (AML)
    techniques.
    Predictions for '[MASK]':
      he is researching advanced machine learning ( cad )
      techniques. (0.1885)
      he is researching advanced machine learning ( ai )
      techniques. (0.1468)
      he is researching advanced machine learning ( ada )
      techniques. (0.0763)
```

*Rule-based contextual expansion*: For structured text like clinical notes, context-aware rules can be defined. For instance, if “BP” occurs alongside “mmHg,” it is likely referring to “blood pressure.”

```
    # Rule-based context-aware abbreviation expansion
    def rule_based_expansion(text):
        if "BP" in text and "mmHg" in text:
            text = text.replace("BP", "blood pressure")
        if "AML" in text and "leukemia" in text:
            text = text.replace("AML", "acute myeloid leukemia")
        return text

    # Example text
    clinical_text = ("BP was recorded as 120/80 mmHg. The leukemia"
    "diagnosis was confirmed as AML type.")

    # Apply rule-based expansion
    expanded_text = rule_based_expansion(clinical_text)
    print(expanded_text)
```

Output:

```
blood pressure was recorded as 120/80 mmHg. The leukemia
diagnosis was confirmed as acute myeloid leukemia type.
```

*Hybrid methods:* A combination of dictionaries, rule-based approaches, and machine learning can be used to handle abbreviations and acronyms more effectively.

### 12.3.5 Sentence Splitting and Paragraph Segmentation

**Notebook: Section 12.3.5. Sentence Splitting and Paragraph Segmentation**
This notebook provides code examples demonstrating sentence splitting and paragraph segmentation.

Link to the GitHub repository:
► https://github.com/sn-code-inside/BioPy
Go to: Chap. 12—► Sect. 12.3.5.

Splitting text into sentences and paragraphs enables downstream tasks such as text summarization, NER, and relationship extraction. By breaking down text into meaningful units, NLP systems can better interpret context and process information more efficiently. Biomedical texts, however, pose challenges for sentence and paragraph segmentation due to features, such as the frequent use of abbreviations, numbered lists, and complex sentence structures.



Sentence splitting, also known as sentence tokenization, involves dividing a text into individual sentences. While this process might appear straightforward, as punctuation marks like periods, exclamation marks, or question marks often indicate the end of a sentence, biomedical texts contain many exceptions that require advanced handling:

Abbreviations and acronyms: Texts often use abbreviations like "e.g.," "Dr.," or specific terms such as "et al." or "Fig." These can be mistaken for sentence boundaries unless carefully handled.

Decimal numbers: Numeric data such as "3.5 mg" or "pH 7.4" should not trigger sentence breaks.

Complex parenthetical expressions: Sentences may include nested parentheses or references, such as "(see Fig. 2b for details)."

Modern sentence tokenization techniques use rule-based methods, machine learning models, or hybrid approaches to address these challenges.

Example code for sentence splitting.

```
import nltk
nltk.download('punkt')
from nltk.tokenize import sent_tokenize

# Example biomedical text
text = ("Dr. Smith reported that p53 mutations were observed. The"
"patient was given 3.5 mg of the drug. See Fig. 2b for details.")

# Sentence splitting
sentences = sent_tokenize(text)
print(sentences)
```

Output:

```
['Dr. Smith reported that p53 mutations were observed.',
'The patient was given 3.5 mg of the drug.', 'See Fig.', '2b
for details.']
```

Paragraph segmentation involves splitting text into paragraphs, which represent higher-level structural units. Unlike sentences, paragraphs provide context and group related information, often used for tasks like topic modeling, summarization, and document-level information extraction. Paragraph segmentation typically relies on:

Line breaks or indentation: In structured documents like journal articles, line breaks or indentations often signal paragraph boundaries.

Section headings: In biomedical texts, paragraphs often begin with headings or keywords like "Introduction" or "Methods."

Explicit markers: Paragraph markers (e.g., <p> in HTML or \n\n in plain text) can aid in segmentation.

Example code for paragraph segmentation.

```
# Example biomedical text with paragraph markers
text = ("Introduction\n\np53 mutations are common in cancer."
"They affect DNA repair pathways.\n\nMethods\n\nThe study"
"analyzed 100 samples using sequencing.")

# Paragraph splitting based on double line breaks
paragraphs = text.split('\n\n')
for i, paragraph in enumerate(paragraphs):
    print(f"Paragraph {i + 1}: {paragraph}")
```

Output:

```
Paragraph 1: Introduction
Paragraph 2: p53 mutations are common in cancer. They affect
DNA repair pathways.
Paragraph 3: Methods
Paragraph 4: The study analyzed 100 samples using sequencing.
```

### 12.3.6 Normalization of Text

**Notebook: Section 12.3.6. Normalization of Text**

This notebook provides code examples demonstrating the normalization of text.

Link to the GitHub repository:

▸ https://github.com/sn-code-inside/BioPy

Go to: Chap. 12—▸ Sect. 12.3.6.

Text normalization aims to standardize textual data for consistency and improved analysis. It includes steps such as converting text to lowercase, expanding contractions, normalizing Unicode characters, handling mixed-case terms, and standardizing formatting. In biomedical texts, normalization becomes more complex due to specific elements such as gene names, chemical formulas, and mixed-case terms, which need careful handling to preserve their scientific meaning while ensuring compatibility with NLP pipelines.



*Converting to lowercase:* Standardizing text to lowercase is a common normalization step to ensure uniformity. However, in biomedical texts, some terms are case-sensitive and carry distinct meanings. For example, gene names like "p53" and "P53" may represent different concepts in certain contexts. Chemical compounds, such as "NaCl" (sodium chloride), require case preservation. A selective approach to lowercasing, where specific terms or patterns are excluded, is often necessary.

*Expanding contractions:* Biomedical texts often avoid contractions, but they may still appear in patient narratives or less formal clinical notes. For instance: "can't" to "cannot," "it's" to "it is," etc. Expanding contractions improves clarity and ensures uniform representation of words.

*Unicode normalization:* Texts may include Unicode characters for symbols (e.g., μg for micrograms) or special accents in author names. Normalizing these characters to a standard form ensures compatibility across systems.

Handling mixed-case terms: Mixed-case terms such as "anti-TNFα" or "non-Small Cell Lung Cancer" (NSCLC) are common in biomedical texts. Simple normalization can obscure their meaning, so domain-specific rules or lexicons may be required to standardize such terms effectively.

Standardizing chemical formulas: Chemical formulas like "$H_2O$" or "$CO_2$" require specific handling to retain their scientific integrity. Normalization may involve preserving their format while ensuring consistency across datasets.

In practice, normalization often involves combining multiple tasks into a pipeline tailored to the text's characteristics.

```
import re
import unicodedata
from contractions import fix

# Function to normalize biomedical text
def normalize_text(text):
    # Expand contractions
    text = fix(text)
    # Normalize Unicode
    text = unicodedata.normalize('NFKD', text)
    # Handle case sensitivity
    words = text.split()
    text = ' '.join(word if word.isupper() or word[0].
    islower() else word.lower() for word in words)
    # Remove unnecessary spaces and standardize chemical formulas
    text = re.sub(r'\s+', ' ', text).strip()
    return text

# Example text
text = ("It's common to use 5 µg of Anti-TNFα therapies in"
"non-Small Cell Lung Cancer.")

# Normalize the text
normalized_text = normalize_text(text)
print(normalized_text)
```

Output:

```
it is common to use 5 µg of anti-tnfα therapies in non-Small
cell lung cancer.
```

## 12.4 Python Libraries for NLP in Biomedicine

Python provides a variety of libraries that make NLP in biomedicine accessible and efficient. These libraries support tasks such as tokenization, text classification, NER, and more. Some are general-purpose NLP tools, while others are specifically designed for the biomedical data. In this section, we will describe some Python libraries commonly used in biomedical NLP, including NLTK, spaCy, scikit-learn, and specialized libraries such as scispaCy, Biopython, and PyMedTermino.

### 12.4.1 Natural Language Toolkit (NLTK)

The Natural Language Toolkit (NLTK) is one of the most popular and widely used Python libraries for general NLP tasks [17]. As described above, it provides an extensive suite of tools for handling various stages of text processing, including tokenization, stemming, lemmatization, stopword removal, part-of-speech tagging, and parsing. While NLTK is not specifically designed for biomedical texts, its modular structure and functionality can be adapted to handle tasks in biomedical NLP. Its components can be customized or integrated with domain-specific resources, such as biomedical ontologies and lexicons, to improve its capabilities for specialized applications.

### 12.4.2 spaCy for Biomedical Text Processing

spaCy is a modern, production-ready NLP library that is designed for high efficiency and performance, particularly when working with large text corpora. It provides state-of-the-art tools for tasks like tokenization, part-of-speech tagging, NER, dependency parsing, and text classification. Unlike older libraries, spaCy emphasizes speed and scalability, making it an excellent choice for deploying NLP models in real-world applications.

While spaCy's default models are optimized for general-domain texts, its flexibility allows easy adaptation for biomedical NLP tasks through the use of specialized models, such as those provided by scispaCy. These models are pretrained on large biomedical corpora and are capable of recognizing domain-specific entities like genes, proteins, diseases, and chemicals, addressing the unique needs of biomedical text processing.



spaCy offers fast and robust tokenization that can handle complex biomedical terms, hyphenations, and abbreviations while maintaining high accuracy. General NER capabilities of spaCy can identify entities like names, organizations, and dates. With biomedical extensions like scispaCy, it can also recognize genes, diseases, and chemicals, facilitating specialized tasks such as entity linking [18]. Furthermore, spaCy supports multiple pretrained models, and biomedical adaptations (e.g., scispaCy's `en_ner_bc5cdr_md` and `en_ner_jnlpba_md`) make it effective for scientific and clinical texts. In addition, spaCy includes tools for analyzing syntactic structure, enabling tasks such as extracting relationships between biomedical entities. spaCy is designed to integrate with machine learning workflows, providing compatibility with libraries like TensorFlow and PyTorch for building custom NLP models.

### 12.4.3 scikit-Learn for Text Classification

scikit-learn is one of the most widely used Python libraries for machine learning, supporting a broad range of tools that make it suitable for various NLP tasks, including text classification. It provides efficient implementations of text vectorization techniques such as term frequency-inverse document frequency (TF-IDF), which converts raw text into numerical features suitable for machine learning models. These features can then be used in classification algorithms like support vector machines (SVMs), Naive Bayes, logistic regression, or random forests to solve diverse NLP problems. In biomedical NLP, scikit-learn is often employed to classify medical documents, such as categorizing research articles based on disease types, predicting treatment options based on clinical notes, or even classifying patient records by diagnosis. Its modular design and compatibility with other Python libraries make it particularly useful for building and evaluating customized classification pipelines for domain-specific applications.

scikit-learn also includes robust tools for model evaluation, allowing researchers to assess the performance of classification models using metrics such as precision, recall, F1 score, and area under the curve (AUC). For instance, in a task where clinical notes are classified to identify patients with specific conditions, scikit-learn can be used to train and test models while measuring their effectiveness in correctly identifying relevant cases versus avoiding false positives.

scikit-learn is not specialized for biomedical data but can be adapted to the domain with the inclusion of custom preprocessing steps or biomedical-specific feature extraction techniques. For example, biomedical text often contains domain-specific terminology, abbreviations, and multiword expressions that require careful handling. By integrating scikit-learn with other libraries such as NLTK, spaCy, or scispaCy for preprocessing, researchers can ensure that the textual features used in machine learning models are both accurate and relevant to the task.

### 12.4.4 Bio-Specific Libraries: scispaCy, Biopython, and PyMedTermino

Several Python libraries are specifically designed to handle the complexities of biomedical text processing, supporting tailored tools for tasks such as NER, terminology standardization, and ontology integration. These bio-specific libraries extend the capabilities of general NLP frameworks by incorporating domain-specific knowledge and functionality to address the unique challenges of biomedical data.

scispaCy is a prominent library for processing biomedical and scientific texts, built as an extension of spaCy [18]. It offers pretrained models trained on biomedical corpora, enabling it to recognize specialized entities such as genes, proteins, diseases, chemicals, and more. scispaCy is particularly useful for tasks like extracting key entities from clinical notes or research articles and linking these entities to biomedical ontologies such as Unified Medical Language System (UMLS) or MeSH. This capability allows researchers to perform more complex analyses, such as mapping recognized entities to their standard identifiers or linking terms to a broader semantic network. With its robust NER capabilities and support for efficient text processing, scispaCy is well suited for large-scale biomedical text mining projects.

Biopython, while traditionally focused on computational biology, includes features that complement text processing tasks in biomedicine [19]. For instance, it provides tools for handling biological sequences, parsing file formats like FASTA and GenBank, and interacting with databases such as NCBI Entrez and PubMed. When paired with NLP libraries, Biopython can facilitate tasks like linking text-derived information to genomic data or integrating text annotations with biological datasets. This makes it a versatile library for interdisciplinary applications where textual and biological data need to be processed together.

PyMedTermino is another specialized library aimed at handling biomedical terminologies and ontologies [20]. It simplifies working with standard biomedical vocabularies like SNOMED-CT, ICD-10, and MeSH by providing programmatic access to these resources. PyMedTermino enables tasks such as terminology normalization, concept mapping, and building hierarchical relationships between biomedical terms. By standardizing terms across different datasets, it helps researchers ensure consistency and improve the interpretability of their analyses.

These bio-specific libraries address the unique requirements of biomedical text processing by providing domain-adapted models, tools for handling biological data, and access to standardized terminologies. They can be used independently or integrated into larger NLP pipelines to increase the performance of biomedical applications in research, clinical decision-making, and data mining.

## 12.5 NER in Biomedical Texts

NER involves identifying and classifying specific terms, or "entities," in unstructured text into predefined categories such as diseases, drugs, genes, proteins, or symptoms [4]. NER is useful for extracting key information from scientific articles, clinical notes, and other textual data, allowing users to analyze large volumes of data efficiently. It can be used for information extraction, building knowledge graphs, and enabling LBD, particularly in areas like drug development, clinical decision support, and precision medicine.



As discussed earlier, entities in biomedical texts often consist of multiword expressions, such as "non-small cell lung cancer," or ambiguous terms, such as "BP," which can refer to either blood pressure or biological process. This complexity demands specialized tools and frameworks, such as scispaCy and spaCy, which are designed to handle the nuances of biomedical texts.

The predefined categories for biomedical NER include diseases (e.g., "diabetes mellitus"), drugs and chemicals (e.g., "ibuprofen" or "cisplatin"), genes and proteins (e.g., "BRCA1" or "p53"), and symptoms (e.g., "headache" or "fever"). Accurate identification and classification of these entities using biomedical NER simplifies the study of their relationships, supporting researchers in making new discoveries and advancing healthcare and life sciences [21].

### 12.5.1 Biomedical NER: Identifying Genes, Proteins, Diseases, and Drugs

Biomedical NER extends traditional NER by focusing on entities specific to medicine, biology, and pharmacology, enabling the extraction of highly specialized information [4]. This includes identifying and linking entities like genes, proteins, diseases, and drugs. Biomedical NER has been used in extracting gene-disease relationships from research papers, such as identifying "BRCA1" as a gene and "breast cancer" as a disease and linking them to study their association [22]. It also supports the identification of drug-drug interactions by recognizing drug names in clinical reports or scientific articles and extracting the relationships between them. Moreover, biomedical NER facilitates the recognition of protein-protein or other molecular interactions, where both the entities and their relationships help elucidating biological mechanisms.

Gene names can overlap with common English words, such as "PAX" (a gene) versus "pax" (a general term). Drug names also can add complexity as different brand names or abbreviations may refer to the same compound, for example, "acetaminophen" and "Tylenol." Abbreviations present further ambiguity, with acronyms like "MS" possibly referring to "multiple sclerosis" or "mass spectrometry," depending on the context. These challenges require specialized approaches to ensure accurate recognition and classification of entities.

To address these issues, biomedical NER models use domain-specific datasets and pretrained models designed for the complexities of biomedical text [23]. They often integrate knowledge from resources such as UMLS, MeSH, or DrugBank, and use machine learning or deep learning techniques. This approach improves precision for tasks like extracting relationships, building knowledge graphs, and facilitating discoveries in medical research and clinical practice.

### 12.5.2 Using scispaCy for Biomedical NER

Using scispaCy for biomedical NER is an effective approach for extracting meaningful entities from scientific and biomedical texts [18]. As an extension of spaCy, scispaCy is designed for the biomedical domain and offers pretrained models trained on large, domain-specific corpora. These models are well suited for recognizing a wide range of biomedical entities, such as genes, diseases, drugs, and chemicals, making them useful for tasks like gene-disease relationship extraction, drug discovery, and clinical text analysis.

scispaCy's main features include pretrained models for biomedical NER, which allow for identification of entities specific to the biomedical field. The library also supports large-scale entity linking to biomedical knowledge bases, such as UMLS. This linking feature allows researchers to not only identify entities but also to connect them to detailed information about their definitions, relationships, and associated research. Furthermore, scispaCy is optimized for processing large corpora quickly and efficiently, making it a practical choice for handling the substantial amount of text.

For example, scispaCy can process text such as "EGFR mutations are common in non-small cell lung cancer. Patients are treated with gefitinib" and identify entities like EGFR (a gene), non-small cell lung cancer (a disease), and gefitinib (a drug). By linking these entities to UMLS concepts, researchers can retrieve detailed information about each entity, including associated biological pathways, clinical guidelines, and ongoing clinical trials. This capability supports complex analyses that rely on integrating biomedical data from diverse sources. scispaCy is also effective at handling ambiguities inherent in biomedical texts.

## 12.6 Text Classification and Categorization

Text classification is the process of assigning predefined categories or labels to text data [24]. It is particularly useful for organizing and categorizing large volumes of text, such as scientific articles, clinical notes, EHRs, or drug labels. This enables researchers and healthcare professionals to quickly retrieve relevant information, make data-driven decisions, and automate time-consuming tasks like literature reviews. We will discuss the importance of text classification in biomedical NLP and review common techniques used for classifying biomedical texts.

### 12.6.1 Classifying Biomedical Documents

Text classification can be applied to a range of tasks to support research, clinical decision-making, and pharmacovigilance [25, 26]. For example, research articles can be automatically categorized based on their topics, such as cancer, diabetes, or neuroscience, or their specific focus areas, such as drug discovery or gene therapy. This automation makes it easier to organize and retrieve scientific literature, helping researchers quickly find relevant information. Similarly, clinical document classification enables the sorting of clinical notes or EHRs by diagnoses, patient symptoms, or treatment plans which can support healthcare providers with contextually relevant data.

Another important application is adverse event reporting, where text classification can identify mentions of adverse drug reactions in patient narratives, case studies, or other textual data. Flagging these reports can support drug safety monitoring and pharmacovigilance efforts which helps to detect and mitigate risks associated with medications.



Biomedical text classification typically involves transforming raw, unstructured text into a structured format, such as a vectorized representation using techniques like term frequency-inverse document frequency (TF-IDF) or word embeddings [27]. These representations capture the semantic and syntactic features of the text, which are then fed into machine learning algorithms like support vector machines (SVMs), logistic regression, or deep learning models to classify the text into predefined categories [25, 28–30]. This structured approach enables high accuracy and scalability, making text classification a core component of biomedical NLP workflows.

### 12.6.2 Supervised Learning for Text Classification

Supervised learning for text classification involves training a model on labeled data, where each text sample is associated with a predefined category [30]. The model examines patterns in the labeled data and learns to map specific features of the text to its corresponding label. Once trained, the model can classify new, unseen texts into the appropriate categories, making supervised learning a reliable method for automating classification tasks in biomedical contexts.

The process of supervised text classification begins with data preprocessing. This step involves cleaning and preparing the text by removing irrelevant elements such as stopwords and punctuation and applying stemming or lemmatization to standardize terms. Preprocessing ensures that the text is in a consistent format, improving the model's ability to learn from the data. Next, the text is converted into a numerical representation that machine learning algorithms can process. Methods such as bag-of-words (BoW) and TF-IDF are commonly used to create vectorized representations of the text [27, 31]. These approaches capture the frequency and importance of words in the text while preserving meaningful patterns.

. Algorithms such as Naive Bayes, SVMs, and logistic regression are popular for text classification due to their ability to handle high-dimensional data effectively [30]. Each algorithm has its strengths; for instance, Naive Bayes is computationally efficient and works well for tasks with strong probabilistic relationships, while SVM is effective at handling sparse data and identifying complex decision boundaries. The labeled biomedical text data is then used to train the model, allowing it to learn the relationships between text features and their associated categories. After training, the model's performance is evaluated using metrics like accuracy, precision, recall, and F1-score. These metrics outline model's ability to correctly classify text samples and point areas for improvement.

Supervised learning is particularly effective in biomedical NLP tasks such as classifying research articles by topic, sorting clinical notes by diagnosis, or identifying adverse drug reaction reports. Its use of labeled data ensures high accuracy and relevance, establishing it as a widely used method for biomedical text classification.

## 12.7 Information Extraction and Relation Extraction

Information extraction and relation extraction are tasks in NLP that support biomedical research. Information extraction focuses on extracting structured and relevant information, such as entities, concepts, or events, from unstructured text. Relation extraction builds upon this by identifying and categorizing meaningful relationships between these extracted entities, such as interactions between proteins, associations between genes and diseases, or links between drugs and adverse effects. Together, information extraction and relation extraction enable researchers to transform biomedical text into structured, actionable knowledge.

### 12.7.1 Extracting Key Entities and Relationships from Biomedical Texts

Information extraction focuses on identifying specific pieces of information, such as entities, attributes, or facts, from unstructured text [6]. In the biomedical domain, these entities often include genes, proteins, diseases, drugs, and symptoms. Extracting this information from sources like scientific papers, clinical notes, and drug labels helps researchers and clinicians distill key features from vast amounts of textual data. Once these entities are identified, relation extraction takes the process further by uncovering relationships between two or more entities, providing structured knowledge that supports biomedical research and healthcare.

Relation extraction identifies meaningful associations, such as gene-disease links, drug-drug interactions, or protein-protein interactions, which are required for understanding complex biological and medical phenomena [32, 33]. For instance, relation extraction can uncover the connection between a gene mutation and a disease, such as "BRCA1 is associated with breast cancer," or point out adverse drug interactions, such as "Aspirin interacts with Warfarin." Similarly, relation extraction can elucidate molecular interactions by extracting relationships between proteins, such as "p53 interacts with MDM2," helping to explain biological pathways and mechanisms.

These tasks help automate knowledge extraction and hypothesis generation. Gene-disease relationships can aid in identifying genetic markers for diagnostic or therapeutic purposes. Drug-drug interaction analysis supports pharmacovigilance, ensuring patient safety by identifying adverse effects of coadministered medications. Protein-protein interaction studies contribute to understanding cellular processes and can help identify drug targets.

In practice, information extraction and relation extraction are implemented using advanced NLP techniques, often with domain-specific models and tools. NER identifies biomedical entities in the text, while relation extraction models analyze the surrounding context to classify the relationships between these entities. Libraries such as scispaCy and frameworks like Hugging Face Transformers support efficient implementation of these tasks. Pretrained models like BioBERT or SciBERT further increase accuracy by providing domain-specific contextual embeddings, which help manage the complexity and ambiguity inherent in biomedical texts. In summary, information extraction and relation extraction convert unstructured biomedical data into structured knowledge by identifying key entities and their relationships, thereby accelerating research and improving decision-making in healthcare.



### 12.7.2 Rule-Based Versus Machine Learning Approaches for Relation Extraction

Relation extraction can be approached using rule-based methods and machine learning-based methods [34, 35]. Each approach has its strengths and limitations, making them suitable for different use cases in biomedical contexts.

Rule-based methods rely on predefined rules or patterns to identify and extract relationships between entities. These rules often use grammatical structures, such as dependency parsing, or lexical cues, such as specific keywords or trigger phrases. For example, a rule-based approach to extract gene-disease relationships might target sentences that follow the pattern "[GENE] is associated with [DISEASE]." While this approach is straightforward and effective for well-defined and consistent relationships, it can be limited by its rigidity. Rule-based methods struggle with complex sentence structures or novel relationships that do not match the predefined patterns, leading to missed extractions. Despite these limitations, rule-based methods are relatively easy to implement for domains where the relationships are well understood and can be explicitly described.

In contrast, machine learning–based methods use supervised or semi-supervised learning models to extract relationships [29, 30, 35]. These methods require a labeled dataset where relationships between entities have already been annotated. The model learns to identify patterns in the data that signal a relationship and can generalize to new instances based on these patterns. For instance, a machine learning model for extracting gene-disease relationships might take entity pairs, such as "BRCA1" and "breast cancer," as input and predict whether a relationship exists based on contextual features like the words and syntax between the entities. These methods are more flexible than rule-based approaches and can adapt to diverse and complex sentence structures. However, their reliance on large, high-quality labeled datasets for training can be a barrier.

Machine learning–based methods often outperform rule-based approaches when dealing with unstructured or highly variable text. They can identify implicit relationships that are not directly stated, making them suitable for large-scale extraction tasks. Moreover, deep learning models like BioBERT and SciBERT can capture semantic details and domain-specific context, further improving the accuracy of relation extraction.

Choosing between rule-based and machine learning–based approaches depends on the specific requirements of the task. Rule-based methods are suitable for tasks with well-defined and consistent patterns, especially when labeled data is limited. On the other hand, machine learning–based methods are better suited for tasks requiring scalability, adaptability, and the ability to generalize to new patterns. In practice, a hybrid approach that combines the precision of rule-based methods with the flexibility of machine learning models often yields the best results.

### 12.7.3 Extracting Drug-Drug Interactions from Literature

Extracting drug-drug interactions (DDIs) from biomedical literature may support safer and more effective treatment regimens [36, 37]. DDIs can influence the efficacy of medications or lead to adverse effects, so their identification is important for drug safety and pharmacovigilance. However, information about DDIs is often buried within unstructured text, such as research articles, clinical notes, or drug labels, which makes manual extraction labor-intensive and error-prone. NLP can automate DDI extraction, which facilitates faster analysis and supporting informed clinical decision-making [38].

The process of extracting DDIs begins with data preprocessing. This involves cleaning the text to remove irrelevant elements, tokenizing it into meaningful units, and filtering out stopwords. Entity recognition is then performed to identify the drugs mentioned in the text. NER models, such as those available in scispaCy, are specifically tailored for biomedical entities and can accurately extract drug names from unstructured text. For instance, in the sentence "Aspirin interacts with Warfarin," an NER model would identify "Aspirin" and "Warfarin" as drug entities.

Once the drugs are identified, relation extraction techniques are applied to determine whether a relationship, specifically an interaction, exists between them. Relation extraction can be approached using rule-based methods, where predefined patterns or keywords, such as "interacts with," are used to identify DDIs. For example, a rule might flag any sentence containing "[DRUG1] interacts with [DRUG2]" as an interaction. While straightforward, this approach may struggle with complex sentence structures or implicit relationships.

Machine learning–based approaches offer more flexibility and adaptability for relation extraction. These models are trained on annotated datasets of drug-drug interactions, learning to classify whether a pair of drugs in a sentence exhibits an interaction based on contextual features like the words and syntactic relationships between them. Advanced methods, such as those employing pretrained biomedical models like BioBERT or SciBERT, can capture the semantic nuances of biomedical text, enabling the identification of DDIs even when they are expressed indirectly or in varied linguistic formats.

Automating the extraction of DDIs from literature provides several benefits. It reduces the time and effort required to analyze large amounts of biomedical text, improves the accuracy of identifying drug interactions, and supports real-time updates to clinical decision support systems. This may ensure that healthcare providers and researchers have access to the up-to-date information about drug interactions, ultimately improving patient safety and treatment outcomes. Integrating NLP techniques with scispaCy and machine learning models can greatly support the extraction of DDIs, thereby advancing drug safety research and clinical practice.

### 12.7.4 Benefits of Relation Extraction in Biomedical Research

Relation extraction brings several benefits in biomedical research by improving the ability to process and understand complex information from unstructured text. One main advantage is speed and efficiency. Automated relation extraction allows researchers to quickly analyze large amounts of biomedical text, such as clinical reports, scientific articles, or drug labels. This capability is useful in scenarios where manually reviewing thousands of documents would be time-consuming and impractical. Using NLP techniques, researchers can extract meaningful relationships at scale, allowing them to focus on interpreting the results and making informed decisions.



Another benefit is the ability to uncover new relationships, such as gene-disease associations, drug-disease interactions, or protein-protein interactions. These discoveries can lead to novel hypotheses, driving new research directions and expanding scientific understanding. For instance, LBD uses relation extraction to identify previously unrecognized connections in the biomedical literature, leading to breakthroughs in areas like drug repurposing, biomarker identification, or understanding disease mechanisms.

Relation extraction also contributes to the structured representation of biomedical knowledge. It enables the systematic organization and retrieval of information by converting unstructured text into structured formats, such as knowledge graphs or relational databases. These structured representations make it easier to integrate diverse data sources, enabling researchers to explore complex relationships across multiple domains. For example, knowledge graphs built through relation extraction can support drug discovery by connecting molecular interactions, clinical trial data, and disease pathways, or assist in clinical decision support by linking patient records with relevant guidelines and treatment options.

Overall, relation extraction supports the accessibility, usability, and integration of biomedical information, driving efficiency, discovery, and organization. Its applications in fields like drug development, precision medicine, and translational research underline its transformative impact on the way biomedical knowledge is analyzed, interpreted, and applied.

## 12.8 Topic Modeling and Text Mining

Topic modeling is an unsupervised machine learning technique designed to uncover hidden thematic structures within large collections of text documents [39]. Identifying and grouping related topics provide a way to summarize and explore unstructured text. In biomedical research, topic modeling has a wide range of applications, including clustering research articles by themes, identifying emerging trends in scientific literature, and analyzing patterns in clinical notes or EHRs. This approach helps researchers to distill key information from extensive datasets and can guide them toward relevant areas of interest.

### 12.8.1 Latent Dirichlet Allocation (LDA) for Topic Modeling in Biomedical Texts

Latent Dirichlet allocation (LDA) is a widely used algorithm for topic modeling, employing a probabilistic approach to uncover hidden thematic structures within text data [40–42]. LDA operates under the assumption that documents are mixtures of topics, and each topic is characterized by a specific distribution of words. LDA identifies the underlying topics that characterize the collection of documents by examining word co-occurrence patterns throughout the corpus. This capability makes it a useful tool for extracting meaningful information from unstructured text.

In the biomedical context, LDA is used for a variety of applications in both research and clinical data [40, 41]. It can be used to identify trends in scientific literature by detecting key topics discussed in research papers, such as advancements in cancer research, drug development strategies, or the emergence of gene therapy techniques. This automated identification of trends helps researchers stay informed about the latest developments in their fields. LDA also facilitates the clustering of documents by topic, organizing large collections of papers into thematic groups. This organization simplifies literature reviews, allowing researchers to quickly locate studies relevant to specific subjects or questions. In clinical settings, LDA can uncover patterns in unstructured clinical notes or EHRs. For instance, it can help identify common symptoms associated with diseases, analyze treatment pathways, or detect recurring themes in patient care.

The main concepts behind LDA are straightforward yet effective. LDA treats documents as collections of words, and topics are represented as distributions over these words. Each document is viewed as a mixture of multiple topics, and each word is assigned a probability of belonging to a particular topic. This probabilistic representation allows LDA to discover topics that frequently occur together in the corpus, even if these topics are not explicitly labeled or predefined. Words with high probabilities within a topic can be interpreted as the defining features, helping to clarify the thematic focus of that topic.

For instance, when applied to a corpus of biomedical research papers, LDA might identify a topic characterized by words such as "tumor," "mutation," and "oncology," which would likely correspond to cancer research. Another topic might be represented by terms like "clinical trial," "dosage," and "adverse effect," indicative of drug development studies. By associating documents with topics, LDA offers a structured way to explore and interpret large collections of text.

The implementation of LDA is facilitated by Python libraries such as gensim and scikit-learn, which provide user-friendly tools for preprocessing text, creating document-term matrices, and training topic models. These libraries make it easier for researchers to apply LDA to biomedical datasets, allowing them to summarize large volumes of text, identify trends, and generate hypotheses for further investigation.

### 12.8.2 Clustering Biomedical Research Papers by Topic

**Notebook: Section 12.8.2. Clustering Biomedical Research Papers by Topic**

This notebook provides code examples demonstrating clustering biomedical research papers by topic.

Link to the GitHub repository:

▶ https://github.com/sn-code-inside/BioPy

Go to: Chap. 12—▶ Sect. 12.8.2.

To demonstrate how LDA works, let's apply it to a simple collection of biomedical research papers to automatically discover the main topics. We can use ensim library to analyze and cluster documents by topic.

Example: Implementing LDA for topic modeling in biomedical texts.

```
import gensim
from gensim import corpora
from nltk.corpus import stopwords
from nltk.tokenize import word_tokenize

# Example biomedical texts (research papers or clinical notes)
documents = [
    "The patient was diagnosed with lung cancer and
    underwent chemotherapy.",
    "This study investigates the genetic mutations in breast
    cancer patients.",
    "The drug metformin is commonly used for treating diabetes.",
    "Hypertension is a risk factor for heart disease and stroke.",
    "The gene BRCA1 is linked to breast cancer risk."
]

# Preprocessing the text: Tokenize and remove stopwords
stop_words = set(stopwords.words('english'))
texts = [[word for word in word_tokenize(doc.lower()) if word.
isalnum() and word not in stop_words] for doc in documents]
```

```
# Create a dictionary and corpus for LDA
dictionary = corpora.Dictionary(texts)
corpus = [dictionary.doc2bow(text) for text in texts]

# Apply LDA for topic modeling
lda_model = gensim.models.LdaModel(corpus, num_topics=3,
id2word=dictionary, passes=15)

# Print the discovered topics
topics = lda_model.print_topics(num_words=5)
for i, topic in topics:
    print(f"Topic {i+1}: {topic}")
```

Output:

```
Topic 1: 0.100*"cancer" + 0.100*"breast" + 0.083*"study" +
0.083*"genetic" + 0.083*"mutations"
Topic 2: 0.111*"diabetes" + 0.111*"metformin" + 0.083*"treating"
+ 0.083*"patients" + 0.056*"commonly"
Topic 3: 0.120*"lung" + 0.120*"cancer" + 0.092*"chemotherapy" +
0.063*"risk" + 0.063*"disease"
```

In this example, we applied LDA to a small collection of biomedical documents. The model discovered three topics that represent the key themes in the documents:

Topic 1: Focuses on genetic mutations and breast cancer.

Topic 2: Relates to diabetes and treatment with metformin.

Topic 3: Discusses lung cancer and chemotherapy.

The LDA model assigns probabilities to each word in the documents for each topic. In the output, you can see the most important words for each topic along with their weights (i.e., the probability that they belong to that topic).

Once topics have been discovered, it is useful to visualize the results to understand the underlying structure of the data. pyLDAvis is a Python package designed to help visualize LDA models and understand the relationships between topics.

Example: Visualizing LDA topics using pyLDAvis. import pyLDAvis

```
import pyLDAvis.gensim_models
# Visualize the topics
lda_visualization = pyLDAvis.gensim_models.prepare(lda_
model, corpus, dictionary)
pyLDAvis.display(lda_visualization)
```

pyLDAvis creates an interactive visualization of the LDA model, where each topic is represented as a circle. The size of the circle indicates the importance of the topic, and the distance between the circles shows the similarity between the topics. By clicking on each topic, you can view the most relevant words associated with that topic, helping to interpret the results and gain a deeper understanding of the discovered themes. LDA can be used to analyze a large corpus of research papers to detect emerging trends in biomedical research.

## 12.9 Sentiment Analysis in Biomedical Research

Sentiment analysis in NLP is used to assess the emotional tone, attitude, or polarity (positive, negative, or neutral) expressed in a piece of text [8, 43]. While it has been widely utilized in domains like marketing and social media analytics, sentiment analysis is now finding useful applications in biomedical research. Sentiment analysis can be used to examine patient reviews, social media discussions, drug feedback, or clinical narratives to learn patient satisfaction, treatment outcomes, public perception of healthcare interventions, and the overall sentiment surrounding medical topics.

### 12.9.1 Sentiment Analysis in Biomedical Texts

Sentiment analysis is used to evaluate emotional tone or attitudes expressed in various contexts, helping better understanding into patient experiences, public health trends, and clinical narratives [8, 44]. It is particularly useful for understanding patient sentiments and experiences by analyzing reviews or feedback on healthcare platforms. This approach can show how patients perceive treatments, medications, or healthcare services, helping providers identify areas of satisfaction or concern. Monitoring public health trends through sentiment analysis of social media posts, blogs, and forums helps to understand societal perceptions of health interventions, public health measures, or emerging diseases. In clinical settings, sentiment analysis of narratives written by doctors or patients can identify emotional expressions related to treatment outcomes, complications, or concerns about diagnoses and drugs.



Biomedical texts often have a formal and technical tone, which differ from the language seen in social media or product reviews. Patients' sentiments may not always be stated directly in emotional terms but can often be inferred from their descriptions of symptoms, treatment effectiveness, or adverse effects. For example, a patient might describe severe headaches and dizziness following a medication, which signals a negative sentiment, even if dissatisfaction is not stated outright. In contrast, a report mentioning improvement of symptoms with no side effects indicates positive sentiment, while a neutral tone might appear in statements describing an absence of change during treatment.

Examples of sentiments in biomedical texts include positive expressions such as "The medication significantly improved my symptoms with no side effects," indicating satisfaction with the treatment. Negative sentiment might be conveyed in a statement like, "I experienced severe headaches and dizziness after starting this treatment," reflecting distress or dissatisfaction. A neutral sentiment could be expressed in observations like, "The patient showed no signs of improvement during the treatment period," which conveys a factual report without emotional judgment. These examples illustrate how sentiment analysis in biomedical contexts requires careful interpretation of both explicit and implicit cues to capture the underlying emotional or attitudinal tones. Addressing the specific challenges of biomedical language, sentiment analysis supports improved patient care, helps refine healthcare services, and can inform public health strategies.

### 12.9.2 Challenges of Sentiment Analysis in Scientific Texts

Sentiment analysis in biomedical texts presents several challenges that set it apart from general sentiment analysis tasks and require specialized approaches to address the nuances of scientific and clinical language [27]. A primary challenge arises from the technical and objective tone prevalent in biomedical literature and clinical notes. These texts often employ domain-specific terminology and precise language to describe medical conditions, treatments, and outcomes. Words that may indicate sentiment in general contexts, such as "progression," carry different implications in biomedical texts. For instance, "progression" in a medical report typically signals a worsening condition (negative sentiment), while in general usage, it may simply mean advancement or development without a sentiment attached.

Another noticeable challenge is the presence of implied sentiment. Biomedical texts, particularly clinical notes, often convey sentiment indirectly. For example, a treatment might be described as "ineffective," which implies negative sentiment about the outcome but does not use overtly emotional language. Similarly, phrases like "no improvement observed" or "the patient did not tolerate the medication well" imply dissatisfaction or concern, even though they are stated in neutral or technical terms. NLP models need to be capable of understanding these nuanced expressions to infer sentiment accurately.

Ambiguity further complicates sentiment analysis in biomedical texts. Discussions of treatments or conditions may reflect mixed sentiments, where the same treatment produces positive results for some patients and adverse effects for others. For example, a study might describe a drug as "effective for reducing symptoms in most patients but associated with severe side effects in some." Such statements can be difficult to classify as purely positive or negative and may require more granular sentiment categorization.

The domain-specific vocabulary in biomedical texts also poses a hurdle. Many terms, such as "tolerated," "responded," or "relapse," carry specialized meanings that differ from their general usage. For instance, "tolerated" in a clinical context often describes a patient's ability to endure a medication without adverse effects, which may suggest neutral or positive sentiment depending on the context. General-purpose sentiment analysis models trained on nonbiomedical corpora are typically ill-equipped to interpret such terms accurately. This makes it necessary to use domain-adapted models, such as those fine-tuned on biomedical datasets, to capture the specialized semantics of the field.

These challenges illustrate the complexity of performing sentiment analysis in biomedical texts and motivate the need for domain-specific tools and models. Addressing these issues involves using annotated biomedical datasets, incorporating contextual embeddings from pretrained models like BioBERT, and developing algorithms capable of interpreting nuanced and domain-specific language.

### 12.9.3 Analyzing Sentiments in Patient Reviews of Drugs

**Notebook: Section 12.9.3. Analyzing Sentiments in Patient Reviews of Drugs**
This notebook provides code examples demonstrating analyzing sentiments in patient reviews of drugs.
Link to the GitHub repository:
► https://github.com/sn-code-inside/BioPy
Go to: Chap. 12—► Sect. 12.9.3.

Analyzing patient reviews of drugs is a valuable application of sentiment analysis that helps to learn, patient satisfaction, treatment effectiveness, and adverse effects [45, 46]. These reviews, often shared on healthcare platforms, may include descriptions of patients' conditions, the treatments they received, and their experiences with the outcomes. By systematically extracting and analyzing the sentiments expressed in these reviews, healthcare providers can identify patterns in patient feedback, understand common concerns, and assess the perceived benefits and risks of medications.

The process typically starts with data collection, where patient reviews are gathered from online platforms, healthcare forums, or other relevant sources. These reviews usually contain information about patient experiences, including drug efficacy, side effects, and overall satisfaction. Following data collection, the text undergoes preprocessing to ensure it is suitable for analysis. This step involves tokenizing the text into meaningful units, removing stopwords to eliminate irrelevant words, and applying stemming or lemmatization to standardize the vocabulary. Preprocessing improves the quality of the data and improves the performance of the sentiment analysis model.



Once the text is preprocessed, sentiment classification is performed using an appropriate model. The model assigns each review a sentiment label such as positive, negative, or neutral, based on the emotional tone or attitude conveyed in the text. Positive sentiment might indicate satisfaction with the treatment, such as "This medication completely resolved my symptoms with no side effects." Negative sentiment could reflect dissatisfaction or adverse effects, such as "I experienced severe nausea and headaches after taking this drug." Neutral sentiment often describes observations without emotional judgment, like "The patient showed no significant change during the treatment."

An example of implementing sentiment analysis in Python involves using the VADER (Valence Aware Dictionary and sEntiment Reasoner) tool. While VADER is primarily designed for social media and general text sentiment analysis, it can be adapted for biomedical contexts, especially for analyzing patient reviews. VADER's lexicon-based approach allows it to identify sentiment scores quickly and efficiently, making it suitable for exploratory analysis or when working with large datasets.

Example code using NLTK.

```
from nltk.sentiment import SentimentIntensityAnalyzer
import nltk

# Download the VADER lexicon
nltk.download('vader_lexicon')

# Initialize VADER sentiment analyzer
sia = SentimentIntensityAnalyzer()

# Example patient reviews of drugs
reviews = [
    "This medication worked wonderfully for my migraine.",
    "I had terrible side effects, including nausea and
    dizziness. Not recommended.",
    "The drug was okay, no side effects, but symptoms
    remain."
]

# Analyze the sentiment for each review
for review in reviews:
    sentiment = sia.polarity_scores(review)
    print(f"Review: {review}")
    print(f"Sentiment Scores: {sentiment}")
    print("-" * 50)
```

Output (output socres may change):

```
Review: This medication worked wonders for my migraine. No
more headaches!
Sentiment Scores: {'neg': 0.0, 'neu': 0.606, 'pos': 0.394,
'compound': 0.5994}
--------------------------------------------------
Review: I had terrible side effects, including nausea and
dizziness. Not recommended.
Sentiment Scores: {'neg': 0.37, 'neu': 0.63, 'pos': 0.0,
'compound': -0.5707}
--------------------------------------------------
Review: The drug was okay, no side effects, but symptoms
remain.
Sentiment Scores: {'neg': 0.145, 'neu': 0.724, 'pos': 0.131,
'compound': -0.0387}
--------------------------------------------------
```

VADER assigns sentiment scores to text based on its emotional tone. The compound score is a single value that represents the overall sentiment of the text. Positive compound scores indicate positive sentiment, negative scores indicate negative sentiment, and values close to zero represent neutral sentiment.

In the first review, “This medication worked wonderfully for my migraine” the compound score is positive, reflecting a positive sentiment about the medication’s effectiveness. The second review, which mentions “terrible side effects,” is classified as negative due to the explicit dissatisfaction expressed. The third review contains a mix of sentiments, with positive elements like “okay and no side effects” and negative elements like “symptoms remain” resulting in a compound score that reflects a somewhat neutral sentiment with a slight negative bias. This ability to handle nuanced texts is one of VADER’s strengths, but fine-tuning or domain-specific tools may be necessary for more complex biomedical texts.

Sentiment analysis of patient reviews moves beyond individual assessments to show broader, aggregated patterns. Aggregating the sentiment scores of multiple reviews, researchers can analyze how patients collectively perceive a specific drug or treatment. Calculating average sentiment scores for a drug, for example, helps to identify whether the overall feedback leans positive, negative, or neutral. This aggregate analysis can inform decisions in drug development, marketing, and patient education.

Another example:

```
from vaderSentiment.vaderSentiment import
SentimentIntensityAnalyzer

# Initialize the VADER sentiment analyzer
analyzer = SentimentIntensityAnalyzer()

# Example patient reviews
reviews = [
    "This medication worked perfectly and completely
    resolved my symptoms.",
    "I experienced severe nausea and headaches after taking
    this drug, with no improvement.",
    "The patient did not show any major changes during the
    treatment."
]

# Analyze the sentiment of each review
for review in reviews:
    sentiment_scores = analyzer.polarity_scores(review)
    print(f"Review: {review}")
    print(f"Sentiment Scores: {sentiment_scores}")
    if sentiment_scores['compound'] >= 0.5:
        print("Overall Sentiment: Positive")
    elif sentiment_scores['compound'] <= -0.5:
        print("Overall Sentiment: Negative")
    else:
        print("Overall Sentiment: Neutral")
    print()
```

Output (output socres may change):

```
Review: This medication worked perfectly and completely
resolved my symptoms.
Sentiment Scores: {'neg': 0.0, 'neu': 0.531, 'pos': 0.469,
'compound': 0.7346}
Overall Sentiment: Positive

Review: I experienced severe nausea and headaches after
taking this drug, with no improvement.
Sentiment Scores: {'neg': 0.316, 'neu': 0.684, 'pos': 0.0,
'compound': -0.6224}
Overall Sentiment: Negative

Review: The patient did not show any major changes during
the treatment.
Sentiment Scores: {'neg': 0.0, 'neu': 1.0, 'pos': 0.0,
'compound': 0.0}
Overall Sentiment: Neutral
```

This example demonstrates how VADER can assign sentiment scores to patient reviews and classify them as positive, negative, or neutral based on the overall sentiment. For more accurate results in biomedical contexts, fine-tuning VADER's lexicon or employing domain-specific sentiment models like BioBERT or SciBERT can capture the nuances of medical terminology and improve sentiment analysis performance.

## 12.10 NLP for Literature-Based Discovery (LBD)

LBD is a transformative application of NLP in biomedical research, allowing the identification of hidden relationships within large collections of scientific texts [7]. LBD is built on the principle that connections between seemingly unrelated concepts can be inferred from existing literature, even if these relationships are not explicitly mentioned in a single document. For example, if concept A (a drug) is linked to concept B (a protein) in one study and concept B is linked to concept C (a disease) in another study, LBD can hypothesize a relationship between concept A and concept C, such as the drug's potential to treat the disease. This approach has been instrumental in generating new hypotheses, identifying opportunities for drug repurposing, and uncovering novel gene-disease or protein-disease associations.

### 12.10.1 Overview of LBD

LBD is designed to analyze large quantities of biomedical literature to uncover new, indirect relationships between concepts that may not be explicitly linked within a single document. This approach provides a systematic way to generate hypotheses by drawing on existing knowledge in the biomedical domain. LBD is particularly important in generating new hypotheses, discovering therapeutic targets, and identifying novel associations between biological entities [47]. For example, researchers can explore connections between genes, proteins, and diseases that have not been previously investigated, opening up new avenues for research and innovation.

A key application of LBD is drug repurposing, where existing drugs are analyzed for possible new uses. Identifying indirect links between drugs and diseases, LBD can uncover opportunities to repurpose medications for conditions beyond their original indications. Similarly, LBD facilitates the discovery of gene-disease associations by integrating knowledge from multiple publications to show previously unexplored relationships. These discoveries can contribute to understanding disease mechanisms, identifying biomarkers, or developing targeted therapies.

The foundation of LBD lies in the Swanson Hypothesis, which states that if two distinct entities are both connected to a common third entity, an undiscovered relationship may exist between the original entities. This principle was demonstrated in Swanson's seminal discovery of the link between fish oil and Raynaud's disease. By observing that fish oil was associated with reduced blood viscosity and that Raynaud's disease involved abnormal blood viscosity, Swanson hypothesized a therapeutic benefit of fish oil for Raynaud's disease—a connection that had not been previously recognized.

LBD builds on this principle by systematically identifying these "hidden" relationships within the biomedical literature. Modern LBD techniques use advanced NLP tools, such as NER and relation extraction, to identify entities and their connections across large text corpora. By constructing knowledge graphs or semantic networks that map these relationships, researchers can visualize indirect links and propose new hypotheses or validate discoveries.



### 12.10.2 Tools and Approaches for LBD in Biomedical Texts

Implementing LBD in biomedical texts involves systematically applying NLP techniques to extract and analyze relationships between entities, which helps the discovery of novel connections. The process relies on multiple steps, each using specialized tools and methods to address the complexity of biomedical data. Together, these methods provide a framework for generating hypotheses and finding hidden patterns from the collections of scientific literature.

The first step in LBD is data collection, which involves gathering large sets of biomedical texts. Sources like PubMed, clinical trial databases, and scientific reviews serve as rich repositories of unstructured data. Tools such as PubMed APIs or web scraping frameworks are commonly used to collect this data. It is important that the dataset is complete and relevant to the research focus to make LBD effective.

The next step is entity extraction, where the goal is to identify and categorize biomedical entities such as genes, drugs, diseases, and proteins. NER tools like scispaCy and BioBERT are particularly suited for this task, as they are pretrained on domain-specific corpora and can recognize complex biomedical terminology. For example, scispaCy can identify entities such as "EGFR" (gene) or "gefitinib" (drug) in scientific text with high accuracy.

Once the entities are identified, relationship extraction is performed to determine how these entities are connected. This step involves analyzing sentences, paragraphs, or entire documents to identify explicit or implicit relationships. Approaches for relationship extraction include rule-based methods, which rely on predefined patterns or triggers (e.g., "inhibits," "is associated with"), and machine learning–based methods, which use annotated datasets to train models capable of generalizing across diverse text structures. Advanced NLP models, such as BioBERT, are good at extracting relationships due to their ability to understand biomedical context and semantics.

After extracting relationships, discovery algorithms are applied to identify indirect connections between entities. Techniques like Latent Semantic Analysis (LSA) or LDA analyze the co-occurrence and distribution of terms across documents, helping to uncover patterns that suggest novel links. Network-based methods, such as constructing knowledge graphs, are also effective. These graphs represent entities as nodes and relationships as edges, allowing algorithms to explore paths that connect previously unlinked nodes.

The final step in LBD is validation, where the discovered relationships are assessed for their credibility and relevance. This can involve cross-referencing findings with established biomedical knowledge bases like UMLS, DrugBank, or MeSH to ensure they are consistent with existing evidence. For novel connections, experimental validation, such as laboratory studies or clinical trials, may be necessary to test the hypotheses generated by LBD.

Using these steps, LBD combines NLP and computational analysis to mine the biomedical literature for actionable discoveries. The use of domain-specific tools and algorithms ensures precision and relevance, making LBD a relevant approach for advancing biomedical research and innovation.

## 12.11 Ethical and Regulatory Considerations in Biomedical NLP

The integration of NLP into biomedical research and healthcare has led to major advances, but it also raises important ethical and regulatory considerations [48, 49]. As NLP is increasingly used to analyze sensitive data such as clinical notes, EHRs, and patient feedback, it is needed to address concerns related to privacy, informed consent, transparency, and the responsible deployment of these technologies [50]. These concerns are particularly relevant in high-stakes areas like clinical decision-making, drug discovery, and patient care, where errors or biases can have serious consequences.

Clinical texts often contain identifiable patient information, which need be anonymized or deidentified before being used in NLP tasks [50]. Even with deidentification, there is a risk of reidentification, particularly if multiple data sources are combined. Strong data protection practices are required to prevent breaches of patient privacy. In the context of machine learning, models trained on sensitive clinical data could inadvertently reveal private information, requiring measures like differential privacy to safeguard data.

Patients must provide informed consent before their data is used for research purposes, including NLP analysis. However, NLP often involves secondary use of data that was not originally collected for research. Therefore, it is important that patients are fully informed about how their data will be used. In situations where it is not feasible to obtain explicit consent from every patient (e.g., large retrospective studies), ethical review boards must carefully consider the balance between the benefits of the research and the risks to patient privacy.

NLP models may reflect or amplify biases present in the data. For example, if a dataset used to train a clinical NLP model is disproportionately representative of a specific demographic group, the model may produce biased outcomes for underrepresented groups. Biomedical NLP models requires careful attention to how data is collected, processed, and used to address such issues. Researchers must evaluate whether their models perform equally well across different demographic groups, diseases, or treatment types.

## 12.12 Exercises and Questions

The following examples and questions are designed to give students practical experience with the techniques and challenges of applying NLP in biomedical research, from text preprocessing and sentiment analysis to more complex tasks like relation extraction and LBD.

### Exercises

1. Tokenizing biomedical text: Write a Python script using spaCy or NLTK to tokenize clinical notes into sentences and words. Explore the effect of tokenization on text containing biomedical abbreviations and acronyms.
2. NER in biomedical text: Implement a biomedical NER model using scispaCy to extract entities like genes, diseases, and drugs from a set of research abstracts. Display the extracted entities along with their labels.
3. Text classification of clinical notes: Use scikit-learn to classify clinical notes based on disease categories (e.g., cancer, diabetes, cardiovascular diseases). Preprocess the text using TF-IDF vectorization, train a classifier, and evaluate its performance.
4. Relation extraction in biomedical literature: Write a rule-based approach to extract relationships between genes and diseases from biomedical texts. For example, identify sentences where a gene is associated with a disease.
5. Sentiment analysis of patient reviews: Use VADER or another sentiment analysis tool to analyze the sentiment of patient reviews of medications. Determine the overall sentiment (positive, negative, or neutral) for each review and aggregate the results.
6. Topic modeling on research papers: Apply Latent Dirichlet Allocation (LDA) to cluster a collection of biomedical research papers into topics. Visualize the topics and interpret the top words for each topic.
7. Building a custom NER model: Use spaCy to train a custom NER model to identify specific biomedical terms (e.g., protein names or drug classes) from a small, labeled dataset.
8. Information Extraction from EHRs: Implement an NLP pipeline to extract and summarize key information (e.g., diagnoses, medications, and treatment outcomes) from deidentified EHRs.

9. Detecting drug-drug interactions: Using a set of biomedical abstracts, create a rule-based or machine learning approach to identify drug-drug interactions. Use extracted relationships to build a drug interaction network.
10. LBD for drug repurposing: Implement a simple LBD system to explore connections between drugs and diseases by analyzing research abstracts. Use indirect associations to propose new hypotheses for drug repurposing.

## Questions to Be Answered

1. What are the key challenges of applying NER in biomedical texts?
2. How does tokenization differ when processing biomedical texts compared to general domain texts?
3. Why is it important to remove stopwords in some biomedical NLP tasks, and in which cases might you choose not to remove them?
4. Explain how scispaCy differs from general-purpose NLP tools like spaCy for handling biomedical texts.
5. How does the Latent Dirichlet Allocation (LDA) algorithm work in the context of biomedical text mining?
6. What is the importance of data privacy and deidentification when using NLP models to process clinical notes or EHRs?
7. Describe the role of sentiment analysis in biomedical research. In what types of texts can sentiment analysis be applied, and what are the limitations?
8. What is LBD, and how can it help in drug repurposing and hypothesis generation in biomedical research?
9. Explain the process of extracting relations between biomedical entities such as genes and diseases. What are the main techniques used?

**Acknowledgement** The language of the human-generated text was corrected with the assistance of artificial intelligence (AI) tools [GPT-3.5 and GTP-4 from OpenAI]. GitHub Co-Pilot was used to check the correctness of the codes. The text underwent subsequent human revision to ensure its accuracy.

## References

1. Aramaki E, Wakamiya S, Yada S, Nakamura Y. Natural language processing: from bedside to everywhere. Yearb Med Inform. 2022;31(1):243–53.
2. Wieland-Jorna Y, van Kooten D, Verheij RA, de Man Y, Francke AL, Oosterveld-Vlug MG. Natural language processing systems for extracting information from electronic health records about activities of daily living. A systematic review. JAMIA Open. 2024;7(2):ooae044.
3. Doan S, Conway M, Phuong TM, Ohno-Machado L. Natural language processing in biomedicine: a unified system architecture overview. Methods Mol Biol. 2014;1168:275–94.
4. Sung M, Jeong M, Choi Y, Kim D, Lee J, Kang J. BERN2: an advanced neural biomedical named entity recognition and normalization tool. Bioinformatics. 2022;38(20):4837–9.
5. Wang Y, Wang Y, Peng Z, Zhang F, Zhou L, Yang F. Medical text classification based on the discriminative pre-training model and prompt-tuning. Digit Health. 2023;9:20552076231193213.

6. Landolsi MY, Hlaoua L, Ben Romdhane L. Information extraction from electronic medical documents: state of the art and future research directions. Knowl Inf Syst. 2023;65(2):463–516.
7. Cheerkoot-Jalim S, Khedo KK. Literature-based discovery approaches for evidence-based healthcare: a systematic review. Health Technol (Berl). 2021;11(6):1205–17.
8. Denecke K, Reichenpfader D. Sentiment analysis of clinical narratives: a scoping review. J Biomed Inform. 2023;140:104336.
9. Murphy RM, Klopotowska JE, de Keizer NF, Jager KJ, Leopold JH, Dongelmans DA, et al. Adverse drug event detection using natural language processing: a scoping review of supervised learning methods. PLoS One. 2023;18(1):e0279842.
10. Wu HY, Chiang CW, Li L. Text mining for drug-drug interaction. Methods Mol Biol. 2014;1159:47–75.
11. Jeynes JCG, Corney M, James T. A large-scale evaluation of NLP-derived chemical-gene/protein relationships from the scientific literature: implications for knowledge graph construction. PLoS One. 2023;18(9):e0291142.
12. Alessa A, Faezipour M. Flu outbreak prediction using twitter posts classification and linear regression with historical Centers for Disease Control and Prevention reports: prediction framework study. JMIR Public Health Surveill. 2019;5(2):e12383.
13. Nadkarni PM, Ohno-Machado L, Chapman WW. Natural language processing: an introduction. J Am Med Inform Assoc. 2011;18(5):544–51.
14. Khurana D, Koli A, Khatter K, Singh S. Natural language processing: state of the art, current trends and challenges. Multimed Tools Appl. 2023;82(3):3713–44.
15. Liu H, Christiansen T, Baumgartner WA Jr, Verspoor K. BioLemmatizer: a lemmatization tool for morphological processing of biomedical text. J Biomed Semantics. 2012;3:3.
16. Sarica S, Luo J. Stopwords in technical language processing. PLoS One. 2021;16(8):e0254937.
17. Bird E, Loper E, Klein E. Natural language processing with python. O'Reilly Media Inc.; 2009.
18. Neumann M, King D, Beltagy I, Ammar BW. ScispaCy: fast and robust models for biomedical natural language processing. ArXiv. 2019:abs/1902.07669.
19. Cock PJ, Antao T, Chang JT, Chapman BA, Cox CJ, Dalke A, et al. Biopython: freely available Python tools for computational molecular biology and bioinformatics. Bioinformatics. 2009;25(11):1422–3.
20. Lamy J-B, Venot A, Duclos C. PyMedTermino: an open-source generic API for advanced terminology services. Stud Health Technol Inform. 2015;210:924–8.
21. Zhang B. Getting to know named entity recognition: better information retrieval. Med Ref Serv Q. 2024;43(2):196–202.
22. Wei CH, Leaman R, Lu Z. SimConcept: a hybrid approach for simplifying composite named entities in biomedical text. IEEE J Biomed Health Inform. 2015;19(4):1385–91.
23. Durango MC, Torres-Silva EA, Orozco-Duque A. Named entity recognition in electronic health records: a methodological review. Healthc Inform Res. 2023;29(4):286–300.
24. Dogra V, Verma S, Kavita, Chatterjee P, Shafi J, Choi J, et al. A complete process of text classification system using state-of-the-art NLP models. Comput Intell Neurosci. 2022;2022:1883698.
25. Srivastava SK, Singh SK, Suri JS. State-of-the-art methods in healthcare text classification system: AI paradigm. Front Biosci (Landmark Ed). 2020;25(4):646–72.
26. Madi IAE, Redjdal A, Bouaud J, Seroussi B. Exploring explainable AI techniques for text classification in healthcare: a scoping review. Stud Health Technol Inform. 2024;316:846–50.
27. Dey RK, Das AK. Modified term frequency-inverse document frequency based deep hybrid framework for sentiment analysis. Multimed Tools Appl. 2023:1–24.
28. Richard E, Reddy B. Text classification for clinical trial operations: evaluation and comparison of natural language processing techniques. Ther Innov Regul Sci. 2021;55(2):447–53.
29. Li Z, Gurgel H, Dessay N, Hu L, Xu L, Gong P. Semi-supervised text classification framework: an overview of dengue landscape factors and satellite earth observation. Int J Environ Res Public Health. 2020;17(12).
30. Kangoo NA, Roy A, editors. Supervised machine learning text classification: a review. Singapore: Springer Nature Singapore; 2023.
31. Juluru K, Shih HH, Keshava Murthy KN, Elnajjar P. Bag-of-words technique in natural language processing: a primer for radiologists. Radiographics. 2021;41(5):1420–6.

32. Fraile Navarro D, Ijaz K, Rezazadegan D, Rahimi-Ardabili H, Dras M, Coiera E, et al. Clinical named entity recognition and relation extraction using natural language processing of medical free text: a systematic review. Int J Med Inform. 2023;177:105122.
33. Huang MS, Han JC, Lin PY, You YT, Tsai RT, Hsu WL. Surveying biomedical relation extraction: a critical examination of current datasets and the proposal of a new resource. Brief Bioinform. 2024;25(3).
34. Milošević N, Thielemann W. Comparison of biomedical relationship extraction methods and models for knowledge graph creation. J Web Semantics. 2023;75:100756.
35. Zhao Y, Yuan X, Yuan Y, Deng S, Quan J. Relation extraction: advancements through deep learning and entity-related features. Soc Netw Anal Min. 2023;13(1):92.
36. Kim S, Liu H, Yeganova L, Wilbur WJ. Extracting drug-drug interactions from literature using a rich feature-based linear kernel approach. J Biomed Inform. 2015;55:23–30.
37. Zhu Y, Li L, Lu H, Zhou A, Qin X. Extracting drug-drug interactions from texts with BioBERT and multiple entity-aware attentions. J Biomed Inform. 2020;106:103451.
38. Zhang T, Leng J, Liu Y. Deep learning for drug-drug interaction extraction from the literature: a review. Brief Bioinform. 2020;21(5):1609–27.
39. Albalawi R, Yeap TH, Benyoucef M. Using topic modeling methods for short-text data: a comparative analysis. Front Artif Intell. 2020;3:42.
40. Danler M, Hackl WO, Neururer SB, Huber L, Pfeifer B. Visualizing nursing narratives: an evaluation of latent Dirichlet allocation topic modeling for care reports. Stud Health Technol Inform. 2024;316:1709–13.
41. Inoue M, Fukahori H, Matsubara M, Yoshinaga N, Tohira H. Latent Dirichlet allocation topic modeling of free-text responses exploring the negative impact of the early COVID-19 pandemic on research in nursing. Jpn J Nurs Sci. 2023;20(2):e12520.
42. Rasiwasia N, Vasconcelos N. Latent Dirichlet allocation models for image classification. IEEE Trans Pattern Anal Mach Intell. 2013;35(11):2665–79.
43. Zunic A, Corcoran P, Spasic I. Sentiment analysis in health and well-being: systematic review. JMIR Med Inform. 2020;8(1):e16023.
44. Greaves F, Ramirez-Cano D, Millett C, Darzi A, Donaldson L. Use of sentiment analysis for capturing patient experience from free-text comments posted online. J Med Internet Res. 2013;15(11):e239.
45. Nair AB, Abhinand K, Anamika U, Jaison DT, Ajitha V, Anoop VS. "Hey..! This medicine made me sick": sentiment analysis of user-generated drug reviews using machine learning techniques. ArXiv. 2024:abs/2404.13057.
46. Korkontzelos I, Nikfarjam A, Shardlow M, Sarker A, Ananiadou S, Gonzalez GH. Analysis of the effect of sentiment analysis on extracting adverse drug reactions from tweets and forum posts. J Biomed Inform. 2016;62:148–58.
47. Bhasuran B, Murugesan G, Natarajan J. Literature Based Discovery (LBD): towards hypothesis generation and knowledge discovery in biomedical text mining. ArXiv. 2023:abs/2310.03766.
48. Bear Don't Walk OJt, Reyes Nieva H, Lee SS, Elhadad N. A scoping review of ethics considerations in clinical natural language processing. JAMIA Open. 2022;5(2):ooac039.
49. Wang C, Liu S, Yang H, Guo J, Wu Y, Liu J. Ethical considerations of using ChatGPT in health care. J Med Internet Res. 2023;25:e48009.
50. Rabbani N, Bedgood M, Brown C, Steinberg E, Goldstein RL, Carlson JL, et al. A natural language processing model to identify confidential content in adolescent clinical notes. Appl Clin Inform. 2023;14(3):400–7.

# Single-Cell RNA Sequencing Data Analysis

Contents




J. U. Kazi, *Python Essentials for Biomedical Data Analysis: An Introductory Textbook*,
https://doi.org/10.1007/978-3-031-85600-6_13

Single-cell RNA sequencing (scRNA-seq) allows researchers to investigate the gene expression profiles of individual cells, determining the heterogeneity of cell populations that cannot be captured by bulk RNA sequencing. This ability to analyze thousands or even millions of cells at single-cell resolution has driven major advances in fields such as developmental biology, immunology, cancer research, and neuroscience. By capturing the unique transcriptional signatures of individual cells, scRNA-seq enables the discovery of new cell types, characterization of complex tissues, and understanding of dynamic biological processes like differentiation and disease progression. However, working with scRNA-seq data involves a series of computational steps, from raw data preprocessing to advanced analyses, including clustering, cell type identification, and trajectory inference. Python has become a widely used language in this area due to its libraries for bioinformatics, including scanpy and AnnData, which offer tools for processing, visualizing, and analyzing scRNA-seq data. In this chapter, we will explore a brief workflow for analyzing scRNA-seq data using Python.

**Learning Goals**

In this chapter, we will learn how to use Python-based tools to analyze scRNA-seq data. We will explore key steps in the analysis workflow, including data preprocessing, quality control and interpretation methods such as gene regulatory network inference analysis. By working through each part of the workflow, we will learn how to analyze cellular heterogeneity and gene expression regulation at the single-cell level.

## 13.1 Basics of scRNA-seq

scRNA-seq is a high-throughput method used to profile gene expression at the single-cell level [1]. Unlike traditional bulk RNA sequencing, which measures average gene expression across a population of cells, scRNA-seq captures the gene expression signature of each individual cell, revealing cellular heterogeneity that might be masked in bulk experiments.

### 13.1.1 The scRNA-seq Technology

scRNA-seq was developed to address the limitations of bulk RNA sequencing, which measures average gene expression across a population of cells and cannot resolve differences between individual cell types or subpopulations within heterogeneous tissues. By isolating single cells, capturing their RNA, converting it into complementary DNA (cDNA), and sequencing it, scRNA-seq allows us to quantify gene expression at the resolution of individual cells [2]. This level of detail shows patterns of cellular diversity and function that were previously inaccessible.

One of the key applications of scRNA-seq is identifying rare cell types within a sample, such as stem cells, immune subsets, or tumor cells, which might be masked in bulk sequencing due to their low abundance [3]. This capability is particularly important in studying diseases like cancer, where rare cell populations may play a role in disease progression or resistance to treatment. Furthermore, scRNA-seq allows us to explore cellular heterogeneity within tissues or organs by comparing the gene expression profiles of individual cells, revealing differences that underlie complex biological functions or disease states [4]. Another use of scRNA-seq is tracking cellular development and differentiation [5]. Capturing snapshots of gene expression at various stages, it is possible to reconstruct dynamic processes, such as stem cell differentiation, immune response activation, or tumor evolution. This temporal resolution allows to study processes in developmental biology and the mechanisms of disease progression.

The scRNA-seq workflow involves several key steps. First, tissues are dissociated into single cells. These cells are then isolated, commonly using microfluidics-based platforms like 10x Genomics Chromium or droplet-based methods such as Drop-seq [6, 7]. RNA from each cell is captured and reverse-transcribed into cDNA, which is subsequently amplified to generate sufficient material for sequencing. High-throughput sequencing is then performed to measure gene expression levels. Each of these steps is important for preserving the integrity of the single-cell data and minimizing technical noise. The 10x Genomics Chromium system is known for its high throughput and cost efficiency, making it suitable for large-scale studies. Drop-seq offers scalability and affordability for experiments that require many cells, while Smart-seq provides higher sensitivity and full-length transcript coverage, which is ideal for studying transcript isoforms and alternative splicing. Overall, scRNA-seq has transformed the field of genomics by enabling the study of cellular diversity at single-cell resolution.

### 13.1.2 Importance and Applications in Biomedical Research

scRNA-seq has become a transformative tool in biomedical research, enabling the exploration of cellular heterogeneity and dynamics across a wide range of tissues and conditions. Its ability to profile gene expression at single-cell resolution makes it indispensable in fields where understanding tissue complexity and cellular interactions is necessary.



scRNA-seq has been widely used in tumor biology to study distinct subpopulations of cancer cells within a single tumor [8, 9]. These subpopulations often show different behaviors, such as proliferative capacity, metastatic potential, or resistance to therapy [10]. These heterogeneous populations can be used to explain tumor evolution and the mechanisms of metastasis, which in turn guide the development of targeted therapies to overcome treatment resistance. Furthermore, scRNA-seq allows us to identify rare immune cell subsets, such as specialized T-cell or B-cell populations, and to understand their roles in immune responses [11, 12].

scRNA-seq enables tracing the lineage and differentiation pathways of cells during embryonic development [13]. Profiling cells at different developmental stages allows us to reconstruct trajectories that describe how progenitor cells differentiate into specialized cell types. This approach has led to the identification of key regulators of cell fate decisions and improved our understanding of developmental processes, such as organogenesis and tissue regeneration. In neuroscience, scRNA-seq has been instrumental in mapping the diverse cell types of the brain, including neurons, glial cells, and other supporting cells [14]. Moreover, scRNA-seq guides to identify cellular changes associated with neurological disorders, such as Alzheimer's disease, Parkinson's disease, and autism spectrum disorders.

The versatility of scRNA-seq in addressing diverse biological questions emphasizes its importance in biomedical research. Its applications extend beyond these fields to areas such as regenerative medicine, infectious disease, and personalized medicine, where the ability to understand cellular diversity and dynamics has far-reaching implications for advancing health and disease management.

### 13.1.3 Differences Between Bulk and scRNA-seq

Bulk RNA sequencing and scRNA-seq differ fundamentally in the resolution and information they provide about gene expression. While bulk RNA sequencing measures the average gene expression profile of a population of cells, scRNA-seq captures the gene expression profiles of individual cells, that can be used to study cellular diversity within a tissue or sample. These differences make each technique suited for distinct research questions (◘ Table 13.1).

Choosing between bulk RNA-seq and scRNA-seq depends on the research question and the biological context. Bulk RNA-seq is well suited for studies where population-level gene expression is sufficient, such as comparing gene expression between treated and untreated samples or assessing overall transcriptomic changes in

**◘ Table 13.1** Differences between bulk RNA sequencing and scRNA-seq

| Aspect | Bulk RNA sequencing | scRNA-seq |
|---|---|---|
| Resolution | Measures the average gene expression profile of a population of cells, providing a population-level overview | Captures gene expression profiles of individual cells, offering much higher resolution that reveals cellular diversity within a tissue or sample |
| Ability to address heterogeneity | Treats the sample as homogeneous; cannot distinguish between different cell types or states within the population | Uncovers cellular heterogeneity; enables identification of distinct cell types, dynamic cellular states, and transcriptional changes underlying biological processes |
| Data complexity | Data is less complex and generally less noisy; relies on simpler methods for differential expression analysis | Data is inherently noisy with high variability and dropouts due to the small amount of RNA in individual cells; requires advanced computational and statistical methods |
| Computational methods required | Simpler computational methods are sufficient for data processing and analysis | Requires advanced techniques such as dimensionality reduction, clustering, and trajectory inference to interpret complex data |
| Appropriate applications | Well suited for studies where population-level gene expression is sufficient, like comparing treated vs. untreated samples or assessing overall changes | Preferred for research involving cellular heterogeneity, rare cell populations, or dynamic changes at the single-cell level, despite higher complexity and cost |

a tissue. However, for questions involving cellular heterogeneity, rare cell populations, or dynamic changes at the single-cell level, scRNA-seq is the preferred approach despite its higher complexity and cost.

## 13.2 Experimental Design and Data Acquisition

The success of an scRNA-seq experiment depends on careful experimental design and careful planning of data acquisition. Due to the complexity and cost of scRNA-seq, we need to make strategic decisions to ensure that the experiment is optimized for the specific biological questions. This involves selecting the right sample types, preparing high-quality single-cell suspensions, and choosing an appropriate sequencing platform to achieve reliable and meaningful results.

### 13.2.1 Experimental Workflow for scRNA-seq

The experimental workflow for scRNA-seq involves multiple steps to isolate individual cells, capture their RNA, and generate sequencing data that reflects the transcriptomic profiles of those cells [15].

Cell isolation: This step is important for ensuring that each cell can be profiled independently [16]. Cell dissociation methods vary depending on the tissue type and include enzymatic digestion, mechanical disruption, or a combination of both to create a single-cell suspension. Once dissociated, techniques like fluorescence-activated cell sorting (FACS) or microfluidics are commonly used to isolate individual cells. FACS sorts cells based on fluorescent markers, enabling the selection of specific cell populations, while microfluidics-based systems encapsulate cells in droplets for downstream processing. For fragile or rare cell types, gentle handling and optimized protocols help to preserve cell viability and functionality.



RNA capture and amplification: After cell isolation, the mRNA from each cell is captured and reverse-transcribed into cDNA. This step is necessary because mRNA is unstable and needs to be converted into a more stable form for downstream analysis. Since the amount of RNA in a single cell is extremely small, amplification is necessary to generate sufficient cDNA for sequencing. Amplification methods vary by platform, with some emphasizing full-length transcript coverage (e.g., Smart-seq) and others prioritizing high-throughput processing (e.g., 10x Genomics Chromium).

Sequencing: The amplified cDNA is sequenced using high-throughput sequencing technologies, typically generating millions of short reads that correspond to gene expression levels. Illumina sequencing is the most commonly used technology due to its high accuracy, scalability, and cost-effectiveness. The sequencing depth, or the number of reads per cell, is a parameter that influences the resolution of the data. Deeper sequencing provides more coverage of the transcriptome but increases costs, so the depth must be balanced with the number of cells being profiled and the specific goals of the study.

### 13.2.2 Key Considerations in scRNA-seq Experimentation

Designing an scRNA-seq experiment requires careful planning and attention to several factors to ensure high-quality, reproducible results [17]. The success of the experiment depends on decisions that optimize the resolution, accuracy, and biological relevance of the data.

Sampling: The choice of tissue or cell type should align with the specific biological question being addressed. For example, profiling tumor tissues might focus on identifying rare cancer subpopulations, while immune studies may target blood-derived immune cells. The number of cells to be sequenced is another important factor. Capturing a sufficient number of cells increases the likelihood of identifying rare populations and capturing the full heterogeneity of the sample. Moreover, the number of cells must be balanced with the sequencing depth and budget constraints.

Cell viability: Dead or dying cells can release degraded RNA or intracellular contents, contaminating the surrounding environment and introducing artifacts into the sequencing data. Ensuring proper sample collection, processing, and storage protocols can minimize cell death. Fresh samples are ideal, but when immediate processing is not possible, cryopreservation protocols should be employed to maintain cell integrity. Quality control at the dissociation step helps to avoid overstressing cells, which can lead to altered gene expression profiles.

Sequencing depth: The sequencing depth, defined as the number of reads allocated per cell, determines the sensitivity of the experiment. High sequencing depth allows for the detection of low-abundance transcripts and provides a more complete view of the transcriptome, but it comes at an increased cost. The optimal depth depends on the study's goals. For instance, studies focusing on rare transcripts or subtle gene expression changes may require deeper sequencing, while exploratory studies targeting broad population-level analysis may prioritize profiling more cells over sequencing depth.

Biological replicates: Incorporating biological replicates such as samples from different individuals, conditions, or time points ensure the robustness and generalizability of the findings. Replicates help account for natural variability between samples and reduce the likelihood of drawing conclusions based on artifacts or anomalies in a single dataset. Proper experimental controls, such as untreated versus treated samples or diseased versus healthy tissues, further improve the interpretability of the results.

Experimental design and technical considerations: A well-structured experimental design minimizes technical batch effects and maximizes reproducibility. For example, processing all samples in parallel, using standardized protocols, and randomizing sample preparation steps can reduce technical variability. scRNA-seq experiments often require advanced computational tools for data integration and analysis, particularly when working with multiple replicates or datasets. Careful planning for downstream analysis, including cell clustering, differential expression, and trajectory inference, ensures that the experimental design aligns with the computational needs.

## 13.3 Processing Raw scRNA-seq Data

Once scRNA-seq data has been generated, the first step in the analysis pipeline is to process the raw data. This typically involves converting the raw sequencing reads (in FASTA or FASTQ format) into formats suitable for downstream analysis, such as aligned reads or gene expression matrices. In this section, we will briefly describe the procedures involved in handling raw sequencing data.

### 13.3.1 FASTA and FASTQ Files: Structure and Content

The primary output from sequencing platforms like 10x Genomics and Smart-seq are FASTA or FASTQ files. These files contain the raw nucleotide sequences (reads) generated from the experiment, along with metadata and quality information. FASTA and FASTQ formats were previously discussed in different contexts in Chap. 3 and Chap. 10; here, we provide a brief further review of these file formats.

FASTA: This format stores nucleotide sequences but does not include quality information. Each sequence is represented by a header line starting with >, followed by the nucleotide sequence.

Example of a FASTA entry:

```
>seq1
ATGCAGTGCTGACTGAGTGC
```

FASTQ: A more common format in RNA sequencing, FASTQ files contain both the sequence information and a corresponding quality score for each nucleotide. Each entry consists of four lines:

1. Header line starting with @ and containing the read identifier.
2. Nucleotide sequence (e.g., ACTG).
3. Optional header (can be a + sign or additional information).
4. Quality score, representing the confidence in each base call.

Example of a FASTQ entry:

```
@seq1
ATGCAGTGCTGACTGAGTGC
+
IIIIIIIIIIIIIIIIIII
```

### 13.3.2 Quality Control and Preprocessing of Raw Data

Before performing any downstream analysis, it is important to ensure the quality of the raw sequencing data. Poor-quality reads can introduce noise and artifacts, reducing the accuracy of the analysis. Several standalone command-line programs such as FastQC, Trimmomatic, and cutadapt can be used to assess the quality of scRNA-seq data.

Steps in quality control:

*Assessing read quality:* FastQC, a quality control tool for high-throughput sequencing data, such as FASTQ files, can be used to generate reports on the overall quality of the raw sequencing reads. FastQC provides visualizations of various metrics, such as per-base sequence quality, GC content, and the presence of adapter sequences.

Example command for running FastQC:

```
fastqc sample.fastq
```

*Trimming low-quality reads:* Low-quality bases and adapter sequences are often removed using tools like Trimmomatic or cutadapt [18, 19]. This ensures that only high-quality reads are used in the downstream analysis.

Example of trimming reads with cutadapt:

```
cutadapt -q 20 -o trimmed_output.fastq sample.fastq
```

### 13.3.3 Mapping Reads to a Reference Genome

Once the quality of the raw sequencing data has been confirmed, the next step is to align the reads to a reference genome or transcriptome. Alignment converts the raw nucleotide sequences into meaningful information by mapping them to known gene locations.

Key steps in alignment include the following:

*Indexing the reference genome:* The first step in alignment is to create an index of the reference genome, which allows the alignment tool to efficiently map reads to the genome. This is typically done using tools like STAR, kallisto, or HISAT2 [20–22].

*Aligning reads:* Once the reference genome is indexed, the sequencing reads can be aligned to it. The alignment process generates output files in SAM or BAM format, which store the mapped reads along with their alignment positions.

*Generating gene expression matrices:* After alignment, gene expression matrices are generated, where rows represent genes, and columns represent cells. The values in the matrix correspond to the expression level of each gene in each cell.

### 13.3.4 Python Tools for Preprocessing and Alignment

Several Python-based libraries simplify the scRNA-seq preprocessing and analysis workflow, enabling us to work with data within a single environment. Notable Python tools include:

scanpy: A popular library for single-cell data analysis, providing preprocessing, normalization, clustering, and visualization functions [23]. scanpy can be used to process 10× Genomics datasets and handle gene expression matrices.

Example of reading 10× Genomics data in scanpy:

```
import scanpy as sc

# Read 10x Genomics data
adata = sc.read_10x_mtx('path_to_data_directory/', var_
names='gene_symbols', cache=True, prefix='your_file_name_prefix')
```

anndata: A library that works with scanpy for handling annotated data matrices, which are commonly used to store scRNA-seq data [24].

HTSeq: A Python library for processing high-throughput sequencing data, particularly RNA-seq. It includes functions for counting aligned reads and generating gene expression matrices [25].

Overall, Python offers an ecosystem for preprocessing and aligning scRNA-seq data that can be used to manage data analysis pipeline within a Python environment.

## 13.4 Quality Control and Filtering of scRNA-seq Data

After processing and aligning the raw scRNA-seq data, quality control (QC) is an important step to ensure the reliability and accuracy of the results. scRNA-seq can generate noisy data, and without careful QC, poor-quality cells or technical artifacts can distort downstream analyses. This section describes common QC practices, such as identifying and removing low-quality cells, detecting doublets, normalizing gene expression data, and filtering the data to retain biologically meaningful cells.

### 13.4.1 Identifying and Removing Low-Quality Cells



Not all cells captured in an scRNA-seq experiment are of sufficient quality to be included in the final analysis. Cells with degraded RNA, low RNA content, or high proportions of mitochondrial gene expression often represent stressed or dying cells, which should be excluded. Identifying low-quality cells is one of the first steps in the quality control (QC) process. Key quality metrics include:

*Number of detected genes:* Cells with an abnormally low number of detected genes may indicate poor-quality or dying cells, as healthy cells should express a sufficient number of genes.

*Total read counts:* Cells with very low total read counts (UMIs) may also be indicative of low RNA content or technical issues during sequencing.

*Percentage of mitochondrial reads:* High proportions of mitochondrial gene expression can signal damaged cells or cellular stress, as mitochondria play a role in cellular respiration.

Using these metrics, cells that fall outside acceptable thresholds can be removed from the dataset.

Example in Python using scanpy:

```
import scanpy as sc
import numpy as np

# Load the data (assuming 10x Genomics format, we need both
# matrix.mtx.gz and barcodes.tsv.gz files )
adata = sc.read_10x_mtx('path_to_data_directory/',
var_names='gene_symbols', cache=True,
prefix='your_file_name_prefix')

# Annotate mitochondrial genes
adata.var['mt'] = adata.var_names.str.startswith('MT-')
# For human data, mitochondrial genes start with 'MT-'

# Calculate QC metrics
sc.pp.calculate_qc_metrics(adata, qc_vars=['mt'],
percent_top=None, log1p=False, inplace=True)

# Plot the distribution of total counts, detected genes, and
# mitochondrial content
sc.pl.violin(adata, ['n_genes_by_counts', 'total_counts',
'pct_counts_mt'], jitter=0.4)

# Apply filters
# Filter out cells with fewer than 200 genes expressed
sc.pp.filter_cells(adata, min_genes=200)

# Filter out cells with more than 20% mitochondrial gene
# expression
adata = adata[adata.obs['pct_counts_mt'] < 20, :]
```

In this example, scanpy is used to calculate the percentage of mitochondrial gene expression and filter out cells that either have fewer than 200 detected genes or high mitochondrial content.

### 13.4.2 Detection of Doublets and Empty Drops

Artifacts such as doublets (two or more cells captured together in a single droplet) or empty droplets (droplets containing only ambient RNA) can introduce bias and noise into the scRNA-seq dataset. It is important to identify and remove these artifacts to prevent their inclusion in downstream analysis.

Doublets: Doublets occur when two or more cells are captured in a single droplet or reaction chamber, resulting in combined gene expression profiles that do not represent a single cell. If not detected, doublets can lead to the incorrect identification of new cell types or skewed gene expression profiles.

Tools for doublet detection: Scrublet is a Python tool that estimates the likelihood that a particular cell is a doublet by simulating artificial doublets and comparing their gene expression profiles to those of the observed cells [26].

Example: Doublet detection using scrublet:

In this example, we will use example data from scanpy.

```
import scanpy as sc
import scrublet as scr
import numpy as np
import matplotlib.pyplot as plt
from scipy import sparse

# Load the preprocessed and filtered data (AnnData object)
adata = sc.read_h5ad('filtered_data.h5ad')
# file name here is a place holder

# Use raw counts for Scrublet
    counts_matrix = adata.raw.X

# Ensure counts_matrix is dense
if sparse.issparse(counts_matrix):
counts_matrix = counts_matrix.A # Convert to NumPy array

# Ensure counts_matrix contains integer counts
counts_matrix = counts_matrix.astype(np.int64)

# Initialize Scrublet
scrub = scr.Scrublet(counts_matrix)

# Run doublet detection
doublet_scores, predicted_doublets = scrub.scrub_doublets()

# Add doublet scores and predictions to the AnnData object
adata.obs['doublet_score'] = doublet_scores
adata.obs['predicted_doublet'] = predicted_doublets

# Visualize the doublet scores
plt.figure(figsize= (4, 6))
plt.hist(doublet_scores, bins=50, density=True)
plt.xlabel('Doublet Score')
plt.ylabel('Frequency')
plt.title('Distribution of Doublet Scores')
plt.show()
```

Empty drops: Empty droplets are droplets that do not contain cells but may capture ambient RNA present in the solution. Including empty droplets in the analysis can introduce background noise and affect the accuracy of gene expression measurements. To identify and remove empty droplets, we can analyze the total counts per cell and apply a threshold. However, more sophisticated methods consider the distribution of ambient RNA and the expected cell-containing droplets. Tools like DropletUtils (R package), Cell Ranger (10× Genomics) and SoupX (R package) can estimate and remove empty droplets and ambient RNA contamination.

Example: Filtering empty droplets based on total counts.

```
import scanpy as sc
import matplotlib.pyplot as plt
import numpy as np

# Load the raw data (unfiltered)
adata = sc.read_10x_mtx('path_to_raw_feature_bc_matrix/',
var_names='gene_symbols', cache=True)

# Calculate QC metrics
sc.pp.calculate_qc_metrics(adata, inplace=True)

# Visualize the distribution of total counts per cell
plt.figure(figsize= (4, 6))
plt.hist(adata.obs['total_counts'], bins=100)
plt.xlabel('Total Counts')
plt.ylabel('Number of Cells')
plt.title('Total Counts per Cell')
plt.show()
# Determine a threshold to filter out empty droplets
# For example, use the inflection point or knee of the
# distribution
# Here, we use a simple threshold (adjust based on data)
min_counts_threshold = 500

# Filter out cells with total counts less than the threshold
sc.pp.filter_cells(adata, min_counts=min_counts_threshold)

# Save the filtered dataset
adata.write('adata_filtered_no_empty_drops.h5ad')
```

The threshold for total counts should be chosen carefully. Too high a threshold may exclude low RNA content cells, while too low a threshold may retain empty droplets. Visual inspection of the histogram and methods like the knee or inflection point detection can aid in selecting an appropriate cutoff. For more precise detection of empty droplets, consider using the `emptyDrops` function from the DropletUtils package in R. This method models the ambient RNA profile to distinguish real cells from empty droplets. Cell Ranger emptyDrops() function is also available as a separate package called EmptyDrops.

### 13.4.3 Normalization and Scaling of Gene Expression Data

After filtering out low-quality cells, the next step in the preprocessing pipeline is to normalize and scale the gene expression data. Normalization adjusts the gene expression values across cells to account for differences in sequencing depth and capture efficiency. Scaling standardizes the values, ensuring that genes contribute equally to analyses like clustering and dimensionality reduction.

Normalization: Normalization ensures that gene expression counts from each cell are comparable by adjusting for differences in total counts (library size) across cells. For instance, some cells might have more reads due to technical variations, but this does not necessarily indicate higher gene expression levels. Normalization methods like counts per million (CPM) or log-normalization are used to adjust for these discrepancies.

In single-cell RNA-seq data, a common normalization approach involves scaling the total counts per cell to a fixed target (e.g., 10,000 reads per cell) and then applying a log transformation. This process mitigates the effects of varying sequencing depths and reduces the impact of highly expressed genes.

Example in Python Using scanpy:

```
import scanpy as sc
# Assuming 'adata' is your AnnData object containing raw counts
# Normalize total counts per cell to 10,000
sc.pp.normalize_total(adata, target_sum=1e4)
# Logarithmize the data
sc.pp.log1p(adata)
```

This example, `sc.pp.normalize_total(adata, target_sum = 1e4)`, scales each cell's total counts to 10,000. This means that after normalization, the sum of counts for each cell will be 10,000. `sc.pp.log1p(adata)` performs a natural logarithm transformation after adding 1 to each value `(log(1 + x))`, which helps to stabilize variance and reduce the influence of high-count genes.

Scaling: Scaling involves centering and standardizing the gene expression values so that each gene has a mean of zero and a standard deviation of one across all cells. This step is needed for algorithms that assume normally distributed data, such as Principal Component Analysis (PCA), which is commonly used in dimensionality reduction.



```
# Scale the data to unit variance and zero mean
sc.pp.scale(adata, max_value=10)
```

In this example, sc.pp.scale(adata, max_value = 10) centers and scales the expression values for each gene by centering subtracts the mean expression of each gene, so that the gene's expression across all cells has a mean of zero and scaling divide by the standard deviation so that the expression has a unit variance. The `max_value = 10` parameter caps the scaled values at 10 (and – 10). This prevents extreme outliers from skewing the results of downstream analyses by limiting the effect of very highly expressed genes.

## 13.5 Dimensionality Reduction Techniques

After performing quality control on the scRNA-seq dataset, the next step in the analysis pipeline is to reduce the high-dimensional gene expression data into a lower-dimensional space, where patterns and relationships between cells become easier to visualize and analyze. Dimensionality reduction techniques are often used in scRNA-seq data analysis because each cell can express thousands of genes, making the data highly complex and difficult to interpret in its raw form. By reducing the dimensionality, we can preserve the most important information while simplifying the analysis. Here we will apply commonly apply dimensionality reduction techniques in scRNA-seq analysis such as PCA, t-Distributed Stochastic Neighbor Embedding (t-SNE), and Uniform Manifold Approximation and Projection (UMAP) to visualize cellular heterogeneity, clustering cells, and identifying subpopulations.

### 13.5.1 PCA for scRNA-seq Data

PCA is one of the basic dimensionality reduction methods used in scRNA-seq analysis. It transforms the high-dimensional gene expression data into a smaller number of principal components (PCs). Each principal component captures the directions in which the variance in the data is greatest, with the first few components typically explaining the majority of the variance.

PCA is particularly useful in the early stages of scRNA-seq analysis for identifying major sources of variability emphasizing the axes of greatest variance and helping us to understand the main sources of heterogeneity in the dataset. Furthermore, PCA often serves as a preprocessing step before clustering cells, as it reduces the noise and redundancy in the data.

Example of PCA in Python Using scanpy:

```
import scanpy as sc

# Load and preprocess the data (assuming the data is already
# normalized and filtered)
adata = sc.read_h5ad('path_to_preprocessed_data.h5ad')

# Perform PCA
sc.tl.pca(adata, svd_solver='arpack')

# Plot the variance explained by each principal component
sc.pl.pca_variance_ratio(adata, log=True)

# Visualize cells in the first two principal components
sc.pl.pca(adata, color=['cell_type_annotation'])
# Assuming cell type annotations exist
```

In this example, PCA projects the data into a lower-dimensional space where each cell is represented by its values along the principal components. The plot of variance explained by each principal component helps determine how many PCs to use for further analysis (e.g., clustering).

### 13.5.2 t-SNE for scRNA-seq Data

t-SNE is widely used for visualizing scRNA-seq data. Unlike PCA, which captures linear relationships, t-SNE is designed to preserve the local structure of the data, making it particularly effective for visualizing complex datasets where the relationships between cells are not easily captured by linear methods [27]. t-SNE maps cells into a two-dimensional space, where similar cells (in terms of gene expression) are grouped together, making it easier to identify clusters or subpopulations. We often use t-SNE to explore the results of clustering methods by visualizing how well cells are separated into groups. However, t-SNE has some limitations, such as being computationally intensive for large datasets and sometimes producing unstable results depending on the parameters used (e.g., perplexity).

Example of t-SNE in Python using scanpy:

```
import scanpy as sc

# Load and preprocess the data (assuming PCA has been per-
formed)
adata = sc.read_h5ad('path_to_preprocessed_data.h5ad')

# Perform PCA if not already done
sc.tl.pca(adata, svd_solver='arpack')

# Run t-SNE
sc.tl.tsne(adata, n_pcs=40, perplexity=30)

# Visualize the result
sc.pl.tsne(adata, color=['cell_type_annotation',
'gene_of_interest'])
```



Perplexity is a key parameter in t-SNE that influences how the algorithm balances the local versus global structure of the data. Adjusting it may yield different visualizations. t-SNE is not recommended for quantitative comparisons between clusters or for inferring distances between cells but is useful for exploratory visualizations.

### 13.5.3 UMAP for scRNA-seq Data

UMAP is similar to t-SNE but has some key advantages. UMAP tends to preserve both local and global structures in the data better than t-SNE and is typically faster and more scalable, making it suitable for large single-cell datasets [28, 29]. Like t-SNE, UMAP reduces high-dimensional data to two dimensions, providing an intuitive visualization of cellular heterogeneity. UMAP preserves both local and global relationships between cells better than t-SNE, making it more interpretable for understanding larger-scale structures in the data.

Example of UMAP in Python using scanpy:

```
import scanpy as sc

# Load and preprocess the data (assuming PCA has been
# performed)
adata = sc.read_h5ad('path_to_preprocessed_data.h5ad')

# Compute neighbors (required for UMAP)
sc.pp.neighbors(adata, n_neighbors=10, n_pcs=40)
# Run UMAP
sc.tl.umap(adata)
# Visualize the result
sc.pl.umap(adata, color=['cell_type_annotation',
'gene_of_interest'])
```

UMAP tends to be more stable than t-SNE, and its parameters (such as `n_neighbors` and `min_dist`) offer flexibility in preserving local versus global structures in the data. UMAP is widely used in scRNA-seq analysis for clustering, visualization, and exploring cellular differentiation pathways.

### 13.5.4 Visualizing Single-Cell Clusters

One of the most important outcomes of dimensionality reduction is the ability to visualize cell clusters (discussed further below), which represent distinct subpopulations or cell types within the dataset. After reducing the data using PCA, t-SNE, or UMAP, cells can be visualized in a 2D or 3D plot, where each point represents a single cell, and its position reflects its gene expression profile relative to other cells.

In scRNA-seq, clustering is often used in combination with dimensionality reduction to identify cell types, allowing us to interpret clusters as distinct cell types or subtypes. Furthermore, it enables exploration of developmental trajectories; in certain biological contexts, clusters may represent different stages of cell differentiation.

## 13.6 Clustering and Cell Type Identification

Once the dimensionality of scRNA-seq data has been reduced, the next step in the analysis pipeline is clustering. Clustering groups cells based on their gene expression profiles, allowing us to identify distinct subpopulations within a tissue or sample. These clusters often correspond to different cell types or states, enabling the discovery of novel cell types and the characterization of heterogeneous tissues. We will explore common clustering methods for identifying cell types based on gene expression markers, and Python tools used for clustering and annotation.

### 13.6.1 Clustering Methods

Clustering is a form of unsupervised learning where the goal is to partition cells into groups such that cells within each group are more similar to each other than to cells in other groups. Several clustering algorithms can be applied to scRNA-seq data, each with its strengths and limitations. Commonly used clustering methods include K-means clustering, hierarchical clustering, Louvain clustering, and Leiden clustering.

K-means clustering is a partitioning method that divides the data into a predefined number of clusters [30]. The algorithm works by initializing cluster centroids and iteratively assigning each cell to the nearest centroid based on its gene expression profile. The centroids are updated in each iteration to minimize the within-cluster variance. However, one limitation of K-means is that it requires the number of clusters to be specified beforehand, which can be challenging for complex scRNA-seq datasets where the number of cell types is not known. Furthermore, it assumes spherical cluster shapes that may not capture the complex structures inherent in scRNA-seq data.

Example of K-means clustering:

```
import scanpy as sc
from sklearn.cluster import KMeans

# Assuming PCA has been performed and stored in adata.obsm['X_
pca']
# Perform K-means clustering with 5 clusters (change as
needed)
kmeans = KMeans(n_clusters=5, random_state=0).fit(adata.
obsm['X_pca'])
adata.obs['kmeans_clusters'] = kmeans.labels_.astype(str)

# Visualize the clusters in UMAP space
sc.pl.umap(adata, color='kmeans_clusters')
```



Hierarchical clustering builds a hierarchy of clusters by recursively merging (agglomerative) or splitting (divisive) clusters. It is useful for visualizing relationships between clusters in a tree-like structure called a dendrogram. However, the method can be computationally intensive for large datasets and it requires a distance metric and linkage criterion, which can affect the results.

Louvain and Leiden are community detection algorithms commonly used in scRNA-seq analysis. These algorithms treat the cells as nodes in a graph, where edges represent similarities between cells based on their gene expression profiles. Louvain or Leiden clustering detects communities (or clusters) in the graph, making it particularly well suited for finding complex cell subpopulations in scRNA-seq data. Louvain and Leiden clustering are widely used because they do not require the number of clusters to be specified beforehand, making them ideal for exploratory analysis. They can capture complex, nonspherical cluster structures. The Leiden algorithm is an improved version of Louvain that shows better performance and faster convergence.

Example of Louvain clustering:

```
import scanpy as sc

# Compute the neighborhood graph (required for clustering)
sc.pp.neighbors(adata, n_neighbors=10, n_pcs=40)

# Run UMAP
sc.tl.umap(adata)

# Run Louvain clustering
sc.tl.louvain(adata, resolution=0.5)
# Adjust the resolution for more or fewer clusters

# Visualize the Louvain clusters in UMAP space
sc.pl.umap(adata, color='louvain')
```

Example of Leiden clustering:

```
import scanpy as sc

# Compute the neighborhood graph (required for clustering)
sc.pp.neighbors(adata, n_neighbors=10, n_pcs=40)

# Run UMAP
sc.tl.umap(adata)

# Run Leiden clustering
sc.tl.leiden(adata, resolution=0.5)
# Adjust resolution for more or fewer clusters

# Visualize the Leiden clusters in UMAP space
sc.pl.umap(adata, color='leiden')
```

### 13.6.2 Cell Type Annotation and Marker Gene Detection

After clustering cells, the next important task is to annotate the clusters by identifying which cell types they represent. This is usually done by examining the expression of known marker genes—genes that are specifically expressed in certain cell types.

Marker genes are identified by comparing the gene expression profiles of cells within a cluster to those of cells in other clusters. Common approaches for marker gene detection include the identification of genes that are significantly overexpressed in one cluster compared to others. Tools such as scanpy's `rank_genes_groups` function can be used for this purpose. We can also use existing knowledge about known marker genes for different cell types to help annotate clusters.

Example: detecting marker genes using scanpy:

```
 import scanpy as sc
 # Find marker genes for each cluster using the 'leiden'
 # clustering
 sc.tl.rank_genes_groups(adata, groupby='leiden',
method='wilcoxon')
 # Plot the top marker genes for each cluster
 sc.pl.rank_genes_groups(adata, n_genes=5, sharey=False)
```

In this example, marker genes are identified for each cluster using the Wilcoxon rank-sum test, and the top genes are visualized. These marker genes can then be compared to known cell-type markers from the literature to annotate the clusters.

Once marker genes have been identified, the clusters can be annotated by comparing the marker gene expression to known reference datasets or published studies. Several reference databases, such as CellMarker and PanglaoDB, provide lists of marker genes for various human and mouse cell types [31, 32].

Example of annotating clusters manually:

```
# Manually annotate clusters based on known marker genes
cell_type_map = {
    '0': 'T cells',
    '1': 'B cells',
    '2': 'Monocytes',
    '3': 'Dendritic cells',
    # Add mappings for all clusters
}
adata.obs['cell_type'] = adata.obs['leiden'].map(cell_type_map)

# Visualize the annotated cell types in UMAP space
sc.pl.umap(adata, color='cell_type')
sc.pl.umap(adata, color='cell_type')
```



In this example, clusters are manually annotated based on their marker gene expression profiles. The results are then visualized in UMAP space to show the distribution of cell types.

### 13.6.3 Python Tools for Clustering

Several Python libraries facilitate clustering and cell type identification in scRNA-seq analysis. Some of the most widely used tools include:

scanpy: A widely used library for scRNA-seq analysis, which includes functions for clustering (Louvain, Leiden), marker gene detection, and visualization.

Harmony: A tool for integrating data from multiple scRNA-seq experiments, accounting for batch effects, and improving clustering across datasets. Harmony can be used within scanpy through its API.

Example of using harmony in scanpy:

```
import harmonypy as hm

# Assuming PCA has been performed
ho = hm.run_harmony(adata.obsm['X_pca'], adata.obs, 'batch')
adata.obsm['X_pca_harmony'] = ho.Z_corr.T

# Use the corrected PCA embeddings for clustering
sc.pp.neighbors(adata, use_rep='X_pca_harmony')
sc.tl.leiden(adata)
sc.tl.umap(adata)

# Visualize the clusters
sc.pl.umap(adata, color=['leiden', 'batch'])
```

scikit-learn: A general-purpose machine learning library that includes clustering algorithms such as K-means and hierarchical clustering.

## 13.7 Differential Gene Expression Analysis

Differential gene expression (DGE) analysis is a key step in scRNA-seq studies. It aims to identify genes that show significant differences in expression between two or more cell populations. These differences may correspond to changes in cell states, responses to treatments, or intrinsic differences between cell types. DGE analysis helps to identify genes that drive functional differences between cell types or conditions, determining underlying biological processes. We will discuss DGE analysis, common statistical approaches, and demonstrate how to perform DGE using Python libraries such as scanpy and mention integration with R packages when necessary.

### 13.7.1 Differential Expression in scRNA-seq

Differential gene expression analysis in scRNA-seq is different from bulk RNA sequencing due to the nature of the data. In bulk RNA-seq, gene expression is measured as the average across a large number of cells, whereas scRNA-seq measures gene expression in individual cells. As a result, scRNA-seq data are sparse (containing many zeros) and displays high variability between cells, requiring specialized statistical approaches.

Applications of differential gene expression in scRNA-seq include identifying genes that are uniquely expressed in specific cell types or subpopulations, which can serve as markers for those cell types. It is also used to detect changes in gene expression during transitions between cell states, such as differentiation. Furthermore, comparing gene expression profiles between treated and untreated cells helps identify genes involved in drug response or disease progression.

### 13.7.2 Identifying Differentially Expressed Genes Across Cell Populations

DGE analysis typically involves comparing gene expression levels between two or more groups of cells. These groups can be defined based on clustering results, experimental conditions, or other criteria. The goal is to determine which genes are significantly upregulated or downregulated in one group compared to others. Various methods have been proposed for DGE analysis [33–35]. Here, we describe a simple model for identifying differentially expressed genes (DEGs). Steps in DGE analysis include defining groups, calculating differential expression, and adjusting for multiple testing.

*Define groups:* The first step is to define the groups of cells you want to compare. These groups could be based on the clusters identified in previous analyses (e.g., comparing cluster 1 vs. cluster 2) or experimental conditions (e.g., treated vs. untreated cells).

*Calculate differential expression:* Use statistical tests to compare the expression levels of each gene between the groups. Common statistical methods include the Wilcoxon rank-sum test, t-tests, and generalized linear models. For scRNA-seq data, nonparametric tests are often preferred due to the sparsity and noise in the data.

*Adjust for multiple testing:* Since thousands of genes are tested simultaneously, it is important to adjust the p-values for multiple testing to control the false discovery rate (FDR). This can be done using methods like the Benjamini-Hochberg procedure.

### 13.7.3 Statistical Approaches for DEG in scRNA-seq Analysis



DGE analysis in scRNA-seq employs statistical methodologies tailored to handle the distinctive features of scRNA-seq data, such as sparsity, high variability, and zero inflation [36]. Approaches can be broadly categorized into those adapted from bulk RNA-seq and those designed specifically for single-cell data. The Wilcoxon rank-sum test, a nonparametric method, compares gene expression distributions between two groups without assumptions about normality, making it robust to the high proportion of zeros and sparsity inherent in scRNA-seq data [33]. Conversely, the t-test compares group means and assumes normality and equal variances, limiting its applicability to normalized or transformed scRNA-seq data.

Logistic regression identifies genes discriminating between groups by modeling group membership probabilities, incorporating covariates to adjust for confounders [37, 38]. The likelihood ratio test (LRT) evaluates gene contributions to group differences and is useful in generalized linear models (GLMs) [39]. Specialized single-cell methods include MAST (model-based analysis of single-cell transcriptomics), which employs a hurdle model to manage sparsity and bimodal expression, and SCDE (single-cell differential expression), which accounts for dropout events and technical noise using mixture models [40, 41].

Other widely used methods, such as DESeq2 and edgeR, originally developed for bulk RNA-seq, can analyze scRNA-seq data with appropriate adaptations. These methods model read counts using the negative binomial distribution but face challenges like overdispersion and high dropout rates when directly applied to scRNA-seq. Other single-cell-specific approaches, like DECENT and DEsingle, leverage zero-inflated negative binomial models to address the excess zeros and variability [42, 43]. DECENT integrates technical variations using a hierarchical model, while DEsingle distinguishes between biological and technical zeros using constrained maximum likelihood estimation.

### 13.7.4 Python Tools for Differential Expression

Python libraries like scanpy provide functions for performing DGE analysis in scRNA-seq data. In addition, packages like diffxpy offer advanced statistical models tailored for single-cell data. In the following example, we will perform DGE analysis between two clusters using scanpy's rank_genes_groups function, which implements the Wilcoxon rank-sum test by default.

```
import scanpy as sc

# Load the processed single-cell data (assuming data has
# been normalized and clustered)
adata = sc.read_h5ad('path_to_preprocessed_data.h5ad')

# Perform differential expression analysis between clusters
sc.tl.rank_genes_groups(adata, groupby='leiden',
method='wilcoxon')

# Visualize the top differentially expressed genes
sc.pl.rank_genes_groups(adata, n_genes=10, sharey=False)
```

In this example, the `rank_genes_groups` function compares the gene expression between clusters defined by Louvain clustering. The top differentially expressed genes are visualized, displaying which genes are uniquely upregulated in each cluster.

Alternative statistical tests in Scanpy:

t-Test:

```
sc.tl.rank_genes_groups(adata, groupby='leiden', method='t-test')
```

Logistic regression:

```
# Perform differential expression analysis using logistic
# regression
sc.tl.rank_genes_groups(adata, groupby='leiden',
method='logreg')

# Visualize the top genes
sc.pl.rank_genes_groups(adata, n_genes=10, sharey=False)
```

## 13.8 Trajectory Inference and Pseudotime Analysis

Trajectory inference and pseudotime analysis are key methods used in scRNA-seq to study dynamic biological processes, such as cell differentiation, development, and disease progression [44–46]. Through analysis of gene expression changes over "pseudotime", which represents an inferred timeline of cell progression, we can uncover the pathways that cells take as they transition between different states or types. Unlike real time, pseudotime is an abstract representation of how far along a trajectory a cell lies, which is determined by its gene expression profile relative to other cells. We will discuss the basic concepts behind trajectory inference, key applications in scRNA-seq, and the tools used to implement pseudotime analysis in Python, including scanpy and Monocle.

### 13.8.1 Developmental Trajectories

Developmental trajectories describe the paths cells follow as they change gene expression over time, transitioning between different cell types or states. In numerous biological processes, including embryonic development, immune responses, and cancer progression, cells evolve from one state to another, with gradual shifts in gene expression during these transitions [47–50]. For instance, during embryonic development, stem cells differentiate into specialized cell types like neurons, muscle cells, and skin cells. scRNA-seq enables us to study these transitions at a granular level by capturing gene expression profiles of individual cells [50]. Applying trajectory inference and pseudotime analysis to these profiles, we can reconstruct developmental or differentiation pathways based on gene expression patterns.



### 13.8.2 Pseudotime Estimation in Single-Cell Data

Pseudotime estimation in single-cell data involves inferring the relative progression of cells along a biological process based on changes in their gene expression patterns, rather than measuring time chronologically [51]. Cells at the beginning of a process—such as early developmental stages—display distinct gene expression profiles compared to cells that have progressed further along a trajectory. To estimate pseudotime, we first apply dimensionality reduction techniques to reduce the complexity of the gene expression data and to visualize the relationships between cells [52]. With these simplified representations, a "root" or starting cell is identified; this cell is believed to be at the beginning of the biological process, such as a stem cell in a differentiation pathway. An algorithm is then used to infer the trajectory of cells through the reduced-dimensional space, assigning lower pseudotime values to cells closer to the root in terms of gene expression, and higher pseudotime values to those further along the trajectory. The inferred trajectory is visualized, often using methods like UMAP or t-SNE, with cells colored according to their pseudotime values to illustrate the progression of the biological process.

### 13.8.3 Python Tools for Trajectory Analysis

Several Python-based tools are available for performing trajectory inference and pseudotime analysis. These tools make it easy to analyze and visualize developmental trajectories based on scRNA-seq data.

**scanpy** includes functions for pseudotime estimation and trajectory inference. In particular, scanpy can integrate with the Palantir package for trajectory inference, which uses diffusion maps to model cell differentiation.

Example of pseudotime analysis using scanpy:

```
import scanpy as sc
import numpy as np

# Load the preprocessed single-cell data
adata = sc.read_h5ad('path_to_preprocessed_data.h5ad')

# Compute neighbors (required for pseudotime)
sc.pp.neighbors(adata, n_neighbors=15, use_rep='X_pca')


# Perform clustering using the Leiden algorithm
sc.tl.leiden(adata)

# Plot UMAP to visualize clusters
sc.pl.umap(adata, color='leiden')

# Select a root cell for pseudotime analysis
# Here, we choose a cell from cluster '0' as the root cell
root_cluster = '0'
root_cells = adata.obs names[adata.obs['leiden'] == root_cluster]
root_cell = root_cells[0]

# Set the root cell index in adata
adata.uns['iroot'] = np.flatnonzero(
adata.obs_names == root_cell)[0]

# Perform DPT (diffusion pseudotime) analysis
sc.tl.dpt(adata)

# Plot the UMAP with pseudotime coloring
sc.pl.umap(adata, color='dpt_pseudotime')
```

In this example, scanpy's DPT (Diffusion Pseudotime) method is used to estimate pseudotime for each cell, and the trajectory is visualized in UMAP space, with cells colored by their pseudotime values.

**Monocle** is one of the most widely used tools for pseudotime and trajectory analysis in scRNA-seq. Initially developed in R, Monocle is now also available through Python wrappers and is highly regarded for its ability to model complex trajectories in biological processes, such as differentiation. Monocle fits a principal graph to the data in low-dimensional space, and the pseudotime is calculated based on the projection of cells onto this graph [53]. Monocle can model branching trajectories, where cells diverge into different lineages.

**Palantir** is another Python tool designed for pseudotime and trajectory inference. Palantir uses diffusion maps to model cell fate probabilities and differentiation trajectories [52]. It is particularly well suited for datasets where cells follow multiple differentiation paths. Comprehensive guidelines are available at palantir.readthedocs.io.

## 13.9 Integration of scRNA-seq Datasets

As scRNA-seq studies grow in scale and complexity, the challenge of integrating multiple datasets becomes more common. These datasets may come from different experimental conditions, platforms, or even species, making it necessary to develop methods to harmonize them. Integration makes it possible to compare cell populations across conditions, study shared cellular states, or pool data for larger and relevant analyses. However, such integration must account for technical variability, such as batch effects, that could obscure biological differences.

### 13.9.1 Challenges in Integrating Data Across Conditions and Studies



Integrating scRNA-seq datasets from different sources presents several challenges. Batch effects, technical variations introduced during different experimental runs, such as differences in reagent quality, sequencing depth, or sample handling, can introduce unwanted variability, leading to spurious clustering of cells from the same batch rather than from the same biological group. Differences in sequencing technology, as various platforms like 10x Genomics, Smart-seq, or Drop-seq may have different levels of sensitivity, coverage, and gene detection efficiency, make direct comparisons difficult. Furthermore, differences in cell types or conditions, when datasets come from different tissues, organisms, or experimental conditions, can make it challenging to disentangle technical variability from true biological differences. The goal of integration is to remove these technical artifacts while preserving true biological variation, allowing us to draw meaningful conclusions from combined datasets.

### 13.9.2 Batch Effect Correction Techniques

To integrate scRNA-seq datasets effectively, it is important to correct for batch effects while preserving the underlying biological signals. Several computational methods have been developed to address this challenge [54]. Mutual nearest neighbors (MNNs) correction identifies cells across different datasets that are most similar in gene expression space and aligns these mutual neighbors, thereby removing batch effects while preserving biological variation [55]. This method is particularly effective when datasets contain overlapping cell types. Canonical correlation analysis (CCA) offers another approach by identifying linear combinations of gene expression variables that are maximally correlated across datasets [56]. Aligning datasets based on shared cell types, CCA has proven effective for integrating datasets with common cell populations, as demonstrated in tools like Seurat's CCA implementation in R. Harmony is another widely used method, which iteratively adjusts cell embeddings to eliminate batch effects while maintaining biological diversity [57]. It is computationally efficient and performs well even with complex datasets involving multiple batch effects. BBKNN (batch balanced k-nearest neighbors) offers a different strategy, working directly within the k-nearest neighbors graph [58]. This method ensures that the graph construction is balanced across batches, making it especially compatible with frameworks like scanpy and effective for batch correction while maintaining dataset integrity. Each of these techniques provides unique strengths, allowing us to choose an approach that best suits the characteristics and requirements of the scRNA-seq data.

### 13.9.3 Integrating Data from Different Sequencing Technologies

Integrating scRNA-seq datasets generated from different technologies, such as 10x Genomics and Smart-seq, presents unique challenges due to differences in resolution and sensitivity. For instance, 10x Genomics captures a large number of cells with lower gene detection sensitivity, while Smart-seq detects fewer cells but provides higher sensitivity and full-length transcript coverage. Overcoming these disparities requires careful normalization of sequencing depth, gene expression distributions, and technical variability while preserving shared biological signals [56]. Normalization techniques, such as log-normalization or scaling, are often employed to bring gene expression values from different datasets into a comparable range. Identifying common cell types across datasets is an important step, as these overlapping populations can serve as anchors or reference points for alignment. Batch correction methods, such as Harmony, BBKNN or LIGER, can then be applied to eliminate technical discrepancies and harmonize the datasets [57–59]. Once integration is complete, the aligned dataset enables robust downstream analyses, including clustering, differential expression analysis, and the exploration of biological variation across the combined data.

### 13.9.4 Example Code for Integration

Python tools such as scanpy can effectively integrate several integration methods including BBKNN and Harmony for batch effect correction.

Example of dataset integration using scanpy:

```
import scanpy as sc

# Load multiple datasets
adata1 = sc.read_h5ad('dataset1.h5ad')
adata2 = sc.read_h5ad('dataset2.h5ad')

# Concatenate the datasets
adata_combined = adata1.concatenate(adata2, batch_key='batch')

# Perform BBKNN batch correction
import scanpy.external as sce
sce.pp.bbknn(adata_combined, batch_key='batch')

# Run UMAP to visualize the integrated datasets
sc.tl.umap(adata_combined)
sc.pl.umap(adata_combined, color=['batch', 'cell_type'])
```

In this example, BBKNN is used to correct batch effects in two scRNA-seq datasets, and the integrated data is visualized using UMAP. The resulting plot shows how cells from different batches align, and cell types are color-coded.

Example of using Harmony for batch correction:



```
import harmonypy as hm

# Run PCA on the combined dataset
sc.tl.pca(adata_combined)

# Run Harmony for batch correction
ho = hm.run_harmony(adata_combined.obsm['X_pca'],
adata_combined.obs, 'batch')

# Update the PCA embeddings with Harmony results
adata_combined.obsm['X_pca_harmony'] = ho.Z_corr.T

# Visualize the Harmony-corrected data in UMAP
sc.pp.neighbors(adata_combined, use_rep='X_pca_harmony')
sc.tl.umap(adata_combined)
sc.pl.umap(adata_combined, color=['batch', 'cell_type'])
```

In this example, Harmony is applied to correct batch effects, and the corrected PCA embeddings are used to generate a UMAP plot.

## 13.10 Advanced Analysis and Interpretation of scRNA-seq Data

After completing initial steps of scRNA-seq analysis such as quality control, clustering, and differential expression, advanced methods can be employed to further interpret the data, allowing us to explore functional and regulatory aspects of gene expression at the single-cell level and gain deeper understanding into biological processes like gene regulation, cell-cell communication, and functional enrichment.

### 13.10.1 Gene Regulatory Network Inference

Gene regulatory networks (GRNs) capture complex interactions between genes, often mediated by transcription factors and other regulatory molecules, to control gene expression patterns [60]. Inferring these networks from scRNA-seq data provides overview an key regulators that drive biological processes, such as cell fate decisions and differentiation. GRN inference methods include correlation-based approaches, which calculate pairwise correlations of gene expression levels across cells to suggest possible regulatory relationships. Regression-based methods, such as LASSO regression, model gene expression as a function of regulators, provide a predictive framework for network construction [61]. Moreover, information theory-based methods, like GENIE3 and GRNBoost, use mutual information to identify gene interactions, capturing nonlinear relationships that may otherwise be missed [62, 63].

Among Python tools, pySCENIC stands out as an implementation of the SCENIC pipeline for GRN inference from scRNA-seq data [64]. It leverages co-expression patterns and motif enrichment analysis to pinpoint transcription factors and their target genes. We can use pySCENIC to analyze processed and filtered single-cell data, define a list of transcription factors, and execute the pipeline to construct the regulatory network. The process includes pruning the network based on motif analysis, assessing transcription factor activity in each cell using AUCell, and visualizing transcription factor activity across clusters. This workflow enables the identification of key regulatory elements underlying cellular processes.

### 13.10.2 Functional Enrichment Analysis (GO, KEGG)

Functional enrichment analyses, including Gene Ontology (GO) and KEGG pathway analysis, are approaches to uncover the biological functions and pathways associated with DEGs. These analyses help identify key biological processes, cellular components, and molecular functions enriched in specific gene sets. For example, GO analysis categorizes DEGs based on biological processes, cellular components, or molecular functions, while KEGG pathway analysis maps DEGs to known signaling and metabolic pathways.

Python tools like gseapy facilitate functional enrichment analysis by enabling us to analyze a list of DEGs using GO and KEGG databases. Gseapy automates the process of enrichment testing and provides visualizations such as bar plots or enrichment maps, emphasizing the top enriched terms or pathways [65]. This visualization helps interpret the biological relevance of DEGs in a straightforward manner. For instance, after obtaining a DEG list, we can run gseapy to identify GO terms or

KEGG pathways and visualize results as ranked terms, providing a clear view of the most relevant biological activities.

Combining functional enrichment analysis with complementary approaches, such as gene regulatory network inference or clustering, we can gain an understanding of the biological processes and regulatory mechanisms driving cellular behavior. These tools collectively enhance the interpretation of scRNA-seq datasets, linking gene expression changes to functional outcomes and biological significance.

### 13.10.3 Pathway Analysis and Cell-Cell Communication

Pathway analysis provides a deeper understanding of biological pathways that are active within specific cell populations, determining how cells respond to stimuli or interact with their environment. This approach identifies pathways that are impacted, shedding light on the molecular mechanisms driving cellular processes. Meanwhile, cell-cell communication analysis focuses on the dynamic interactions between cell types, identifying how they communicate through signaling pathways mediated by ligand-receptor interactions. This analysis is particularly useful for studying complex processes like immune responses, tissue development, and cancer progression, where intercellular communication plays a role.

Tools like CellPhoneDB, a Python-based resource for cell-cell communication analysis, enable us to identify ligand-receptor pairs mediating interactions between different cell types in scRNA-seq data [66]. Leveraging curated databases of ligand-receptor interactions, CellPhoneDB allows the exploration of how signaling pathways connect distinct cell populations. The results of these analyses can be visualized as heatmaps, which show interaction intensities between cell types, or as interaction networks, emphasizing key communication pathways and their mediators.

Combining pathway analysis with cell-cell communication studies, we can gain a holistic understanding of cellular behavior, linking intracellular pathways with intercellular signaling networks. This integrated perspective increases the ability to decipher complex biological systems and their underlying regulatory mechanisms.



## 13.11 Exercises and Questions

The following exercises and questions are designed to give you an understanding of scRNA-seq analysis, from data preprocessing and clustering to advanced analyses.

**Exercises**

1. Preprocessing scRNA-seq data: Write a Python script to load a scRNA-seq dataset from 10× Genomics, perform quality control (filter out cells with high mitochondrial gene content), normalize the data, and visualize the results using UMAP.
2. Clustering single cells: Perform Louvain clustering on a scRNA-seq dataset using scanpy. Visualize the clusters in UMAP space and compare them to known cell types.

3. Differential expression analysis: Identify differentially expressed genes between two cell populations (e.g., cluster 1 vs. cluster 2). Use scanpy to rank genes and visualize the top 10 differentially expressed genes.
4. Dimensionality reduction with PCA and UMAP: Apply Principal Component Analysis (PCA) and UMAP to reduce the dimensionality of a scRNA-seq dataset. Compare the two methods by visualizing the cells in 2D space.
5. Batch effect correction: Load two scRNA-seq datasets from different experimental conditions. Use BBKNN or Harmony to correct for batch effects, and visualize the integrated dataset using UMAP.
6. Trajectory inference: Use scanpy's Diffusion Pseudotime (DPT) method to estimate the pseudotime of cells in a differentiation pathway. Visualize the pseudotime trajectory and identify genes that vary along the trajectory.
7. Gene regulatory network inference: Use pyscenic to infer gene regulatory networks from scRNA-seq data. Identify transcription factors that play key roles in regulating gene expression in specific cell types.
8. Functional enrichment analysis: Perform Gene Ontology (GO) enrichment analysis on differentially expressed genes from a scRNA-seq dataset. Use gseapy to visualize the enriched biological processes.
9. Cell-cell communication analysis: Using CellPhoneDB, analyze an scRNA-seq dataset to identify ligand-receptor interactions between different cell populations. Visualize the communication network between cell types.
10. Integrating datasets from multiple sources: Write a Python script to integrate two scRNA-seq datasets using Harmony or BBKNN. Compare how well the integration aligns cell types between the datasets and visualize the results using UMAP.

## ? Questions to Be Answered

1. What are the key steps in preprocessing scRNA-seq data, and why is quality control important?
2. How does Principal Component Analysis (PCA) help in reducing the dimensionality of scRNA-seq data?
3. Explain the difference between Louvain clustering and K-means clustering in the context of single-cell data analysis.
4. What is pseudotime, and how is it used to model developmental trajectories in scRNA-seq data?
5. Describe how batch effects can impact scRNA-seq analysis and name two methods to correct for batch effects.
6. What is the role of gene regulatory network inference in scRNA-seq analysis, and which algorithms are commonly used for this task?
7. How is functional enrichment analysis applied in scRNA-seq, and what information does it provide?
8. How do ligand-receptor interactions reveal insights into cell-cell communication in scRNA-seq studies?
9. What challenges arise when integrating scRNA-seq datasets from different sequencing technologies, and how can they be addressed?
10. Why is it important to account for multiple testing in differential expression analysis, and how is the false discovery rate controlled?

**Acknowledgement** The language of the human-generated text was corrected with the assistance of artificial intelligence (AI) tools [GPT-3.5 and GTP-4 from OpenAI]. GitHub Co-Pilot was used to check the correctness of the codes. The text underwent subsequent human revision to ensure its accuracy.

## References

1. Jovic D, Liang X, Zeng H, Lin L, Xu F, Luo Y. Single-cell RNA sequencing technologies and applications: a brief overview. Clin Transl Med. 2022;12(3):e694.
2. Kolodziejczyk AA, Kim JK, Svensson V, Marioni JC, Teichmann SA. The technology and biology of single-cell RNA sequencing. Mol Cell. 2015;58(4):610–20.
3. Huang D, Ma N, Li X, Gou Y, Duan Y, Liu B, et al. Advances in single-cell RNA sequencing and its applications in cancer research. J Hematol Oncol. 2023;16(1):98.
4. Ke M, Elshenawy B, Sheldon H, Arora A, Buffa FM. Single cell RNA-sequencing: a powerful yet still challenging technology to study cellular heterogeneity. BioEssays. 2022;44(11):e2200084.
5. Fan W, Yang C, Hou X, Wan J, Liao B. Novel insights into the sinoatrial node in single-cell RNA sequencing: from developmental biology to physiological function. J Cardiovasc Dev Dis. 2022;9(11).
6. Gao C, Zhang M, Chen L. The comparison of two single-cell sequencing platforms: BD rhapsody and 10x genomics chromium. Curr Genomics. 2020;21(8):602–9.
7. Salomon R, Kaczorowski D, Valdes-Mora F, Nordon RE, Neild A, Farbehi N, et al. Droplet-based single cell RNAseq tools: a practical guide. Lab Chip. 2019;19(10):1706–27.
8. Alchahin AM, Tsea I, Baryawno N. Recent advances in single-cell RNA-sequencing of primary and metastatic clear cell renal cell carcinoma. Cancers (Basel). 2023;15(19).
9. Yu W, Wang C, Shang Z, Tian J. Unveiling novel insights in prostate cancer through single-cell RNA sequencing. Front Oncol. 2023;13:1224913.
10. Otsuji K, Takahashi Y, Osako T, Kobayashi T, Takano T, Saeki S, et al. Serial single-cell RNA sequencing unveils drug resistance and metastatic traits in stage IV breast cancer. NPJ Precis Oncol. 2024;8(1):222.
11. Wang X, Chen Y, Li Z, Huang B, Xu L, Lai J, et al. Single-cell RNA-Seq of T cells in B-ALL patients reveals an exhausted subset with remarkable heterogeneity. Adv Sci (Weinh). 2021;8(19):e2101447.
12. Li K, Zhang C, Zhou R, Cheng M, Ling R, Xiong G, et al. Single cell analysis unveils B cell-dominated immune subtypes in HNSCC for enhanced prognostic and therapeutic stratification. Int J Oral Sci. 2024;16(1):29.
13. Potter SS. Single-cell RNA sequencing for the study of development, physiology and disease. Nat Rev Nephrol. 2018;14(8):479–92.
14. Armand EJ, Li J, Xie F, Luo C, Mukamel EA. Single-cell sequencing of brain cell transcriptomes and Epigenomes. Neuron. 2021;109(1):11–26.
15. Haque A, Engel J, Teichmann SA, Lonnberg T. A practical guide to single-cell RNA-sequencing for biomedical research and clinical applications. Genome Med. 2017;9(1):75.
16. Arsenio J. Single-cell Transcriptomics of immune cells: cell isolation and cDNA library generation for scRNA-Seq. Methods Mol Biol. 2020;2184:1–18.
17. Nguyen QH, Pervolarakis N, Nee K, Kessenbrock K. Experimental considerations for single-cell RNA sequencing approaches. Front Cell Dev Biol. 2018;6:108.
18. Bolger AM, Lohse M, Usadel B. Trimmomatic: a flexible trimmer for Illumina sequence data. Bioinformatics. 2014;30(15):2114–20.
19. Martin M. Cutadapt removes adapter sequences from high-throughput sequencing reads. EMBnet J. 2011;17:10–2.
20. Dobin A, Davis CA, Schlesinger F, Drenkow J, Zaleski C, Jha S, et al. STAR: ultrafast universal RNA-seq aligner. Bioinformatics. 2013;29(1):15–21.
21. Bray NL, Pimentel H, Melsted P, Pachter L. Near-optimal probabilistic RNA-seq quantification. Nat Biotechnol. 2016;34(5):525–7.

22. Kim D, Paggi JM, Park C, Bennett C, Salzberg SL. Graph-based genome alignment and genotyping with HISAT2 and HISAT-genotype. Nat Biotechnol. 2019;37(8):907–15.
23. Wolf FA, Angerer P, Theis FJ. SCANPY: large-scale single-cell gene expression data analysis. Genome Biol. 2018;19(1):15.
24. Virshup I, Rybakov S, Theis FJ, Angerer P, Wolf FA. anndata: Annotated data. bioRxiv. 2021:2021.12.16.473007.
25. Anders S, Pyl PT, Huber W. HTSeq–a Python framework to work with high-throughput sequencing data. Bioinformatics. 2015;31(2):166–9.
26. Wolock SL, Lopez R, Klein AM. Scrublet: computational identification of cell doublets in single-cell transcriptomic data. Cell Syst. 2019;8(4):281–91 e9.
27. Kobak D, Berens P. The art of using t-SNE for single-cell transcriptomics. Nat Commun. 2019;10(1):5416.
28. McInnes L, Healy J, Melville J. UMAP: Uniform Manifold Approximation and Projection for Dimension Reduction. arXiv. 2020;1802.03426.
29. Becht E, McInnes L, Healy J, Dutertre CA, Kwok IWH, Ng LG, et al. Dimensionality reduction for visualizing single-cell data using UMAP. Nat Biotechnol. 2018;37:38–44.
30. Steinley D. K-means clustering: a half-century synthesis. Br J Math Stat Psychol. 2006;59(Pt 1):1–34.
31. Hu C, Li T, Xu Y, Zhang X, Li F, Bai J, et al. CellMarker 2.0: an updated database of manually curated cell markers in human/mouse and web tools based on scRNA-seq data. Nucleic Acids Res. 2023;51(D1):D870–D6.
32. Franzen O, Gan LM, Bjorkegren JLM. PanglaoDB: a web server for exploration of mouse and human single-cell RNA sequencing data. Database (Oxford). 2019;2019:baz046.
33. Heumos L, Schaar AC, Lance C, Litinetskaya A, Drost F, Zappia L, et al. Best practices for single-cell analysis across modalities. Nat Rev Genet. 2023;24(8):550–72.
34. Chen H, Tian T, Luo H, Jiang Y. Identification of differentially expressed genes at the single-cell level and prognosis prediction through bulk RNA sequencing data in breast cancer. Front Genet. 2022;13:979829.
35. Squair JW, Gautier M, Kathe C, Anderson MA, James ND, Hutson TH, et al. Confronting false discoveries in single-cell differential expression. Nat Commun. 2021;12(1):5692.
36. Das S, Rai A, Merchant ML, Cave MC, Rai SN. A comprehensive survey of statistical approaches for differential expression analysis in single-cell RNA sequencing studies. Genes (Basel). 2021;12(12):1947.
37. Dezem FS, Marcao M, Ben-Cheikh B, Nikulina N, Omotoso A, Burnett D, et al. A machine learning one-class logistic regression model to predict stemness for single cell transcriptomics and spatial omics. BMC Genomics. 2023;24(1):717.
38. Wang T, Nabavi S, editors. Differential gene expression analysis in single-cell RNA sequencing data. 2017 IEEE International Conference on Bioinformatics and Biomedicine (BIBM); 2017; 13–16 Nov. 2017.
39. Nault R, Saha S, Bhattacharya S, Dodson J, Sinha S, Maiti T, et al. Benchmarking of a Bayesian single cell RNAseq differential gene expression test for dose-response study designs. Nucleic Acids Res. 2022;50(8):e48.
40. Finak G, McDavid A, Yajima M, Deng J, Gersuk V, Shalek AK, et al. MAST: a flexible statistical framework for assessing transcriptional changes and characterizing heterogeneity in single-cell RNA sequencing data. Genome Biol. 2015;16:278.
41. Kharchenko PV, Silberstein L, Scadden DT. Bayesian approach to single-cell differential expression analysis. Nat Methods. 2014;11(7):740–2.
42. Ye C, Speed TP, Salim A. DECENT: differential expression with capture efficiency adjustmeNT for single-cell RNA-seq data. Bioinformatics. 2019;35(24):5155–62.
43. Miao Z, Deng K, Wang X, Zhang X. DEsingle for detecting three types of differential expression in single-cell RNA-seq data. Bioinformatics. 2018;34(18):3223–4.
44. Hou W, Ji Z, Chen Z, Wherry EJ, Hicks SC, Ji H. A statistical framework for differential pseudotime analysis with multiple single-cell RNA-seq samples. Nat Commun. 2023;14(1):7286.
45. Cannoodt R, Saelens W, Saeys Y. Computational methods for trajectory inference from single-cell transcriptomics. Eur J Immunol. 2016;46(11):2496–506.
46. Saelens W, Cannoodt R, Todorov H, Saeys Y. A comparison of single-cell trajectory inference methods. Nat Biotechnol. 2019;37(5):547–54.

47. Ueda Y, Nakamura T, Nie J, Solivais AJ, Hoffman JR, Daye BJ, et al. Defining developmental trajectories of prosensory cells in human inner ear organoids at single-cell resolution. Development. 2023;150(12).
48. Farrell JA, Wang Y, Riesenfeld SJ, Shekhar K, Regev A, Schier AF. Single-cell reconstruction of developmental trajectories during zebrafish embryogenesis. Science. 2018;360(6392).
49. Zhu Q, Gao P, Tober J, Bennett L, Chen C, Uzun Y, et al. Developmental trajectory of prehematopoietic stem cell formation from endothelium. Blood. 2020;136(7):845–56.
50. Moiso E, Farahani A, Marble HD, Hendricks A, Mildrum S, Levine S, et al. Developmental deconvolution for classification of cancer origin. Cancer Discov. 2022;12(11):2566–85.
51. Wagner A, Regev A, Yosef N. Revealing the vectors of cellular identity with single-cell genomics. Nat Biotechnol. 2016;34(11):1145–60.
52. Setty M, Kiseliovas V, Levine J, Gayoso A, Mazutis L, Pe'er D. Characterization of cell fate probabilities in single-cell data with Palantir. Nat Biotechnol. 2019;37(4):451–60.
53. Van den Berge K, Roux de Bezieux H, Street K, Saelens W, Cannoodt R, Saeys Y, et al. Trajectory-based differential expression analysis for single-cell sequencing data. Nat Commun. 2020;11(1):1201.
54. Stuart T, Butler A, Hoffman P, Hafemeister C, Papalexi E, Mauck WM 3rd, et al. Comprehensive integration of single-cell data. Cell. 2019;177(7):1888–1902.e21.
55. Haghverdi L, Lun ATL, Morgan MD, Marioni JC. Batch effects in single-cell RNA-sequencing data are corrected by matching mutual nearest neighbors. Nat Biotechnol. 2018;36(5):421–7.
56. Butler A, Hoffman P, Smibert P, Papalexi E, Satija R. Integrating single-cell transcriptomic data across different conditions, technologies, and species. Nat Biotechnol. 2018;36(5):411–20.
57. Korsunsky I, Millard N, Fan J, Slowikowski K, Zhang F, Wei K, et al. Fast, sensitive and accurate integration of single-cell data with harmony. Nat Methods. 2019;16(12):1289–96.
58. Polanski K, Young MD, Miao Z, Meyer KB, Teichmann SA, Park JE. BBKNN: fast batch alignment of single cell transcriptomes. Bioinformatics. 2020;36(3):964–5.
59. Liu J, Gao C, Sodicoff J, Kozareva V, Macosko EZ, Welch JD. Jointly defining cell types from multiple single-cell datasets using LIGER. Nat Protoc. 2020;15(11):3632–62.
60. Badia IMP, Wessels L, Muller-Dott S, Trimbour R, Ramirez Flores RO, Argelaguet R, et al. Gene regulatory network inference in the era of single-cell multi-omics. Nat Rev Genet. 2023;24(11):739–54.
61. Omranian N, Eloundou-Mbebi JM, Mueller-Roeber B, Nikoloski Z. Gene regulatory network inference using fused LASSO on multiple data sets. Sci Rep. 2016;6:20533.
62. Huynh-Thu VA, Irrthum A, Wehenkel L, Geurts P. Inferring regulatory networks from expression data using tree-based methods. PLoS One. 2010;5(9).
63. Moerman T, Aibar Santos S, Bravo Gonzalez-Blas C, Simm J, Moreau Y, Aerts J, et al. GRNBoost2 and Arboreto: efficient and scalable inference of gene regulatory networks. Bioinformatics. 2019;35(12):2159–61.
64. Aibar S, Gonzalez-Blas CB, Moerman T, Huynh-Thu VA, Imrichova H, Hulselmans G, et al. SCENIC: single-cell regulatory network inference and clustering. Nat Methods. 2017;14(11):1083–6.
65. Fang Z, Liu X, Peltz G. GSEApy: a comprehensive package for performing gene set enrichment analysis in Python. Bioinformatics. 2023;39(1)
66. Efremova M, Vento-Tormo M, Teichmann SA, Vento-Tormo R. CellPhoneDB: inferring cell-cell communication from combined expression of multi-subunit ligand-receptor complexes. Nat Protoc. 2020;15(4):1484–506.

# Supplementary Information




J. U. Kazi, *Python Essentials for Biomedical Data Analysis: An Introductory Textbook*,
https://doi.org/10.1007/978-3-031-85600-6

# Appendices

## Additional Resources and Reading

The following additional resources and reading materials are recommended by author for those interested in expanding their knowledge and staying updated on the latest developments:

### Advanced Python Textbooks

*Fluent Python* by Luciano Ramalho: Offers an in-depth exploration of Python's advanced features, ideal for those looking to deepen their understanding. Link: https://www.oreilly.com/library/view/fluent-python-2nd/9781492056348/

*Python for Data Analysis* by Wes McKinney: Suggested reading to help you understand data analysis using Python. Link: https://www.oreilly.com/library/view/python-for-data/9781098104023/

### Specialized Bioinformatics Books

*Practical Bioinformatics* by Michael Agostino: A resource for applying various bioinformatics concepts. Link: https://www.routledge.com/Practical-Bioinformatics/Agostino/p/book/9780815344568 (2nd edition will be available in 2026).

*Bioinformatics Algorithms: An Active Learning Approach* by Phillip Compeau and Pavel Pevzner: Provides an introduction to the algorithmic techniques in bioinformatics. Link: https://www.bioinformaticsalgorithms.org/

### Online Documentation and Tutorials

SciPy and NumPy Documentation: For a deeper understanding of scientific computing in Python.

Jupyter Notebook Tutorials: Interactive tutorials on Jupyter Notebooks, a popular tool for Python coding in scientific research.

### Research Papers and Case Studies

Access to databases like PubMed and Google Scholar to find research papers where Python has been used in groundbreaking biomedical research.

### Blogs and Websites

Towards Data Science on Medium: A platform with numerous articles on Python in data science, many of which are applicable to biomedicine.

KDnuggets: A leading site on AI, analytics, big data, and data science, often featuring Python-related content.

### Online Course Platforms with Specialized Tracks

Udacity and Pluralsight: Platforms provide more specialized and advanced courses in Python, particularly in data science and AI.

## YouTube Channels and Podcasts

Channels like "Corey Schafer" and "Sentdex" for Python tutorials.

Podcasts like "Talk Python To Me" and "Python Bytes" for the latest discussions in the Python community.

## Professional and Academic Journals

Journals such as *Nature Biotechnology* and *Journal of Chemical Information and Modeling* for the latest research at the intersection of Python and biomedicine.

## Python in Emerging Technologies

Books and articles on the application of Python in emerging fields like artificial intelligence in healthcare and blockchain in medicine.

## Networking and Professional Development

Joining professional organizations like the International Society for Computational Biology (ISCB) can provide access to a wealth of resources and networking opportunities in the field of bioinformatics.

These resources provide a foundation for both the theoretical and practical aspects of Python in biomedicine, catering to a range of interests from programming fundamentals to specialized applications in bioinformatics and data analysis.

## Advanced Python Libraries for Biomedical Research

- *Hands-On Machine Learning with Scikit-Learn, Keras, and TensorFlow* by Aurélien Géron: For those interested in applying machine learning and deep learning to biomedical data.
- *Deep Learning for the Life Sciences* by Bharath Ramsundar, Peter Eastman, Patrick Walters, and Vijay Pande: This book focuses on applying deep learning techniques in biology and healthcare.

Specialized Resources for Single-Cell RNA Sequencing (scRNA-seq)

- *Orchestrating Single-Cell Analysis with Bioconductor* by Robert Amezquita, Aaron Lun, Stephanie Hicks, Rafael Gottardo, Alan O'Callaghan: A resource for those interested in the single-cell RNA sequencing, providing practical workflows and analysis strategies. Link: https://bioconductor.org/books/release/OSCA/
- Comprehensive online tutorials: Consider adding curated online platforms such as The Comprehensive scRNA-seq Course by Harvard Medical School (free, available online) which provides hands-on learning on single-cell technologies.

Biostatistics and Computational Biology

- *Computational Biology: A Practical Introduction to BioData Processing and Analysis with Linux, MySQL, and R* by Röbbe Wünschiers: Ideal for students interested in computational biology concepts beyond Python.
- *Biostatistics for Biological and Health Sciences* by Mario Triola, Marc Triola, Jason Roy: This book provides more in-depth biostatistics knowledge useful in biomedical data interpretation. Link: https://eu.pearson.com/biostatistics-for-the-biological-and-health-sciences-global-edition/9781292452036

Bioinformatics-Specific Libraries and Tools

- Biopython Documentation and Tutorials: Suggested for working with biological data in Python, offering extensive examples for sequence analysis and more.
- Galaxy Project: For students who prefer working with biomedical data in a graphical environment, the Galaxy Project platform provides a user-friendly interface for bioinformatics analysis, including Python-based tools.

Workshops and Certification Programs

- Coursera—Applied Data Science with Python: A well-structured course that includes bioinformatics applications, machine learning, and data visualization.
- European Bioinformatics Institute (EMBL-EBI) Training: EMBL-EBI offers a range of workshops and online training sessions for bioinformaticians, including Python-based tools and workflows for genomic and transcriptomic data analysis.

Specialized Resources for Natural Language Processing (NLP) in Biomedicine

- *Natural Language Processing with Python* by Steven Bird, Ewan Klein, and Edward Loper: For those interested in biomedical NLP, this resource is a great primer on NLP techniques.
- Biomedical Text Mining (BioNLP): Many free resources, including BioNLP-ST shared tasks, are available for researchers working on natural language processing in biomedical research.

Python for Data Visualization in Biomedicine

- *Python Data Science Handbook* by Jake VanderPlas: A guide for mastering data visualization in Python, with a focus on libraries like Matplotlib and Seaborn, which are useful for visualizing complex biomedical data.

Conferences and Workshops

- ISMB (Intelligent Systems for Molecular Biology): The ISMB conference is a great platform for researchers working at the intersection of computational biology and bioinformatics to exchange knowledge on Python-based methods and applications.
- RECOMB (Research in Computational Molecular Biology): A leading conference on bioinformatics that covers computational techniques for genome analysis, systems biology, and more, with a strong focus on the latest Python applications.

# Glossary

**Algorithm:** - A step-by-step procedure or formula for solving a problem. In programming, algorithms are implemented in code to perform tasks like searching, sorting, or data manipulation.

**Array:** - A data structure that stores a fixed-size sequential collection of elements of the same type. In Python, arrays are commonly implemented using NumPy arrays rather than built-in lists for efficiency in numerical computations.

**Assertion:** - A statement that tests if a condition is true. In Python, assertions are used to debug code using the assert keyword.

**Class:** - A blueprint for creating objects (a particular data structure), providing initial values for state (member variables or attributes), and implementations of behavior (member functions or methods).

**Class** inheritance: - A mechanism in which a new class (subclass) is derived from an existing class (parent class), inheriting its attributes and methods.

**Comprehension:** - A concise way to create lists, sets, or dictionaries in Python. List comprehensions, set comprehensions, and dictionary comprehensions provide a syntactically compact way to generate new collections from existing ones. Example: [x**2 for x in range(10)] generates a list of squares.

**Conditional** statement: - A block of code that performs different actions based on whether a specified condition is true or false. In Python, these include `if`, `elif`, and `else`.

**Context** manager: - A structure that allows for the proper acquisition and release of resources, like file handling, using the `with` keyword. Example: `with open('file.txt', 'r') as file:`

**DataFrame:** - A 2D, labeled data structure, like a table or spreadsheet, in the pandas library. DataFrames are valuable for handling and analyzing structured data.

**Data** type: - The kind of data that can be stored and manipulated within a program. Common data types in Python include `int`, `float`, `str`, and `bool`.

**Decorator:** - A design pattern in Python that allows a user to add new functionality to an existing object without modifying its structure. Decorators are often used with functions or methods.

**Decorator** function: - A higher-order function in Python that takes another function as an argument and extends its behavior without explicitly modifying it. Example: `@my_decorator`

**Dictionary:** - An unordered collection of key-value pairs. Dictionaries are defined with curly braces `{ }` and allow efficient lookups.

**Docstring:** - A special string used to document a specific segment of code, typically placed immediately after the function or class definition. It provides an easy way to document functionality for future reference or user understanding. Example: `"""This function returns the square of a number."""`

**Enumerate:** - A Python built-in function used to add a counter to an iterable and return it in a form of an enumerate object. Example: `for i, value in enumerate(my_list):`

**Exception:** - An error detected during the execution of a program. Python uses `try-except` blocks to handle exceptions and prevent crashes.

**Expression:** - A piece of code that evaluates to a value, such as `5 + 3` or `x * y`. An expression can be as simple as a literal value or a complex function call.

**Function:** - A reusable block of code that performs a specific task. Functions are defined using the `def` keyword and can accept input parameters and return values.

**Generator:** - A special type of iterator returned by a function that yields results one at a time, as needed, instead of all at once. Generators are memory-efficient for large data sets.

**Immutable:** - An object whose state or content cannot be modified after it is created. Examples in Python include strings (`str`), tuples (`tuple`), and frozensets.

**Iterable:** - Any Python object capable of returning its members one at a time, permitting iteration over it in a loop. Examples include lists, tuples, dictionaries, and strings.

**Iterator:** - An object that contains a countable number of values and can be iterated upon. You can traverse through all the values using an iterator's `__iter__()` and `__next__()` methods.

**Lambda** function: - An anonymous, inline function defined with the `lambda` keyword. Lambda functions can have any number of arguments but only one expression, used primarily for short, simple functions.

**List:** - An ordered, mutable collection of items. Lists are defined with square brackets `[]` and can contain items of different data types.

**Loop:** - A sequence of instructions that is continually repeated until a certain condition is reached. Common loops in Python include `for` and `while` loops.

**Map** function: - A function that applies another function to all items in an input list or iterable. It returns an iterator with the results. Example: `map(lambda x: x**2, [1, 2, 3, 4])`

**Method:** - A function that is associated with an object and typically called using dot notation. For example, `.append()` is a method of list objects.

**Module:** - A file containing Python code (e.g., functions, classes, variables). It can be imported into other scripts or modules to reuse code.

**Mutable:** - Objects that can be modified after creation, such as lists and dictionaries. This is the opposite of immutable, where objects cannot be changed (e.g., tuples, strings).

**Namespace:** - A container where names are mapped to objects. Namespaces prevent conflicts between identifiers by isolating them. Python has different namespaces, such as local, global, and built-in.

**Object:** - An instance of a class, which can contain both data (variables) and functions (methods). In Python, everything is an object, including integers and strings.

**Package:** - A directory of Python modules containing an additional `__init__.py` file, making it a Python package. Packages are a way to organize related modules.

**Pickle:** - A Python module that serializes (converts) Python objects into a byte stream to store them in a file or send them over a network, and deserialize (reconvert) them back into objects.

**Recursion:** - A process in which a function calls itself, directly or indirectly, to solve a problem. Recursion is typically used for problems that can be broken down into similar subproblems.

**Set:** - An unordered collection of unique items. Sets in Python are defined with curly braces `{ }` and do not allow duplicate values. Example: set([1, 2, 3, 3]) results in {1, 2, 3}.

**Slice:** - A mechanism to select a range of items from sequence types like lists, tuples, or strings using the slice notation [start:end:step].

**String:** - A series of characters, surrounded by single quotes `' '`, double quotes `" "`, or triple quotes `''' / """`. Strings are immutable in Python.

**Type** hinting: - A feature in Python that allows specifying the expected type of a function's arguments and return values to improve code readability and reduce errors. Example: `def add(x: int, y: int) -> int:`

**Tuple:** - An ordered, immutable collection of items, similar to a list but cannot be modified after creation. Tuples are defined with parentheses `()`.

**Variable:** - A named location used to store data in memory. Variables in Python are dynamically typed, meaning they can hold data of any type, and their type can change.

**Virtual** environment: - A self-contained Python environment that allows you to install dependencies and packages in isolation from other projects or the global Python environment. Virtual environments are created using tools like `venv` or `virtualenv`.

**Yield:** - A keyword in Python used to return a generator. It allows functions to return intermediate results and resume from where they left off without restarting the function from scratch.

GPSR Compliance

*The European Union's (EU) General Product Safety Regulation (GPSR) is a set of rules that requires consumer products to be safe and our obligations to ensure this.*

*If you have any concerns about our products, you can contact us on ProductSafety@springernature.com*

In case Publisher is established outside the EU, the EU authorized representative is:

Springer Nature Customer Service Center GmbH
Europaplatz 3
69115 Heidelberg, Germany

**Batch number: 09112167**

Printed by Printforce, the Netherlands